VOLUME ONE HUNDRED AND THIRTY FOUR

CURRENT TOPICS IN DEVELOPMENTAL BIOLOGY

Sex Determination in Vertebrates

CURRENT TOPICS IN DEVELOPMENTAL BIOLOGY

"A meeting-ground for critical review and discussion of developmental processes"
A.A. Moscona and Alberto Monroy (Volume 1, 1966)

VOLUME ONE HUNDRED AND THIRTY FOUR

CURRENT TOPICS IN DEVELOPMENTAL BIOLOGY

Sex Determination in Vertebrates

Edited by

BLANCHE CAPEL
Department of Cell Biology,
Duke University Medical Center,
Durham, NC, United States

ELSEVIER

ACADEMIC PRESS
An imprint of Elsevier

Academic Press is an imprint of Elsevier
50 Hampshire Street, 5th Floor, Cambridge, MA 02139, United States
525 B Street, Suite 1650, San Diego, CA 92101, United States
The Boulevard, Langford Lane, Kidlington, Oxford OX5 1GB, United Kingdom
125 London Wall, London, EC2Y 5AS, United Kingdom

First edition 2019

ISBN: 978-0-12-811544-2
ISSN: 0070-2153

For information on all Academic Press publications
visit our website at https://www.elsevier.com/books-and-journals

Publisher: Fiona Pattison
Acquisition Editor: Fiona Pattison
Editorial Project Manager: Shellie Bryant
Production Project Manager: Denny Mansingh
Cover Designer: Alan Studholme

Typeset by SPi Global, India

Contents

Contributors

Arthur P. Arnold
Department of Integrative Biology & Physiology, Laboratory of Neuroendocrinology of the Brain Research Institute, University of California, Los Angeles, CA, United States

Hayk Barseghyan
Center for Genetic Medicine Research, Children's National Medical Center; Department of Genomics and Precision Medicine, The George Washington University, Washington, DC, United States

Josephine Bowles
School of Biomedical Sciences; Institute for Molecular Bioscience, The University of Queensland, Brisbane, QLD, Australia

Tasman Daish
Comparative Genome Biology Laboratory, Department of Molecular and Biomedical Science, School of Biological Sciences, The University of Adelaide, Adelaide, SA, Australia

Emmanuèle Délot
Center for Genetic Medicine Research, Children's National Medical Center; Department of Genomics and Precision Medicine, The George Washington University, Washington, DC, United States

Bruce W. Draper
Molecular and Cellular Biology, University of California, Davis, CA, United States

Neil J. Gemmell
Department of Anatomy, University of Otago, Dunedin, New Zealand

Alexander Goikoetxea
Department of Anatomy, University of Otago, Dunedin, New Zealand

Nitzan Gonen
The Francis Crick Institute, London, United Kingdom

Andy Greenfield
Mammalian Genetics Unit, Medical Research Council Harwell Institute, Oxfordshire, United Kingdom

Frank Grützner
Comparative Genome Biology Laboratory, Department of Molecular and Biomedical Science, School of Biological Sciences, The University of Adelaide, Adelaide, SA, Australia

Ricardo S. Hattori
Sao Paulo Fisheries Institute—Campos do Jordao, APTA/SAA, Campos do Jordão, CP, Brazil

Timothy A. Hore
Department of Anatomy, University of Otago, Dunedin, New Zealand

Michelle E. Kossack
Molecular and Cellular Biology, University of California, Davis, CA, United States

Robin Lovell-Badge
The Francis Crick Institute, London, United Kingdom

Shingo Miyawaki
Institute of Advanced Medical Sciences, Tokushima University, Tokushima; Graduate School of Frontier Biosciences, Osaka University, Osaka, Japan

Serge Nef
Department of Genetic Medicine and Development, University of Geneva Medical School, Geneva, Switzerland

Oscar Ortega-Recalde
Department of Anatomy, University of Otago, Dunedin, New Zealand

Abhinav Parivesh
Center for Genetic Medicine Research, Children's National Medical Center, Washington, DC, United States

Reynaldo Patiño
U.S. Geological Survey, Texas Cooperative Fish and Wildlife Research Unit and Departments of Natural Resources Management and Biological Sciences, Texas Tech University, Lubbock, TX, United States

Cassy Spiller
School of Biomedical Sciences, The University of Queensland, Brisbane, QLD, Australia

Isabelle Stévant
Department of Genetic Medicine and Development, University of Geneva Medical School, Geneva, Switzerland

Carlos A. Strüssmann
Department of Marine Biosciences, Tokyo University of Marine Science and Technology, Tokyo, Japan

Makoto Tachibana
Institute of Advanced Medical Sciences, Tokushima University, Tokushima; Graduate School of Frontier Biosciences, Osaka University, Osaka, Japan

Minoru Tanaka
Division of Biological Science, Nagoya University, Nagoya, Japan

Erica V. Todd
Department of Anatomy, University of Otago, Dunedin, New Zealand

Eric Vilain
Center for Genetic Medicine Research, Children's National Medical Center; Department of Genomics and Precision Medicine, The George Washington University, Washington, DC, United States

Yoji Yamamoto
Department of Marine Biosciences, Tokyo University of Marine Science and Technology, Tokyo, Japan

Preface

Vertebrate sex determination may be the most evolutionarily diverse developmental process known. While the determination of the body axis, or development of the eye or heart employs highly conserved pathways across vertebrates, the problem of how to determine sex seems to have been solved in many different ways across species. In some cases, this involves the appearance of a novel upstream regulator that gains control of the downstream pathway. However, in other cases, the downstream pathway itself seems to involve different genes, or even when the same genes are involved, they do not maintain the same order in the pathway: cassettes of genes display heterochronic expression patterns (Czerwinski, Natarajan, Barske, Looger, & Capel, 2016). What's more, in mammals a cell fate decision in the supporting cell lineage (to differentiate as pre-Sertoli or pre-granulosa cells) controls the process of sex determination and recruits other cells to follow suit (Koopman, 2016). However, in fish, germ cells seem to play the dominant role, with increases in germ cell number leading to female development, and decreases in germ cell number leading to male (Kossack & Draper, 2019; Tanaka, 2019).

The advent of genome and transcriptome sequencing has made molecular investigation of sex determination pathways possible in a vast array of species. Using these tools, many non-classic model organisms are now being studied, which has resulted in a vastly expanded view of evolutionary solutions to this problem. In this volume, we include reviews that consider fish (Kossack & Draper, 2019; Tanaka, 2019; Yamamoto, Hattori, Patiño, & Strüssmann, 2019), platypus (Daish & Grützner, 2019), mouse (Nef, Stévant, & Greenfield, 2019), and human sex determining systems (Parivesh, Barseghyan, Délot, & Vilain, 2019) as models of this diversity. For example, environmental sex determination has been poorly understood until recently, but owing to transcriptome information in temperature-dependent species, we are beginning to crack this mechanism (Ge et al., 2018). Similarly, sex change in fish has fascinated investigators for 50 years, but only now are we beginning to understand this process as an endocrine response to dominance and/or stress (Gemmell, Todd, Goikoetxea, Ortega-Recalde, & Hore, 2019; Yamamoto et al., 2019).

Classically, "primary sex determination" has referred to the process of cell fate commitment of the supporting cells to granulosa or Sertoli cell fate in the early bipotential gonad. Both transcription factors such as SRY and SOX9 (Gonen & Lovell-Badge, 2019) and epigenetic mechanisms (Miyawaki & Tachibana, 2019) are now known to be important in this process. Our detailed understanding of the molecular pathways involved is evolving rapidly due to the advent of single cell sequencing. A review by Nef et al. (2019) describes the arc of cell fate commitment in supporting cells and the subsequent commitment of other gonadal cell types to their differentiation pathways. Several reviews in this volume illustrate how our advancing understanding of the sex determination in mammals is being translated in a manner that will be relevant to patients with Differences in Sex Determination (DSDs) (Gonen & Lovell-Badge, 2019; Parivesh et al., 2019).

In a review entitled, "Rethinking sex determination of non-gonadal tissues" (Arnold, 2019), Arnold challenges the classic use of the term "primary sex determination" to describe what he believes should more rightly be called, "gonadal sex determination." The reason that the process of gonadal sex determination has been called "primary sex determination" is because the decision within the gonad plays a dominant role in the sexual differentiation of the rest of the organism, including differentiation of the sex ducts, the genitalia, musculature, and the brain. There is very strong evidence that this is the case because when the gonad develops in a manner discordant with the sex chromosomes (e.g., as a testis in an XX individual (Koopman, Gubbay, Vivian, Goodfellow, & Lovell-Badge, 1991), or as an ovary in an XY host (Lovell-Badge & Robertson, 1990)), the sex of the gonad as testis or ovary controls the phenotypic sex of the organism, regardless of the sex chromosome constitution. However, as Arnold points out, more and more studies have indicated a cell autonomous role for the sex chromosomes in some behavior patterns and aspects of early embryonic development, independent of the sex of the gonad. Furthermore, in birds, it is clear that sex chromosomes play a more dominant role (Clinton, Zhao, Nandi, & McBride, 2012; Zhao et al., 2010).

An aspect of the sex determination problem that is not often considered is how heteromorphic sex chromosomes are managed by the cell. After all, the evolution of a pair of heteromorphic sex chromosomes means that physical dosage of some genes will not be the same between males and females, and, perhaps of more significance, some regions of the sex chromosomes will not be paired at meiosis (which is a classic signal to abort meiosis and channel the cell into cell death pathways). How these problems are managed by cells is the subject of a review by Daish and Grützner (2019), who consider this problem in many species with special attention to the platypus, which has evolved a system of 10 sex chromosomes that pair in a chain during meiosis.

In contrast to mammals, in which the X and Y chromosomes differ across most of their length, in fish, heteromorphism between the sex chromosomes is often very limited. This may explain why sex change and sex reversals are "functional" in these species—e.g., they lead to production of fertile gametes of the opposite sex (Tanaka, 2019). This is not true in mammals where sex-reversed individuals are rarely fertile (Weber & Capel, 2018). This may be true for at least two reasons: first, the sex chromosomes disrupt meiosis in the opposite sex (for example, the presence of two X chromosomes arrests meiosis during spermatogenesis); and second, the sex chromosomes have acquired genes required for gametogenesis (for example, the Y chromosome in mammals encodes genes required for spermatogenesis (Yamauchi, Riel, Stoytcheva, & Ward, 2014)). Surprisingly in zebrafish, medaka, and other species, the gonads pass through a transient hermaphrodite phase before committing to adult testis or ovary fate (Kossack & Draper, 2019; Tanaka, 2019). In many species, the number of germ cells at this stage dictates the future sex of gonad differentiation. Surprisingly, at some life stages and under some conditions, both developing oocytes and developing sperm are found in the same gonad. Whereas in mammals, the somatic gonad has strict control over whether the germ cells differentiate as oocytes or sperm (Spiller & Bowles, 2019), in medaka, there is a gene expressed in germ cells that controls this decision independent of the sex of the gonad (Tanaka, 2019).

Regardless of the species, one overriding principle does seem to be conserved. The male and female pathways, within germ cells, somatic cells, and across the gonad, seem to be antagonistic on many levels. Activation of one pathway is always coupled with repression of the opposite pathway. It seems likely that the male and female pathways operate as networks in which the order of genes can be shuffled and/or individual genes can be substituted, as long as the net outcome is promotion of one sex and repression of the other (Capel, 2017).

The evolution of new genes that influence this balanced antagonism is common, and there seem to be "back up" pathways within an organism such that if one dominant layer of control is stripped away, another takes its place. This principle was demonstrated in mice by balanced deletion of key genes from the male and female sides of the network, which resulted in a new equilibrium dominated by remaining elements of the pathways (Jameson, Lin, & Capel, 2012; Lavery et al., 2012; Nicol & Yao, 2015). More recently evidence emerged showing that wild zebrafish still have heteromorphic sex chromosomes, but these were lost in domestic strains and replaced with a multigenic system that also seems to be vulnerable to environmental influences (Kossack & Draper, 2019; Wilson et al., 2014). Transition from one system to another was recently captured in bearded dragons, which normally depend on a ZW system in which females are the heterogametic sex. However, in a group of dragons living in Northern Australia this system was dominated by temperature in warmer climes (e.g., ZZ individuals developed as females), which led to loss of the W chromosome and creation of a lineage of homogametic ZZ temperature-dependent animals within one generation (Holleley et al., 2015).

We can only hope that this amazing diversity and capacity of the system to right itself will be resilient in the face of climate change and other man-made environmental challenges.

Blanche Capel
Department of Cell Biology, Duke University Medical Center, Durham, NC, United States

References

Arnold, A. P. (2019). Rethinking sex determination of non-gonadal tissues. *Current Topics in Developmental Biology*, *134*, 289–315.

Capel, B. (2017). Vertebrate sex determination: Evolutionary plasticity of a fundamental switch. *Nature Reviews. Genetics*, *18*, 675–689.

Clinton, M., Zhao, D., Nandi, S., & McBride, D. (2012). Evidence for avian cell autonomous sex identity (CASI) and implications for the sex-determination process? *Chromosome Research: An International Journal on the Molecular, Supramolecular and Evolutionary Aspects of Chromosome Biology*, *20*, 177–190.

Czerwinski, M., Natarajan, A., Barske, L., Looger, L. L., & Capel, B. (2016). A timecourse analysis of systemic and gonadal effects of temperature on sexual development of the red-eared slider turtle Trachemys scripta elegans. *Developmental Biology*, *420*, 166–177.

Daish, T., & Grützner, F. (2019). Evolution and meiotic organization of heteromorphic sex chromosomes. *Current Topics in Developmental Biology*, *134*, 1–48.

Ge, C., Ye, J., Weber, C., Sun, W., Zhang, H., Zhou, Y., et al. (2018). The histone demethylase KDM6B regulates temperature-dependent sex determination in a turtle species. *Science*, *360*, 645–648.

Gemmell, N. J., Todd, E. V., Goikoetxea, A., Ortega-Recalde, O., & Hore, T. A. (2019). Natural sex change in fish. *Current Topics in Developmental Biology*, *134*, 71–117.

Gonen, N., & Lovell-Badge, R. (2019). The regulation of *Sox9* expression in the gonad. *Current Topics in Developmental Biology*, *134*, 223–252.

Holleley, C. E., O'Meally, D., Sarre, S. D., Marshall Graves, J. A., Ezaz, T., Matsubara, K., et al. (2015). Sex reversal triggers the rapid transition from genetic to temperature-dependent sex. *Nature*, *523*, 79–82.

Jameson, S. A., Lin, Y. T., & Capel, B. (2012). Testis development requires the repression of Wnt4 by Fgf signaling. *Developmental Biology*, *370*, 24–32.

Koopman, P. (2016). The curious world of gonadal development in mammals. *Current Topics in Developmental Biology*, *116*, 537–545.

Koopman, P., Gubbay, J., Vivian, N., Goodfellow, P., & Lovell-Badge, R. (1991). Male development of chromosomally female mice transgenic for Sry. *Nature*, *351*, 117–121.

Kossack, M. E., & Draper, B. W. (2019). Genetic regulation of sex determination and maintenance in zebrafish (*Danio rerio*). *Current Topics in Developmental Biology*, *134*, 119–149.

Lavery, R., Chassot, A. A., Pauper, E., Gregoire, E. P., Klopfenstein, M., de Rooij, D. G., et al. (2012). Testicular differentiation occurs in absence of R-spondin1 and Sox9 in mouse sex reversals. *PLoS Genetics*, *8*, e1003170.

Lovell-Badge, R., & Robertson, E. (1990). XY female mice resulting from a heritable mutation in the murine primary testis determining gene, Tdy. *Development*, *109*, 635–646.

Miyawaki, S., & Tachibana, M. (2019). Role of epigenetic regulation in mammalian sex determination. *Current Topics in Developmental Biology*, *134*, 195–221.

Nef, S., Stévant, I., & Greenfield, A. (2019). Characterizing the bipotential mammalian gonad. *Current Topics in Developmental Biology*, *134*, 167–194.

Nicol, B., & Yao, H. H. (2015). Gonadal identity in the absence of pro-testis factor SOX9 and pro-ovary factor beta-catenin in mice. *Biology of Reproduction*, *93*, 35.

Parivesh, A., Barseghyan, H., Délot, E., & Vilain, E. (2019). Translating genomics to the clinical diagnosis of disorders/differences of sex development. *Current Topics in Developmental Biology*, *134*, 317–375.

Spiller, C., & Bowles, J. (2019). Sexually dimorphic germ cell identity in mammals. *Current Topics in Developmental Biology*, *134*, 253–288.

Tanaka, M. (2019). Regulation of germ cell sex identity in medaka. *Current Topics in Developmental Biology*, *134*, 151–165.

Weber, C., & Capel, B. (2018). Sex reversal. *Current Biology: CB*, *28*, R1234–R1236.

Wilson, C. A., High, S. K., McCluskey, B. M., Amores, A., Yan, Y. L., Titus, T. A., et al. (2014). Wild sex in zebrafish: Loss of the natural sex determinant in domesticated strains. *Genetics*, *198*, 1291–1308.

Yamamoto, Y., Hattori, R. S., Patiño, R., & Strüssmann, C. A. (2019). Environmental regulation of sex determination in fishes: Insights from atheriniformes. *Current Topics in Developmental Biology*, *134*, 49–69.

Yamauchi, Y., Riel, J. M., Stoytcheva, Z., & Ward, M. A. (2014). Two Y genes can replace the entire Y chromosome for assisted reproduction in the mouse. *Science*, *343*, 69–72.

Zhao, D., McBride, D., Nandi, S., McQueen, H. A., McGrew, M. J., Hocking, P. M., et al. (2010). Somatic sex identity is cell autonomous in the chicken. *Nature*, *464*, 237–242.

CHAPTER ONE

Evolution and meiotic organization of heteromorphic sex chromosomes

Tasman Daish, Frank Grützner*
Comparative Genome Biology Laboratory, Department of Molecular and Biomedical Science, School of Biological Sciences, The University of Adelaide, Adelaide, SA, Australia
*Corresponding author: e-mail address: frank.grutzner@adelaide.edu.au

Contents

Current Topics in Developmental Biology, Volume 134
ISSN 0070-2153
https://doi.org/10.1016/bs.ctdb.2019.01.009

Abstract

The evolution of heteromorphic sex chromosomes has occurred independently many times in different lineages. The differentiation of sex chromosomes leads to dramatic changes in sequence composition and function and guides the evolutionary trajectory and utilization of genes in pivotal sex determination and reproduction roles. In addition, meiotic recombination and pairing mechanisms are key in orchestrating the resultant impact, retention and maintenance of heteromorphic sex chromosomes, as the resulting exposure of unpaired DNA at meiosis triggers ancient repair and checkpoint pathways. In this review, we summarize the different ways in which sex chromosome systems are organized at meiosis, how pairing is affected, and differences in unpaired DNA responses. We hypothesize that lineage specific differences in meiotic organization is not only a consequence of sex chromosome evolution, but that the establishment of epigenetic changes on sex chromosomes contributes toward their evolutionary conservation.

1. Introduction

The genetic determination of sex often leads to the differentiation of sex chromosomes. The generally accepted scenario is that sexually antagonistic genes that provide an advantage in either males or females will evolve a sex bias supported by the suppression of genetic recombination. This can allow sequences to rapidly diverge on the sex chromosomes or persist stably with varying degrees of differentiation. The extent of sex chromosome heteromorphism can therefore vary from small chromatin differences on otherwise cytologically indistinguishable chromosomes to highly morphologically differentiated sex chromosomes. Sex chromosomes with small sex specific regions may not interfere with the normal progression of meiosis, which provides greater flexibility to acquire new sex determining systems prior to the establishment of meiotic silencing or pairing mechanisms that mediate passage of unpaired sequences through meiosis. The sex chromosomes of avians and mammals in particular are highly differentiated with extensive X and Y or Z and W specific regions. In addition to the different levels of sex chromosome differentiation some lineages turn over sex chromosomes and sex determination mechanisms rapidly while in some lineages the sex chromosomes are evolutionarily conserved. In this review, we explore the possibility that the conservation of sex chromosomes may be associated with the mechanisms that enable the passage of such highly differentiated sex chromosomes through vertebrate meiosis.

The generation of haploid gametes in meiosis followed by the syngamy of parental genomes are key elements of sexual reproduction. In the first meiotic division the correct alignment and segregation of homologous chromosomes is complex and subject to rigorous surveillance checks to ensure correct pairing and segregation. Consequently, multiple strategies have emerged through evolution to protect genomes from aberrant homolog synapsis, translocations, and unpaired DNA (which arises when heteromorphic sex chromosomes are present). These protective response programmes and checkpoints function to prevent the transmission of gametes that have chromosome losses or gains or to initiate the destruction by programmed cell death of those cells that register changes beyond certain thresholds. Such programmes include both transcriptional and post transcriptional silencing or suppression mechanisms and can involve the establishment and maintenance of epigenetic modifications and also the recruitment of structural histone variants to the unpaired, translocated, or foreign DNA or chromatin. The process of meiotic homolog synapsis is therefore a fundamental process of gamete production. However, due to the complexity of its regulation and wide range of checkpoint mechanisms that have evolved, many aspects remain poorly understood.

While great advances in our understanding of chromosome synapsis, recombination and chromosome segregation have been achieved in model organisms (e.g., mouse, yeast) a more complete appreciation of the species-specific differences in the meiotic organization of autosomes and sex chromosomes is vital to understand how the components and mechanisms underlying gametogenesis have shaped sex chromosome evolution. It is in these contexts that we compare and contrast the emergence and transmission of the heteromorphic sex chromosomes of the major evolutionary lineages through the complexities and obstacles of meiosis and gametogenesis. We also consider the consequences of these fundamental biological processes on reproduction.

2. Differentiation of autosomes into heteromorphic sex chromosomes

In textbooks sex chromosomes are commonly depicted as a large X and a small Y and described as a way of achieving both sex determination and an equal ratio of males and females in a population. Decades of painstaking research has revealed impressive combinations of environmental and genetic sex determining systems, male and female heterogamety and various degrees of sex chromosome heteromorphisms, hemizygosity and monozygosity. This mixed bag is peppered with additional oddities such as meiotic multiples and intraspecies sex chromosome instabilities which confound the formulation of broad mechanistic rules. With more and more sex chromosome systems discovered, scientists have questioned how and why this has occurred, and whether the laws of sex chromosome evolution are applicable across species (Abbott, Norden, & Hansson, 2017; Marshall Graves, 2008; Peichel, 2017; Schartl, Schmid, & Nanda, 2016). Finding "common ground" when describing sex chromosomes and sex determining systems outside mammals and birds is a challenge due to the combination of mechanisms, spectrum of systems even within some populations, and the dynamism and adaptability of the genetic machinery (Schartl et al., 2016). The sex chromosomes of therian mammals, despite the spectacular range and diversity of species within both marsupial and placental lineages, are remarkably stable and conserved in both gene content and origin, presenting an ordered acquisition of strata or "generations of change." This is unlike the situation in amphibians, reptiles, fishes and insects.

All extant heteromorphic sex chromosome pairs originated as autosomal homologs following the random acquisition of a mutation that transformed a gene into a sex determining locus on one of the homolog (Bergero & Charlesworth, 2009; Charlesworth, 1990; Rice, 1996). For mammals this must have occurred back around the split with monotremes but prior to the marsupial divergence, between 180 and 140 Mya, presumably superseding whatever sex determining system was functioning in the primordial mammalian lineage at the time. Theory and observation then describe a period of evolutionary time within which the regions adjacent to this new locus become refractory to recombinatory interactions with the homolog due to the additional acquisition, recruitment or importation of additional sexually antagonistic genes or gene blocks providing reproductive advantage (Fig. 1, Upper panel). The maintenance of recombination inhibition is fostered by inversions, the accumulation of heterochromatin, and additional mutations that gradually or acutely promote the loss of genes that lack similar sex specific functions or advantage. This results in significant loss of homology and formation of morphologically divergent chromosomes, a mechanistic process termed Mullers' Ratchet (Muller, 1964).

The seemingly inevitable gene loss and subsequent size reduction of the chromosome carrying the sex determining gene (W or Y) has invited predictions about the terminal fates of extant mammalian Y chromosomes including the human Y (Charlesworth & Charlesworth, 2000; Graves, 2006, 2016). This has been a topical issue given that numerous examples exist where Y chromosome loss has occurred in some vertebrate species,

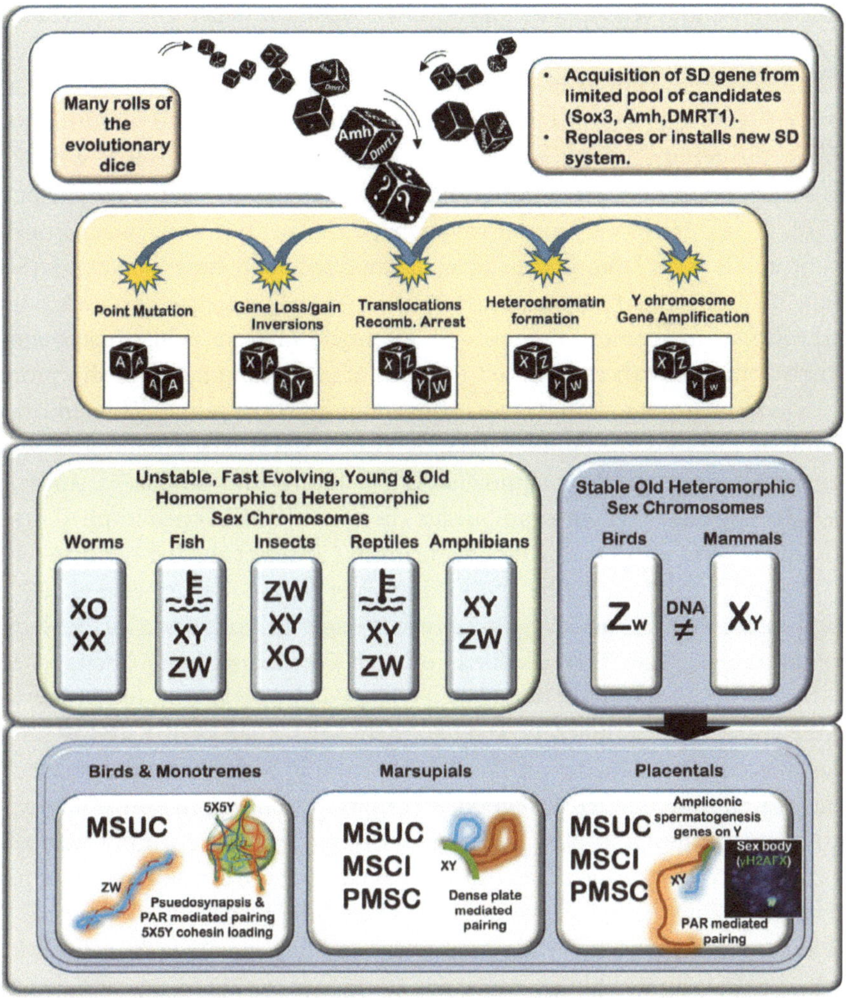

Fig. 1 Amniote sex chromosome evolution and the management of heteromorphic sex chromosomes in meiosis. Upper panel: Heteromorphic sex chromosomes can be formed through recruitment by mutation from an autosomal pair or existing homomorphic sex chromosomes. Initiation of the evolutionary path to heteromorphic sex chromosomes can involve the co-opting of a limited cohort of potential sex determining genes followed by (1) the preservation of a limited number of genes which function in sex specific reproduction in a non-recombining block, (2) maintenance of this gene block through recombination suppression via inversion/translocations, (3) acquisition of heterochromatin and loss or export of sequences/genes not under selective pressure and (4) chromosome specific gene recruitment and/or amplification. Middle panels: Spectrum of sex chromosome systems of major amniote lineages (left) and the relatively stable heteromorphic avian and mammalian sex chromosomes indicating

(Continued)

such as two mole vole species and another rodent in which the Y, including *Sry*, has been lost and replaced with another autosomal *Sox9* regulator, CBX2 (Bagheri-Fam et al., 2012; Kuroiwa et al., 2011; Mulugeta et al., 2016). Part of this reasoning also stems from the extremely limited number of Y genes required for initiation and progression past Meiosis I, which in the case of mice is only two, *Sry* and *Eif2s3y* (Yamauchi, Riel, Stoytcheva, & Ward, 2014). However, the relatively large mouse Y is uncharacteristically euchromatic with large populations of amplified testis expressed genes (Soh et al., 2014). In fact, the long arm is populated with lineage specific ampliconic and palindromic copies of genes with essential functions in spermatogenesis which contrasts with the limited number of genes originally on the proto-Y, called gametologs, which have survived the genetic "purge." The inflation of lineage specific gene families with functions specific to spermatogenesis is a common theme for the mammalian Y, including the human Y, and also includes large fast evolving and lineage specific miRNAs with implied functions in meiotic sex chromosome regulation (Cortez et al., 2014; Costa et al., 2006; Guo, Su, Zhou, & Sha, 2009; Hughes et al., 2012; Hughes et al., 2010; Hughes, Skaletsky, Koutseva, Pyntikova, & Page, 2015; Li et al., 2013; Naqvi, Bellott, Lin, & Page, 2018; Skaletsky et al., 2003; Zhang, Peng, Wang, & Su, 2007; Zhang, Vibranovski, Landback, Marais, & Long, 2010). However, the repeat content and ampliconic and heterochromatic nature of Y and W chromosomes has been a major hurdle in understanding Y chromosome gene content and biology due to its sequence composition being incompatible with orthodox shotgun and next generation sequencing and assembly methods. Fortunately major progress has been achieved on this front allowing comparative Y chromosome assessments and the identification of maintained conserved Y regions in certain amphibian and fish species indicating Y loss may not be an inevitable consequence of the gene attrition pathway on the

Fig. 1—Cont'd lack of common evolutionary origin. Lower panel: Management of the avian, monotreme, marsupial and therian heteromorphic sex chromosomes in prophase I indicating silencing and pairing status. Left: Monotremes and chicken sex chromosomes lack sex chromosome specific silencing (MSCI) but utilize self-association strategies, undergo a period of general transcriptional suppression and show cohesin loading (monotremes only). Middle: marsupial sex chromosomes lack a PAR and maintain proximity via the dense plate and undergo MSCI. Right: Therian sex chromosomes have PAR pairing, are transcriptionally suppressed at pachytene in concert with sex body (inset) formation whereby the sex chromatin accumulates phosphorylated H2AFX (γH2AFX). SD, sex determining; PAR; pseudoautosomal region; MSUC, meiotic silencing of unpaired chromatin; MSCI, meiotic sex chromosome inactivation; PMSC, post meiotic sex chromatin; ≠, no sequence homology.

sex determining chromosome (Bellott et al., 2014; Cortez et al., 2014; Hughes & Page, 2015; Hughes & Rozen, 2012; Soh et al., 2014). In addition to the influx to and amplification of Y/W sequences during differentiation, the maintained Y genes that resided on the neosex chromosomes either reduced expression in tissues other than testis, as occurred in therians, or maintained more ancestral profiles as in birds and monotremes (Cortez et al., 2014). This point of difference between the therian Y and the avian W/monotreme Y was recently supported by data showing the W did not have gene cohorts with oocyte specific expression (Bellott et al., 2017). This work, however, does not include monotremes, whose consideration is integral to the full appreciation of mammalian sex chromosome evolution.

A superficial sketch of amniote sex chromosome evolution reveals contrasting clusters of dynamism and diversity juxtaposed against stable conservation, the latter being represented by birds and therian mammals (Fig. 1, Middle panel). Both represent the emergence and maintenance over evolutionary time of highly differentiated sex chromosomes. Interestingly, in mammals we see the origin of the therian XY common to all extant mammalian lineages except the monotremes which represent a unique multiple sex chromosome system unrelated in origin and sequence to all other mammals, yet with significant homology to bird sex chromosomes (Grutzner et al., 2004; Rens et al., 2007; Veyrunes et al., 2008). Unlike the sex chromosomes of reptiles which show extensive diversity in both sex determining and sex chromosome systems, the sex chromosomes of practically all extant bird species (neognathes) are uniformly and highly differentiated like therian sex chromosomes. Considering how similar the therian XY are in terms of the degree of heteromorphism, the sequence allocated to mediating pairing, the pseudoautosomal region (PAR), can be dramatically different. For example, the marsupials have done away with the PAR completely and while the mouse has a relatively large Y its PAR of ~0.5 Mb is dwarfed by that of the 2.5 Mb human PAR (Helena Mangs & Morris, 2007; Perry, Palmer, Gabriel, & Ashworth, 2001). The functional consequences of PAR size for pairing and potential aneuploidy risk are discussed below. Just as intriguing as the PAR absence in marsupials is the complex arrangement of the platypus 10 sex chromosomes (5X5Y) which, although not all identified, must have 9 PARs to mediate assembly of the meiotic chain. Contrasting the relative stability of extant avian and mammalian heteromorphic sex chromosome systems is the vast spectrum of diverse examples observed in an array of lower vertebrates. Recent work in snakes and amphibians show highly dynamic and even merged sex determining systems within phylogenetically close species, and also new discoveries overturning established sex determining paradigms (Emerson, 2017; Gamble et al., 2017; Miura, 2017). It is clear that different orders of snakes have sex chromosomes representing different stages in the differentiation pathway, from phenotypically identical ZW (Henophidians) to advanced heteromorphic ZWs (Caenophidians); however, this understanding has recently been challenged with work proposing a Henophidian XY system (Gamble et al., 2017) and even more recent work showing the same sex chromosomes have been recruited into both ZW and XY systems (Augstenova et al., 2018). Amphibian species on the other hand show near complete sex chromosome homomorphy and even examples of ZW and XY systems in the same species as observed for *Rana rugosa* (Miura, Ohtani, & Ogata, 2012). This allows a highly flexible system in which genes can be recruited into or disposed of from the sex determining pathway without restrictions arising from the involvement of heteromorphic chromosomes (Miura, 2017). These examples provide fertile ground for the discovery of unknown mechanisms of sex chromosome evolution that can be applied to mammals. We discuss here consequences of such systems in terms of reproduction and gametogenesis in mammals, and the pathways of such systems through key evolutionary crossroads with contrasting perspectives from avian and unusual vertebrate examples.

Advances in molecular cytology and deep sequencing have led to the discovery of sex chromosomes in many vertebrates and invertebrates, challenging previous views on sex chromosome evolution but also providing novel insights and testable hypotheses. The discovery of heteromorphic sex chromosomes also marks the starting point to investigate numerous questions regarding the genetic content, gene expression patterns as well as wholesale epigenetic changes that occur exclusively on sex chromosomes. Such changes include dosage compensation and meiotic organization and meiotic silencing which is observed on sex chromosomes in some lineages. The evolutionary differentiation of sex chromosomes seems in direct contravention with the fundamental purpose and mechanisms essential to the production of genetically diverse gametes. In the following paragraphs we discuss the meiotic fate of heteromorphic sex chromosomes and how the mechanisms that have been specifically established on sex chromosomes may shape sex chromosome evolution.

3. Friend or foe: Pairing and synapsis of heteromorphic sex chromosomes

Meiotic chromosome pairing and recombination are key events ensuring successful sexual reproduction. Once sexual antagonism is established on the proto-sex chromosomes these mechanisms are counterproductive to sex chromosome function. On the other hand, chromosomal pairing is also essential for segregation at meiotic anaphase I. The synaptonemal complex (SC) is a highly ordered and complex structure assembled in prophase I. It serves to mediate and maintain the pairing process between homologous sequences until crossovers (COs) or chiasmata are established and the diplotene stage has commenced in preparation for metaphase I and segregation. Because the SC is intimately linked to the process of recombination, DSB resolution, and more recently CO interference and CO machinery distribution, mutations in the core components negatively impact gamete production and fertility (Libuda, Uzawa, Meyer, & Villeneuve, 2013; Rog, Kohler, & Dernburg, 2017).

The gross arrangement of the SC is a tripartite structure bridging the aligned chromosomal axes and composed of (1) axial elements (AE) which are termed lateral elements (LEs) following SC assembly, (2) transverse filaments (TFs) which are tethered to the LEs and (3) the central elements (CEs) which serve to link the TFs and complete SC assembly. SC assembly, maintenance and disassembly is a highly dynamic and fluid process comprising highly specialized structural and non-structural components, the regulation of which remains unclear (Biswas, Hempel, Llano, Pendas, & Jessberger, 2016; Rog et al., 2017). In terms of evolutionary conservation there is a striking incongruence between the high conservation of the structural components and their arrangements among plants, insects, worms, yeast, birds and mammals, and the low conservation of the underlying amino acid sequences of the key SC proteins (Casey, Daish, Barbero, & Grutzner, 2017; Fraune et al., 2016). Despite this, certain structural motifs and secondary structures are conserved, such as the coiled coil domains of the CEs.

Interestingly SC mutant mice reveal differences in the level of synaptic errors required to induce a checkpoint arrest and apoptosis between the sexes (Bolcun-Filas et al., 2009; Yang et al., 2006). Loss of Synaptonemal complex protein 2 (*Sycp2*) or 3 (*Sycp3*) in males results in general reductions in synapsis and spermatocyte loss at pachytene by cell death, while in oocytes this rigor in pairing monitoring is reduced, resulting in a high incidence of segregation errors and aneuploidy (Yang et al., 2006; Yuan et al., 2002, 2000). This indicates sex differences in core meiotic checkpoint pathways which may account for the observed higher frequency of chromosomal errors in female gametes.

While the general structure and role of the SC in meiotic chromosome organization is highly conserved there are significant variations in the assembly schedule and dependence on the DNA double strand break (DSB) pathway activation for synapsis. Such diversity is shown in female *D. melanogaster* and *C. elegans* where pairing and synapsis can occur independent of recombination, and in male *D. melanogaster* which lacks recombination altogether and also in *S. pombe* and *A. nidulans* which have recombination independent and dependent pairing, respectively, which can occur in the absence of an SC (Bahler, Wyler, Loidl, & Kohli, 1993; Cahoon & Hawley, 2013; Egel-Mitani, Olson, & Egel, 1982; Rog & Dernburg, 2013). Three striking examples involving heteromorphic sex chromosomes highlight this flexibility and structural adaptability of the SC core proteins, one in marsupials involving a structure called the dense plate, one in female chicken involving non-homologous synapsis during ZW equalization, and another in platypus male meiosis involving massive accumulations of a structural cohesin (Casey et al., 2017).

Other sources of variation in the SC include numerous post translational modifications necessary for faithful orchestration of SC assembly and disassembly in the context of DNA damage pathway activation, homology searching and crossover formation. Such essential modifications include SUMOylation, acetylation and phosphorylation and can involve cell cycle regulators such as CDK1-CyclinB and polo-like kinases (PLKs). More recently the 26S proteasome has demonstrated roles in pairing of homologous sequences in budding yeast (Ahuja et al., 2017; Rog et al., 2017).

4. Pairing without homology: The dense plate of the marsupial XY

Unlike the other major mammalian lineages, marsupials lack any region of sex chromosome homology and therefore do not undergo recombination at pseudoautosomal regions prior to their segregation (Page et al., 2003; Solari & Bianchi, 1975). This is observed in all marsupial species studied indicating establishment of this pairing mechanism early in marsupial evolution. Early reports on the mechanism of sex chromosome pairing clearly showed through transmission electron microscopy (TEM) the absence of a tripartite SC structure linking the X and Y; however, it became clear through immunostaining that components of the SC were utilized or recruited from the axial elements to form a proteinaceous structure which notably lacked cohesins and linked the XY through prophase I to allow faithful segregation (Page et al., 2005, 2006). Interestingly, in addition to the axial/lateral element component SYCP3, the dense plate was also constructed from the central element protein SYCP1 indicating multiple SC components were co-opted to assemble the dense plate (Page et al., 2006). DP assembly initiates through mid-pachytene with SYCP3 showing marked association with the still separate X and Y chromosomes. As autosomal SYCP3 begins to be lost from the completely synapsed autosomes from mid-pachytene, the dense plate SYCP3 accumulation increases, and the sex chromosomes become "paired" and move to the nuclear envelope by diplotene similar to the murine sex body (Fig. 1 Lower panel). This trajectory departs from that seen in mouse where the XY pair earlier via their PARs and also exhibit a looping configuration prior to DP formation (Page et al., 2005).

5. The dynamic nature of heteromorphic sex chromosome generation: Meiotic multiples

Yet another variation in the sex chromosome world are complex multiple sex chromosome systems. This curious presence of sex chromosome multiples appears to have independently emerged many times throughout amniote evolution; however, the identification of any defining causative feature has yet to be established. Considering the possible deleterious consequences of having to manage a multi-sex-chromosome system during prophase I synapsis and segregation, it is surprising that meiotic multiples have independently arisen many times in different species. Just as diverse as the lineages in which they are found is the stability of the multivalents, which are observed to form chains through meiosis if composed of sex chromosomes, or more commonly in plants, rings for autosomal multivalents (Gruetzner, Ashley, Rowell, & Marshall Graves, 2006).

The functional advantage or contribution to speciation through which multivalents are preserved within a lineage is highlighted by some recent investigations into primate, amphibian, insect and fish karyotypes (Bueno, Moresco, Konerat, Moreira-Filho, & Margarido, 2016; Gazoni et al., 2018; Gross et al., 2009; Kral, Korinkova, Forman, & Krkavcova, 2011; Mattos, Carvalho, Carvalho, & Schneider, 2018; Roco et al., 2015; Steinberg, Nieves, & Mudry, 2014). Some fish species display enormous meiotic chains of up to 20 elements which are not sex specific. There also are examples of the sex chromosome multiples exceeding the number of the autosomes such as the case with the 6X6Y smokey jungle frog (Gazoni et al., 2018; Gross et al., 2009). Highly dynamic and heterogeneous chain configurations within the same population of scorpions show strikingly high fidelity at segregation with negligible aneuploidy, and involve management of these complex arrangements using pseudosynapsis (pairing of heterologous sequences) and achiasmate bivalents (Mattos et al., 2018). Examples in spiders describe premeiotic X1X2 pairing and inactivation in females and the association of a male univalent X with a centrosome, presumably to allow reliable segregation (Kral et al., 2011). A mammalian example implicating underlying genomic traits conducive to sex chromosome–autosome translocations is the proposed independent formation of the multivalent sex chromosomes of the Meso and South American Howler monkeys involving both Y- and X-autosomal translocations leading to meiotic chain expansions (Steinberg et al., 2014). These examples clearly demonstrate the tendency for chromosome systems to undergo dynamic and rapid rearrangements, and the wide parameters existing to tolerate, stably accommodate and reliably passage them through generations in the face of stringent regulatory checkpoints and potential reproductive failure.

Factors thought to influence the composition of multiple sex chromosomes include (1) the frequency of metacentric (bi-armed) or acrocentric chromosomes in the particular karyotype, and (2) which sex has the heteromorphic sex chromosomes (Pokorna, Altmanova, & Kratochvil, 2014; Yoshida & Kitano, 2012). An observed non-random segregation occurring in female meiosis during formation of the first polar body is thought to be contingent on centromere numbers and therefore a stronger force pulling toward either the polar body, as in humans, or the egg as in mice (Pardo-Manuel de Villena & Sapienza, 2001b; Pardo-Manuel de Villena & Sapienza, 2001c). Correlating centromere positions with multiple sex chromosome status for hundreds of amniote species established that X to autosomal translocations (resulting in XY1Y2 sex chromosomes) were biased toward species with low acrocentric frequencies. In contrast, those with Y to autosomal translocations (resulting in X1X2Y sex chromosomes) were biased toward species with high acrocentric chromosome frequencies (Yoshida & Kitano, 2012). The causative mechanism underlying the female Meiosis I centromere-based segregation bias described above is termed meiotic drive, and cannot influence the evolution or fixation of Y:autosomal Robertsonian fusions as these will not experience passage through the meiotic drive environment of the female (Pokorna et al., 2014). While these are reasonable explanations for the described observations, it remains unclear why mammals have numerous examples of species with sex chromosome multiples, and why avians, which are thought to also have meiotic drive, have not provided any examples of sex chromosome multiples (Axelsson et al., 2010; Pala et al., 2012; Pardo-Manuel de Villena & Sapienza, 2001a). An additional curiosity is the transition from female ZW heterogamety to male monotreme 5X5Y heterogamety, given that the chicken and platypus share significant homology between the Z and X5, and that monotremes have the largest and most stable sex chromosome system in mammals (Grutzner et al., 2004; Rens et al., 2007; Veyrunes et al., 2008).

6. A role for SMCs in sex chromosome organization: Cohesins are key players in the regulation of meiotic pairing dynamics

Cohesins and their close functional and structural relative, the condensins, are multi-component molecules with diverse functions in genome regulation including mitotic and meiotic division, DNA repair, transcriptional control, and chromosome condensation. Mutations affecting cohesins result in reproductive and fertility deficits and cancer (Litwin & Wysocki, 2018; Rankin, 2015; Remeseiro & Losada, 2013). The primary and best characterized roles for cohesin are homolog and sister chromatid tethering during mitotic and meiotic cell divisions. In humans and mice, the meiotic-specific cohesin complex assembles in a tri-partite configuration composed of the structural maintenance of chromosomes (SCC) proteins SMC1β and SMC3, the bridging or linker α-kleisin subunit Rad21, which is also bound by the stromal antigen protein STAG3. This structure forms a ring in which the DNA strands of meiotic homologs and sister chromatids are held until their separations in Meiosis I and II, respectively. Consequently, these proteins are essential for successful gametogenesis (Biswas et al., 2016; Hopkins et al., 2014; Ward, Hopkins, McKay, Murray, & Jordan, 2016). The cohesins and especially SMC3 are integral components of the SC where they participate in chromosome condensation and pairing. Recent evidence has implicated a possible role in sex chromosome specific regulation in the platypus (Casey et al., 2017). Another meiosis specific cohesin variant, RAD21L, is associated with the sex body in mouse and there are functional links between cohesin components, the nucleolus, and heterochromatin formation. The observed massive accumulation of SMC3 and STAG3 specifically to the unpaired regions of the platypus sex chromosome chain through pachytene is an intriguing observation (Casey et al., 2017; Herran et al., 2011; Ishiguro, Kim, Fujiyama-Nakamura, Kato, & Watanabe, 2011). The marked accumulation of these two cohesin components specifically to meiotic sex chromosomes has not previously been observed in any species, and the fact that it is restricted to unpaired DNA and is not associated with the multiple PARs, implicates a role specific to the

10 sex chromosomes. Furthermore, accumulation occurs concomitant with the transient contraction and coalescing of the 10 elements into a heterochromatic paranucleolar mass (Fig. 1 Lower panel). These observations require further investigation, including the determination of the cohesin status of the equalized ZW in pachytene chicken oocytes, to identify the degree of conservation of strategies utilized to manage heteromorphic sex chromosomes.

7. Nucleolar association and its role in heterochromatin formation

The role of heterochromatin in the orchestration of bivalents through prophase I seems to be species specific in terms of chromocenter localization, centromere distributions, ribosomal DNA (NORs) locations and number (Berrios, 2017). One curious and less well understood feature of mammalian spermatocyte heterochromatin dynamics is the relationship and functional interplay between heteromorphic sex chromosomes and the nucleolus and more specifically the role of the nucleolus in heterochromatinization of unpaired DNA and sex chromosome silencing in both gametogenesis and X chromosome inactivation (XCI). The position of the centromere relative to the telomere is accepted as being fundamental to the homology search and synaptic dynamics during and following the bouquet stage. This process also involves the distribution of heterochromatin and associated nucleolar dynamics which occurs in mammalian spermatocytes and has long been a curious feature of meiosis. While its significance remains elusive the role of the nucleolus in sex chromosome silencing in general is gaining more attention.

It has been convincingly shown that the distribution of nucleoli through prophase I is non-random in terms of their central or peripheral location, but is dependent on NOR location(s) relative to the centromeric heterochromatin and the telomeres (Berrios, Ayarza, Moreno, Paulos, & Fernandez-Donoso, 1999; Berrios et al., 2004; Hofgartner, Schmid, Krone, Zenzes, & Engel, 1979; Schmid, Loser, Schmidtke, & Engel, 1982). This process occurs through the early prophase I bouquet stage when nuclear envelope (NE) telomere attachment and heterochromatin coalescence begins as the necessary prelude to homolog synapsis. The observation that heteromorphic sex chromosomes that lack a NOR show transient or persistent nucleolar associations through meiotic prophase I raises questions regarding the functional links between the nucleolar machinery, facultative heterochromatin formation and sex chromosome silencing.

There has been a wealth of work done characterizing the nucleolar-bivalent-XY associations in spermatocytes during prophase I in multiple vertebrate and insect species using techniques such as silver staining and TEM (Beguelini, Marchesin, Azeredo-Oliveira, & Morielle-Versute, 2011; Berrios et al., 2004; Castanhole et al., 2008; Hofgartner et al., 1979; Kierszenbaum & Tres, 1974; Morielle-Souza & de Azeredo-Oliveira, 2008; Sampaio, Pires, da Rosa, & Dias, 2014; Schmid et al., 1982; Solari & Tres, 1970; Stahl, Hartung, Devictor, & Berge-Lefranc, 1984). The coalescence of heterochromatin involved formation of at least one large nucleolus, normally peripherally located, which originated from the union of multiple autosomal/NOR associated smaller nucleoli (Berrios, 2017; Tres, 2005). In insects with multivalent sex chromosomes such as Triatomes, sex chromosome-nucleolar association occurs; however, the authors speculate this is due to sex chromosome located NORs (Madeira et al., 2016; Severi-Aguiar, Lourenco, Bicudo, & Azeredo-Oliveira, 2006). Apart from the meiotic persistence of nucleoli in heteropteran insects, the more commonly observed nucleolar dynamics involve prophase I associations with heterochromatin masses which then ends mid-pachytene with nucleolar component dispersal when ribosomal DNA transcription is suspended (Morielle-Souza & de Azeredo-Oliveira, 2008; Peruquetti, Assis, Taboga, & de Azeredo-Oliveira, 2008). In this context the XY-nucleolar associations were shown to involve granular component invaginations into the XY body, a loss of telomere-NE attachment during this association (Berrios & Fernandez-Donoso, 1990; Solari, 1969a, 1969b; Solari & Tres, 1970), and localization of proteins with likely chromatin modifying or structural roles such as XMR to both the unpaired XY regions and nucleolar domains (Calenda, Allenet, Escalier, Bach, & Garchon, 1994; Escalier & Garchon, 2005).

It remains an open question if sex chromosome-nucleolar associations, whether transient or stable, are required for successful meiotic progression, management of unpaired DNA or maintenance of heterochromatic states, or whether the relationship is collateral to the general heterochromatin-nucleolar interactions. One possible clue to the biological meaning of heterochromatin-nucleolar associations comes from observations in female mammalian cells which undergo XCI. The association of the inactivated X with the nucleolus in female somatic cells has been observed previously; however, only recently has it been demonstrated that this linkage is essential for maintenance of specific repressive marks on the silenced X chromatin and that this interaction involves repeats, CTCF, and lncRNAs (Yang et al., 2015; Zhang, Huynh, & Lee, 2007).

Most relevant to the behavior of heteromorphic sex chromosomes in meiosis is the link between CTCF and cohesins in the process of nucleolar tethering and management of heterochromatin localization (Yang et al., 2015; Yusufzai, Tagami, Nakatani, & Felsenfeld, 2004). In addition to the well documented transient sex body-nucleolar interactions occurring from pachytene in diverse species, recent work in monotremes by Daish, Casey, and Grutzner (2015) show a persistent and tight tethering of the multivalent meiotic chain to a large, central singular nucleolar structure concomitant with massive cohesin accumulations to the unpaired chromatin loops. This suggests that a mechanism of sex chromosome specific heterochromatin maintenance occurs through prophase I in platypus mediated by a nucleolar-cohesin axis. This provides a speculative model of unpaired DNA management through critical meiotic stages involving synapsis and segregation. However, despite the sex chromosome homology between the chicken Z and the platypus sex chromosomes, the nucleolar size and number are highly variable and there is no reported nucleolar-ZW association in the chicken and quail, demonstrating the flexibility of sex chromosome meiotic behavior, even when involving similar gene sets (Andraszek, Gryzinska, Knaga, Wojcik, & Smalec, 2012; Rens et al., 2007).

Another intriguing observation made in platypus prophase I spermatocytes was the tethering of chromosome 6 to the large singular nucleolus (Daish et al., 2015). Not only is this the "future" therian sex chromosome but it also contains the only NOR in platypus. Unlike meiotic nucleolar-NOR association dynamics in mouse, where the multiple nucleoli detach from the NOR bivalent and migrate to the nuclear periphery and the sex body, chromosome 6 remains nucleolar bound through the synapsis programme and into pachytene, while the rDNA activity is presumed to be suppressed due to repressive histone modifications at the site of contact. Such observations invite conclusions, while only speculative, on the potential integration of an autosome into the heteromorphic sex chromosome programme of unpaired DNA checkpoint avoidance mechanisms involving epigenetic and nuclear localization regulations.

8. Meiotic consequences of sex chromosome differentiation: Activation of pathways and mechanisms to identify and manage unpaired DNA

Meiosis functions to divide a diploid genome into haploid gametes in preparation for reconstitution at fertilization into a diploid cell combining the genetic material of two parental genomes. The complex events comprising gametogenesis must occur faithfully in both sexes, however, the presence of heteromorphic sex chromosomes in males results in a distinct programme of sequestration and transcriptional silencing of the XY due to the presence of non-homologous sequences not present in XX females. This silencing occurs in the context of the elaborate schedule of pairing and synapsis of homologous sequences, the concurrent induction of genome wide DNA DSBs, their resolution and resulting recombination and chiasmata formation which enables proper segregation at anaphase I (Page & Hawley, 2003; Zickler & Kleckner, 2015). The response to the presence of unpaired heterologous sequences through both female and male meiosis is a general evolutionarily conserved response called meiotic silencing of unpaired chromatin (MSUC) and involves the silencing and epigenetic modification of any unpaired DNA. However, its manifestation specifically on sex chromosomes in heterogametic species is termed meiotic sex chromosome inactivation (MSCI) (McKee & Handel, 1993; Turner, 2007). While the MSUC response is an ancient eukaryotic defence and meiotic checkpoint mechanism, MSCI is restricted to most therian mammals and is not automatically applied to heteromorphic sex chromosomes during meiosis. MSCI results in the formation of a sex chromosome specific heterochromatic domain, the "sex body," which locates to the nuclear periphery and is easily visualized using antibodies against a well characterized cohort of signature repressive chromatin marks (Fig. 1 Lower panel) and the conspicuous absence of transcriptional activity markers (Chadwick & Willard, 2003; Handel, 2004; Peters et al., 2001; Turner, 2015). Failure of MSCI results in spermatocyte elimination by apoptosis in response to the activation of a pairing checkpoint, and therefore this phenomenon has direct relevance to the cause of genetic disease and fertility (Royo et al., 2010; Turner, Mahadevaiah, Ellis, Mitchell, & Burgoyne, 2006).

Why would there evolve a mechanism to remove spermatocytes that express genes on sex chromosomes during prophase I? On face value this does seem puzzling and requires extension of our current understanding of MSCI and the underlying machinery effecting its establishment. This involves observations in oocytes and spermatocytes with induced pairing errors (Turner, 2005; Turner et al., 2005), identification of gene movements between the X chromosome and autosomes over evolutionary time using large scale comparative genomics approaches (Chen, Krinsky, & Long, 2013; Long & Emerson, 2017; Potrzebowski, Vinckenbosch, & Kaessmann, 2010; Potrzebowski et al., 2008), generation of model organism mutants with impaired MSCI (Baarends et al., 2007; Baudat, Manova, Yuen, Jasin, & Keeney, 2000; Bellani, Romanienko, Cairatti, & Camerini-Otero, 2005; Cloutier et al., 2015; Cloutier, Mahadevaiah, Ellnati, Toth, & Turner, 2016), and characterization of heteromorphic sex chromosome meiotic behavior in multiple species over many years (Daish et al., 2015; Jaramillo-Lambert & Engebrecht, 2010; Jaramillo-Lambert, Harigaya, Vitt, Villeneuve, & Engebrecht, 2010; Moens et al., 1997; Namekawa, VandeBerg, McCarrey, & Lee, 2007; Solari, 1969a, 1974; Solari & Tres, 1967). Some reasonable explanations for MSCI include the prevention of inappropriate recombination between heterologous sequences during the process of DNA DSB repair in late prophase I (Inagaki, Schoenmakers, & Baarends, 2010; McKee & Handel, 1993), the prevention of expression of genes only beneficial to the opposite sex and harmful to gametogenesis progression in the heterogametic sex (Meiklejohn & Tao, 2010), or the precise application of the more general MSUC response to the heteromorphic sex chromosomes to avoid checkpoint detection (Turner, 2007). Fundamentally, the numerous proposed explanations for mammalian MSCI put forward over recent years must be considered alongside the fact that MSCI emerged independently within the context of existing mechanisms which regulate and monitor the pairing status of all meiotic chromosomes in both males and females (Schimenti, 2005; Turner, 2005; Turner et al., 2005). The fact that meiotic silencing factors such as phosphorylated H2AFX (γH2AX) and BRCA1 are also associated with the marsupial XY prior to their union indicates not only that silencing of unpaired DNA in prophase I preceded the marsupial-eutherian split but that the X and Y enter pachytene in a pre-inactivated state prior to formation of a sex body (Franco, Sciurano, & Solari, 2007; Namekawa et al., 2007).

9. Conserved mechanisms act to silence unpaired DNA in meiosis

The most ancient form of a pairing checkpoint was identified in a strain of the bread mold *Neurospora crassa* which had an unpaired copy of the *Asm-1*$^{+}$ allele (Shiu, Raju, Zickler, & Metzenberg, 2001). This seminal Cell paper demonstrated that not only was the unpaired copy silenced but all other copies of the DNA were transcriptionally repressed even if they were paired. Consequently this phenomenon was called meiotic silencing of unpaired DNA (MSUD) and replaced the previous transvection terminology. Subsequent efforts aimed to characterize the MSUD mechanism in terms of unpaired DNA size limits, the nature of the silencing signal dissemination, and the unpaired DNA detection machinery (Kutil, Seong, & Aramayo, 2003; Pratt, Lee, & Aramayo, 2004; Shiu & Metzenberg, 2002). The latter is particularly sophisticated as it was shown that allele polymorphisms comprising as little as 6% could be detected and silenced. The major departure from other demonstrated mechanisms of unpaired DNA silencing is that *N. Crassa* uses a small RNA mediated pathway involving DICER and ARGONAUTE family proteins, that small RNAs are detected matching the unpaired DNA, and that these are lost in small RNA biogenesis pathway mutants (Alexander et al., 2008; Hammond, Spollen, et al., 2013; Hammond, Xiao, et al., 2013; Lee, Pratt, McLaughlin, & Aramayo, 2003). For a thorough recent review of MSUC research see Hammond (2017).

As mentioned, MSUC occurs in prophase I and also represents an ancient and genome-wide response to aberrant pairing functioning in both male and female gonads (Schimenti, 2005). This response is expressed via the transcriptional silencing of any unpaired DNA and accumulation of repressive epigenetic marks and histone variants also seen in the XY sex body. Such pairing checkpoints presumably function to reduce the production of aneuploid gametes and therefore maximize fertility. As explained more fully below the pairing checkpoint is intimately linked to the process of recombination and DNA DSB resolution leading to selective chiasmata formation. The machinery participates in the marking of both heteromorphic sex chromosomes and unpaired autosomes. How directly the initial genome wide programme of DSB formation and resolution is linked to the sex chromatin associated DSB machinery has remained unclear. In both male and female mice the T(X;16)16H translocation shows both the consequences of non-pairing, which is meiotic loss through apoptosis of a proportion of germ cells, but also the "escape" from checkpoint detection via synaptic quadrivalents and accumulation of repressive marks on any remaining unpaired sequences (Odorisio, Rodriguez, Evans, Clarke, & Burgoyne, 1998; Turner et al., 2005). Similarly, the unpaired X chromosome in X0 female oocytes can, in a majority of the oocyte pool, evade checkpoint arrest and elimination by a process of self-association or "pseudosynapsis." The less robust checkpoint operating in female meiosis was recently more thoroughly confirmed and characterized using mutant strains that also revealed incomplete or variable silencing of genes on unpaired sex chromatin based on γH2AFX accumulations (Cloutier et al., 2016). Importantly it was demonstrated that most of these self-associating X chromosome regions also failed to accumulate the silencing factors similar to the other non-synapsing X chromosome oocytes. Pairing with any sequences was shown to mask checkpoint detection as the single X could evade silencing marks by partial pairing with autosomes. In fact this response to unpaired chromosomes or chromosome segments, and consequent recruitment of DNA damage repair (DDR) pathway components, contribute to the viability of gametes with multiple Robertsonian translocations through complex pairing configurations (Manterola et al., 2009). Furthermore the use of mice mutant for components of the silencing machinery (such as γH2AFX, BRCA1, ATR and H2A ubiquitination) shown to locate to the unpaired autosome, or to the X in XO females, XY oocytes and XY in males, prevent the MSUC response, resulting in arrest and elimination by apoptosis (Baarends et al., 2005; Turner et al., 2005). While these meiotic responses to unpaired DNA appear essential to successful gamete production in both sexes in mammals, the fact that similar responses to unpaired Z regions occur in chicken oocytes suggests the MSUC response predates the emergence of the therian heteromorphic XY. However, this fails to account for the seemingly greater reliance on pseudosynapsis between the heteromorphic sex chromosome sequences in chicken oocytes and platypus spermatocytes nor explain why this mechanism was not carried over to the therian XY during its differentiation.

10. The recombination pathway machinery is utilized in the management of the unpaired DNA response

Heteromorphic sex chromosomes must negotiate prophase I in the face of a complex programme of DNA DSBs, homology searching, synapsis of homologous sequences and SC formation, chromosome scale pairing, and resolution of a restricted set of DSBs into crossovers to enable proper segregation at anaphase I. How these unpaired or heterologous sex chromosome sequences are integrated into recombination machinery pathways determines the success of the meiotic cell in producing viable gametes, a process that is dynamic and adaptive, depending on the sex and species. In all species studied, DSBs are induced by the topoisomerase-like transesterase Spo11 and, depending on the status or extent of recombination occurring in a particular species, can differ in number from approximately a dozen in *Drosophila* and yeast to hundreds in mouse (de Massy, 2013; Keeney, 2001; Keeney, Giroux, & Kleckner, 1997). DSBs are initially processed by the MRN complex (MRE11, RAD50, and NBS1) before the resected 3′ ssDNA ends are stabilized by replication protein A (RPA) prior to replacement by the recombinase Rad51 or DMC1 (Gaines et al., 2015; San Filippo, Sung, & Klein, 2008; Sugiyama, Zaitseva, & Kowalczykowski, 1997; Suwaki, Klare, & Tarsounas, 2011; Tsubouchi & Ogawa, 1998; Wang & Haber, 2004). However, the number of DSBs resolved into crossovers is restricted to 1–2 per chromosome to enable genetic exchanges to occur and establish tensioned alignment at metaphase prior to segregation. Meiosis specific bias toward non-homologous crossovers exists in the context of controls over DSB distributions and CO interference, which also restrict chiasmata number and general location (Anderson, Reeves, Webb, & Ashley, 1999; Cole, Kauppi, et al., 2012; Cole, Keeney, & Jasin, 2012; Hollingsworth, 2010; Lao & Hunter, 2010; Martini, Diaz, Hunter, & Keeney, 2006). The DSB programme is intimately linked to the consequent process of pairing and SC formation, but also to the way in which sequences which share limited homology are able to "survive" meiosis in parallel to this schedule of homologous pairing. In most species studied, the sex chromosomes maintain a limited region of homology, the pseudoautosomal region (PAR), which enables restricted pairing and therefore inclusion in the schedule of pairing and crossover resolution, and importantly the production of chiasmata ensuring amphitelic spindle attachments and proper segregation due to this bi-orientation. XY PAR pairing and generation of the single chiasma is observed to occur later than the autosomal pairing schedule and remarkably may involve a Spo11 variant specific to the XY PAR chiasma (Kauppi et al., 2011). DDR components such as Rad51 persist on the unpaired XY sequences into pachytene and diplotene at which time the DNA DSBs must be resolved by the usage of a sister chromatid template as homologous recombination (HR) is not possible due to the absence of homologous sequences (Moens et al., 1997). The sites and number of DSBs and resolved crossovers is a highly regulated and complex process that acts as a key driver of evolution by promoting genetic variation. It has therefore been the focus of considerable research efforts over many years (Baudat, Imai, & de Massy, 2013; Bolcun-Filas & Schimenti, 2012; Zickler & Kleckner, 2015).

After the early prophase I wave of Spo11 mediated DSBs in mouse, there follows a closely monitored and regulated process of break repair that preferentially utilizes homologous chromosomes, a process possibly influenced by assembly of SC components or HORMA domain (HORMAD) containing proteins along the chromatin axes (Li, Bolcun-Filas, & Schimenti, 2011; Shin et al., 2010; Wojtasz et al., 2012). Recent work in *C. elegans* also supports a role for SC components in regulating final crossover distribution per chromosome (Libuda et al., 2013). The SC may also function to locate the initial DSBs to DNA at the chromatin axes, after which the DNA damage signaling machinery migrates and populates the chromatin loops, thus filling the whole XY body as described for γH2AFX, MDC1 and RNF8, and additional modifications such as ubiquitination and sumoylation that lead to transcriptional silencing (Huen et al., 2007; Ichijima et al., 2011; La Salle, Sun, Zhang, Matunis, & Handel, 2008; Lu & Matunis, 2013; Lu, Xiong, Kuang, Korakavi, & Yu, 2013; Mailand et al., 2007). Key players in the strand invasion and consequent homology search and repair process, such as Rad51 and MLH1, or the DDR sensor kinases ATR and ATM, can be used as markers to identify the timing of PAR pairing or degree of asynapsis, the latter identified by localization of BRCA1 and TOPBP1 specifically on unpaired chromatin (Burgoyne, Mahadevaiah, & Turner, 2009; Keeney & Neale, 2006; MacQueen & Hochwagen, 2011; Neale & Keeney, 2006; Royo et al., 2013). As the leptotene wave of DSBs is required for proper pairing and synapsis, the persistence of DDR proteins mark asynaptic homologs, translocations, or the heterologous sex chromosome arms (Burgoyne et al., 2009; Mahadevaiah et al., 2008). One key difference between accumulated proteins on asynaptic autosomes and unpaired sex chromosomes, such as HORMAD 1/2, BRCA1 and TOPBP1, is the kinase targeting them for phosphorylation. Both the early wave of DSBs and the later sex chromosome specific wave both accumulate the DDR components resulting in phosphorylation of H2AFX. However, the latter wave requires the specific recruitment of ATR instead of ATM (Barchi et al., 2005; Bellani et al., 2005; Mahadevaiah et al., 2001). Unlike the persistence of asynaptic autosomes, the presence and subsequent marking of heteromorphic sex chromosomes with DDR machinery fails to induce a checkpoint or result in spermatocyte elimination by apoptosis (Burgoyne et al., 2009; Manterola et al., 2009).

The dependency of pairing on DNA DSBs is revealed in mutants for components such as Spo11 or MDC1 which show blocking of meiotic progression beyond early prophase I and disrupted or incomplete synapsis (Baudat et al., 2000; Pittman et al., 1998; Romanienko & Camerini-Otero, 2000; Yoshida et al., 1998). Interestingly Spo11 mutant mice show Spo11-independent DNA DSBs and the presence of an XY "pseudo body" containing asynaptic DNA but not sex chromatin (Carofiglio et al., 2013). Recent work in mice on Spo11 isoform mutants shows XY PAR pairing deficits result in apoptosis and infertility, and that this cell death is contingent on the breaching of thresholds of XY PAR non-crossovers (Faisal & Kauppi, 2016; Kauppi et al., 2011). The Spo11-independent DSBs are possibly carryovers from the premeiotic cell cycle. However, the degree of disconnect between the DSB programme and ability to initiate the pairing process remains obscure as these mutants appear to continue a schedule of "blind" synapsis not contingent on homology. Thus it appears the spermatocyte can maintain an MSUC checkpoint in concert with these pairing attempts. The persistence of unpaired DNA, represented by their association with DDR and repair machinery, is easily visualized by immunostaining for γH2AFX which spreads and coats the unpaired sex chromatin in addition to any other heterologous or unsynapsed sequences (Burgoyne et al., 2009; Perera et al., 2004; Refolio, Cavero, Marcon, Freire, & San-Segundo, 2011; Turner et al., 2005). In fact the unpaired XY accumulates a host of repressive changes in addition to γH2AFX such as histone methylation and deacetylation, ubiquitination and sumoylation, the presence of which are usually dependent on the presence of different DDR components (Inagaki et al., 2010; Royo et al., 2013; Turner, 2015; Wojtasz et al., 2012).

During therian prophase I the chromatin loops of the XY, as described above, attracts the phosphorylated form of the H2AFX histone variant, a process mediated by BRCA1 and ATR during the second wave of sex chromosome specific γH2AFX deposition. The Turner group has revealed that a key component in ATR guidance and sex body epigenetic regulation is TOPBP1, which is involved in the regulation of both sex chromosomes and autosomes (ElInati et al., 2017). In addition to this hallmark feature is an ensemble of classical repressive epigenetic modifications, but recent work surprisingly also showed changes normally associated with gene activation (Hasegawa et al., 2015; Maezawa, Hasegawa, Yukawa, et al., 2018) The Namekawa group demonstrated open chromatin signatures at intergenic and intronic sequences specific to sex chromosomes during MSCI using a mouse mutant for the PRC1 component SCML2. SCML2 is thought to be recruited to the sex chromatin by interacting with γH2AFX-MDC1, and is subsequently responsible for suppressing H2AK119 ubiquitination on pachytene sex chromatin, therefore playing roles in both opening sex chromatin during MSCI and then closing it in spermatids (see below) by the removal of ubiquitin on H2AK119 (Adams et al., 2018; Hasegawa et al., 2015; Maezawa, Hasegawa, Alavattam, et al., 2018; Sin, Kartashov, Hasegawa, Barski, & Namekawa, 2015). This demonstrates that multilayered epigenetic regulatory pathways exist both effecting repressive epigenetic modifications (MSCI), and setting up specific gene cohorts for the ensuing post MSCI period when spermatogenesis specific gene expression is required. Another key meiotic regulator is the bromodomain containing epigenetic reader protein BRDT that regulates gene expression by nucleosome and protein associations and by recruiting other remodeling proteins to acetylated histones (Miller et al., 2016). This protein is essential for male gametogenesis. Its loss disrupts MSCI and synapsis of the sex chromosome PAR; however, the schedule of DSBs (γH2AFX), their repair (Rad51), and autosomal synapsis remain unaffected (Manterola et al., 2018). Furthermore it was shown that BRDT loss disrupted the establishment and distribution of both repressive (H3K9me3 and H3K4me) and activating (H3K9ac) chromatin modifications specifically on the sex chromosomes, resulting in loss of transcriptional suppression (MSCI) visualized by the presence of RNA PolII over the XY. This recent work by the Namekawa and Wolgemuth groups highlights the complexity of the sex chromatin specific regulatory pathways following the DDR resolution and its intimate relationship to MSCI maintenance, chromatin modulation, and post meiotic gene regulation.

11. The evolutionary history of meiotic silencing

While there may be conserved MSUD and/or MSUC responses to unpaired DNA present in diverse vertebrate species and even variant forms in insects, yeast and fungi, the expression of this repressive response mechanism specific to the heterologous sequences of differentiated sex chromosomes (MSCI) appears as flexible and dynamic as the emergence and decay of the sex chromosomes themselves. This sex chromosome specific transcriptional silencing is centered at pachytene, at which time it is markedly contrasted by the enhanced state of transcription of the autosomes (Namekawa et al., 2006; Turner, 2015). This differential transcriptional activity, possibly a consequence of co-opting MSUC specifically to heteromorphic sex chromosomes, induced the transposition of genes required for spermatogenesis from the X (Emerson, Kaessmann, Betran, & Long, 2004; Zhang et al., 2010). Interestingly this has been shown as having also occurred in XY species which have ambiguous MSCI status such as insects (Betran, Thornton, & Long, 2002; Vibranovski, Lopes, Karr, & Long, 2009; Vibranovski, Zhang, & Long, 2009). Presumably the persistence of sequestering repair pathway components to asynaptic autosomal sites over the protracted pachytene period may reduce the efficacy of MSCI, resulting in aberrant and inappropriate X chromosome gene transcription and death. The fact that both chicken and platypus sex chromosomes do not exhibit therian like sex body formation (MSCI) through female and male prophase I, respectively, and also lack a differential state of transcription between autosomes and sex chromosomes is intriguing.

MSUC and MSCI also function in the germline of hemizygous XO male worms and include similar cohorts of repressive chromatin changes to effect transcriptional silencing (Checchi & Engebrecht, 2011a, 2011b; Kelly et al., 2002; Strome, Kelly, Ercan, & Lieb, 2014). See Maine (2010) for a comprehensive review of meiotic silencing in *C. elegans* (Maine, 2010). This most extreme form of heteromorphism results in a similar schedule of checkpoint avoidance as that observed to occur to the single unpaired accessory chromosome of Zebrafinch males and also in hemizygous XO male grasshoppers (Cabrero, Teruel, Carmona, & Camacho, 2007; Cabrero, Teruel, Carmona, Jimenez, & Camacho, 2007; Pigozzi & Solari, 1998, 2005; Viera, Calvente, et al., 2004; Viera, Santos, et al., 2004). While the single X in male worms undergoes meiotic condensation and transcriptional repression, there are differences in the utilized effector pathways between closely related *Caenorhabditis* species in terms of specific histone 3 lysine 9 methyltransferases, even though they all have MSCI (Larson, Van, Nakayama, & Engebrecht, 2016). While the meiotic unpaired DNA response checkpoints of nematodes appear to be similar to those occurring in mammals, the consequence of its failure may be less significant in terms of gamete viability, inviting questions about the reasons for its maintenance in the germline (Bessler, Andersen, & Villeneuve, 2010; Checchi & Engebrecht, 2011b; Jaramillo-Lambert & Engebrecht, 2010; Larson et al., 2016).

The response to heteromorphic sex chromosomes in insect male meiosis and the requirement for an MSUC/MSCI response remains surprisingly less clear in the case of the genetic and developmental model organism *Drosophila melanogaster*. Unlike the hemizygous XO male grasshopper, which as described has a condensed and transcriptionally silenced meiotic X chromosome, the XY chromosomes of male *Drosophila* are yet to be unambiguously characterized as undergoing meiotic silencing despite the application of numerous modern genomic assessment technologies (Deng et al., 2011; Hense, Baines, & Parsch, 2007; Kemkemer, Hense, & Parsch, 2011; Landeen, Muirhead, Wright, Meiklejohn, & Presgraves, 2016; Meiklejohn, Landeen, Cook, Kingan, & Presgraves, 2011; Mikhaylova & Nurminsky, 2011; Vibranovski, 2014; Vibranovski, Lopes, et al., 2009; Vibranovski et al., 2012). A curious feature of *D. melanogaster* male meiosis is the absence of recombination, which may argue against the function of "orthodox" MSCI, due to the non-requirement of mechanisms preventing exchange of genetic material between heterologous sequences (McKee, 2009; Morgan, 1912). However, the fact that many genes have been transposed from the X to autosomes supports the idea that relocation maintains expression of genes important for spermatogenesis in the face of transcriptional quelling or MSCI as proposed for the mammalian X (Betran et al., 2002; Betran, Emerson, Kaessmann, & Long, 2004; Long & Emerson, 2017; Potrzebowski et al., 2008; Vibranovski, Zhang, & Long, 2009). The fact that neither expression analyses in testis of transgenes inserted on the X (Hense et al., 2007; Hoyle, Hutchens, Turner, & Raff, 1995) nor X:autosomal global transcriptional output ratios (Meiklejohn et al., 2011; Vibranovski, Lopes, et al., 2009) provide convincing evidence either way suggests that the *Drosophila* XY does not experience a typical MSCI programme. Cytological studies using surface spreads, in combination with immunofluorescence against MSCI proteins such as γH2AFX, ATR, H3K27me3, and RNA FISH, may illuminate the MSCI status in flies. Studies also demonstrate a role for cohesin axis components in the pairing and segregation of achiasmatic *Drosophila* homologs. These studies suggest a role in the regulation of the alternate segregation of the 5X and 5Y sex chromosomes in the platypus, which show late prophase I cohesin coating of the unsynapsed regions (Casey et al., 2017; Gyuricza et al., 2016).

12. Birds and monotremes use alternative meiotic silencing mechanisms to deal with the unpaired DNA of heteromorphic sex chromosomes

Using physical mapping with BAC clones, it was surprisingly shown that the multiple monotreme sex chromosomes had no homology to the eutherian XY but had significant regions of homology between the X3X5 of platypus and the chicken Z (Rens et al., 2007; Veyrunes et al., 2008). Monotreme chromosome 6 in fact was identified as the mammalian autosome from which the therian sex chromosomes were recruited. This presents the intriguing scenario whereby homologous sex chromosome sequences, while being maintained in sex determining sex chromosome systems in both chicken and monotremes, transition from the female ZW to the male XY system. This therefore presented unique systems in which to assess the evolution, not only of sex determining systems, but relevant to this review, the emergence of mammalian meiotic silencing mechanisms, and the dependence on MSCI and MSUC as response programmes to prevent meiotic progression and reduce nondisjunction and aneuploidy in the face of aberrant synapsis. We now compare and contrast the avian and monotreme sex chromosome meiotic dynamics with each other and the therian XY and track the emergence of mammalian MSCI.

Unlike the typical therian response to unpaired DNA in male meiosis described above, birds and monotremes appear to be different in terms of the selective recruitment of silencing factors to heterologous sex chromosome DNA, and the manner in which the sex chromosomes self-associate through prophase I (Casey et al., 2017; Pigozzi, 2016). Three key papers have investigated MSCI in chicken and platypus and shown some similarities in the sex chromosome behavior in these species (Daish et al., 2015; Guioli, Lovell-Badge, & Turner, 2012; Schoenmakers et al., 2009). The other aspect relevant to the passage of heterologous sequences through prophase I is the aforementioned process of sex chromosome self-synapsis or self-association, which invites speculation on its role in unpaired DNA checkpoint avoidance. One significant difference between platypus and chicken is the fidelity of this self-association or "pseudo-synapsis" process, which appears less efficient in chicken as revealed by the accumulation of DDR components to the Z when self-synapsis fails to occur in a minority of oocytes (Guioli et al., 2012). Furthermore the higher occurrence of synaptic errors in birds and the corresponding lack of a meiotic arrest response, even in the presence of translocations and aberrant pairing, is highlighted by the maintenance of a large heterochromatic univalent in the Bengalese and Zebra Finch (del Priore & Pigozzi, 2014; Pigozzi & Solari, 1998, 2005). This large univalent chromosome, which undergoes synapsis with its homolog in oocytes, enters male prophase in an inactive state and displays repressive histone marks or variants such as methylation of histone 3 lysine 9, methylation of histone 4 lysine 20, and HP1β (del Priore & Pigozzi, 2014; Goday & Pigozzi, 2010; Inagaki et al., 2010). In the heterogametic females, hallmark MSCI marks such as γH2AFX, fail to accumulate on the heterologous sequences, and typical heterochromatic marks are restricted to the W that appears, similar to the abovementioned accessory chromosome in males, to enter prophase I in a pre-inactivated state termed meiotic silencing pre-synapsis (MSPS) (del Priore & Pigozzi, 2014; Guioli et al., 2012).

The prototherian platypus similarly has sex chromosomes that fail to accumulate γH2AFX, but whose prophase I chromosomes accumulate a wave of DNA damage repair pathway components (RPA32). However, unlike therian mammals, these marks do not persist on the heterologous sex chromosome sequences possibly due to the analogous process of self-association or pseudo-pairing (Daish et al., 2015). Two notable differences between the behavior of heteromorphic sex chromosome in birds and platypus is both the absence of the SC transverse filament SYCP1 at points of sex chromosome pseudosynapsis, and nucleolar tethering of the coalesced sex chromosome chain in platypus prophase I. However, the limited analyses on monotreme sex chromosomes through male meiosis indicate a similar lack of sex chromosome specific silencing marks like γH2AFX, an equivalent period of general transcriptional repression, and the described highly regulated sex chromosome self-association indicating an avian-like schedule of sex chromosome dynamics through prophase I (Pigozzi, 2016). Unfortunately, the monotremes are currently restricted in terms of functional in vivo testing that might reveal and dissect additional important meiotic pairing pathways relevant to heteromorphic sex chromosome behavior through meiosis to allow adequate comparative analyses between avians and therian mammals.

A curious feature of avian sex chromosome meiotic association involves an initial orthodox PAR pairing and consequent single recombination nodule formation, but also an accompanying process of heteromorphic synapsis or "pseudo-synapsis" between the nonhomologous ZW sequences termed synaptic adjustment and equalization (Rahn & Solari, 1986). This was observed in numerous carinate species whereby a schedule of W lengthening and Z shortening occurred to allow a form of full length sex chromosome alignment (Hogan, Jarrell, Ryder, & Greenbaum, 1992; Solari & Moses, 1977; Solari & Pigozzi, 1993). This phenomenon was also observed in the more primitive ratite *Rhea americana,* which has less differentiated ZW chromosomes and, interestingly, displayed multiple recombination nodules (Pigozzi & Solari, 1997). Not only is the ZW observed to align, albeit in a twisted configuration with the Z wrapping around the W, an SYCP1 junction forms between the equalized ZW length at mid-pachytene that completely overlaps with the LE SYCP3 that is then lost upon ZW desynapsis during late pachytene (Schoenmakers et al., 2009). Two excellent high resolution analyses of this process in the context of meiotic silencing in chicken were conducted by the Baarends and Turner groups demonstrating that pseudo synapsis of the heterologous Z and W sequences serves to transiently protect the non-homologous sequences from unpaired DNA checkpoint machinery and a consequent silencing response.

13. Anything goes: MSCI of pseudo-heteromorphic sex chromosomes in an XX male mammal

As described above there exists remarkable variation in sex chromosome management strategies through evolution including which sex evolves heteromorphic sex chromosomes, the requirement for sex chromosome specific silencing, generation of complex sex chromosome multiples, and the adaptive utilization of structural synapsis components to enable sex chromosome interactions such as the marsupial dense plate. The recent molecular characterization of closely related mole vole species further highlights the extant variety existing within the heteromorphic sex chromosome and meiotic management/checkpoint space, by demonstrating MSCI on an isomorphic X chromosome pair in a male mammal (Matveevsky, Bakloushinskaya, & Kolomiets, 2016; Matveevsky, Kolomiets, Bogdanov, Hakhverdyan, & Bakloushinskaya, 2017; Mulugeta et al., 2016). Furthermore, the sequencing of two of these species, *Ellobius talpinus* and *Ellobius lutescens*, revealed that not only the loss of the Y chromosome, including *Sry*, occurred independently on two occasions in this lineage, but that a limited number of Y chromosome genes were retained and present in both males and females. While *E. lutescens* has male and female X0 status, *E. talpinus and E. tancrei* have both male and female XX sex chromosomes. What is particularly intriguing is the fact that in oocytes the two X chromosomes, similar to autosomes, undergo full synapsis and SC formation along the entire chromosome axis while in testis the same two X chromosomes have pairing limited to terminal telomeric regions (Bakloushinskaya et al., 2012; Baklushinskaia et al., 2010; Matveevsky et al., 2016, 2017). Not only do the two Xs in males fail to fully pair, but they also form a peripheral sex body and accumulate typical silencing factors such as H2A ubiquitination, sumoylation, ATR (*E. tancrei* only) and H2AFX phosphorylation similar to a typical therian XY sex body experiencing MSCI (Matveevsky et al., 2016). Quite apart from the question of the sex determination genes, how the same homologs are selected for recruitment into the MSCI/MSUC pathway in one sex and not the other remains unknown. However, the analysis of *E. tancrei and E. talpinus* hybrids demonstrates hybrid sterility even though all meiotic chromosomes appear aligned, adjusted and undergoing full synapsis, revealing an underlying genetic incompatibility. How this relates to the case of male XX MSCI and female fully paired XX chromosomes is not understood, however, *E. lutescens*, which have a single X in males and females, show an inactivated single X (Matveevsky et al., 2017; Mulugeta et al., 2016). Another remarkable feature of *Ellobius* sex chromosomes is the maintenance of an intact X inactivation center (Xic) resulting in inactivation of one X in both the XX female and the XX male (Mulugeta et al., 2016). However, *E. lutescens* are non-viable when their single X is present in two copies, and fail to initiate XCI, meaning the ability to silence one of the 2 Xs has been lost in this species (Lyapunova, Vorontsov, & Zakarjan, 1975).

14. Escape from meiotic silencing and the transcriptional regulation of sex chromosome linked genes beyond meiosis I

Following the initial molecular and epigenetic characterizations of MSCI during murine meiosis it was observed that the sex chromosomes retained repressive histone marks present in the XY body beyond meiosis I that persisted even until mature sperm production (Greaves, Rangasamy, Devoy, Marshall Graves, & Tremethick, 2006; Khalil, Boyar, & Driscoll, 2004; Namekawa et al., 2006; Turner et al., 2006). Prior to these studies it was thought that silenced genes on the XY were reactivated due to their requirement in spermatogenesis, with some sex chromosome linked genes previously observed to be expressed in round spermatids (Hendriksen, 1999; Hendriksen et al., 1995). Using both immunostaining for epigenetic marks and RNA PolII and Cot-1 RNA FISH, it was clearly demonstrated that the XY maintained a state of transcriptional suppression, and included some repressive epigenetic marks present in the XY body such as HP1β, HP1γ and H3K9me2. However, the two hallmark inactive X chromosome histone variants, γH2AFX and macroH2A, were not associated with PMSC (Namekawa et al., 2006). Gene expression analyses indicated a significant sex chromosome specific maintenance of transcriptional silencing occurred in round spermatids, obviously occurring after XY segregation. However, this was incomplete and seemingly selective as localized groups of genes were reactivated after MSCI. Soon after it was shown that a specific class of X linked genes, cohorts of multi-copy or ampliconic genes with high or exclusive testis expression (comprising ~18% of the X chromosome gene content), were transcriptionally reactivated in post-meiotic spermatids (Mueller et al., 2008). This was in contrast to the reactivation of single copy genes that showed limited or incomplete reactivation, suggesting that amplification of essential spermatogenesis genes made them refractory to PMSC mechanisms. Aspects of the PMSC regulatory pathway were revealed in

Y chromosome deletion mutants and by selectively knocking down the multi-copy *Sly* gene using an siRNA transgenic approach (Cocquet et al., 2009; Ellis et al., 2005). This revealed a striking de-repression of XY linked genes normally experiencing PMSC in spermatids, and is thought to involve an H3K9 methylation-CBX1-SLY axis to maintain repressive chromatin environments on the sex chromosomes (Cocquet et al., 2009).

15. How do genes on the silenced sex chromosomes reactivate after MSCI while in the PMSC environment?

Insight into the mechanism responsible for the selective recruitment of genes for transcriptional de-repression within the PMSC came with the identification of the germline specific polycomb protein, Sex comb on midleg-like 2 (SCML2), and the DNA damage response protein Ring finger protein 8 (RNF8), which cooperatively regulate the ubiquitination of regulatory regions on the sex chromosomes to promote gene expression (Adams et al., 2018; Maezawa, Hasegawa, Alavattam, et al., 2018; Maezawa, Hasegawa, Yukawa, et al., 2018; Sin et al., 2012). RNF8 activation, in partnership with MDC1, leads to sex chromosome specific ubiquitination and epigenetic changes promoting gene expression (Lu et al., 2010; Sin et al., 2012). The demonstration that RNF8 mediated H2AK119ub leads to the establishment of activating marks such as H3K4 di- and tri-methylation, the recruitment of the H2AF2 histone variant, and the recently characterized crotonylation of histone lysines specifically on the post meiotic sex chromosomes, identifies RNF8 as a key player in the regulation of spermatogenesis (Adams et al., 2018; Liu et al., 2017; Tan et al., 2011). RNF8 is a component of the PRC1 complex and has known functions in gene repression during development (Wang, 2004). SCML2 is a polycomb protein and also a PRC1 component, acting to inhibit the mono-ubiquitination of H2AK119, resulting in sex chromosome specific gene activation. More recently SCML2 also has been shown to act as a pericentromeric heterochromatin regulator (Hasegawa et al., 2015). How these two proteins function to co-operatively regulate sex chromosome gene expression in spermatids was recently investigated using a double knockout mouse model (Adams et al., 2018). Interestingly, SCML2 selectively removed RNF8-dependent and independent H2AK119ub, but was not involved in RNF8-mediated H2AK119 polyubiquitination. Furthermore, genes that are regulated by RNF8 and SCML2 in spermatid PMSC also were shown to have increased enrichment for the transcriptional activation mark H3K27ac (Adams et al., 2018). Such regulation by DDR components usually involved in sex chromosome silencing following MDC1-γH2AFX pathway activation demonstrates the utilization of MSCI machinery to effect the post meiotic expression of sex linked genes by priming gene enhancers with activating marks. These genes are proven critical for successful spermatogenesis and this work therefore provides valuable insight into sex chromosome biology and epigenetic regulation with direct relevance to human fertility.

While there are similarities between the inactive X in female somatic cells and the inactive XY in males in terms of establishment and maintenance of repressed chromatin, the extent and gene classes escaping silencing differ markedly. The Xi has up to 25% transcriptional activity from its protein coding genes while the meiotic XY "escapees" are restricted to the post-MSCI period as outlined above (Carrel & Willard, 2005). However, while MSCI prevents mRNA production from the sex chromosomes, the surprising presence of X linked miRNA transcripts in pachytene spermatocytes has been shown by multiple groups (Buchold et al., 2010; Liu et al., 2015; Marcon, Babak, Chua, Hughes, & Moens, 2008; Song, Fang, Li, Liu, & Thomas Chao, 2009; Sosa, Flores, Yan, & McCarrey, 2015). While the role of MSUC escapees can be confidently predicted to contribute specifically to spermatogenesis, apart from targeting genes for post transcriptional suppression, the specific function of these miRNAs in prophase I, and why they reside on an otherwise silenced X chromosome, was unclear. Sosa et al. (2015) extended this work using RNA-FISH (and RNA PolII localization) to directly visualize nascent miRNA transcript production (pri-miRNAs) identifying cohorts of X linked miRNAs which escaped MSCI, PMSC, or both (Sosa et al., 2015). It was proposed that MSCI avoidance was in part achieved by the non-random localization of the miRNA cluster X regions to the periphery of the XY body to escape the repressive sex body chromatin environment. The escape of certain X linked miRNAs from MSCI made sense, as their requirement for MSCI and spermatogenic progression had been demonstrated in *Ago4* knockout mice (Modzelewski, Holmes, Hilz, Grimson, & Cohen, 2012). Using similar approaches to visualize nascent miRNA transcripts, Royo et al. (2015) came to the opposite conclusion that X linked miRNAs were, similar to coding genes, silenced by MSCI (Royo et al., 2015). However, they demonstrated the consequences of miRNAs being expressed through pachytene by employing a miR-BAC transgene approach which resulted in elevated programmed cell death at the mid-pachytene stage (Royo et al., 2015). While similar miRNAs were visualized by RNA-FISH between the Turner and McCarrey groups leading to differing results, the fact that miRNAs are required for spermatogenesis (*Ago4* mutant), but also kill spermatocytes if actively expressed during pachytene (miR-BAC expression), and that mature miRNA cohorts have been detected by multiple groups throughout all meiotic and post-meiotic stages, indicates additional dissection of the critical MSCI/PMSC windows are required using alternative methods. Quite apart from the specific question of whether miRNA clusters are actively silenced by MSCI at pachytene, is the puzzle of why mature miRNAs present due to their perdurance from preceding stages do not induce apoptosis, while overexpression of the same miRNA at a similar meiotic stage does induce cell death.

16. Outstanding questions and future perspectives

Amniote sex chromosome systems can be highly variable due to rapid differentiation or can be stably maintained as genetically and morphologically conserved structures. Rapid changes in gene acquisition and loss interact with ancient checkpoint and surveillance responses associated with non-pairing, and even the complete loss of homologous regions comprising the PARs, associated with fundamental programmes such as DSB formation, repair, and recombination. As some sex chromosome systems have evolved exclusive silencing responses while others have not, such changes may be context dependent contributors to the evolutionary trajectory of these sex chromosomes. This process occurs in parallel to the specific selection pressures of the sequences involved, which are to varying degrees tolerated, negotiated, or blocked by the employment of gross or fine scale genomic changes consisting of transposition, heterochromatinization or translocation. For this reason, future meiotic work requires analyses to be conducted in a much broader range of species featuring different or unorthodox sex chromosome systems.

Great advances have been made toward the understanding of sex chromosome gene expression regulation in meiosis, and sex chromosome evolution in general, by the application of transcriptomics. For example, large scale comparative RNA-seq datasets identified a lack of backup sex chromosome retrogenes on autosomes in monotremes, which suggested an absence of MSCI, later confirmed and characterized through male platypus meiosis. Such approaches have also contributed to an understanding of the dosage compensation status of monotremes and birds which, while unlike the wholesale transcriptional suppression involved in MSCI and

XCI, offers opportunities for greater appreciation of more subtle gene or region-specific regulation.

The observed interactions between the nucleolus and sex chromosomes have been documented in many evolutionarily distinct species. However, the necessity or nature of this interaction in terms of silencing causation or consequence, and the resultant epigenetic contributions, remain major questions in meiotic heteromorphic sex chromosome dynamics. Whether nucleolar association was an instrumental step leading to therian MSCI and sex body formation requires further exploration in key lineage contexts prior to marsupial and therian divergence, such as bird oogenesis and monotreme spermatogenesis, which passage not only similar sex chromosome genes and require sex chromosome pseudosynapsis, but also lack sex chromosome specific silencing. Consequently, nucleolar contributions to sex chromosome epigenetic establishment or maintenance, especially for examples such as the cohesin-laden multiple sex chromosomes of platypus, remain unresolved.

Some major outstanding questions remain in areas such as the contribution of noncoding RNAs and post-transcriptional regulatory mechanisms acting on gene networks, including sex chromosomes in gametogenesis, and their possible contributing roles in idiopathic infertility. Also the functional relevance to gamete formation and reproduction of the fast evolving X linked lncRNAs with lineage specific transcriptional complexity remains unknown, while the mechanisms leading to and functionality of the ampliconic testis specific Y gene cohorts is also incomplete. In terms of the physical management of heteromorphic sex chromosomes, the molecular details of the alternate segregation of complex meiotic multiples, or how segregation bias is achieved as well as a greater understanding of sex differences in meiotic pairing checkpoint stringencies, are either incomplete or unknown. Finally, appreciation of the evolutionary or functional rules governing specific unusual examples such as the XX male and female mole voles which have MSCI, warrants attention to fully understand the degree to which they depart from those forces shaping the majority of mammalian sex chromosomes.

References

Abbott, J. K., Norden, A. K., & Hansson, B. (2017). Sex chromosome evolution: Historical insights and future perspectives. *Proceedings of the Biological Sciences*, *284*(1854), 1–9.

Adams, S. R., Maezawa, S., Alavattam, K. G., Abe, H., Sakashita, A., Shroder, M., et al. (2018). RNF8 and SCML2 cooperate to regulate ubiquitination and H3K27 acetylation for escape gene activation on the sex chromosomes. *PLoS Genetics*, *14*, e1007233.

Ahuja, J. S., Sandhu, R., Mainpal, R., Lawson, C., Henley, H., Hunt, P. A., et al. (2017). Control of meiotic pairing and recombination by chromosomally tethered 26S proteasome. *Science*, *355*, 408–411.

Alexander, W. G., Raju, N. B., Xiao, H., Hammond, T. M., Perdue, T. D., Metzenberg, R. L., et al. (2008). DCL-1 colocalizes with other components of the MSUD machinery and is required for silencing. *Fungal Genetics and Biology*, *45*, 719–727.

Anderson, L. K., Reeves, A., Webb, L. M., & Ashley, T. (1999). Distribution of crossing over on mouse synaptonemal complexes using immunofluorescent localization of MLH1 protein. *Genetics*, *151*, 1569–1579.

Andraszek, K., Gryzinska, M., Knaga, S., Wojcik, E., & Smalec, E. (2012). Number and size of nucleoli in the spermatocytes of chicken and Japanese quail. *Folia Biologica*, *60*, 121–127.

Augstenova, B., Johnson Pokorna, M., Altmanova, M., Frynta, D., Rovatsos, M., & Kratochvil, L. (2018). ZW, XY, and yet ZW: Sex chromosome evolution in snakes even more complicated. *Evolution*, *72*(8), 1701–1707.

Axelsson, E., Albrechtsen, A., van, A. P., Li, L., Megens, H. J., Vereijken, A. L., et al. (2010). Segregation distortion in chicken and the evolutionary consequences of female meiotic drive in birds. *Heredity (Edinburgh)*, *105*, 290–298.

Baarends, W. M., Wassenaar, E., Hoogerbrugge, J. W., Schoenmakers, S., Sun, Z. W., & Grootegoed, J. A. (2007). Increased phosphorylation and dimethylation of XY body histones in the Hr6b-knockout mouse is associated with derepression of the X chromosome. *Journal of Cell Science*, *120*, 1841–1851.

Baarends, W. M., Wassenaar, E., van der Laan, R., Hoogerbrugge, J., Sleddens-Linkels, E., Hoeijmakers, J. H., et al. (2005). Silencing of unpaired chromatin and histone H2A ubiquitination in mammalian meiosis. *Molecular and Cellular Biology*, *25*, 1041–1053.

Bagheri-Fam, S., Sreenivasan, R., Bernard, P., Knower, K. C., Sekido, R., Lovell-Badge, R., et al. (2012). Sox9 gene regulation and the loss of the XY/XX sex-determining mechanism in the mole vole *Ellobius lutescens*. *Chromosome Research*, *20*, 191–199.

Bahler, J., Wyler, T., Loidl, J., & Kohli, J. (1993). Unusual nuclear structures in meiotic prophase of fission yeast: A cytological analysis. *The Journal of Cell Biology*, *121*, 241–256.

Bakloushinskaya, I. Y., Matveevsky, S. N., Romanenko, S. A., Serdukova, N. A., Kolomiets, O. L., Spangenberg, V. E., et al. (2012). A comparative analysis of the mole vole sibling species *Ellobius tancrei* and *E. talpinus* (Cricetidae, Rodentia) through chromosome painting and examination of synaptonemal complex structures in hybrids. *Cytogenetic and Genome Research*, *136*, 199–207.

Baklushinskaia, I., Romanenko, S. A., Grafodatskii, A. S., Matveevskii, S. N., Liapunova, E. A., & Kolomiets, O. L. (2010). The role of chromosome rearrangements in evolution of mole voles of genus Ellobius (Rodentia, Mammalia). *Genetika*, *46*, 1290–1293.

Barchi, M., Mahadevaiah, S., Di Giacomo, M., Baudat, F., de Rooij, D. G., Burgoyne, P. S., et al. (2005). Surveillance of different recombination defects in mouse spermatocytes yields distinct responses despite elimination at an identical developmental stage. *Molecular and Cellular Biology*, *25*, 7203–7215.

Baudat, F., Imai, Y., & de Massy, B. (2013). Meiotic recombination in mammals: Localization and regulation. *Nature Reviews. Genetics*, *14*, 794–806.

Baudat, F., Manova, K., Yuen, J. P., Jasin, M., & Keeney, S. (2000). Chromosome synapsis defects and sexually dimorphic meiotic progression in mice lacking Spo11. *Molecular Cell*, *6*, 989–998.

Beguelini, M. R., Marchesin, S. R., Azeredo-Oliveira, M. T., & Morielle-Versute, E. (2011). Nucleolar behavior during meiosis in four species of phyllostomid bats (Chiroptera, Mammalia). *Genetics and Molecular Research*, *10*, 552–565.

Bellani, M. A., Romanienko, P. J., Cairatti, D. A., & Camerini-Otero, R. D. (2005). SPO11 is required for sex-body formation, and Spo11 heterozygosity rescues the prophase arrest of Atm-/- spermatocytes. *Journal of Cell Science*, *118*, 3233–3245.

Bellott, D. W., Hughes, J. F., Skaletsky, H., Brown, L. G., Pyntikova, T., Cho, T. J., et al. (2014). Mammalian Y chromosomes retain widely expressed dosage-sensitive regulators. *Nature*, *508*, 494–499.

Bellott, D. W., Skaletsky, H., Cho, T. J., Brown, L., Locke, D., Chen, N., et al. (2017). Avian W and mammalian Y chromosomes convergently retained dosage-sensitive regulators. *Nature Genetics*, *49*, 387–394.

Bergero, R., & Charlesworth, D. (2009). The evolution of restricted recombination in sex chromosomes. *Trends in Ecology & Evolution*, *24*, 94–102.

Berrios, S. (2017). Nuclear architecture of mouse spermatocytes: Chromosome topology, heterochromatin, and nucleolus. *Cytogenetic and Genome Research*, *151*, 61–71.

Berrios, S., Ayarza, E., Moreno, M., Paulos, A., & Fernandez-Donoso, R. (1999). Non-random distribution of the pericentromeric heterochromatin in meiotic prophase nuclei of mammalian spermatocytes. *Genetica*, *106*, 187–195.

Berrios, S., & Fernandez-Donoso, R. (1990). Nuclear architecture of human pachytene spermatocytes: Quantitative analysis of associations between nucleolar and XY bivalents. *Human Genetics*, *86*, 103–116.

Berrios, S., Fernandez-Donoso, R., Pincheira, J., Page, J., Manterola, M., & Cerda, M. C. (2004). Number and nuclear localisation of nucleoli in mammalian spermatocytes. *Genetica*, *121*, 219–228.

Bessler, J. B., Andersen, E. C., & Villeneuve, A. M. (2010). Differential localization and independent acquisition of the H3K9me2 and H3K9me3 chromatin modifications in the *Caenorhabditis elegans* adult germ line. *PLoS Genetics*, *6*, e1000830.

Betran, E., Emerson, J. J., Kaessmann, H., & Long, M. (2004). Sex chromosomes and male functions: Where do new genes go? *Cell Cycle*, *3*, 873–875.

Betran, E., Thornton, K., & Long, M. (2002). Retroposed new genes out of the X in Drosophila. *Genome Research*, *12*, 1854–1859.

Biswas, U., Hempel, K., Llano, E., Pendas, A., & Jessberger, R. (2016). Distinct roles of meiosis-specific cohesin complexes in mammalian spermatogenesis. *PLoS Genetics*, *12*, e1006389.

Bolcun-Filas, E., Hall, E., Speed, R., Taggart, M., Grey, C., de Massy, B., et al. (2009). Mutation of the mouse Syce1 gene disrupts synapsis and suggests a link between synaptonemal complex structural components and DNA repair. *PLoS Genetics*, *5*, e1000393.

Bolcun-Filas, E., & Schimenti, J. C. (2012). Genetics of meiosis and recombination in mice. *International Review of Cell and Molecular Biology*, *298*, 179–227.

Buchold, G. M., Coarfa, C., Kim, J., Milosavljevic, A., Gunaratne, P. H., & Matzuk, M. M. (2010). Analysis of microRNA expression in the prepubertal testis. *PLoS One*, *5*, e15317.

Bueno, V., Moresco, R. M., Konerat, J. T., Moreira-Filho, O., & Margarido, V. P. (2016). Instability of multiple sex chromosomes systems in fish: The case of Erythrinus erythrinus (Bloch & Schneider, 1801) (Characiformes, Erythrinidae). *Zebrafish*, *13*, 26–32.

Burgoyne, P. S., Mahadevaiah, S. K., & Turner, J. M. (2009). The consequences of asynapsis for mammalian meiosis. *Nature Reviews. Genetics*, *10*, 207–216.

Cabrero, J., Teruel, M., Carmona, F. D., & Camacho, J. P. (2007). Histone H2AX phosphorylation is associated with most meiotic events in grasshopper. *Cytogenetic and Genome Research*, *116*, 311–315.

Cabrero, J., Teruel, M., Carmona, F. D., Jimenez, R., & Camacho, J. P. (2007). Histone H3 lysine 9 acetylation pattern suggests that X and B chromosomes are silenced during entire male meiosis in a grasshopper. *Cytogenetic and Genome Research*, *119*, 135–142.

Cahoon, C. K., & Hawley, R. S. (2013). Flies get a head start on meiosis. *PLoS Genetics*, *9*, e1004051.

Calenda, A., Allenet, B., Escalier, D., Bach, J. F., & Garchon, H. J. (1994). The meiosis-specific Xmr gene product is homologous to the lymphocyte Xlr protein and is a component of the XY body. *The EMBO Journal*, *13*, 100–109.

Carofiglio, F., Inagaki, A., de Vries, S., Wassenaar, E., Schoenmakers, S., Vermeulen, C., et al. (2013). SPO11-independent DNA repair foci and their role in meiotic silencing. *PLoS Genetics, 9*, e1003538.

Carrel, L., & Willard, H. F. (2005). X-inactivation profile reveals extensive variability in X-linked gene expression in females. *Nature, 434*, 400–404.

Casey, A. E., Daish, T. J., Barbero, J. L., & Grutzner, F. (2017). Differential cohesin loading marks paired and unpaired regions of platypus sex chromosomes at prophase I. *Scientific Reports*, 7, 4217.

Castanhole, M. M., Pereira, L. L., Souza, H. V., Bicudo, H. E., Costa, L. A., & Itoyama, M. M. (2008). Heteropicnotic chromatin and nucleolar activity in meiosis and spermiogenesis of Limnogonus aduncus (Heteroptera, Gerridae): A stained nucleolar organizing region that can serve as a model for studying chromosome behavior. *Genetics and Molecular Research*, 7, 1398–1407.

Chadwick, B. P., & Willard, H. F. (2003). Chromatin of the Barr body: Histone and non-histone proteins associated with or excluded from the inactive X chromosome. *Human Molecular Genetics, 12*, 2167–2178.

Charlesworth, B. (1990). Mutation-selection balance and the evolutionary advantage of sex and recombination. *Genetical Research, 55*, 199–221.

Charlesworth, B., & Charlesworth, D. (2000). The degeneration of Y chromosomes. *Philosophical Transactions of the Royal Society of London. Series B, Biological Sciences, 355*, 1563–1572.

Checchi, P. M., & Engebrecht, J. (2011a). *Caenorhabditis elegans* histone methyltransferase MET-2 shields the male X chromosome from checkpoint machinery and mediates meiotic sex chromosome inactivation. *PLoS Genetics*, 7, e1002267.

Checchi, P. M., & Engebrecht, J. (2011b). Heteromorphic sex chromosomes: Navigating meiosis without a homologous partner. *Molecular Reproduction and Development, 78*, 623–632.

Chen, S., Krinsky, B. H., & Long, M. (2013). New genes as drivers of phenotypic evolution. *Nature Reviews. Genetics, 14*, 645–660.

Cloutier, J. M., Mahadevaiah, S. K., ElInati, E., Nussenzweig, A., Toth, A., & Turner, J. M. (2015). Histone H2AFX links meiotic chromosome asynapsis to prophase I oocyte loss in mammals. *PLoS Genetics, 11*, e1005462.

Cloutier, J. M., Mahadevaiah, S. K., ElInati, E., Toth, A., & Turner, J. (2016). Mammalian meiotic silencing exhibits sexually dimorphic features. *Chromosoma, 125*, 215–226.

Cocquet, J., Ellis, P. J., Yamauchi, Y., Mahadevaiah, S. K., Affara, N. A., Ward, M. A., et al. (2009). The multicopy gene Sly represses the sex chromosomes in the male mouse germline after meiosis. *PLoS Biology*, 7, e1000244.

Cole, F., Kauppi, L., Lange, J., Roig, I., Wang, R., Keeney, S., et al. (2012). Homeostatic control of recombination is implemented progressively in mouse meiosis. *Nature Cell Biology, 14*, 424–430.

Cole, F., Keeney, S., & Jasin, M. (2012). Preaching about the converted: How meiotic gene conversion influences genomic diversity. *Annals of the New York Academy of Sciences, 1267*, 95–102.

Cortez, D., Marin, R., Toledo-Flores, D., Froidevaux, L., Liechti, A., Waters, P. D., et al. (2014). Origins and functional evolution of Y chromosomes across mammals. *Nature, 508*, 488–493.

Costa, Y., Speed, R. M., Gautier, P., Semple, C. A., Maratou, K., Turner, J. M., et al. (2006). Mouse MAELSTROM: The link between meiotic silencing of unsynapsed chromatin and microRNA pathway? *Human Molecular Genetics, 15*, 2324–2334.

Daish, T. J., Casey, A. E., & Grutzner, F. (2015). Lack of sex chromosome specific meiotic silencing in platypus reveals origin of MSCI in therian mammals. *BMC Biology, 13*, 106.

de Massy, B. (2013). Initiation of meiotic recombination: How and where? Conservation and specificities among eukaryotes. *Annual Review of Genetics, 47*, 563–599.

del Priore, L., & Pigozzi, M. I. (2014). Histone modifications related to chromosome silencing and elimination during male meiosis in Bengalese finch. *Chromosoma, 123*, 293–302.

Deng, X., Hiatt, J. B., Nguyen, D. K., Ercan, S., Sturgill, D., Hillier, L. W., et al. (2011). Evidence for compensatory upregulation of expressed X-linked genes in mammals, *Caenorhabditis elegans* and *Drosophila melanogaster*. *Nature Genetics, 43*, 1179–1185.

Egel-Mitani, M., Olson, L. W., & Egel, R. (1982). Meiosis in *Aspergillus nidulans*: Another example for lacking synaptonemal complexes in the absence of crossover interference. *Hereditas, 97*, 179–187.

Ellnati, E., Russell, H. R., Ojarikre, O. A., Sangrithi, M., Hirota, T., de Rooij, D. G., et al. (2017). DNA damage response protein TOPBP1 regulates X chromosome silencing in the mammalian germ line. *Proceedings of the National Academy of Sciences of the United States of America, 114*, 12536–12541.

Ellis, P. J., Clemente, E. J., Ball, P., Toure, A., Ferguson, L., Turner, J. M., et al. (2005). Deletions on mouse Yq lead to upregulation of multiple X- and Y-linked transcripts in spermatids. *Human Molecular Genetics, 14*, 2705–2715.

Emerson, J. J. (2017). Evolution: A paradigm shift in snake sex chromosome genetics. *Current Biology, 27*, R800–R803.

Emerson, J. J., Kaessmann, H., Betran, E., & Long, M. (2004). Extensive gene traffic on the mammalian X chromosome. *Science, 303*, 537–540.

Escalier, D., & Garchon, H. J. (2005). XMR, a dual location protein in the XY pair and in its associated nucleolus in mouse spermatocytes. *Molecular Reproduction and Development, 72*, 105–112.

Faisal, I., & Kauppi, L. (2016). Sex chromosome recombination failure, apoptosis, and fertility in male mice. *Chromosoma, 125*, 227–235.

Franco, M. J., Sciurano, R. B., & Solari, A. J. (2007). Protein immunolocalization supports the presence of identical mechanisms of XY body formation in eutherians and marsupials. *Chromosome Research, 15*, 815–824.

Fraune, J., Brochier-Armanet, C., Alsheimer, M., Volff, J. N., Schucker, K., & Benavente, R. (2016). Evolutionary history of the mammalian synaptonemal complex. *Chromosoma, 125*, 355–360.

Gaines, W. A., Godin, S. K., Kabbinavar, F. F., Rao, T., VanDemark, A. P., Sung, P., et al. (2015). Promotion of presynaptic filament assembly by the ensemble of *S. cerevisiae* Rad51 paralogues with Rad52. *Nature Communications, 6*, 7834.

Gamble, T., Castoe, T. A., Nielsen, S. V., Banks, J. L., Card, D. C., Schield, D. R., et al. (2017). The discovery of XY sex chromosomes in a boa and python. *Current Biology, 27*, 2148–2153.e4.

Gazoni, T., Haddad, C. F. B., Narimatsu, H., Cabral-de-Mello, D. C., Lyra, M. L., & Parise-Maltempi, P. P. (2018). More sex chromosomes than autosomes in the Amazonian frog *Leptodactylus pentadactylus*. *Chromosoma, 127*, 269–278.

Goday, C., & Pigozzi, M. I. (2010). Heterochromatin and histone modifications in the germline-restricted chromosome of the zebra finch undergoing elimination during spermatogenesis. *Chromosoma, 119*, 325–336.

Graves, J. A. (2006). Sex chromosome specialization and degeneration in mammals. *Cell, 124*, 901–914.

Graves, J. A. (2016). Did sex chromosome turnover promote divergence of the major mammal groups?: De novo sex chromosomes and drastic rearrangements may have posed reproductive barriers between monotremes, marsupials and placental mammals. *BioEssays, 38*, 734–743.

Greaves, I. K., Rangasamy, D., Devoy, M., Marshall Graves, J. A., & Tremethick, D. J. (2006). The X and Y chromosomes assemble into H2A.Z-containing [corrected] facultative heterochromatin [corrected] following meiosis. *Molecular and Cellular Biology, 26*, 5394–5405.

Gross, M. C., Feldberg, E., Cella, D. M., Schneider, M. C., Schneider, C. H., Porto, J. I., et al. (2009). Intriguing evidence of translocations in Discus fish (Symphysodon, Cichlidae) and a report of the largest meiotic chromosomal chain observed in vertebrates. *Heredity (Edinburgh), 102*, 435–441.

Gruetzner, F., Ashley, T., Rowell, D. M., & Marshall Graves, J. A. (2006). How did the platypus get its sex chromosome chain? A comparison of meiotic multiples and sex chromosomes in plants and animals. *Chromosoma, 115*, 75–88.

Grutzner, F., Rens, W., Tsend-Ayush, E., El-Mogharbel, N., O'Brien, P. C., Jones, R. C., et al. (2004). In the platypus a meiotic chain of ten sex chromosomes shares genes with the bird Z and mammal X chromosomes. *Nature, 432*, 913–917.

Guioli, S., Lovell-Badge, R., & Turner, J. M. (2012). Error-prone ZW pairing and no evidence for meiotic sex chromosome inactivation in the chicken germ line. *PLoS Genetics, 8*, e1002560.

Guo, X., Su, B., Zhou, Z., & Sha, J. (2009). Rapid evolution of mammalian X-linked testis microRNAs. *BMC Genomics, 10*, 97.

Gyuricza, M. R., Manheimer, K. B., Apte, V., Krishnan, B., Joyce, E. F., McKee, B. D., et al. (2016). Dynamic and stable cohesins regulate synaptonemal complex assembly and chromosome segregation. *Current Biology, 26*, 1688–1698.

Hammond, T. M. (2017). Sixteen years of meiotic silencing by unpaired DNA. *Advances in Genetics, 97*, 1–42.

Hammond, T. M., Spollen, W. G., Decker, L. M., Blake, S. M., Springer, G. K., & Shiu, P. K. (2013). Identification of small RNAs associated with meiotic silencing by unpaired DNA. *Genetics, 194*, 279–284.

Hammond, T. M., Xiao, H., Boone, E. C., Decker, L. M., Lee, S. A., Perdue, T. D., et al. (2013). Novel proteins required for meiotic silencing by unpaired DNA and siRNA generation in *Neurospora crassa*. *Genetics, 194*, 91–100.

Handel, M. A. (2004). The XY body: A specialized meiotic chromatin domain. *Experimental Cell Research, 296*, 57–63.

Hasegawa, K., Sin, H. S., Maezawa, S., Broering, T. J., Kartashov, A. V., Alavattam, K. G., et al. (2015). SCML2 establishes the male germline epigenome through regulation of histone H2A ubiquitination. *Developmental Cell, 32*, 574–588.

Helena Mangs, A., & Morris, B. J. (2007). The human pseudoautosomal region (PAR): Origin, function and future. *Current Genomics, 8*, 129–136.

Hendriksen, P. J. (1999). Do X and Y spermatozoa differ in proteins? *Theriogenology, 52*, 1295–1307.

Hendriksen, P. J., Hoogerbrugge, J. W., Themmen, A. P., Koken, M. H., Hoeijmakers, J. H., Oostra, B. A., et al. (1995). Postmeiotic transcription of X and Y chromosomal genes during spermatogenesis in the mouse. *Developmental Biology, 170*, 730–733.

Hense, W., Baines, J. F., & Parsch, J. (2007). X chromosome inactivation during Drosophila spermatogenesis. *PLoS Biology, 5*, e273.

Herran, Y., Gutierrez-Caballero, C., Sanchez-Martin, M., Hernandez, T., Viera, A., Barbero, J. L., et al. (2011). The cohesin subunit RAD21L functions in meiotic synapsis and exhibits sexual dimorphism in fertility. *The EMBO Journal, 30*, 3091–3105.

Hofgartner, F. J., Schmid, M., Krone, W., Zenzes, M. T., & Engel, W. (1979). Pattern of activity of nucleolus organizers during spermatogenesis in mammals as analyzed by silver-staining. *Chromosoma, 71*, 197–216.

Hogan, K. M., Jarrell, G. H., Ryder, E. J., & Greenbaum, I. F. (1992). Heterosynapsis and axial equalization of the sex chromosomes of the northern bobwhite quail. *Cytogenetics and Cell Genetics, 60*, 131–134.

Hollingsworth, N. M. (2010). Phosphorylation and the creation of interhomolog bias during meiosis in yeast. *Cell Cycle*, *9*, 436–437.

Hopkins, J., Hwang, G., Jacob, J., Sapp, N., Bedigian, R., Oka, K., et al. (2014). Meiosis-specific cohesin component, Stag3 is essential for maintaining centromere chromatid cohesion, and required for DNA repair and synapsis between homologous chromosomes. *PLoS Genetics*, *10*, e1004413.

Hoyle, H. D., Hutchens, J. A., Turner, F. R., & Raff, E. C. (1995). Regulation of beta-tubulin function and expression in Drosophila spermatogenesis. *Developmental Genetics*, *16*, 148–170.

Huen, M. S., Grant, R., Manke, I., Minn, K., Yu, X., Yaffe, M. B., et al. (2007). RNF8 transduces the DNA-damage signal via histone ubiquitylation and checkpoint protein assembly. *Cell*, *131*, 901–914.

Hughes, J. F., & Page, D. C. (2015). The biology and evolution of mammalian Y chromosomes. *Annual Review of Genetics*, *49*, 507–527.

Hughes, J. F., & Rozen, S. (2012). Genomics and genetics of human and primate y chromosomes. *Annual Review of Genomics and Human Genetics*, *13*, 83–108.

Hughes, J. F., Skaletsky, H., Brown, L. G., Pyntikova, T., Graves, T., Fulton, R. S., et al. (2012). Strict evolutionary conservation followed rapid gene loss on human and rhesus Y chromosomes. *Nature*, *483*, 82–86.

Hughes, J. F., Skaletsky, H., Koutseva, N., Pyntikova, T., & Page, D. C. (2015). Sex chromosome-to-autosome transposition events counter Y-chromosome gene loss in mammals. *Genome Biology*, *16*, 104.

Hughes, J. F., Skaletsky, H., Pyntikova, T., Graves, T. A., van Daalen, S. K., Minx, P. J., et al. (2010). Chimpanzee and human Y chromosomes are remarkably divergent in structure and gene content. *Nature*, *463*, 536–539.

Ichijima, Y., Ichijima, M., Lou, Z., Nussenzweig, A., Camerini-Otero, R. D., Chen, J., et al. (2011). MDC1 directs chromosome-wide silencing of the sex chromosomes in male germ cells. *Genes & Development*, *25*, 959–971.

Inagaki, A., Schoenmakers, S., & Baarends, W. M. (2010). DNA double strand break repair, chromosome synapsis and transcriptional silencing in meiosis. *Epigenetics*, *5*, 255–266.

Ishiguro, K., Kim, J., Fujiyama-Nakamura, S., Kato, S., & Watanabe, Y. (2011). A new meiosis-specific cohesin complex implicated in the cohesin code for homologous pairing. *EMBO Reports*, *12*, 267–275.

Jaramillo-Lambert, A., & Engebrecht, J. (2010). A single unpaired and transcriptionally silenced X chromosome locally precludes checkpoint signaling in the *Caenorhabditis elegans* germ line. *Genetics*, *184*, 613–628.

Jaramillo-Lambert, A., Harigaya, Y., Vitt, J., Villeneuve, A., & Engebrecht, J. (2010). Meiotic errors activate checkpoints that improve gamete quality without triggering apoptosis in male germ cells. *Current Biology*, *20*, 2078–2089.

Kauppi, L., Barchi, M., Baudat, F., Romanienko, P. J., Keeney, S., & Jasin, M. (2011). Distinct properties of the XY pseudoautosomal region crucial for male meiosis. *Science*, *331*, 916–920.

Keeney, S. (2001). Mechanism and control of meiotic recombination initiation. *Current Topics in Developmental Biology*, *52*, 1–53.

Keeney, S., Giroux, C. N., & Kleckner, N. (1997). Meiosis-specific DNA double-strand breaks are catalyzed by Spo11, a member of a widely conserved protein family. *Cell*, *88*, 375–384.

Keeney, S., & Neale, M. J. (2006). Initiation of meiotic recombination by formation of DNA double-strand breaks: Mechanism and regulation. *Biochemical Society Transactions*, *34*, 523–525.

Kelly, W. G., Schaner, C. E., Dernburg, A. F., Lee, M. H., Kim, S. K., Villeneuve, A. M., et al. (2002). X-chromosome silencing in the germline of *C. elegans*. *Development*, *129*, 479–492.

Kemkemer, C., Hense, W., & Parsch, J. (2011). Fine-scale analysis of X chromosome inactivation in the male germ line of *Drosophila melanogaster. Molecular Biology and Evolution, 28*, 1561–1563.

Khalil, A. M., Boyar, F. Z., & Driscoll, D. J. (2004). Dynamic histone modifications mark sex chromosome inactivation and reactivation during mammalian spermatogenesis. *Proceedings of the National Academy of Sciences of the United States of America, 101*, 16583–16587.

Kierszenbaum, A. L., & Tres, L. L. (1974). Nucleolar and perichromosomal RNA synthesis during meiotic prophase in the mouse testis. *The Journal of Cell Biology, 60*, 39–53.

Kral, J., Korinkova, T., Forman, M., & Krkavcova, L. (2011). Insights into the meiotic behavior and evolution of multiple sex chromosome systems in spiders. *Cytogenetic and Genome Research, 133*, 43–66.

Kuroiwa, A., Handa, S., Nishiyama, C., Chiba, E., Yamada, F., Abe, S., et al. (2011). Additional copies of CBX2 in the genomes of males of mammals lacking SRY, the Amami spiny rat (*Tokudaia osimensis*) and the Tokunoshima spiny rat (Tokudaia tokunoshimensis). *Chromosome Research, 19*, 635–644.

Kutil, B. L., Seong, K. Y., & Aramayo, R. (2003). Unpaired genes do not silence their paired neighbors. *Current Genetics, 43*, 425–432.

La Salle, S., Sun, F., Zhang, X. D., Matunis, M. J., & Handel, M. A. (2008). Developmental control of sumoylation pathway proteins in mouse male germ cells. *Developmental Biology, 321*, 227–237.

Landeen, E. L., Muirhead, C. A., Wright, L., Meiklejohn, C. D., & Presgraves, D. C. (2016). Sex chromosome-wide transcriptional suppression and compensatory cis-regulatory evolution mediate gene expression in the Drosophila male germline. *PLoS Biology, 14*, e1002499.

Lao, J. P., & Hunter, N. (2010). Trying to avoid your sister. *PLoS Biology, 8*, e1000519.

Larson, B. J., Van, M. V., Nakayama, T., & Engebrecht, J. (2016). Plasticity in the meiotic epigenetic landscape of sex chromosomes in Caenorhabditis species. *Genetics, 203*, 1641–1658.

Lee, D. W., Pratt, R. J., McLaughlin, M., & Aramayo, R. (2003). An argonaute-like protein is required for meiotic silencing. *Genetics, 164*, 821–828.

Li, X. C., Bolcun-Filas, E., & Schimenti, J. C. (2011). Genetic evidence that synaptonemal complex axial elements govern recombination pathway choice in mice. *Genetics, 189*, 71–82.

Li, G., Davis, B. W., Raudsepp, T., Pearks Wilkerson, A. J., Mason, V. C., Ferguson-Smith, M., et al. (2013). Comparative analysis of mammalian Y chromosomes illuminates ancestral structure and lineage-specific evolution. *Genome Research, 23*, 1486–1495.

Libuda, D. E., Uzawa, S., Meyer, B. J., & Villeneuve, A. M. (2013). Meiotic chromosome structures constrain and respond to designation of crossover sites. *Nature, 502*, 703–706.

Litwin, I., & Wysocki, R. (2018). New insights into cohesin loading. *Current Genetics, 64*, 53–61.

Liu, Y., Niu, M., Yao, C., Hai, Y., Yuan, Q., Liu, Y., et al. (2015). Fractionation of human spermatogenic cells using STA-PUT gravity sedimentation and their miRNA profiling. *Scientific Reports, 5*, 8084.

Liu, S., Yu, H., Liu, Y., Liu, X., Zhang, Y., Bu, C., et al. (2017). Chromodomain protein CDYL acts as a crotonyl-CoA hydratase to regulate histone crotonylation and spermatogenesis. *Molecular Cell, 67*, 853–866.e5.

Long, M., & Emerson, J. J. (2017). Meiotic sex chromosome inactivation: Compensation by gene traffic. *Current Biology, 27*, R659–R661.

Lu, J., & Matunis, M. J. (2013). A mediator methylation mystery: JMJD1C demethylates MDC1 to regulate DNA repair. *Nature Structural & Molecular Biology, 20*, 1346–1348.

Lu, L. Y., Wu, J., Ye, L., Gavrilina, G. B., Saunders, T. L., & Yu, X. (2010). RNF8-dependent histone modifications regulate nucleosome removal during spermatogenesis. *Developmental Cell, 18*, 371–384.

Lu, L. Y., Xiong, Y., Kuang, H., Korakavi, G., & Yu, X. (2013). Regulation of the DNA damage response on male meiotic sex chromosomes. *Nature Communications*, *4*, 2105.
Lyapunova, E. A., Vorontsov, N. N., & Zakarjan, G. G. (1975). Zygotic mortality in *Ellobius lutescens* (Rodentia: Microtinae). *Experientia*, *31*, 417–418.
MacQueen, A. J., & Hochwagen, A. (2011). Checkpoint mechanisms: The puppet masters of meiotic prophase. *Trends in Cell Biology*, *21*, 393–400.
Madeira, F. F., Borsatto, K. C., Lima, A. C., Ravazi, A., de Oliveira, J., da Rosa, J. A., et al. (2016). Nucleolar persistence: Peculiar characteristic of spermatogenesis of the vectors of Chagas disease (Hemiptera, Triatominae). *The American Journal of Tropical Medicine and Hygiene*, *95*, 1118–1120.
Maezawa, S., Hasegawa, K., Alavattam, K. G., Funakoshi, M., Sato, T., Barski, A., et al. (2018). SCML2 promotes heterochromatin organization in late spermatogenesis. *Journal of Cell Science*, *131*, 1–12.
Maezawa, S., Hasegawa, K., Yukawa, M., Kubo, N., Sakashita, A., Alavattam, K. G., et al. (2018). Polycomb protein SCML2 facilitates H3K27me3 to establish bivalent domains in the male germline. *Proceedings of the National Academy of Sciences of the United States of America*, *115*, 4957–4962.
Mahadevaiah, S. K., Bourc'his, D., de Rooij, D. G., Bestor, T. H., Turner, J. M., & Burgoyne, P. S. (2008). Extensive meiotic asynapsis in mice antagonises meiotic silencing of unsynapsed chromatin and consequently disrupts meiotic sex chromosome inactivation. *The Journal of Cell Biology*, *182*, 263–276.
Mahadevaiah, S. K., Turner, J. M., Baudat, F., Rogakou, E. P., de Boer, P., Blanco-Rodriguez, J., et al. (2001). Recombinational DNA double-strand breaks in mice precede synapsis. *Nature Genetics*, *27*, 271–276.
Mailand, N., Bekker-Jensen, S., Faustrup, H., Melander, F., Bartek, J., Lukas, C., et al. (2007). RNF8 ubiquitylates histones at DNA double-strand breaks and promotes assembly of repair proteins. *Cell*, *131*, 887–900.
Maine, E. M. (2010). Meiotic silencing in *Caenorhabditis elegans*. *International Review of Cell and Molecular Biology*, *282*, 91–134.
Manterola, M., Brown, T. M., Oh, M. Y., Garyn, C., Gonzalez, B. J., & Wolgemuth, D. J. (2018). BRDT is an essential epigenetic regulator for proper chromatin organization, silencing of sex chromosomes and crossover formation in male meiosis. *PLoS Genetics*, *14*, e1007209.
Manterola, M., Page, J., Vasco, C., Berrios, S., Parra, M. T., Viera, A., et al. (2009). A high incidence of meiotic silencing of unsynapsed chromatin is not associated with substantial pachytene loss in heterozygous male mice carrying multiple simple robertsonian translocations. *PLoS Genetics*, *5*, e1000625.
Marcon, E., Babak, T., Chua, G., Hughes, T., & Moens, P. B. (2008). miRNA and piRNA localization in the male mammalian meiotic nucleus. *Chromosome Research*, *16*, 243–260.
Marshall Graves, J. A. (2008). Weird animal genomes and the evolution of vertebrate sex and sex chromosomes. *Annual Review of Genetics*, *42*, 565–586.
Martini, E., Diaz, R. L., Hunter, N., & Keeney, S. (2006). Crossover homeostasis in yeast meiosis. *Cell*, *126*, 285–295.
Mattos, V. F., Carvalho, L. S., Carvalho, M. A., & Schneider, M. C. (2018). Insights into the origin of the high variability of multivalent-meiotic associations in holocentric chromosomes of Tityus (Archaeotityus) scorpions. *PLoS One*, *13*, e0192070.
Matveevsky, S., Bakloushinskaya, I., & Kolomiets, O. (2016). Unique sex chromosome systems in *Ellobius*: How do male XX chromosomes recombine and undergo pachytene chromatin inactivation? *Scientific Reports*, *6*, 29949.
Matveevsky, S., Kolomiets, O., Bogdanov, A., Hakhverdyan, M., & Bakloushinskaya, I. (2017). Chromosomal evolution in mole voles *Ellobius* (Cricetidae, Rodentia): Bizarre sex chromosomes, variable autosomes and meiosis. *Genes (Basel)*, *8*(11), 306–326.

McKee, B. D. (2009). Homolog pairing and segregation in Drosophila meiosis. *Genome Dynamics*, *5*, 56–68.
McKee, B. D., & Handel, M. A. (1993). Sex chromosomes, recombination, and chromatin conformation. *Chromosoma*, *102*, 71–80.
Meiklejohn, C. D., Landeen, E. L., Cook, J. M., Kingan, S. B., & Presgraves, D. C. (2011). Sex chromosome-specific regulation in the Drosophila male germline but little evidence for chromosomal dosage compensation or meiotic inactivation. *PLoS Biology*, *9*, e1001126.
Meiklejohn, C. D., & Tao, Y. (2010). Genetic conflict and sex chromosome evolution. *Trends in Ecology & Evolution*, *25*, 215–223.
Mikhaylova, L. M., & Nurminsky, D. I. (2011). Lack of global meiotic sex chromosome inactivation, and paucity of tissue-specific gene expression on the Drosophila X chromosome. *BMC Biology*, *9*, 29.
Miller, T. C., Simon, B., Rybin, V., Grotsch, H., Curtet, S., Khochbin, S., et al. (2016). A bromodomain-DNA interaction facilitates acetylation-dependent bivalent nucleosome recognition by the BET protein BRDT. *Nature Communications*, 7, 13855.
Miura, I. (2017). Sex determination and sex chromosomes in Amphibia. *Sexual Development*, *11*, 298–306.
Miura, I., Ohtani, H., & Ogata, M. (2012). Independent degeneration of W and Y sex chromosomes in frog Rana rugosa. *Chromosome Research*, *20*, 47–55.
Modzelewski, A. J., Holmes, R. J., Hilz, S., Grimson, A., & Cohen, P. E. (2012). AGO4 regulates entry into meiosis and influences silencing of sex chromosomes in the male mouse germline. *Developmental Cell*, *23*, 251–264.
Moens, P. B., Chen, D. J., Shen, Z., Kolas, N., Tarsounas, M., Heng, H. H., et al. (1997). Rad51 immunocytology in rat and mouse spermatocytes and oocytes. *Chromosoma*, *106*, 207–215.
Morgan, T. H. (1912). Special Articles. *Science*, *36*, 718–720.
Morielle-Souza, A., & de Azeredo-Oliveira, M. T. (2008). Study of the nucleolar cycle and ribosomal RNA distribution during meiosis in triatomines (Triatominae, Heteroptera). *Micron*, *39*, 1020–1026.
Mueller, J. L., Mahadevaiah, S. K., Park, P. J., Warburton, P. E., Page, D. C., & Turner, J. M. (2008). The mouse X chromosome is enriched for multicopy testis genes showing postmeiotic expression. *Nature Genetics*, *40*, 794–799.
Muller, H. J. (1964). The relation of recombination to mutational advance. *Mutation Research*, *106*, 2–9.
Mulugeta, E., Wassenaar, E., Sleddens-Linkels, E., van, I. W. F., Heard, E., Grootegoed, J. A., et al. (2016). Genomes of Ellobius species provide insight into the evolutionary dynamics of mammalian sex chromosomes. *Genome Research*, *26*, 1202–1210.
Namekawa, S. H., Park, P. J., Zhang, L. F., Shima, J. E., McCarrey, J. R., Griswold, M. D., et al. (2006). Postmeiotic sex chromatin in the male germline of mice. *Current Biology*, *16*, 660–667.
Namekawa, S. H., VandeBerg, J. L., McCarrey, J. R., & Lee, J. T. (2007). Sex chromosome silencing in the marsupial male germ line. *Proceedings of the National Academy of Sciences of the United States of America*, *104*, 9730–9735.
Naqvi, S., Bellott, D. W., Lin, K. S., & Page, D. C. (2018). Conserved microRNA targeting reveals preexisting gene dosage sensitivities that shaped amniote sex chromosome evolution. *Genome Research*, *28*, 474–483.
Neale, M. J., & Keeney, S. (2006). Clarifying the mechanics of DNA strand exchange in meiotic recombination. *Nature*, *442*, 153–158.
Odorisio, T., Rodriguez, T. A., Evans, E. P., Clarke, A. R., & Burgoyne, P. S. (1998). The meiotic checkpoint monitoring synapsis eliminates spermatocytes via p53-independent apoptosis. *Nature Genetics*, *18*, 257–261.

Page, J., Berrios, S., Parra, M. T., Viera, A., Suja, J. A., Prieto, I., et al. (2005). The program of sex chromosome pairing in meiosis is highly conserved across marsupial species: Implications for sex chromosome evolution. *Genetics, 170*, 793–799.

Page, J., Berrios, S., Rufas, J. S., Parra, M. T., Suja, J. A., Heyting, C., et al. (2003). The pairing of X and Y chromosomes during meiotic prophase in the marsupial species *Thylamys elegans* is maintained by a dense plate developed from their axial elements. *Journal of Cell Science, 116*, 551–560.

Page, S. L., & Hawley, R. S. (2003). Chromosome choreography: The meiotic ballet. *Science, 301*, 785–789.

Page, J., Viera, A., Parra, M. T., de la Fuente, R., Suja, J. A., Prieto, I., et al. (2006). Involvement of synaptonemal complex proteins in sex chromosome segregation during marsupial male meiosis. *PLoS Genetics, 2*, e136.

Pala, I., Naurin, S., Stervander, M., Hasselquist, D., Bensch, S., & Hansson, B. (2012). Evidence of a neo-sex chromosome in birds. *Heredity (Edinburgh), 108*, 264–272.

Pardo-Manuel de Villena, F., & Sapienza, C. (2001a). Female meiosis drives karyotypic evolution in mammals. *Genetics, 159*, 1179–1189.

Pardo-Manuel de Villena, F., & Sapienza, C. (2001b). Nonrandom segregation during meiosis: The unfairness of females. *Mammalian Genome, 12*, 331–339.

Pardo-Manuel de Villena, F., & Sapienza, C. (2001c). Transmission ratio distortion in offspring of heterozygous female carriers of Robertsonian translocations. *Human Genetics, 108*, 31–36.

Peichel, C. L. (2017). Convergence and divergence in sex-chromosome evolution. *Nature Genetics, 49*, 321–322.

Perera, D., Perez-Hidalgo, L., Moens, P. B., Reini, K., Lakin, N., Syvaoja, J. E., et al. (2004). TopBP1 and ATR colocalization at meiotic chromosomes: Role of TopBP1/Cut5 in the meiotic recombination checkpoint. *Molecular Biology of the Cell, 15*, 1568–1579.

Perry, J., Palmer, S., Gabriel, A., & Ashworth, A. (2001). A short pseudoautosomal region in laboratory mice. *Genome Research, 11*, 1826–1832.

Peruquetti, R. L., Assis, I. M., Taboga, S. R., & de Azeredo-Oliveira, M. T. (2008). Meiotic nucleolar cycle and chromatoid body formation during the rat (*Rattus novergicus*) and mouse (*Mus musculus*) spermiogenesis. *Micron, 39*, 419–425.

Peters, A. H., O'Carroll, D., Scherthan, H., Mechtler, K., Sauer, S., Schofer, C., et al. (2001). Loss of the Suv39h histone methyltransferases impairs mammalian heterochromatin and genome stability. *Cell, 107*, 323–337.

Pigozzi, M. I. (2016). The chromosomes of birds during meiosis. *Cytogenetic and Genome Research, 150*, 128–138.

Pigozzi, M. I., & Solari, A. J. (1997). Extreme axial equalization and wide distribution of recombination nodules in the primitive ZW pair of Rhea americana (Aves, Ratitae). *Chromosome Research, 5*, 421–428.

Pigozzi, M. I., & Solari, A. J. (1998). Germ cell restriction and regular transmission of an accessory chromosome that mimics a sex body in the zebra finch, Taeniopygia guttata. *Chromosome Research, 6*, 105–113.

Pigozzi, M. I., & Solari, A. J. (2005). The germ-line-restricted chromosome in the zebra finch: Recombination in females and elimination in males. *Chromosoma, 114*, 403–409.

Pittman, D. L., Cobb, J., Schimenti, K. J., Wilson, L. A., Cooper, D. M., Brignull, E., et al. (1998). Meiotic prophase arrest with failure of chromosome synapsis in mice deficient for Dmc1, a germline-specific RecA homolog. *Molecular Cell, 1*, 697–705.

Pokorna, M., Altmanova, M., & Kratochvil, L. (2014). Multiple sex chromosomes in the light of female meiotic drive in amniote vertebrates. *Chromosome Research, 22*, 35–44.

Potrzebowski, L., Vinckenbosch, N., & Kaessmann, H. (2010). The emergence of new genes on the young therian X. *Trends in Genetics, 26*, 1–4.

Potrzebowski, L., Vinckenbosch, N., Marques, A. C., Chalmel, F., Jegou, B., & Kaessmann, H. (2008). Chromosomal gene movements reflect the recent origin and biology of therian sex chromosomes. *PLoS Biology*, *6*, e80.

Pratt, R. J., Lee, D. W., & Aramayo, R. (2004). DNA methylation affects meiotic trans-sensing, not meiotic silencing, in Neurospora. *Genetics*, *168*, 1925–1935.

Rahn, M. I., & Solari, A. J. (1986). Recombination nodules in the oocytes of the chicken, *Gallus domesticus*. *Cytogenetics and Cell Genetics*, *43*, 187–193.

Rankin, S. (2015). Complex elaboration: Making sense of meiotic cohesin dynamics. *The FEBS Journal*, *282*, 2426–2443.

Refolio, E., Cavero, S., Marcon, E., Freire, R., & San-Segundo, P. A. (2011). The Ddc2/ATRIP checkpoint protein monitors meiotic recombination intermediates. *Journal of Cell Science*, *124*, 2488–2500.

Remeseiro, S., & Losada, A. (2013). Cohesin, a chromatin engagement ring. *Current Opinion in Cell Biology*, *25*, 63–71.

Rens, W., O'Brien, P. C., Grutzner, F., Clarke, O., Graphodatskaya, D., Tsend-Ayush, E., et al. (2007). The multiple sex chromosomes of platypus and echidna are not completely identical and several share homology with the avian Z. *Genome Biology*, *8*, R243.

Rice, W. R. (1996). Sexually antagonistic male adaptation triggered by experimental arrest of female evolution. *Nature*, *381*, 232–234.

Roco, A. S., Olmstead, A. W., Degitz, S. J., Amano, T., Zimmerman, L. B., & Bullejos, M. (2015). Coexistence of Y, W, and Z sex chromosomes in *Xenopus tropicalis*. *Proceedings of the National Academy of Sciences of the United States of America*, *112*, E4752–E4761.

Rog, O., & Dernburg, A. F. (2013). Chromosome pairing and synapsis during *Caenorhabditis elegans* meiosis. *Current Opinion in Cell Biology*, *25*, 349–356.

Rog, O., Kohler, S., & Dernburg, A. F. (2017). The synaptonemal complex has liquid crystalline properties and spatially regulates meiotic recombination factors. *eLife*, *6*, 1–26.

Romanienko, P. J., & Camerini-Otero, R. D. (2000). The mouse Spo11 gene is required for meiotic chromosome synapsis. *Molecular Cell*, *6*, 975–987.

Royo, H., Polikiewicz, G., Mahadevaiah, S. K., Prosser, H., Mitchell, M., Bradley, A., et al. (2010). Evidence that meiotic sex chromosome inactivation is essential for male fertility. *Current Biology*, *20*, 2117–2123.

Royo, H., Prosser, H., Ruzankina, Y., Mahadevaiah, S. K., Cloutier, J. M., Baumann, M., et al. (2013). ATR acts stage specifically to regulate multiple aspects of mammalian meiotic silencing. *Genes & Development*, *27*, 1484–1494.

Royo, H., Seitz, H., Ellnati, E., Peters, A. H., Stadler, M. B., & Turner, J. M. (2015). Silencing of X-linked microRNAs by meiotic sex chromosome inactivation. *PLoS Genetics*, *11*, e1005461.

Sampaio, T. R., Pires, L. B., da Rosa, R., & Dias, A. L. (2014). Activity patterns of nucleolar organizer region during spermatogenesis of different curimatid species (Characiformes: Curimatidae). *Genome*, *57*, 119–124.

San Filippo, J., Sung, P., & Klein, H. (2008). Mechanism of eukaryotic homologous recombination. *Annual Review of Biochemistry*, *77*, 229–257.

Schartl, M., Schmid, M., & Nanda, I. (2016). Dynamics of vertebrate sex chromosome evolution: From equal size to giants and dwarfs. *Chromosoma*, *125*, 553–571.

Schimenti, J. (2005). Synapsis or silence. *Nature Genetics*, *37*, 11–13.

Schmid, M., Loser, C., Schmidtke, J., & Engel, W. (1982). Evolutionary conservation of a common pattern of activity of nucleolus organizers during spermatogenesis in vertebrates. *Chromosoma*, *86*, 149–179.

Schoenmakers, S., Wassenaar, E., Hoogerbrugge, J. W., Laven, J. S., Grootegoed, J. A., & Baarends, W. M. (2009). Female meiotic sex chromosome inactivation in chicken. *PLoS Genetics*, *5*, e1000466.

Severi-Aguiar, G. D., Lourenco, L. B., Bicudo, H. E., & Azeredo-Oliveira, M. T. (2006). Meiosis aspects and nucleolar activity in Triatoma vitticeps (Triatominae, Heteroptera). *Genetica, 126*, 141–151.

Shin, Y. H., Choi, Y., Erdin, S. U., Yatsenko, S. A., Kloc, M., Yang, F., et al. (2010). Hormad1 mutation disrupts synaptonemal complex formation, recombination, and chromosome segregation in mammalian meiosis. *PLoS Genetics, 6*, e1001190.

Shiu, P. K., & Metzenberg, R. L. (2002). Meiotic silencing by unpaired DNA: Properties, regulation and suppression. *Genetics, 161*, 1483–1495.

Shiu, P. K., Raju, N. B., Zickler, D., & Metzenberg, R. L. (2001). Meiotic silencing by unpaired DNA. *Cell, 107*, 905–916.

Sin, H. S., Barski, A., Zhang, F., Kartashov, A. V., Nussenzweig, A., Chen, J., et al. (2012). RNF8 regulates active epigenetic modifications and escape gene activation from inactive sex chromosomes in post-meiotic spermatids. *Genes & Development, 26*, 2737–2748.

Sin, H. S., Kartashov, A. V., Hasegawa, K., Barski, A., & Namekawa, S. H. (2015). Poised chromatin and bivalent domains facilitate the mitosis-to-meiosis transition in the male germline. *BMC Biology, 13*, 53.

Skaletsky, H., Kuroda-Kawaguchi, T., Minx, P. J., Cordum, H. S., Hillier, L., Brown, L. G., et al. (2003). The male-specific region of the human Y chromosome is a mosaic of discrete sequence classes. *Nature, 423*, 825–837.

Soh, Y. Q., Alfoldi, J., Pyntikova, T., Brown, L. G., Graves, T., Minx, P. J., et al. (2014). Sequencing the mouse Y chromosome reveals convergent gene acquisition and amplification on both sex chromosomes. *Cell, 159*, 800–813.

Solari, A. J. (1969a). Changes in the sex chromosomes during meiotic prophase in mouse spermatocytes. *Genetics, 61*(Suppl), 113–120.

Solari, A. J. (1969b). The evolution of the ultrastructure of the sex chromosomes (sex vesicle) during meiotic prophase in mouse spermatocytes. *Journal of Ultrastructure Research, 27*, 289–305.

Solari, A. J. (1974). The behavior of the XY pair in mammals. *International Review of Cytology, 38*, 273–317.

Solari, A. J., & Bianchi, N. O. (1975). The synaptic behaviour of the X and Y chromosomes in the marsupial *Monodelphis dimidiata*. *Chromosoma, 52*, 11–25.

Solari, A. J., & Moses, M. J. (1977). Synaptonemal complexes in a tetraploid mouse spermatocyte. *Experimental Cell Research, 108*, 464–467.

Solari, A. J., & Pigozzi, M. I. (1993). Recombination nodules and axial equalization in the ZW pairs of the Peking duck and the guinea fowl. *Cytogenetics and Cell Genetics, 64*, 268–272.

Solari, A. J., & Tres, L. L. (1967). The ultrastructure of the human sex vesicle. *Chromosoma, 22*, 16–31.

Solari, A. J., & Tres, L. L. (1970). The three-dimensional reconstruction of the XY chromosomal pair in human spermatocytes. *The Journal of Cell Biology, 45*, 43–53.

Song, C., Fang, J., Li, X., Liu, H., & Thomas Chao, C. (2009). Identification and characterization of 27 conserved microRNAs in citrus. *Planta, 230*, 671–685.

Sosa, E., Flores, L., Yan, W., & McCarrey, J. R. (2015). Escape of X-linked miRNA genes from meiotic sex chromosome inactivation. *Development, 142*, 3791–3800.

Stahl, A., Hartung, M., Devictor, M., & Berge-Lefranc, J. L. (1984). The association of the nucleolus and the short arm of acrocentric chromosomes with the XY pair in human spermatocytes: Its possible role in facilitating sex-chromosome acrocentric translocations. *Human Genetics, 68*, 173–180.

Steinberg, E. R., Nieves, M., & Mudry, M. D. (2014). Multiple sex chromosome systems in howler monkeys (Platyrrhini, Alouatta). *Comparative Cytogenetics, 8*, 43–69.

Strome, S., Kelly, W. G., Ercan, S., & Lieb, J. D. (2014). Regulation of the X chromosomes in *Caenorhabditis elegans*. *Cold Spring Harbor Perspectives in Biology, 6*, 1–21.

Sugiyama, T., Zaitseva, E. M., & Kowalczykowski, S. C. (1997). A single-stranded DNA-binding protein is needed for efficient presynaptic complex formation by the *Saccharomyces cerevisiae* Rad51 protein. *The Journal of Biological Chemistry, 272*, 7940–7945.

Suwaki, N., Klare, K., & Tarsounas, M. (2011). RAD51 paralogs: Roles in DNA damage signalling, recombinational repair and tumorigenesis. *Seminars in Cell & Developmental Biology, 22*, 898–905.

Tan, M., Luo, H., Lee, S., Jin, F., Yang, J. S., Montellier, E., et al. (2011). Identification of 67 histone marks and histone lysine crotonylation as a new type of histone modification. *Cell, 146*, 1016–1028.

Tres, L. L. (2005). XY chromosomal bivalent: Nucleolar attraction. *Molecular Reproduction and Development, 72*, 1–6.

Tsubouchi, H., & Ogawa, H. (1998). A novel mre11 mutation impairs processing of double-strand breaks of DNA during both mitosis and meiosis. *Molecular and Cellular Biology, 18*, 260–268.

Turner, J. M. (2005). Sex chromosomes make their mark. *Chromosoma, 114*, 300–306.

Turner, J. M. (2007). Meiotic sex chromosome inactivation. *Development, 134*, 1823–1831.

Turner, J. M. (2015). Meiotic silencing in mammals. *Annual Review of Genetics, 49*, 395–412.

Turner, J. M., Mahadevaiah, S. K., Ellis, P. J., Mitchell, M. J., & Burgoyne, P. S. (2006). Pachytene asynapsis drives meiotic sex chromosome inactivation and leads to substantial postmeiotic repression in spermatids. *Developmental Cell, 10*, 521–529.

Turner, J. M., Mahadevaiah, S. K., Fernandez-Capetillo, O., Nussenzweig, A., Xu, X., Deng, C. X., et al. (2005). Silencing of unsynapsed meiotic chromosomes in the mouse. *Nature Genetics, 37*, 41–47.

Veyrunes, F., Waters, P. D., Miethke, P., Rens, W., McMillan, D., Alsop, A. E., et al. (2008). Bird-like sex chromosomes of platypus imply recent origin of mammal sex chromosomes. *Genome Research, 18*, 965–973.

Vibranovski, M. D. (2014). Meiotic sex chromosome inactivation in Drosophila. *Journal of Genomics, 2*, 104–117.

Vibranovski, M. D., Lopes, H. F., Karr, T. L., & Long, M. (2009). Stage-specific expression profiling of Drosophila spermatogenesis suggests that meiotic sex chromosome inactivation drives genomic relocation of testis-expressed genes. *PLoS Genetics, 5*, e1000731.

Vibranovski, M. D., Zhang, Y. E., Kemkemer, C., Lopes, H. F., Karr, T. L., & Long, M. (2012). Re-analysis of the larval testis data on meiotic sex chromosome inactivation revealed evidence for tissue-specific gene expression related to the drosophila X chromosome. *BMC Biology, 10*, 49. author reply 50.

Vibranovski, M. D., Zhang, Y., & Long, M. (2009). General gene movement off the X chromosome in the Drosophila genus. *Genome Research, 19*, 897–903.

Viera, A., Calvente, A., Page, J., Parra, M. T., Gomez, R., Suja, J. A., et al. (2004). X and B chromosomes display similar meiotic characteristics in male grasshoppers. *Cytogenetic and Genome Research, 106*, 302–308.

Viera, A., Santos, J. L., Page, J., Parra, M. T., Calvente, A., Cifuentes, M., et al. (2004). DNA double-strand breaks, recombination and synapsis: The timing of meiosis differs in grasshoppers and flies. *EMBO Reports, 5*, 385–391.

Wang, P. J. (2004). X chromosomes, retrogenes and their role in male reproduction. *Trends in Endocrinology and Metabolism, 15*, 79–83.

Wang, X., & Haber, J. E. (2004). Role of Saccharomyces single-stranded DNA-binding protein RPA in the strand invasion step of double-strand break repair. *PLoS Biology, 2*, E21.

Ward, A., Hopkins, J., McKay, M., Murray, S., & Jordan, P. W. (2016). Genetic interactions between the meiosis-specific cohesin components, STAG3, REC8, and RAD21L. *G3: Genes, Genomes, Genetics, 6*, 1713–1724.

Wojtasz, L., Cloutier, J. M., Baumann, M., Daniel, K., Varga, J., Fu, J., et al. (2012). Meiotic DNA double-strand breaks and chromosome asynapsis in mice are monitored by distinct HORMAD2-independent and -dependent mechanisms. *Genes & Development*, *26*, 958–973.

Yamauchi, Y., Riel, J. M., Stoytcheva, Z., & Ward, M. A. (2014). Two Y genes can replace the entire Y chromosome for assisted reproduction in the mouse. *Science*, *343*, 69–72.

Yang, F., De La Fuente, R., Leu, N. A., Baumann, C., McLaughlin, K. J., & Wang, P. J. (2006). Mouse SYCP2 is required for synaptonemal complex assembly and chromosomal synapsis during male meiosis. *The Journal of Cell Biology*, *173*, 497–507.

Yang, F., Deng, X., Ma, W., Berletch, J. B., Rabaia, N., Wei, G., et al. (2015). The lncRNA Firre anchors the inactive X chromosome to the nucleolus by binding CTCF and maintains H3K27me3 methylation. *Genome Biology*, *16*, 52.

Yoshida, K., & Kitano, J. (2012). The contribution of female meiotic drive to the evolution of neo-sex chromosomes. *Evolution*, *66*, 3198–3208.

Yoshida, K., Kondoh, G., Matsuda, Y., Habu, T., Nishimune, Y., & Morita, T. (1998). The mouse RecA-like gene Dmc1 is required for homologous chromosome synapsis during meiosis. *Molecular Cell*, *1*, 707–718.

Yuan, L., Liu, J. G., Hoja, M. R., Wilbertz, J., Nordqvist, K., & Hoog, C. (2002). Female germ cell aneuploidy and embryo death in mice lacking the meiosis-specific protein SCP3. *Science*, *296*, 1115–1118.

Yuan, L., Liu, J. G., Zhao, J., Brundell, E., Daneholt, B., & Hoog, C. (2000). The murine SCP3 gene is required for synaptonemal complex assembly, chromosome synapsis, and male fertility. *Molecular Cell*, *5*, 73–83.

Yusufzai, T. M., Tagami, H., Nakatani, Y., & Felsenfeld, G. (2004). CTCF tethers an insulator to subnuclear sites, suggesting shared insulator mechanisms across species. *Molecular Cell*, *13*, 291–298.

Zhang, L. F., Huynh, K. D., & Lee, J. T. (2007). Perinucleolar targeting of the inactive X during S phase: Evidence for a role in the maintenance of silencing. *Cell*, *129*, 693–706.

Zhang, R., Peng, Y., Wang, W., & Su, B. (2007). Rapid evolution of an X-linked microRNA cluster in primates. *Genome Research*, *17*, 612–617.

Zhang, Y. E., Vibranovski, M. D., Landback, P., Marais, G. A., & Long, M. (2010). Chromosomal redistribution of male-biased genes in mammalian evolution with two bursts of gene gain on the X chromosome. *PLoS Biology*, *8*(10), 1–13.

Zickler, D., & Kleckner, N. (2015). Recombination, pairing, and synapsis of homologs during meiosis. *Cold Spring Harbor Perspectives in Biology*, *7*, 1–26.

CHAPTER TWO

Environmental regulation of sex determination in fishes: Insights from Atheriniformes

Yoji Yamamoto[a,*], Ricardo S. Hattori[b], Reynaldo Patiño[c], Carlos A. Strüssmann[a]

[a]Department of Marine Biosciences, Tokyo University of Marine Science and Technology, Tokyo, Japan
[b]Sao Paulo Fisheries Institute—Campos do Jordao, APTA/SAA, Campos do Jordão, CP, Brazil
[c]U.S. Geological Survey, Texas Cooperative Fish and Wildlife Research Unit and Departments of Natural Resources Management and Biological Sciences, Texas Tech University, Lubbock, TX, United States
*Corresponding author: e-mail address: yoji@m.kaiyodai.ac.jp

Contents

Abstract

Sex determination is the first step toward the establishment of phenotypic sex in most vertebrates. Aquatic poikilotherms such as teleost fishes exhibit a high diversity of sex-determination mechanisms and gonadal phenotypes that are remarkably plastic and responsive to a variety of environmental factors (e.g., water temperature, pH, salinity, photoperiod, population density). This chapter reviews current knowledge of genotypic and environmental sex determination systems in fishes with special reference to Atheriniformes—one of the best-characterized taxa in this field—and offers perspectives to guide and stimulate further research.

1. Introduction

Teleost fishes are a functionally and morphologically diverse group of vertebrates distributed across equally diverse environmental landscapes. Although sexual reproduction is a taxonomically conserved trait, sex determination in fishes exhibits a high degree of plasticity. As in many other vertebrates,

Current Topics in Developmental Biology, Volume 134
ISSN 0070-2153
https://doi.org/10.1016/bs.ctdb.2019.02.003

sex in gonochoristic teleosts was originally thought to be genetically predetermined at fertilization (i.e., to exhibit genotypic sex determination; GSD). It has long been known, however, that a variety of environmental factors also can influence the sex determination process (Baroiller, D'Cotta, & Saillant, 2009; Devlin & Nagahama, 2002; Luckenbach & Yamamoto, 2018; Ospina-Álvarez & Piferrer, 2008; Strüssmann & Patiño, 1999). Those species whose sex phenotype is uniquely sensitive to ambient conditions had indeed been cited as examples of environmental sex determination (ESD) and were assumed to lack genetic sex determinants. Results of recent studies have questioned the existence of "true ESD," however, and instead support a scenario where genotypic and environmental sex determinants coexist along a continuum where, as the strength of GSD decreases, that of ESD increases (e.g., Yamamoto, Zhang, Sarida, Hattori, & Strüssmann, 2014). Hence, in many species GSD can be overridden by the environmental milieu that larvae experience during early gonadal development (Fig. 1). The environmental factors that influence sex determination vary among species, and such differences are thought to influence the evolution of reproductive strategies. The best-characterized and prevalent environmental variable affecting sex in fishes is temperature (temperature-dependent sex determination, TSD). To facilitate discussion, this chapter defines TSD in a broad sense to include cases where any measurable effect of temperature on sex determination is observed regardless of the presence or absence of GSD, or whether the effect of temperature is ecologically relevant or not.

Earth is in the midst of a global warming trend that started in the 1950s (IPCC, 2014). Projections of further increases in ambient temperature could directly lead to imbalanced sex ratios in species exhibiting TSD, thus jeopardizing their survival. Conservation of such species requires a full understanding of TSD mechanisms in relation to GSD. Changes in thermal regimes of aquatic habitats, however, are not the only anticipated water quality consequence of global warming. For instance, a retrospective study of managed reservoirs in the south-central United States concluded that continued climate change could also lead to increased salinity and decreased dissolved oxygen concentrations (Gelca et al., 2016). At depths ≥ 3 m, increased temperature could also lead to decreased pH (Gelca et al., 2016) although positive associations between temperature and pH have been reported in soft-water lakes (Bates, Kundzewicz, Wu, & Palutikof, 2008). Similar projections of future changes in aquatic environments, both oceanic and inland, caused by global warming have been made elsewhere including

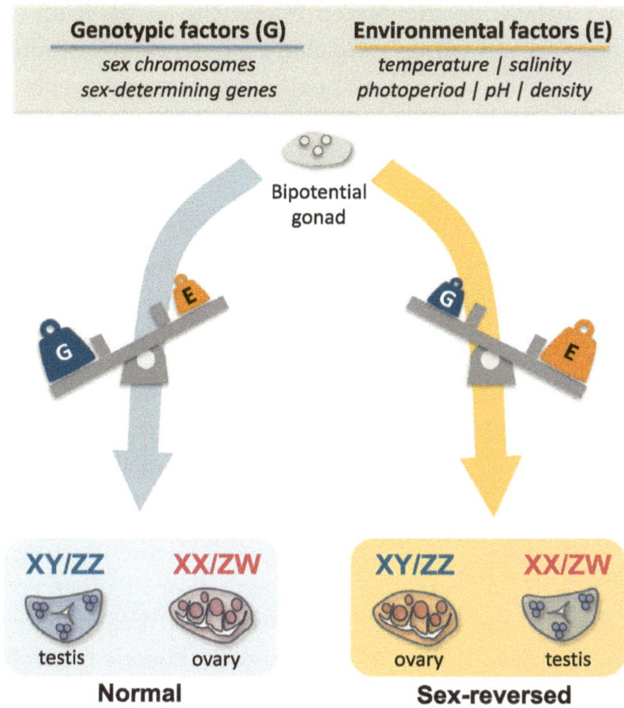

Fig. 1 Schematic representation of sex determination as a balance between environmental (E) and genotypic (G) factors that ultimately defines the development of gonad primordia. Concordance between the genotypic and phenotypic sex is observed when G is heavier (stronger) than E, but when E becomes heavier than G, sex-reversed individuals develop. *Note*: the left pattern also applies when the environmental cue acts in the same direction as genotypic factors, such as masculinizing temperature in male genotype or, conversely, feminizing temperature in female genotype.

Asia and South America (e.g., Japan Meteorological Agency, https://www.data.jma.go.jp/cpdinfo/index_temp.html; Barros & Albernaz, 2014; Marengo, 2008). As discussed later in the chapter, these secondary consequences of global warming also have the potential to influence sex determination in fishes. Additionally, global warming could aggravate the historically preceding impacts of land-use change on aquatic habitats and organisms (Janse et al., 2015; Palmer et al., 2008) and, in turn, the resulting high levels of environmental stress could influence sex determination (see Section 2). The fast pace of climate change-driven habitat change has led to concerns that organismal stress responses could become maladaptive, particularly in species that are slow or unable to acclimate or adapt (Angelier & Wingfield, 2012).

The order Atheriniformes, often referred to as silversides or rainbow fishes, comprise up to 9 families and about 350 species distributed in diverse habitats worldwide, from freshwater to marine (Campanella et al., 2015). An interesting feature of this taxon is that many of its member species, especially those belonging to the family Atherinopsidae, possess the TSD system and may be particularly sensitive to the effects of global warming and climate change (Strüssmann, Conover, Somoza, & Miranda, 2010). The purposes of this chapter are to provide a summary and synthesis of GSD and ESD systems in fishes and of recent developments in this field with special focus on atheriniform fishes, and to offer perspectives for guiding and stimulating further research.

2. Environmental and temperature-dependent sex determination in fishes

The best-characterized environmental factor affecting sex determination in vertebrates is temperature, but a host of other environmental variables including several other factors associated with climate change can affect sex determination/differentiation in gonochoristic fishes. For instance, salinity influences sex determination of European seabass *Dicentrarchus labrax*. In this species, transfers from low to high salinity during the sexually labile period produce male-skewed sex ratios (Saillant, Fostier, Haffray, Menu, & Chatain, 2003). Effects of pH on sex determination have been demonstrated in some cichlids and in swordtail (Reddon & Hurd, 2013; Rubin, 1985). In the West African cichlid *Pelvicachromis pulcher*, for example, more males are produced at lower than higher pH (Reddon & Hurd, 2013). Low dissolved oxygen concentration (hypoxia) may also skew population sex ratios toward males, as shown for zebrafish *Danio rerio* (Shang, Yu, & Wu, 2006) and medaka *Oryzias latipes* (Cheung, Chiu, & Wu, 2014). Population density can modulate sex ratios in anguillids, for which both field and laboratory investigations indicate that high densities promote the formation of male-biased sex ratios (Davey & Jellyman, 2005). As discussed below, the light regime can also influence the direction of gonadal development in some species (Brown, Baumann, & Conover, 2014; Corona-Herrera et al., 2018).

Among fishes, the influence of temperature on phenotypic sex has been particularly well investigated in Atheriniformes (Brown et al., 2014; Corona-Herrera et al., 2016; Strüssmann & Patiño, 1999) (Fig. 2). Atlantic silverside *Menidia menidia* (family Atherinopsidae) was in fact the first teleost where the influence of temperature on sex ratios was demonstrated

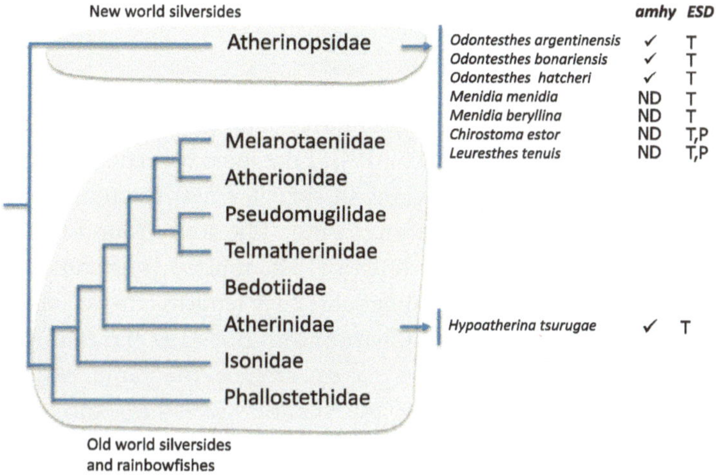

Fig. 2 Scheme showing an overview of available information on sex determination (existence of *amhy* and environmental sex determination, ESD) within the Atheriniformes families and their phylogenetic relationship. The type of ESD is also indicated. T: temperature; P: photoperiod; ND: not determined. The size of branches does not represent divergence time between taxonomic groups. Most studies are concentrated in atherinopsids or New World silversides and knowledge on sex determination in Old World silversides is limited to an atherinid species, *Hypoatherina tsurugae. Modified from Campanella, D., Hughes, L. C., Unmack, P. J., Bloom, D. D., Piller, K. R., et al. (2015). Multi-locus fossil-calibrated phylogeny of Atheriniformes (Teleostei, Ovalentaria).* Molecular Phylogenetics and Evolution, *86, 8–23. https://doi.org/10.1016/j.ympev.2015.03.001.*

(Conover & Kynard, 1981). In this species, young recruits are female-biased at cooler temperatures (at the beginning of the breeding season) but male-biased at warmer temperatures (later in the breeding season). According to a model proposed by Charnov and Bull (1977), the resulting longer growing season for female fish would yield larger-than-average adult individuals at the end of their first year of life, and thus confer a reproductive advantage compared to smaller females born later in the year. An even stronger form of TSD was later reported for the South American silverside pejerrey *Odontesthes bonariensis* (Atherinopsidae). In this species (referred to as pejerrey in this chapter), sex ratios are 100% female or 100% male at the environmentally relevant temperatures of 17 °C (female-promoting temperature) and 29 °C (male-promoting temperature), respectively; at intermediate temperatures mixed-sex populations are produced. These outcomes are reliably obtained when thermal manipulations are applied during the critical period of sex determination, which spans from 1 to 5 weeks after hatching

(Ito, Yamashita, Takashima, & Strüssmann, 2005; Strüssmann, Saito, Usui, Yamada, & Takashima, 1997).

A recent study of California grunion *Leuresthes tenuis* (Atherinopsidae) reported that in addition to temperature, photoperiod (day length) also influences sex ratios (Brown et al., 2014). Like most silversides, the spawning season of California grunion begins in spring when temperatures are still relatively cool and day length is increasing. The study found that the proportion of females was greater at longer (15 h, approx. June summer solstice value) than shorter (12 h, September value) day lengths (Brown et al., 2014), a finding also consistent with the Charnov and Bull (1977) model. California grunion, however, do not seem to experience the same magnitude of seasonal temperature fluctuation as Atlantic silverside, and changes in photoperiod were proposed to represent a more reliable environmental cue for spawning initiation in spring (Brown et al., 2014). An effect of photoperiod on sex determination was recently reported in another atherinopsid, pike silverside *Chirostoma estor*, where continuous illumination produced male-skewed populations compared to control fish reared at 12L: 12D (Corona-Herrera et al., 2018). While the ecological relevance of an effect of constant illumination on sex determination is questionable for a tropical species, the study does confirm that photoperiod can function as environmental sex determinant, at least in silversides. The study also revealed that male-skewed populations had increased expression of male sex-related (*sox9*), stress axis-related (*crf*), and oxidative stress-related (*cat*) genes (Corona-Herrera et al., 2018). The primary mechanism of photoperiod-dependent sex determination in silversides is unknown but is likely associated with seasonal changes in the production of melatonin, a conveyor of circadian signals, which is influenced by the length of the scotophase (melatonin production increases at night) and by temperature (Oliveira & Sánchez-Vásquez, 2010).

The mechanisms of high temperature-induced masculinization have been explored in Atheriniformes and a few other teleosts with reported TSD. Importantly, the condition of being "stressed" seems to be a relevant aspect of the masculinization process in these species. In pejerrey, for example, larvae exposed to masculinizing temperature during the critical period of sex determination consistently showed increased titers of whole-body cortisol (Hattori et al., 2009), which is the primary glucocorticoid produced by the interrenal gland of teleosts and regarded as an important stress hormone (Barton, 2002; Wendelaar Bonga, 1997). In addition, animals treated with exogenous cortisol showed elevated 11-ketotestosterone (11-KT; a potent androgen in teleosts) and testosterone levels, and showed typical molecular signatures of masculinization such as up-regulation of the testicular marker *amh* (anti-müllerian hormone) and down-regulation of the ovarian marker *cyp19a1a* (gonadal aromatase) (Hattori et al., 2009). The steroidogenic enzyme 11β-hydroxysteroid dehydrogenase (HSD11B2) seems to be a key player in heat-induced masculinization of pejerrey (Fernandino, Hattori, Kishii, Strüssmann, & Somoza, 2012). In teleosts, HSD11B2 is important for steroidogenesis within the stress and reproductive axes and is responsible for the inactivation of glucocorticoids and the synthesis of 11-oxygenated androgens like 11-KT, respectively (Perry & Grober, 2003). Hence, in pejerrey, production of 11-KT is induced by the cortisol-mediated activation of *hsd11b2* (Fernandino et al., 2012). These observations strongly support a key role for stress and cortisol in warm temperature-induced masculinization of pejerrey and provide a broader schema for evaluating the impacts of environmental stressors (in addition to temperature) on sex ratios and reproductive development in fishes.

In regards to mechanisms of photoperiod-dependent sex determination, a neuroendocrine path between photoperiod-dependent melatonin production and the stress axis remains to be established but there is experimental evidence for teleosts suggesting interactions between melatonin and cortisol production (Azpeleta, Martínez-Alvarez, Delgado, & De Pedro, 2010; Gesto et al., 2016). In addition, masculinizing photoperiod regimes in pike silverside enhanced *crf* expression, a key hypothalamic regulator of the stress axis (Corona-Herrera et al., 2018), which suggests that prolonged light exposure could enhance the cortisol-masculinization crosstalk during the critical period of sex determination as proposed by Fernandino, Hattori, Moreno Acosta, Strüssmann, and Somoza (2013).

There is also evidence for a direct inhibitory effect of cortisol on the feminization process of fishes (Hayashi et al., 2010; Yamaguchi, Yoshinaga, Yazawa, Gen, & Kitano, 2010). Gonadal estrogen production is regulated by the aromatase gene *cyp19a1a*, which in turn is regulated by follicle-stimulating hormone (FSH) from the pituitary via activation of the cAMP second messenger pathway (Guiguen, Fostier, Piferrer, & Chang, 2010). As in pejerrey, in Japanese flounder *Paralichthys olivaceus* high temperature induces cortisol production (Yamaguchi et al., 2010). Cortisol binds to its receptor, the hormone-receptor complex binds to a cAMP-responsive element (CRE) on the *cyp19a1a* promoter to suppress gene expression, and the consequent reduction in estrogen production impairs the feminization process (Yamaguchi et al., 2010). Down-regulation of *cyp19a1a* by cortisol/high water temperature also has been reported in pejerrey (Hattori et al., 2009) and Atlantic silverside (Duffy et al., 2010), indicating that similar cortisol-dependent suppression of estrogen synthesis may be ubiquitous among Atheriniformes.

The mechanisms of low temperature-dependent "feminization" in silversides (e.g., XY-individuals developing as females) are largely unknown. A possible scenario to explain this feminization process was proposed for pejerrey that involves an early decline of the genotypic-male sex determinant (Zhang et al., 2018) in the context of a "female default" hypothesis (Strüssmann & Ito, 2005). The testis-determining gene, Y-linked anti-müllerian hormone (*amhy*) is present in pejerrey (Yamamoto et al., 2014) and in the congeneric but strongly GSD species, Patagonian pejerrey *Odontesthes hatcheri* (Hattori et al., 2012) (see Section 3). Curiously, in pejerrey but not Patagonian pejerrey, *amhy* expression declines significantly during the critical period of sex determination independently of temperature (Zhang et al., 2018). Regardless of sex genotype or temperature, however, gonadal aromatase *cyp19a1a* is maintained at relatively high levels during the critical period in pejerrey (Zhang et al., 2018). These observations suggest that an ontogenetically programmed ovarian development driven by *cyp19a1a* is the default state that becomes self-sustaining in the absence of a timely and potent masculinizing stimulus (*amhy* and/or high water temperature). Thus, the critical period of TSD in pejerrey might be defined as the time window between the decrease in *amhy* expression and the attainment of a self-sustaining and sex-specific endocrine profile, at which time sex commitment occurs. More extensive molecular pathway analysis using next generation sequencing would be useful to unravel other key players involved in low temperature-dependent feminization process.

A scenario where female is the default sex and testicular formation is triggered by an external factor (high temperature) conceivably renders the gonads prone to locally ambiguous differentiation and the occurrence of intersex condition (co-occurrence of ovarian and testicular tissues within the same individual). Interestingly, this is not the case for pejerrey. On the contrary, intersex condition in this species is extremely rare not only in wild populations but also in laboratory settings where individuals are subjected to complex thermal manipulations during the period of gonadal sex determination/differentiation (Ito et al., 2005; Strüssmann & Ito, 2005). Possible explanations for the unexpected rarity of intersex pejerrey include extragonadal control of gonadal sex differentiation by the central nervous system (Miranda, Strobl-Mazzulla, Strüssmann, Parhar, & Somoza, 2003;

Miranda, Strüssmann, & Somoza, 2001) or the existence of a histological gradient of gonadal sex differentiation (Strüssmann & Ito, 2005) and apoptosis (Yamamoto et al., 2013). Histologically, gonadal sex differentiation in pejerrey begins in the anterior region of the left gonad and proceeds caudally until 10–30% of this gonad has differentiated. Only then will differentiation begin in the anterior region of the contralateral (right) gonad. Interestingly, gonadal apoptosis in the early stages of sex differentiation is observed in the anterior region of the right gonad (Yamamoto et al., 2013). These observations are currently interpreted as a mechanism to prevent discordant sexual differentiation of the paired gonadal lobes (left and right) and may be a characteristic feature of highly temperature-sensitive (TSD) species.

Epigenetic regulation of sex-related genes during the critical period of gonadal differentiation has recently gained attention in fishes. DNA methylation seems to influence sex determination in European sea bass *D. labrax*. In this species, high temperature is able to modify the promoter region of the *cyp19a1* gene and inactivate its expression, leading to testicular development in genotypic females (Navarro-Martín et al., 2011). In half-smooth tongue sole *Cynoglossus semilaevis*, a ZW/ZZ species, differential patterns of *dmrt1* methylation are observed during sex reversal of ZW individuals (pseudomales) that can be passed on to the ZW progeny (Shao et al., 2014). No information is currently available concerning epigenetic regulation of genes associated with TSD in Atheriniformes. Research in this area would yield useful information for a better understanding of TSD mechanisms and of the impacts of environmental stressors on natural populations.

3. Genotypic sex determination and sex-determining genes in fishes

In species with GSD systems, gonadal sex is prescribed by a sex-determining gene located on the Y (male heterogamety) or W (female heterogamety) chromosomes. This gene must be expressed in the primordial gonad during the early stages of sex determination and be relatively insensitive to environmental influences for the phenotypic sex to be manifested correctly. Sex-determining genes are being identified at a rapid pace in recent years thanks to the development of Next Generation Sequencing platforms combined with analytical pipelines that can cover the whole genome and transcriptome with high accuracy, enabling the generation and processing of large amounts of genetic information within a short time.

Despite the variety of sex-determining genes described to date in fishes (Hattori, Strüssmann, Fernandino, & Somoza, 2013; Kikuchi & Hamaguchi, 2013), some trends are becoming apparent. Transcription factors such as *dmrt1* and *sox3* and various members of the TGF-beta superfamily (e.g., *amh*, *gsdf*, and *amhr2*), previously associated with various aspects of gonadal differentiation, are now known as the primary switches of the sex determination pathway in a number of species. Some sex-determining genes, however, had no known function in gonadal development prior to their discovery, such as the immune-related gene *sdY* (Yano et al., 2012). Among Atheriniformes, the *amhy* gene is the sex determinant of at least two atherinopsid species, Patagonian pejerrey and pejerrey (Hattori et al., 2012; Yamamoto et al., 2014), and more recently was also reported as the likely sex-determining gene of cobaltcap silverside *Hypoatherina tsurugae*, a marine atherinid from the Northwest Pacific Ocean (Bej, Miyoshi, Hattori, Strüssmann, & Yamamoto, 2017). Taxonomically, atherinids are fairly removed from atherinopsids within the order Atheriniformes. It is not clear whether the *amhy* genes in these two families are homologous because, with the exception of the TGF-beta domain, their structure differs considerably (Bej et al., 2017; Hattori et al., 2012; Yamamoto et al., 2014). The presence of *amhy* in these families, however, suggests that this gene may be a conserved master switch for sex determination in other atheriniforms (Fig. 2). On the other hand, sex-determining genes appear to have had a high turnover rate within the genus *Oryzias* (medaka species), as evidenced by the existence of *DMY/dmrt1bY*, *sox3bY* and *gsdfY* in this group (Matsuda et al., 2002; Myosho et al., 2012; Nanda et al., 2002; Takehana et al., 2014). For medaka species in which the sex-determining gene is not yet identified, analysis of linkage groups with sex-linked *loci* also supports the poor conservation of sex-determining genes and sex chromosomes in these fishes (Matsuda & Sakaizumi, 2016).

As described in Section 2, the presence of genotypic sex determinants in a given species does not guarantee faithful concordance between its sex genotype and phenotype, in the same manner as mild or even strong ESD mechanisms (e.g., TSD) do not preclude the existence of genetic determinants of gonadal sex. Based on current knowledge, it is possible to suggest that most gonochoristic fishes possess a sex-determining gene or genes regardless of their primary mode of sex determination; thus, sex determination systems would range from strong GSD to strong ESD but "pure" forms of either one would be rare or nonexistent, especially for ESD. The examples of pejerrey and zebrafish illustrate how this paradigm has evolved.

In pejerrey, the remarkable strength and consistency of ESD (TSD), the absence of a thermal plateau with balanced sex ratios, and the fact that sex-bias occurs at environmentally relevant temperatures led to the earlier view that this species lacks genetic determinants (Ospina-Álvarez & Piferrer, 2008; Strüssmann, Moriyama, Hanke, Calsina, & Takashima, 1996). The recent discovery of *amhy* as the potential sex-determining gene in pejerrey, however, put an end to this idea (Yamamoto et al., 2014). In zebrafish, one of the most intensively studied model teleosts, multiple lines of evidence including, most recently high-throughput genetic analysis had also led to the conclusion that sex chromosomes were absent in this species. The natural GSD system of this species was in fact widely believed to be polygenic (Liew & Orbán, 2013); that is, until recently. It now appears that earlier studies failed to detect sex chromosomes in zebrafish because of genetic anomalies present in the domesticated strains used. In a recent study comparing natural and domesticated strains, Wilson et al. (2014) uncovered a strong sex-linked *locus* in chromosome 4 of several wild and natural-strain specimens that is absent or weakened in domesticated strains. As of the time of this writing, the sex-determining gene of zebrafish is still unknown but the discovery of the sex-linked *locus* should facilitate its identification. These observations with zebrafish serve as a cautionary tale of the potential pitfalls of using domesticated fishes in genetic studies with ecological or evolutionary applications. This is not an argument against the well-established polygenic systems of sex determination reported among fishes, but for the need to consider degree of domestication of experimental specimens in data interpretations.

The large repertoire of sex-determining switches found in fishes also implies the existence of diversity or flexibility in the molecular cascade of sex determination, with several alternative pathways resulting in the same sex phenotype. Rewiring of the gene hierarchy during speciation, however, seems to have required adjustments in downstream pathways in some cases. For instance, the *amh* gene is differentially expressed during testicular differentiation of most teleosts (Pfennig, Standke, & Gutzeit, 2015) but in medaka, *amh* shows no sexually dimorphic expression (Klüver et al., 2007). In addition, while the *dmrt1* gene is reportedly an important player in sex determination of fishes and other vertebrates (Koopman & Loffler, 2003), in Patagonian pejerrey it does not show sex-specific expression (Hattori et al., 2008). The rise of *dmy/dmrt1bY* as sex-determining genes in medaka and *amhy* in Patagonian pejerrey seems to have been accompanied by rearrangements in regulatory mechanisms of downstream genes, thus rendering medaka *amh* and Patagonian pejerrey *dmrt1* unnecessary for testicular differentiation. The relatively high expression of sex-determining genes observed during gonad differentiation could well be the driving force for the apparent loss of function of their autosomal paralogs such as *amh* and *dmrt1* in these species. Whether similar processes occur in other fish groups is not yet clear, but a comparative analysis of paralogs of sex-determining genes may shed more light on the plasticity of the gonadal differentiation cascade at the gene level.

The rise of new sex determinants is presumably accompanied by genetic and functional alterations that allow them to be at the top of the sex differentiation cascade. Sequence alterations in sex-determining genes, especially in their regulatory regions, may be necessary to correctly time their expression period as proposed for *dmy* after duplication from *dmrt1* in medaka (Otake, Shinomiya, Kawaguchi, Hamaguchi, & Sakaizumi, 2008) and as observed for *amhy* in relation to the autosomal *amha* in Patagonian pejerrey (Hattori et al., 2012; Yamamoto et al., 2014). One possible mechanism for these genetic changes may be related to mobile elements, such as retrotransposons. Signatures of these elements have been detected in the medaka *dmy* promoter region (Herpin et al., 2010). There are also some cases where the sex-determining gene has lost fragments and become truncated, such as the *amhy* of cobaltcap silverside (Bej et al., 2017) and *sdY* in rainbow trout (Yano et al., 2012). Regardless of how the change occurred, however, an emerging sex-determining gene needs to confer a reproductive advantage in order to become fixed in the population. These genetic changes (loss of fragments or accumulation of repetitive elements) in master sex-determining genes and the sex chromosome where they are located are also likely to offer protection against genetic recombination.

4. Environmental effects on sex determination in wild populations

While temperature modulation of sex determination has been clearly demonstrated for many species under laboratory conditions (Fernandino et al., 2013; Strüssmann & Patiño, 1999), temperature-dependent sex reversal in nature is still controversial. Evidence of its occurrence is based primarily on findings of biased sex ratios in wild populations of silversides from the genera *Menidia* and *Odontesthes* (Conover & Fleisher, 1986; Middaugh & Hemmer, 1987; Strüssmann et al., 2010). Wild populations of Atlantic silverside also show geographical (latitudinal) variation in TSD responses when measured in the laboratory (Conover, 2004). Interestingly, none of the populations examined displayed pure TSD or GSD; the level of TSD was uniformly high in their southern range of distribution and declined toward the northern range, where GSD dominated (Conover, 2004). However, the level of TSD rebounded to a moderate level in the northern-most populations. The length of the growing season likely drives this latitudinal variation in the TSD response and such variations seem to be maintained by contemporary selection (Conover, 2004; Duffy, Hice, & Conover, 2015). Notably, continuous breeding of wild populations of Atlantic silverside with marked TSD at extreme temperatures in the laboratory initially yielded highly skewed sex ratios but the effects gradually vanished, with the minority sex proportionally increasing from one generation to the next until balanced sex ratios were achieved in 8–10 generations (Conover, Van Voorhees, & Ehtisham, 1992). In other words, transition from TSD to GSD was observed over a few generations. Shifts in TSD response among generations/families have been observed also in pejerrey (Hattori et al., 2018; Zhang et al., 2018).

The availability of molecular markers of genetic sex (or of the sex-determining gene itself) in species with TSD facilitates studies of its occurrence in the field, including surveys of the potential effects of global warming on population sex ratios. For example, based on the presence of the *amhy* gene on the Y-chromosome of some Atheriniformes (Bej et al., 2017; Yamamoto et al., 2014), genotypic females (XX) can be distinguished from genotypic males (XY and YY) and the concordance between genotypic and phenotypic sex can be easily evaluated. In fact, a field survey of an invasive, self-propagating pejerrey population in Lake Kasumigaura, Japan, revealed the presence of both male- and female-sex reversals (Yamamoto et al., 2014). Sex-reversed pejerrey of both sexes are viable and fertile (Hattori et al., 2018; Yamamoto et al., 2014) and thus likely to mate in the wild. Progenies of sex-reversed XY females crossed with normal XY males would yield YY offspring (supermales), which also are viable and fertile (Hattori et al., 2018). It would be of interest to determine if YY males are found in nature. Curiously, YY offspring develop as males even when reared at female-inducing temperatures in the laboratory (Hattori et al., 2018), supporting the notion that *amhy* is the primary male-sex determinant in this species and that the presence of two *amhy* copies erases the influence of temperature.

The identification of *amhy* as the potential sex-determining gene of cobaltcap silverside (Bej et al., 2017) has made it possible to search for mismatches between genetic and phenotypic sex in wild populations of this species. A preliminary examination of individuals from Tokyo Bay in Japan revealed the presence of male- and female-sex reversals (Bej et al., 2017). Potential associations between ambient water temperature and sex determination were assessed in 3 year-classes of cobaltcap silverside by estimating their birth dates by otolith analysis and the average water temperature recorded at the time of their birth (unpublished data). One of the year-classes examined showed a high occurrence of male sex reversal, and otolith analysis indicated these fish were born when the temperature was higher than the corresponding birth-time temperatures for the two other year-classes (unpublished data). While further examination is clearly needed, these observations are consistent with the occurrence of natural sex reversal in wild populations of cobaltcap silverside in relation to ambient temperatures. As the global warming trend continues, this approach might prove useful to monitor the effects of climate change on fish sex ratio and population health.

5. Conclusions and perspectives

Recent years have brought much progress in our understanding of sex-determining genes and their diversity among fishes. The identification of *amhy* as the sex-determining gene of pejerrey (Yamamoto et al., 2014), a silverside with the strongest-known TSD among fishes, has led to questioning of the previous notion that this species represents a "pure" form of ESD and, in turn, suggests that true ESD or TSD may not exist in this group of vertebrates (Fig. 1). Conversely, in species with strong GSD systems where sex ratios are normally balanced and stable, exposure to extreme ambient conditions also can influence sex determination, as shown for channel catfish *Ictalurus punctatus* (Patiño, 2019; Patiño et al., 1996). The idea that gonadal fate in gonochoristic teleosts is determined by an inverse relationship (continuum) between genotypic and environmental factors is of increasing appeal (Baroiller & D'Cotta, 2016; Yamamoto et al., 2014).

Cortisol is a major stress hormone in fishes that also seems to be a key player in gonadal sex differentiation of ESD species. Thus, increased cortisol production is required for the masculinizing effect of warm temperatures in a variety of species (Hattori et al., 2008; Hayashi et al., 2010; Yamaguchi et al., 2010). This knowledge is of relevance to an understanding of potential impacts of global warming on fish gonadal development and reproduction, as sex ratios are an important demographic parameter with implications on population dynamics and sustainability. Various environmental conditions (stressors) besides temperature, however, can also elicit changes in cortisol production (Barton, 2002; Wendelaar Bonga, 1997) and therefore could influence sex determination and population sex ratios. An appropriate way to conceptualize the association between environmental challenges and sex determination may be in the context of physiological responses to multiple stressors, which in terms of cortisol production are cumulative in fishes (Barton, 2002). For example, a stressor exposure that by itself does not prompt a cortisol response of sufficient magnitude to influence gonadal fate could become effective in the presence of other challenges. Studies of how multiple (and different types of) stressors affect sex determination are necessary to clarify whether ESD can be seen through the lens of cumulative stress responses.

Knowledge of the mechanisms by which environmental factors induce or allow feminization is very limited in fishes. In most species with ESD, low temperature results in female-biased populations (Strüssmann & Patiño, 1999) but there are exceptions such as channel catfish, where low temperature had no effect on sex ratios and high temperature resulted in the preponderance of females (Patiño et al., 1996). Further studies on environmentally prescribed feminization could bring new insights into the hypothesis of female as the default sex. A novel and potentially productive line of research deals with epigenetic modifications (e.g., methylation) of sex-related genes induced by environmental factors, the involvement of these modifications in gonadal sex differentiation, and an understanding of how these epigenetic imprints are transmitted and are effective in the modulation of offspring sex ratios. The role of hormones outside the sex and stress steroid systems also deserves more research attention. Thyroid hormone, for example, has potent masculinizing activity and is able to induce male-skewed populations in zebrafish (Mukhi, Torres, & Patiño, 2007; Sharma & Patiño, 2013; Sharma, Tang, Mayer, & Patiño, 2016). In addition, in a TSD species of turtle, more males are produced in the presence than the absence of thyroid hormone (Sun et al., 2016).

Atheriniformes include the first example of TSD among fishes and as made evident by the information presented in this chapter, they are poised to continue in their position as prominent models for field studies of sex determination. Pejerrey in particular seems to have the traits of an ideal bioindicator species—it has a strong TSD system, its sex-determining gene has been identified, and preliminary field observations suggest it is possible to detect natural mismatches between genotypic and phenotypic sex.

Long-term monitoring of climate anomalies coupled with field surveys of phenotypic/genotypic sex in pejerrey can contribute to a better understanding and forecasting of the biological impacts of global change. This information would be useful to the development of conservation, management, and mitigation strategies.

Acknowledgments

We thank Dr. Adam Luckenbach for his useful critique of an early draft of this manuscript. Work on this chapter was facilitated by FAPESP-SPRINT grant 2015/50465-5 to R.S.H. and R.P. and by JSPS KAKENHI grants to C.A.S. (26241018) and Y.Y. (15K18728 and 15KK0277). The Texas Cooperative Fish and Wildlife Research Unit is jointly supported by U.S. Geological Survey, Texas Tech University, Texas Parks and Wildlife Department, The Wildlife Management Institute, and U.S. Fish and Wildlife Service.

References

Angelier, F., & Wingfield, J. C. (2012). Importance of the glucocorticoid stress response in a changing world: Theory, hypotheses and perspectives. *General and Comparative Endocrinology*, *190*, 118–128. https://doi.org/10.1016/j.ygcen.2013.05.022.

Azpeleta, C., Martínez-Alvarez, R. M., Delgado, M. J., & De Pedro, N. (2010). Melatonin reduces locomotor activity and circulating cortisol in goldfish. *Hormones and Behavior*, *57*, 323–329. htpps://doi:10.1016/j.yhbeh.2010.01.001.

Baroiller, J. F., & D'Cotta, H. (2016). The reversible sex of gonochoristic fish: Insights and consequences. *Sexual Development*, *10*, 242–266. https://doi.org/10.1159/000452362.

Baroiller, J. F., D'Cotta, H., & Saillant, E. (2009). Environmental effects on fish sex determination and differentiation. *Sexual Development*, *3*, 118–135. https://doi.org/10.1159/000223077.

Barros, D. F., & Albernaz, A. L. M. (2014). Possible impacts of climate change on wetlands and its biota in the Brazilian Amazon. *Brazilian Journal of Biology*, *74*, 810–820.

Barton, B. A. (2002). Stress in fishes: A diversity of responses with particular reference to changes in circulating corticosteroids. *Integrative and Comparative Biology*, *42*, 517–525. https://doi.org/10.1093/icb/42.3.517.

Bates, B. C., Kundzewicz, Z. W., Wu, S., & Palutikof, J. P. (Eds.), (2008). *Climate change and water. Technical paper of the intergovernmental panel on climate change* (p. 210). Geneva: IPCC Secretariat. https://www.ipcc.ch/pdf/technical-papers/climate-change-water-en.pdf.

Bej, D. K., Miyoshi, K., Hattori, R. S., Strüssmann, C. A., & Yamamoto, Y. (2017). A duplicated, truncated *amh* gene is involved in male sex determination in an Old World silverside. *Genes Genomes Genetics*, 7, 2489–2495. https://doi.org/10.1534/g3.117.042697.

Brown, E. E., Baumann, H., & Conover, D. O. (2014). Temperature and photoperiod effects on sex determination in a fish. *Journal of Experimental Marine Biology and Ecology*, *461*, 39–43. https://doi.org/10.1016/j.jembe.2014.07.009.

Campanella, D., Hughes, L. C., Unmack, P. J., Bloom, D. D., Piller, K. R., & Ortí, G. (2015). Multi-locus fossil-calibrated phylogeny of Atheriniformes (Teleostei, Ovalentaria). *Molecular Phylogenetics and Evolution*, *86*, 8–23. https://doi.org/10.1016/j.ympev.2015.03.001.

Charnov, E. L., & Bull, J. (1977). When is sex environmentally determined? *Nature*, *266*, 828.

Cheung, C. H. Y., Chiu, J. M. Y., & Wu, R. S. S. (2014). Hypoxia turns genotypic female medaka fish into phenotypic males. *Ecotoxicology*, *23*, 1260–1269. https://doi.org/10.1007/s10646-014-1269-8.

Conover, D. O. (2004). Temperature-dependent sex determination in fishes. In N. Valenzuela & V. A. Lance (Eds.), *Temperature-dependent sex determination in vertebrates* (pp. 11–20). Washington: Smithsonian Books.

Conover, D., & Fleisher, M. H. (1986). Temperature-sensitive priod of sex determination in the Atlantic silverside, *Menidia menidia*. *Canadian Journal of Fisheries and Aquatic Sciences*, *43*, 514–520. https://doi.org/10.1139/f86-061.

Conover, D. O., & Kynard, B. E. (1981). Environmental sex determination: Interaction of temperature and genotype in a fish. *Science*, *213*, 577–579. https://doi.org/10.1126/science.213.4507.577.

Conover, D. O., Van Voorhees, D. A., & Ehtisham, A. (1992). Sex ratio selection and the evolution of environmental sex determination in laboratory populations of *Menidia menidia*. *Evolution*, *46*, 1722–1730. https://doi.org/10.1111/j.1558-5646.1992.tb01164.x.

Corona-Herrera, G. A., Arranz, S. E., Martínez-Palacios, C. A., Navarrete-Ramírez, P., Toledo-Cuevas, E. M., Valdez-Alarcón, J. J., et al. (2018). Experimental evidence of masculinization by continuous illumination in a temperature sex determination teleost (Atherinopsidae) model: Is oxidative stress involved? *Journal of Fish Biology*, *93*, 229–237. https://doi.org/10.1111/jfb.13651.

Corona-Herrera, G. A., Tello-Ballinas, J. A., Hattori, R. S., Martínez-Palacios, C. A., Strüssmann, C. A., Cárdenas-Reygadas, R., et al. (2016). Gonadal differentiation and temperature effects on sex determination in the freshwater pike silverside *Chirostoma estor* Jordan 1880. *Environmental Biology of Fishes*, *99*, 463–471. https://doi.org/10.1007/s10641-016-0491-z.

Davey, A. J. H., & Jellyman, D. J. (2005). Sex determination in freshwater eels and management options for manipulation of sex. *Reviews in Fish Biology and Fisheries*, *15*, 37–52. https://doi.org/10.1007/s11160-005-7431-x.

Devlin, R. H., & Nagahama, Y. (2002). Sex determination and sex differentiation in fish: An overview of genetic, physiological, and environmental influences. *Aquaculture*, *208*, 191–364. https://doi.org/10.1016/S0044-8486(02)00057-1.

Duffy, T. A., Hice, L. A., & Conover, D. O. (2015). Pattern and scale of geographic variation in environmental sex determination in the Atlantic silverside, *Menidia menidia*. *Evolution*, *69*, 2187–2195. https://doi.org/10.1111/evo.12724.

Duffy, T. A., Picha, M. E., Won, E. T., Borski, R. J., Mcelroy, A. E., & Conover, D. O. (2010). Ontogenesis of gonadal aromatase gene expression in Atlantic silverside (*Menidia menidia*) populations with genetic and temperature-dependent sex determination. *Journal of Experimental Zoology. Part A, Ecological Genetics and Physiology*, *313*, 421–431. https://doi.org/10.1002/jez.612.

Fernandino, J. I., Hattori, R. S., Kishii, A., Strüssmann, C. A., & Somoza, G. M. (2012). The cortisol and androgen pathways cross talk in high temperature-induced masculinization: The 11β-hydroxysteroid dehydrogenase as a key enzyme. *Endocrinology*, *153*, 6003–6011. https://doi.org/10.1210/en.2012-1517.

Fernandino, J. I., Hattori, R. S., Moreno Acosta, O. D., Strüssmann, C. A., & Somoza, G. M. (2013). Environmental stress-induced testis differentiation: Androgen as a by-product of cortisol inactivation. *General and Comparative Endocrinology*, *192*, 36–44. https://doi.org/10.1016/j.ygcen.2013.05.024.

Gelca, R., Hayhoe, K., Scott-Fleming, I., Crow, C., Dawson, D., & Patiño, R. (2016). Climate-water quality relationships in Texas reservoirs. *Hydrological Processes*, *30*, 12–29. https://doi.org/10.1002/hyp.10545.

Gesto, M., Álvarez-Otero, R., Conde-Sieira, M., Otero-Rodiño, C., Usandizaga, S., Soengas, J. L., et al. (2016). A simple melatonin treatment protocol attenuates the response to acute stress in the sole *Solea senegalensis*. *Aquaculture*, *452*, 272–282. https://doi.org/10.1016/j.aquaculture.2015.11.006.

Guiguen, Y., Fostier, A., Piferrer, F., & Chang, C. F. (2010). Ovarian aromatase and estrogens: A pivotal role for gonadal sex differentiation and sex change in fish. *General and Comparative Endocrinology*, *165*, 352–366. https://doi.org/10.1016/j.ygcen.2009.03.002.

Hattori, R. S., Fernandino, J. I., Kishil, A., Kimura, H., Kinno, T., Oura, M., et al. (2009). Cortisol-induced masculinization: Does thermal stress affect gonadal fate in pejerrey, a teleost fish with temperature-dependent sex determination? *PLoS One*, *4*, e6548. https://doi.org/10.1371/journal.pone.0006548.

Hattori, R. S., Fernandino, J. I., Strüssmann, C. A., Somoza, G. M., Yokota, M., & Watanabe, S. (2008). Characterization and expression profiles of DMRT1, AMH, SF1 and P450aro genes during gonadal sex differentiation in Patagonian pejerrey *Odontesthes hatcheri*. *Cybium*, *32*, 95–96.

Hattori, R. S., Murai, Y., Oura, M., Masuda, S., Majhi, S. K., Sakamoto, T., et al. (2012). A Y-linked anti-Mullerian hormone duplication takes over a critical role in sex determination. *Proceedings of the National Academy of Sciences of the United Statesof America*, *109*, 2955–2959. https://doi.org/10.1073/pnas.1018392109.

Hattori, R. S., Strüssmann, C. A., Fernandino, J. I., & Somoza, G. M. (2013). Genotypic sex determination in teleosts: Insights from the testis-determining *amhy* gene. *General and Comparative Endocrinology*, *192*, 55–59. https://doi.org/10.1016/j.ygcen.2013.03.019.

Hattori, R. S., Tashiro, S., Zhang, Y., Kakuta, N., Yokota, M., Strüssmann, C. A., et al. (2018). Demonstration of viability and fertility and development of a molecular tool to identify YY supermales in a fish with both genotypic and environmental sex determination. *Ecology and Evolution*, *8*, 7522–7528. https://doi.org/10.1002/ece3.4148.

Hayashi, Y., Kobira, H., Yamaguchi, T., Shiraishi, E., Yazawa, T., Hirai, T., et al. (2010). High temperature causes masculinization of genetically female medaka by elevation of cortisol. *Molecular Reproduction and Development*, *77*, 679–686. https://doi.org/10.1002/mrd.21203.

Herpin, A., Braasch, I., Kraeussling, M., Schmidt, C., Thoma, E. C., Nakamura, S., et al. (2010). Transcriptional rewiring of the sex determining *dmrt1* gene duplicate by transposable elements. *PLoS Genetics*, *6*(2), e1000844. https://doi.org/10.1371/journal.pgen.1000844.

IPCC. (2014). Climate change 2014: Mitigation of climate change. Summary for policymakers and technical summary, climate change 2014. In *Mitigation of climate change. Part of the working group III contribution to the fifth assessment report of the intergovernmental panel on climate change*. Cambridge University Press. https://doi.org/10.1017/CBO9781107415416.005.

Ito, L. S., Yamashita, M., Takashima, F., & Strüssmann, C. A. (2005). Dynamics and histological characteristics of gonadal sex differentiation in pejerrey (*Odontesthes bonariensis*) at feminizing and masculinizing temperatures. *Journal of Experimental Zoology. Part A, Comparative Experimental Biology*, *303*, 504–514. https://doi.org/10.1002/jez.a.159.

Janse, J. H., Kuiper, J. J., Weijters, M. J., Westerbeek, E. P., Jeuken, M. H. J. L., Bakkenes, M., et al. (2015). GLOBIO-aquatic, a global model of human impact on the biodiversity of inland aquatic ecosystems. *Environmental Science & Policy*, *48*, 99–114. https://doi.org/10.1016/j.envsci.2014.12.007.

Kikuchi, K., & Hamaguchi, S. (2013). Novel sex-determining genes in fish and sex chromosome evolution. *Developmental Dynamics*, *242*, 339–353. https://doi.org/10.1002/dvdy.23927.

Klüver, N., Pfennig, F., Pala, I., Storch, K., Schlieder, M., Froschauer, A., et al. (2007). Differential expression of anti-Müllerian hormone (*amh*) and anti-Müllerian hormone receptor type II (*amhrII*) in the teleost Medaka. *Developmental Dynamics*, *236*, 271–281. https://doi.org/10.1002/dvdy.20997.

Koopman, P., & Loffler, K. A. (2003). Sex determination: The fishy tale of *Dmrt1*. *Current Biology*, *13*, 177–179. https://doi.org/10.1016/S0960-9822(03)00117-9.
Liew, W. C., & Orbán, L. (2013). Zebrafish sex: A complicated affair. *Briefings in Functional Genomics*, *13*, 172–187. https://doi.org/10.1093/bfgp/elt041.
Luckenbach, J. A., & Yamamoto, Y. (2018). *Genetic & environmental sex determination in cold-blooded vertebrates: Fishes, amphibians, and reptiles*. In M. K. Skinner (Ed.), *Vol. 6. Encyclopedia of reproduction* (pp. 176–183). San Diego: Academic Press. https://doi.org/10.1016/B978-0-12-809633-8.20553-0.
Marengo, J. A. (2008). Water and Climate Change. *Estudos Avançados*, *22*, 83–96.
Matsuda, M., Nagahama, Y., Shinomiya, A., Sato, T., Matsuda, C., Kobayashi, T., et al. (2002). *DMY* is a Y-specific DM-domain gene required for male development in the medaka fish. *Nature*, *417*, 559–563. https://doi.org/10.1038/nature751.
Matsuda, M., & Sakaizumi, M. (2016). Evolution of the sex-determining gene in the teleostean genus *Oryzias*. *General and Comparative Endocrinology*, *239*, 80–88. https://doi.org/10.1016/j.ygcen.2015.10.004.
Middaugh, D. P., & Hemmer, M. J. (1987). Influence of environmental temperature on sex ratios in the tidewater silverside, *Menidia peninsulae* (Pisces: Atherinidae). *Copeia*, *4*, 958–964. https://doi.org/10.2307/1445559.
Miranda, L. A., Strobl-Mazzulla, P. H., Strüssmann, C. A., Parhar, I., & Somoza, G. M. (2003). Gonadotropin-releasing hormone neuronal development during the sensitive period of temperature sex determination in the pejerrey fish, *Odontesthes bonariensis*. *General and Comparative Endocrinology*, *132*, 444–453. https://doi.org/10.1016/S0016-6480(03)00117-5.
Miranda, L. A., Strüssmann, C. A., & Somoza, G. M. (2001). Immunocytochemical identification of GtH1 and GtH2 cells during the temperature-sensitive period for sex determination in pejerrey, *Odontesthes bonariensis*. *General and Comparative Endocrinology*, *124*, 45–52. https://doi.org/10.1006/gcen.2001.7687.
Mukhi, S., Torres, L., & Patiño, R. (2007). Effects of larval-juvenile treatment with perchlorate and co-treatment with thyroxine on zebrafish sex ratios. *General and Comparative Endocrinology*, *150*, 486–494. https://doi.org/10.1016/j.ygcen.2006.11.013.
Myosho, T., Otake, H., Masuyama, H., Matsuda, M., Kuroki, Y., Fujiyama, A., et al. (2012). Tracing the emergence of a novel sex-determining gene in medaka, *Oryzias luzonensis*. *Genetics*, *191*, 163–170. https://doi.org/10.1534/genetics.111.137497.
Nanda, I., Kondo, M., Hornung, U., Asakawa, S., Winkler, C., Shimizu, A., et al. (2002). A duplicated copy of *DMRT1* in the sex-determining region of the Y chromosome of the medaka, *Oryzias latipes*. *Proceedings of the National Academy of Sciences of the United States of America*, *99*, 11778–11783. https://doi.org/10.1073/pnas.182314699.
Navarro-Martín, L., Viñas, J., Ribas, L., Díaz, N., Gutiérrez, A., Di Croce, L., et al. (2011). DNA methylation of the gonadal aromatase (*cyp19a*) promoter is involved in temperature-dependent sex ratio shifts in the European sea bass. *PLoS Genetics*, 7(12), e1002447. https://doi.org/10.1371/journal.pgen.1002447.
Oliveira, C., & Sánchez-Vásquez, F. J. (2010). Reproduction rhythms in fish. In E. Kulczykowska, W. Popek, & B. Kapoor (Eds.), *Biological clock in fish* (pp. 185–215). Enfield, NH: Science Publishers.
Ospina-Álvarez, N., & Piferrer, F. (2008). Temperature-dependent sex determination in fish revisited: Prevalence, a single sex ratio response pattern, and possible effects of climate change. *PLoS One*, *3*, 2–4. https://doi.org/10.1371/journal.pone.0002837.
Otake, H., Shinomiya, A., Kawaguchi, A., Hamaguchi, S., & Sakaizumi, M. (2008). The Medaka sex-determining gene *DMY* acquired a novel temporal expression pattern after duplication of *DMRT1*. *Genesis*, *46*, 719–723. https://doi.org/10.1002/dvg.20431.
Palmer, M. A., Reidy Liermann, C. A., Nilsson, C., Flörke, M., Alcamo, J., Lake, P. S., et al. (2008). Climate change and the world's river basins: Anticipating management options. *Frontiers in Ecology and the Environment*, *6*, 81–89. https://doi.org/10.1890/060148.

Patiño, R. (2019). Sex determination, gonadal sex differentiation, and sex control in channel catfish. In H. Wang, F. Piferrer, S. Chen, & Z. Shen (Eds.), *Sex control in aquaculture* (pp. 477–494). Wiley-Blackwell. https://doi.org/10.1002/9781119127291.ch23.

Patiño, R., Davis, K. B., Schoore, J. E., Uguz, C., Strüssmann, C. A., Parker, N. C., et al. (1996). Sex differentiation of channel catfish gonads: Normal development and effects of temperature. *The Journal of Experimental Zoology*, *276*, 209–218. https://doi.org/10.1002/(SICI)1097-010X(19961015)276:3<209::AID-JEZ5>3.0.CO;2-R.

Perry, A. N., & Grober, M. S. (2003). A model for social control of sex change: Interactions of behavior, neuropeptides, glucocorticoids, and sex steroids. *Hormones and Behavior*, *43*, 31–38. https://doi.org/10.1016/S0018-506X(02)00036-3.

Pfennig, F., Standke, A., & Gutzeit, H. O. (2015). The role of *Amh* signaling in teleost fish—Multiple functions not restricted to the gonads. *General and Comparative Endocrinology*, *223*, 87–107. https://doi.org/10.1016/j.ygcen.2015.09.025.

Reddon, A. R., & Hurd, P. L. (2013). Water pH during early development influences sex ratio and male morph in a West African cichlid fish, *Pelvicachromis pulcher*. *Zoology*, *116*, 139–143. https://doi.org/10.1016/j.zool.2012.11.001.

Rubin, D. A. (1985). Effect of pH on sex ratio in cichlids and a poecilliid (Teleostei). *Copeia*, *1985*, 233–235. https://doi.org/10.2307/1444818.

Saillant, E., Fostier, A., Haffray, P., Menu, B., & Chatain, B. (2003). Saline preferendum for the European sea bass, *Dicentrarchus labrax*, larvae and juveniles: Effect of salinity on early development and sex determination. *Journal of Experimental Marine Biology and Ecology*, *287*, 103–117. https://doi.org/10.1016/S0022-0981(02)00502-6.

Shang, E. H. H., Yu, R. M. K., & Wu, R. S. S. (2006). Hypoxia affects sex differentiation and development leading to a male-dominated population in zebrafish (*Danio rerio*). *Environmental Science & Technology*, *40*, 3118–3122. https://doi.org/10.1021/es0522579.

Shao, C., Li, Q., Chen, S., Zhang, P., Lian, J., Hu, Q., et al. (2014). Epigenetic modification and inheritance in sexual reversal of fish. *Genome Research*, *24*, 604–615. https://doi.org/10.1101/gr.162172.113.

Sharma, P., & Patiño, R. (2013). Regulation of gonadal sex ratios and pubertal development by the thyroid endocrine system in zebrafish (Danio rerio). *General and Comparative Endocrinology*, *184*, 111–119. https://doi.org/10.1016/j.ygcen.2012.12.018.

Sharma, P., Tang, S., Mayer, G. D., & Patiño, R. (2016). Effects of thyroid endocrine manipulation on sex-related gene expression and population sex ratios in zebrafish. *General and Comparative Endocrinology*, *235*, 38–47. https://doi.org/10.1016/j.ygcen.2016.05.028.

Strüssmann, C. A., Conover, D. O., Somoza, G. M., & Miranda, L. A. (2010). Implications of climate change for the reproductive capacity and survival of New World silversides (family Atherinopsidae). *Journal of Fish Biology*, *77*, 1818–1834. https://doi.org/10.1111/j.1095-8649.2010.02780.x.

Strüssmann, C. A., & Ito, L. S. (2005). Where does gonadal sex differentiation begin? Gradient of histological sex differentiation in the gonads of pejerrey, *Odontesthes bonariensis* (Pisces, Atherinidae). *Journal of Morphology*, *265*, 190–196. https://doi.org/10.1002/jmor.10351.

Strüssmann, C. A., Moriyama, S., Hanke, E., Calsina, J., & Takashima, F. (1996). Evidence of thermolabile sex determination in pejerrey. *Journal of Fish Biology*, *48*, 643–651. https://doi.org/10.1006/jfbi.1996.0064.

Strüssmann, C. A., & Patiño, R. (1999). Sex determination, environmental. In E. Knobil & J. D. Neil (Eds.), *Encyclopedia of reproduction* (pp. 402–409). San Diego: Academic Press.

Strüssmann, C. A., Saito, T., Usui, M., Yamada, H., & Takashima, F. (1997). Thermal thresholds and critical period of thermolabile sex determination in two atherinid fishes, *Odontesthes bonariensis* and *Patagonina hatcheri*. *The Journal of Experimental Zoology*, *278*, 167–177. https://doi.org/10.1002/(SICI)1097-010X(19970615)278:3<167::AID-JEZ6>3.0.CO;2-M.

Sun, B.-J., Li, T., Mu, Y., McGlashan, J. K., Georges, A., Shine, R., et al. (2016). Thyroid hormone modulates offspring sex ratio in a turtle with temperature-dependent sex determination. *Proceedings of the Royal Society B*, *283*, 20161206. https://doi.org/10.1098/rspb.2016.1206.

Takehana, Y., Matsuda, M., Myosho, T., Suster, M. L., Kawakami, K., Shin-I, T., et al. (2014). Co-option of *Sox3* as the male-determining factor on the y chromosome in the fish *Oryzias dancena*. *Nature Communications*, *5*, 1–10. https://doi.org/10.1038/ncomms5157.

Wendelaar Bonga, S. E. (1997). The stress reponse in fish. *Physiological Reviews*, 77, 591–625. https://doi.org/10.1152/physrev.1997.77.3.591.

Wilson, C. A., High, S. K., McCluskey, B. M., Amores, A., Yan, Y. L., Titus, T. A., et al. (2014). Wild sex in zebrafish: Loss of the natural sex determinant in domesticated strains. *Genetics*, *198*, 1291–1308. https://doi.org/10.1534/genetics.114.169284.

Yamaguchi, T., Yoshinaga, N., Yazawa, T., Gen, K., & Kitano, T. (2010). Cortisol is involved in temperature-dependent sex determination in the Japanese flounder. *Endocrinology*, *151*, 3900–3908. https://doi.org/10.1210/en.2010-0228.

Yamamoto, Y., Hattori, R. S., Kitahara, A., Kimura, H., Yamashita, M., & Strüssmann, C. A. (2013). Thermal and endocrine regulation of gonadal apoptosis during sex differentiation in pejerrey *Odontesthes bonariensis*. *Sexual Development*, 7, 316–324. https://doi.org/10.1159/000353506.

Yamamoto, Y., Zhang, Y., Sarida, M., Hattori, R. S., & Strüssmann, C. A. (2014). Coexistence of genotypic and temperature-dependent sex determination in pejerrey *Odontesthes bonariensis*. *PLoS One*, *9*, 1–8. https://doi.org/10.1371/journal.pone.0102574.

Yano, A., Guyomard, R., Nicol, B., Jouanno, E., Quillet, E., Klopp, C., et al. (2012). An immune-related gene evolved into the master sex-determining gene in rainbow trout, *Oncorhynchus mykiss*. *Current Biology*, *22*, 1423–1428. https://doi.org/10.1016/j.cub.2012.05.045.

Zhang, Y., Hattori, R. S., Sarida, M., García, E. L., Strüssmann, C. A., & Yamamoto, Y. (2018). Expression profiles of *amhy* and major sex-related genes during gonadal sex differentiation and their relation with genotypic and temperature-dependent sex determination in pejerrey *Odontesthes bonariensis*. *General and Comparative Endocrinology*, *265*, 196–201. https://doi.org/10.1016/j.ygcen.2018.03.013.

CHAPTER THREE

Natural sex change in fish

Neil J. Gemmell*, Erica V. Todd, Alexander Goikoetxea, Oscar Ortega-Recalde, Timothy A. Hore
Department of Anatomy, University of Otago, Dunedin, New Zealand
*Corresponding author: e-mail address: neil.gemmell@otago.ac.nz

Contents

Abstract

Sexual fate can no longer be considered an irreversible deterministic process that once established during early embryonic development, plays out unchanged across an organism's life. Rather, it appears to be a dynamic process, with sexual phenotype determined through an ongoing battle for supremacy between antagonistic male and female developmental pathways. That sexual fate is not final and is actively regulated via the suppression or activation of opposing genetic networks creates the potential for flexibility in sexual phenotype in adulthood. Such flexibility is seen in many fish, where sex change is a usual and adaptive part of the life cycle. Many fish are sequential hermaphrodites, beginning life as one sex and changing sometime later to the other. Sequential hermaphrodites include species capable of female-to-male (protogynous), male-to-female (protandrous), or bidirectional (serial) sex change. These natural forms of sex change involve coordinated transformations across multiple biological systems,

Current Topics in Developmental Biology, Volume 134
ISSN 0070-2153
https://doi.org/10.1016/bs.ctdb.2018.12.014

including behavioral, anatomical, neuroendocrine and molecular axes. Here we review the biological processes underlying this amazing transformation, focusing particularly on the molecular aspects, where new genomic technologies are beginning to help us understand how sex change is initiated and regulated at the molecular level.

1. Introduction

Sexual phenotype in most sexually reproducing organisms is dichotomous and static, a sexual mode termed gonochorism (Policansky, 1982). Most vertebrates are gonochoristic, for whom sexual fate is initiated during embryonic development through "sex determination," which triggers the initially bipotential gonad to develop as an ovary or a testis, and "sexual differentiation" the subsequent differentiation of the gonad and associated physiological traits (Capel, 2017; DeFalco & Capel, 2009). The primary sex of an individual may be determined chromosomally (as is seen in most mammals and birds), environmentally (e.g., temperature, pH, population density), or a combination of the two (as observed in many reptiles and fishes) (reviewed in Barske & Capel, 2008; Capel, 2017; Devlin & Nagahama, 2002; Gamble & Zarkower, 2012; Liu et al., 2017). This diversity in mechanism is startling when one considers the critical importance of sex determination for sexual reproduction and species persistence (Capel, 2017). However, despite the obvious diversity of sex-determining mechanisms, sexual fate was for many years considered to be fixed following downstream sexual differentiation cascades, which rely heavily on a few transcription factors and steroid hormones that are evolutionarily conserved (Barske & Capel, 2008; Guiguen, Fostier, Piferrer, & Chang, 2010; Herpin & Schartl, 2011a, 2011b; Kobayashi, Nagahama, & Nakamura, 2013; Matson & Zarkower, 2012; Piferrer & Guiguen, 2008). However, we increasingly recognize that gonadal sexual fate is not only established by competition for primacy between two sexes via antagonistic signaling pathways during embryonic development (Ungewitter & Yao, 2013; Windley & Wilhelm, 2015), but that it requires active maintenance to suppress the opposite sex during adulthood (Lindeman et al., 2015; Matson et al., 2011; Uhlenhaut et al., 2009). We are further aware that while the same genes may participate in sex determination and differentiation in multiple species, they are often expressed in different orders (Capel, 2017) or acquire completely different roles (Liu et al., 2017; Todd, Liu, Muncaster, & Gemmell, 2016).

These new studies establish a framework through which significant plasticity in phenotypic sex can readily emerge. Sex can no longer be viewed as a static feature resulting from a deterministic switch, but rather must be viewed as a reaction norm, with sex attributes being phenotypically plastic (Ah-King & Nylin, 2010). To produce such phenotypic plasticity, the developmental process requires flexible regulatory systems, with switches that may be temporarily active that augment, inhibit or attenuate intrinsic and extrinsic signals (Capel, 2017; Liu et al., 2017; Todd, Liu, et al., 2016; West-Eberhard, 2003). In short, sexual development in vertebrates is actually a very plastic process, during which various genetic and/or environmental inputs can be integrated into the developmental scheme to influence the final sexual expression (Barske & Capel, 2008; Capel, 2017; DeFalco & Capel, 2009; Liu et al., 2017; Todd, Liu, et al., 2016).

This new perspective is strongly supported by an accumulation of evidence in non-mammalian vertebrates, particularly teleost fishes, that display huge plasticity in both gonadal development and sexual expression (Avise & Mank, 2009; Baroiller & D'Cotta, 2001; Devlin & Nagahama, 2002; Kobayashi et al., 2013; Liu et al., 2017; Thomas, Liu, Todd, & Gemmell, 2018; Todd, Liu, et al., 2016). Among teleosts sex is determined by genetics (e.g., XY, ZW, polygenic), environment (e.g., temperature, pH, population density) or their interaction (see Devlin & Nagahama, 2002; Kobayashi et al., 2013). However, in many gonochoristic fishes, the gonadal sexual fate established via primary sex determination can be interrupted, or even reversed, by endogenous and exogenous signals applied either during or after the embryonic period (Baroiller & D'Cotta, 2001; Bhandari, Nakamura, Kobayashi, & Nagahama, 2006; Kobayashi & Iwamatsu, 2005; Paul-Prasanth et al., 2013; Sato, Endo, Yamahira, Hamaguchi, & Sakaizumi, 2005). This plasticity, while incredible, looks tame in comparison to those fishes able to express both sexes at the same time (simultaneous hermaphroditism) or change their sexual expression (sequential hermaphroditism or sex change) according to environmental cues throughout their lifetime (reviewed in Avise & Mank, 2009; Devlin & Nagahama, 2002; Lamm, Liu, Gemmell, & Godwin, 2015; Liu et al., 2017; Sadovy de Mitcheson & Liu, 2008; Todd, Liu, et al., 2016).

Sequential hermaphroditism, where individuals begin life as one sex, changing sex sometime later to the other sex, is among the more stunning examples of phenotypic plasticity in response to environmental influences thus far described (Godwin, 2009; Lamm et al., 2015; Liu et al., 2017; Munday, Buston, & Warner, 2006; Nakamura, Kobayashi, Miura, Alam, & Bhandari, 2005; Thomas et al., 2018; Todd, Liu, et al., 2016; Vega-Frutis, Macías-Ordóñez, Guevara, & Fromhage, 2014; Warner, 1975, 1984). Among vertebrates, sequential hermaphroditism is restricted to the Teleostei, but is phylogenetically widespread among that infraclass occurring in at least 27 families spread across nine teleost orders, which suggests multiple evolutionary origins (Avise & Mank, 2009; Devlin & Nagahama, 2002; Sadovy de Mitcheson & Liu, 2008). Sequential hermaphroditism occurs in three forms (Avise & Mank, 2009): protogynous (female-to-male), protandrous (male-to-female) and serial bidirectional sex change. In all cases, sex change is a usual event in the reproductive cycle of these fishes, is typically cued by changes in social structure or attainment of a critical age or size (Godwin, 2009; Kobayashi et al., 2013; Warner, 1984), and results in changes in reproductive behavior, gonadal anatomy, and external morphology, often within very short time frames (days to weeks) (Godwin, 2009; Lamm et al., 2015; Liu et al., 2017; Munday, Buston, et al., 2006; Nakamura et al., 2005; Thomas et al., 2018; Todd, Liu, et al., 2016; Vega-Frutis et al., 2014; Warner, 1975).

The broad phylogenetic occurrence, diversity of expression and speed of natural sex change make sequentially hermaphroditic fish very powerful models for studying the functioning and evolvability of sex determination and differentiation systems in vertebrates (Herpin et al., 2013; Herpin & Schartl, 2015; Matson & Zarkower, 2012; Wilhelm & Koopman, 2006). Knowledge of how sex change occurs would enhance our understanding of cellular differentiation, reprogramming and developmental commitment (Holmberg & Perlmann, 2012; Koopman, 2008; Orban, Sreenivasan, & Olsson, 2009), which are important for understanding both normal gonadal development in vertebrates and atypical sexual development in humans. Sequential hermaphroditism is also an extraordinary example of phenotypic plasticity in response to environment (see Aubin-Horth & Renn, 2009; Moczek, 2015; West-Eberhard, 2003). This linkage between gene-by-environment interactions and plastic developmental responses (Jaenisch & Bird, 2003; Zhang & Ho, 2011) is strengthening daily through our growing understanding of epigenetic regulation of gene expression (Feinberg, 2007), and its effect on sexual phenotype now emerging (Deveson et al., 2017; Ge et al., 2018; Holleley et al., 2015). Here we review our current understanding of sex change in sequentially hermaphroditic fish.

2. Why change sex?

The biological processes and adaptive benefits of natural sex change have fascinated scientists for decades (Ghiselin, 1969; Munday, Buston, et al., 2006; Robertson, 1972; Warner, 1975, 1984), with the ecological and evolutionary contexts in which sequential hermaphroditism occurs in fishes well-studied and understood (reviewed by Avise & Mank, 2009; Godwin, 2009; Lamm et al., 2015; Nakamura et al., 2005; Warner, 1984).

2.1 Size advantage and the evolution of sequential hermaphroditism

The leading explanation for the adaptive benefit of sequential hermaphroditism, as well as the direction and timing of sex change in teleost fishes, is the Size-Advantage Model (SAM) (Ghiselin, 1969; Munday, Buston, et al., 2006; Warner, 1975). The SAM posits that sex change is adaptive when the reproductive value is greater as one sex when small and the other sex when larger (older), with sex change timed to maximize an individual's expected lifetime reproductive success. The benefits of sex change are tightly associated with a species' mating system and social structure, but in general if the reproductive value for females increases more with body size than for males, then male-first hermaphroditism is most adaptive, while female-first hermaphroditism predominates when the reproductive value for males increases more with body size than for females (Munday, Buston, et al., 2006). Evolutionary transitions between reproductive modes, for example, shifts from protogyny to gonochorism, and vice versa, are associated with changes in the mating system, as predicted by sex allocation theory (Erisman, Petersen, Hastings, & Warner, 2013). New work also demonstrates that the direction of sex change strongly affects effective population sizes, because the populations of protogynous systems are much smaller and more localized than those observed in protandrous systems (Benvenuto, Coscia, Chopelet, Sala-Bozano, & Mariani, 2017).

2.1.1 Protogyny

Protogyny is the most common form of sequential hermaphroditism in fish, occurring in 15 families, and is especially pervasive in social species with polygynous mating systems. In these species, large males aggressively defend territories to monopolize matings with females (Warner, 1984). Protogyny is ubiquitous in the wrasses (Labridae), of which the bluehead wrasse (*Thalassoma bifasciatum*) is among the best studied (Godwin, 2009; Lamm et al., 2015; Liu et al., 2017). Bluehead wrasse are small, polygamous reef fish with a lek-like mating system, in which dominant (terminal-phase, TP) males defend spawning sites to which females have high fidelity (Fig. 1A). Loss of a TP male stimulates sex change in (typically) the largest female of a social group, who then undergoes dramatic changes in behavior,

Fig. 1 Ecological context of (A) protogynous, (B) protandrous and (C) serial bidirectional sex-change in social fishes. (A) The bluehead wrasse (*Thalassoma bifasciatum*) is typical of many protogynous hermaphrodites. A brightly-colored terminal-phase (TP) male defends and courts a harem of smaller females on a tropical reef in the Caribbean sea, one of which will eventually change sex to replace him. (B) A colony of cinnamon clownfish (*Amphiprion melanopus*) inhabiting a bubble anemone. These protandrous hermaphrodites form monogamous breeding pairs consisting of a dominant female and a smaller male, who share shelter space with several immature subordinates. (C) A pair of Maori coral gobies (*Gobiodon histrio*), a species widespread on tropical reefs of the Indo-West Pacific. Monogamous coral gobies live a sessile lifestyle occupying spatially isolated coral colonies, where serial bidirectional sex change enables any two fish to form a heterosexual breeding pair. *Panel A: Image author: Kevin Bryant. Panel B: Image author: Nick Hopgood. Panel C: Image author: Mike Wesemann. Reused with permission from Todd, E. V., Liu, H., Muncaster, S., & Gemmell, N. J. (2016). Bending genders: The biology of natural sex change in fish.* Sexual Development, 10, *223–241.*

anatomy and coloration (Warner & Swearer, 1991). Most juvenile bluehead wrasse develop as female, but a few develop directly as small, initial-phase (IP) males, that mimic the female phenotype and illicit "sneak" matings to achieve fertilization, albeit with a lower level of fertilization success than dominant TP males (Semsar & Godwin, 2004). IP male development is also under social control: more IP males develop on high-density reefs where

TP males can less-effectively monopolize mating opportunities (Munday, Wilson White, & Warner, 2006). In this and other diandric fishes, it remains unknown whether the same or different molecular pathways are involved in sex determination and differentiation of IP and TP males. Recent work shows that female mimicry by IP males has a transcriptional signature in both the brain and the gonad. Sneaker males shared striking similarity in neural gene expression with females, supporting the idea that males with alternative reproductive phenotypes have "female-like" brains (Todd et al., 2018), and because the trigger for sex change is undoubtedly neural it may well be that neural processes involved in the perception and transduction of social cues that initiate sex reversal in females and role change in IP males is similar.

The activators of sex change are less obvious in group-spawning species where social structure is not well-defined, such as the Groupers (Epinephelidae). These fish periodically form sizeable breeding groups where, typically, a single male and several females will break-off to spawn. In such species, whether an individual will undertake protogynous sex change likely depends on attaining a threshold age and/or size, and the sex-ratio and population density of the spawning group (Bhandari, Higa, Nakamura, & Nakamura, 2004; Shapiro, Sadovy, & McGehee, 1993).

2.1.2 Protandry

Protandry is rarer among the teleosts, occurring sporadically across six families. Protandry is usually associated with monogamous or random mating systems, where male territorial defense or intense sperm competition is absent (Munday, Buston, et al., 2006). Under such conditions protandry is adaptive because of the direct positive relationship between female fecundity and body size (Warner, 1975).

Protandrous sex change is socially controlled in monogamous anemonefish (*Amphiprion* and *Premnas* spp.) (Casas et al., 2016; Fricke & Fricke, 1977; Godwin, 1994; Hattori, 2012) (Fig. 1B). Anemonefish live in small social groups among sea anemones. Classically these consist of a large dominant female and a small male, plus smaller subordinate non-breeders. Loss of the dominant female prompts protandrous sex change in her partner and maturation of the most-dominant immature fish as the new breeding male (Hattori, 2012).

Many commercially valuable aquaculture species are protandrous and there is considerable interest in controlling sex ratios by manipulating sex change in farmed fish (Budd, Banh, Domingos, & Jerry, 2015). Australian barramundi (*Lates calcarifer*, Latidae) and gilthead seabream (*Sparus aurata*, Sparidae) both sexually mature as male before becoming female at an older age and larger size (Guiguen, Cauty, Fostier, Fuchs, & Jalabert, 1994; Liarte et al., 2007) and there are significant efforts ongoing to understand sex change in these species (Pauletto et al., 2018). However, most research on protandry has focussed on black porgy (*Acanthopagrus schlegeli*, Sparidae), which reproduce as male for the first 2 years of life before approximately 50% of fish change sex to female (Lee et al., 2001; Wu & Chang, 2013; Wu et al., 2010). Whether a threshold age and/or size also triggers sex inversion in protandrous group-spawning species remains unclear (Guiguen et al., 1994).

2.1.3 Bidirectional sex change

Bidirectional hermaphrodites have the capacity for sex change in either direction, potentially repeatedly during their lifetime. Bidirectional hermaphroditism is known to occur in 10 species spanning five families (Kuwamura, Kadota, & Suzuki, 2015; Manabe et al., 2013; Munday, Kuwamura, & Kroon, 2010). True serial sex change is observed in the monogamous coral-dwelling gobies (Fig. 1C) (e.g., *Gobiodon* and *Paragobiodon*). Sex change in gobies is poorly explained by the SAM—the popular explanation is that bidirectional sex change provides reproductive assurance for these fish in the face of niche specialization and a sessile lifestyle (Munday, Caley, & Jones, 1998; Nakashima, Kuwamura, & Yogo, 2010). Coral gobies experience limited mating opportunities and experience significant risks if moving between spatially isolated coral colonies. Thus, the ability to repeatedly change sex, in either direction, allows any two fish to form a heterosexual breeding pair, reducing both the effort to find a mate and the associated predation risk (Munday et al., 2010).

3. The anatomy of sex change

Detailed histological descriptions of gonadal sex change have been made for representative protandrous (Godwin, 1994; Guiguen et al., 1994; Lee et al., 2001; Wu et al., 2010), protogynous (Lo Nostro, Grier, Andreone, & Guerrero, 2003; Muncaster, Norberg, & Andersson, 2013; Nakamura, Hourigan, Yamauchi, Nagahama, & Grau, 1989) and bidirectional (Cole, 2011; Kuwamura et al., 2015; Sunobe, Nakamura, Kobayashi, Kobayashi, & Nagahama, 2005) sequential hermaphrodites.

In all cases, the process involves radical restructuring of the gonad, transforming a functional gonad of one sexual phenotype into that of the opposite sex. However, sex change follows diverse ontogenetic pathways in different sequentially hermaphroditic lineages (Kobayashi et al., 2013; Liu et al., 2017; Todd, Liu, et al., 2016; Warner & Robertson, 1978) reflecting its multiple independent evolutionary origins (Avise & Mank, 2009; Erisman et al., 2013; Sadovy de Mitcheson & Liu, 2008).

Gonadal restructuring is complete in protogynous wrasses; no testicular tissues are detectable in the ovary before sex change, and only a remnant ovarian lumen and lamellae structure is retained in the secondary testes following sex change (Bhandari et al., 2004; Muncaster et al., 2013; Warner & Robertson, 1978). Protogyny has a common evolutionary origin in wrasse and the key histological events during gonadal sex change are broadly comparable across species (Nakamura et al., 1989). The onset of sex change is heralded by ovarian follicle atresia and oocyte degeneration, followed by the proliferation of spermatogonia and Leydig cells in the peripheral ovarian lamellae before commencement of spermatogenesis characterizes the fully functional testis of a male fish capable of fertilizing eggs (Fig. 2A). This process can be completed rapidly. In tropical bluehead wrasse, which breed year-round and can be induced to change sex at almost any time point, the transformation can take just 8 days (Warner & Swearer, 1991). However, temperate species are slower taking several weeks or months to complete the process (Muncaster et al., 2013) and sex change is typically seasonal starting after spawning from a regressed ovary with only previtellogenic follicles (Bhandari et al., 2004; Jones, 1980; Muncaster et al., 2013).

Precisely where and from what gonadal cell type new testicular tissues derive in protogynous wrasse remains unresolved (Liu et al., 2017). However, in the protogynous seabass and grouper it is thought that crypts of resting spermatogenic tissues occur within the ovarian germinal epithelia prior to sex change (Bhandari et al., 2004; Sadovy de Mitcheson & Liu, 2008; Shapiro et al., 1993).

In other sequential hermaphrodites, sex change proceeds from a bisexual gonad, or "ovotestis" (Liu et al., 2017; Sadovy de Mitcheson & Liu, 2008; Todd, Liu, et al., 2016). In porgies (Sparidae), which include both protandrous and protogynous species, connective tissues strongly demarcate male and female regions in the ovotestis (Sadovy de Mitcheson & Liu, 2008). In the protandrous black porgy, the ovotestis is dominated by active testicular tissue during the first and second spawning seasons when fish are functionally male (Wu et al., 2010). At this time, the ovarian portion of the

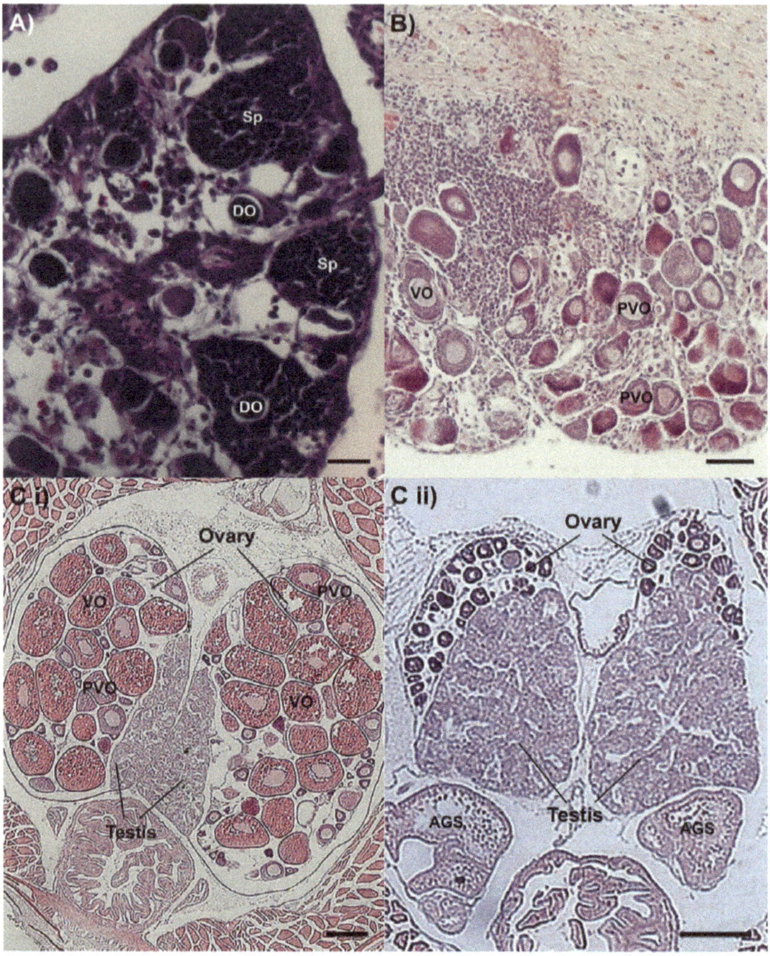

Fig. 2 Histological details of gonads from (A) protogynous, (B) protandrous and (C) bidirectional sex-changing fish. (A) A transitional bluehead wrasse (*Thalassoma bifasciatum*) gonad in the later stages of protogynous sex change (stage 5 of Nakamura et al., 1989), showing degenerating oocytes (DO) and crypts of proliferating spermatocytes (Sp). Scale bar: 20 μm. (B) The ventral portion of a transitional barramundi (*Lates calcarifer*) gonad in the later stages of protandrous sex change, showing developing ovarian tissues including previtellogenic oocytes (PVO). Scale bar 20 μm. (C) The paired ovotestis of a (i) female-phase and a (ii) male-phase individual of the bidirectional sex-changing goby, *Trimma kudoi*. In the female phase, the ovotestis consists of an enlarged ovarian portion containing many vitellogenic (VO) and previtellogenic (PVO) oocytes, and a small testicular portion. In the male phase, testicular tissue in active spermatogenesis dominates the gonad and a small ovarian portion contains mostly early-stage oocytes. The male-phase ovotestis also contains an accessory gonadal structure (AGS) not present in the female-phase. Scale bar 20 μm. *Panel A: Image author: Hui Liu. Panel B: Image author: Quyen Bahn. Panel C: Image authors: Tomoki Sunobe and Hisaya Manabe. Reused with permission from Todd, E. V., Liu, H., Muncaster, S., & Gemmell, N. J. (2016). Bending genders: The biology of natural sex change in fish.* Sexual Development, 10, *223–241*.

ovotestis is small, containing oogonia and a few primary oocytes, but subsequently matures and expands during non-spawning (intersex) periods (Lee, Huang, & Chang, 2008). As the third breeding season approaches, testicular tissues redevelop in those fish that remain functional males but are completely replaced by ovarian tissue in those fish that change sex to female (Lee et al., 2008). Male anemonefish also possess a bisexual gonad that contains both active spermatogenic tissue and previtellogenic oocytes, but in this protandrous species the tissues, while topographically distinct, are not separated by boundary tissues (Godwin, 1994; Shapiro, 1992).

Bidirectional sex-changing gobies also possess an ovotestis that contains ovarian and testicular portions simultaneously (Fig. 2C), but either portion can be reproductively fully functional in a given season (Cole, 2011; Kuwamura et al., 2015; Sunobe et al., 2005). Maintaining a bisexual gonad affords these species enormous flexibility to rapidly adjust their sexual phenotype and gain maximum benefit from reproductive opportunity (Sunobe et al., 2005). Sequential hermaphroditism has multiple evolutionary origins in gobiid fishes and gonadal configurations show substantial taxon-specific variation; from the undelimited ovotestis of *Eviota* gobies, where male and female cells are intermixed, to an increasingly complex and compartmentalized organ (as in *Lythrypnus*, *Trimma*, *Gobiodon* and *Paragobiodon* spp.) (Cole, 2010).

While sex change is an extraordinary phenotype, a bisexual gonadal (Orban et al., 2009), or an immature ovary (Devlin & Nagahama, 2002) is a common feature of the juvenile stage for many non-sex changing fish, before they develop fully as either males or females. In zebrafish the gonad initially develops as a bipotential organ, with a few primordial germ cells in all juveniles initially developing as oocytes (Orban et al., 2009). In some individuals oocytes survive leading to ovary formation, while in others the oocytes are eliminated leading to the formation of a testis (Rodríguez-Marí et al., 2010). Fancl, a member of the Fanconi anemia/BRCA DNA repair pathway appears to be critical to oocyte retention; *fancl* mutation leads to female-to-male sex reversal through an increase in Tp53-mediated germ cell apoptosis (Rodríguez-Marí et al., 2010).

That germ cells of both sexes coexist at least transiently in the gonad of many gonochoristic fishes, and that sexual fate can be altered through the knock out of a single gene (Rodríguez-Marí et al., 2010) implies considerable evolutionary scope for plasticity in reproductive mode (Todd, Liu, et al., 2016). Given this potential for plasticity a reasonable question is why more species do not change sex naturally?

4. The possible cellular origins of gonadal restructuring in protogynous species

Precisely from where and from what gonadal cell type new gonadal tissues derive remains unresolved for species where no germinal tissues of the secondary sex are discernible prior to sex change (Nakamura et al., 1989). For most protogynous species, no testicular tissues can be found in the ovaries before sex change and very few ovarian tissues remain in the secondary testes after sex change (Muncaster et al., 2013; Nakamura et al., 1989; Nozu, Kojima, & Nakamura, 2009). Tracing cellular origins and fates during gonadal sex change is complicated because correctly identifying and staging germ cells in these sexually plastic species is difficult when based solely on histology (Lo Nostro et al., 2003). In protogynous hermaphrodites, early male tissues are often first observed at the periphery of the ovarian lamellae in the vicinity of the germinal epithelium (Lo Nostro et al., 2003). However, establishing whether proliferating spermatogonia arise from a dormant, but sexually differentiated, germ cell population (i.e., co-existence of oogonia and spermatogonia) or from a population of latent sexually bipotent primordial germ cells is hard to establish (Lo Nostro et al., 2003).

In protogynous species, it is currently assumed that testicular construction begins with bipotential germ (gonia) and somatic (epithelial) cells residing within the ovarian germinal epithelium (Lo Nostro et al., 2003; Nakamura et al., 1989; Nozu et al., 2009). Gonial cells clearly proliferate and differentiate into spermatocytes during later stages of sex change, yet it remains to be confirmed whether these are oogonia, spermatogonia or undifferentiated stem cells (Liu et al., 2017). However, the bipotentiality of transplanted spermatogonia and oogonia in rainbow trout (Okutsu, Suzuki, Takeuchi, Takeuchi, & Yoshizaki, 2006; Yoshizaki et al., 2010), a strictly gonochoristic species with an XY sex-determining system, indicates that gonial cells routinely retain sexual plasticity, and are reprogrammable depending on the somatic microenvironment (Todd, Liu, et al., 2016; Yoshizaki et al., 2010).

The sources of the somatic cells that cause testicular restructuring are equally obscure. The somatic cells surrounding the gonial cells near the periphery of ovarian lamellae appear a likely source of presumed Sertoli cells because *gsdf* (gonadal somatic derived factor), a genetic marker involved in testicular differentiation, has been detected in these supporting cells at low levels prior to sex change (Shibata et al., 2010). Furthermore, during ovarian atresia, all vitellogenic and previtellogenic oocytes degenerate, but relatively few follicle cells appear to go through apoptosis (Nozu, Horiguchi, Murata, Kobayashi, &

Nakamura, 2013). Instead the near stages of sex change are characterized by the proliferation of the granulosa cells that surround and support the degenerating oocyte (Nozu et al., 2013). These remaining follicle layers also show positive signals of *cyp11b2* (11β-hydroxylase) a key enzyme responsible for synthesis of 11-ketotestosterone (11-KT) the key male sex steroid in fish (Alam et al., 2006; Li et al., 2013). Whether these cells can be reprogrammed into Sertoli or Leydig cells (steroidogenic cells in testes) to form testicular tubules warrants further investigation (Liu et al., 2017).

Following ovarian atresia, the space left by degenerated oocytes is invaded by the remaining follicle cells and vascularized connective tissues, which appear to give rise to the yellow-brown bodies and cell mass in the central areas of ovarian lamellae (Muncaster et al., 2013; Nakamura et al., 1989). Various types of cells are observed in this cell mass, including macrophages that clean up the degenerating oocytes and connective tissues that form testicular lobules (Muncaster et al., 2013; Nakamura et al., 1989). At this stage, Leydig cells cluster and proliferate around loose connective tissues (Nakamura et al., 1989). The origin of these Leydig cells is not clear, but a likely source is the inner layer of the gonadal wall, which becomes thicker with active cell proliferation as gonadal sex change proceeds (Liu et al., 2017). In the testis of bluehead wrasse this layer contains all somatic cell types, is continuous with the interstitial tissue layer, and may contribute to gonadal restructuring by cell invasion or exchange (Koulish, Kramer, & Grier, 2002). Studies in honeycomb grouper also show that androgen-producing cells form clusters in the inner layer of the gonadal wall near blood vessels of both ovaries and testes as well as transitioning gonads (Alam et al., 2006; Nakamura et al., 2005). Whether and how these cells are reprogrammed or relocated to form Leydig cells during the gonadal transition remains unknown (Liu et al., 2017).

The lack of detailed knowledge about cellular origins and fates in transitioning or sex changing gonads remains a major gap in our understanding of sexual differentiation and sex change in fishes. While tracking cell fates remains challenging, it should become increasingly more feasible with the continued development of cellular markers for fishes, likely strongly empowered by comparative transcriptomic studies in sex-changing species (Liu et al., 2015; Todd, Black, & Gemmell, 2016; Todd et al., 2018) and new single-cell approaches (Stévant et al., 2018). Understanding how the developmental bipotentiality of gonial and somatic cell populations is retained into adulthood is important not only for understanding sex change in sequential hermaphrodites, but also the evolvability of sexual plasticity in teleosts generally (Todd, Liu, et al., 2016).

5. The endocrine regulation of sex change

Sexual differentiation and gonadal development are directed by the balance between gonadal oestrogen and androgen production. In teleosts, 11-ketotestosterone (11-KT) and 17β-estradiol (E2) are the principal androgen and oestrogen that respectively promote testicular and ovarian function (Fig. 3). The relationship between them is especially close, as production of either 11-KT or E2 depends on the bioconversion of testosterone (T) via the 11β-hydroxylase (*cyp11b*)/11β-hydroxysteroid dehydrogenase (11β-HSD, coded by *hsd11b*) and aromatase (*cyp19a1a*) pathways (see Guiguen et al., 2010). The relative expression of these opposing pathways in the gonad determines the sex steroid balance and ultimately decides gonadal fate (Guiguen et al., 2010; Liu et al., 2017; Todd, Liu, et al., 2016).

In sex-changing fishes, dramatic shifts in plasma sex steroids accompany gonadal transition (Devlin & Nagahama, 2002; Godwin, 2010) (Fig. 3),

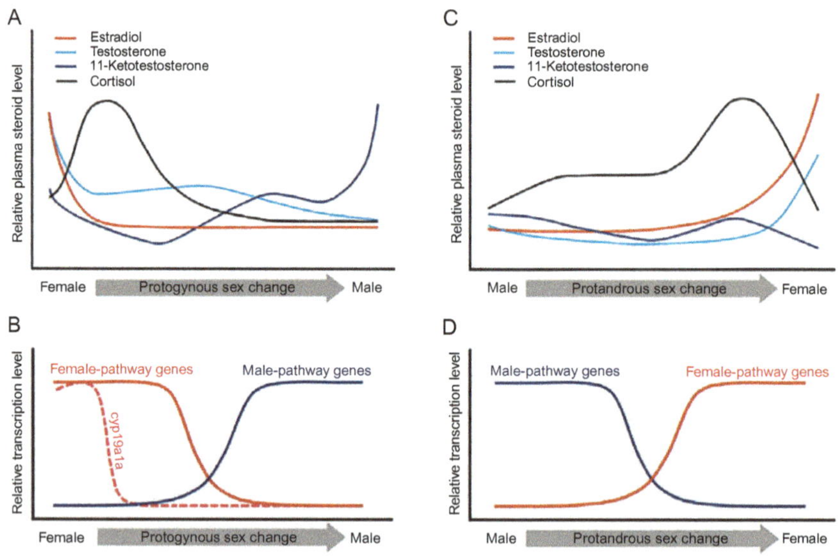

Fig. 3 Shifts in steroid profiles and sex-specific gene expression accompany sex change in each direction. During protogynous sex change, (A) steroid profiles shift from an oestrogenic to an androgenic environment. (B) Following shutdown of the aromatase gene (*Cyp19a1a*), female-specific gene expression is down-regulated before expression profiles become increasingly male-biased. The opposite pattern is observed during protandrous sex change: (C) there is transition from an androgenic to an oestrogenic environment and (D) from male-specific to female-specific expression. *Adapted with Permission from Todd, E. V., Liu, H., Muncaster, S., & Gemmell, N. J. (2016). Bending genders: The biology of natural sex change in fish.* Sexual Development, 10, *223–241.*

yet what tips the balance in favor of the opposing hormonal environment remains unclear. A sharp drop in plasma E2 heralds ovarian degeneration and protogynous sex change, which leads onto a gradual increase in 11-KT production at the onset of spermatogenesis (Bhandari et al., 2006; Bhandari, Komuro, Nakamura, Higa, & Nakamura, 2003; Muncaster et al., 2013; Nakamura et al., 1989; Ohta, Mine, Yamaguchi, & Matsuyama, 2008). The opposite pattern characterizes protandrous sex change; 11-KT levels fall and subsequently plasma E2 increases (Godwin & Thomas, 1993; Lee et al., 2001). In bidirectionally sex changing gobies, E2 levels follow this sexually dimorphic pattern, but there is no apparent association between 11-KT concentrations and gonadal development (Kroon, Munday, & Pankhurst, 2003; Kroon, Munday, & Westcott, 2009; Lorenzi, Earley, & Grober, 2012). The role of 11-KT in gobies is unclear, but routinely low 11-KT levels may reflect the lack of secondary male characteristics in gobies and enable rapid switching between sexual phenotypes (Godwin, 2010; Kroon et al., 2009).

Exogenous manipulation of sex steroids can be used to induce masculinization or feminization in fish (e.g., Chang, Lau, & Lin, 1995; Higa, Ogasawara, Sakaguchi, Nagahama, & Nakamura, 2003; Yeh, Kuo, Ting, & Chang, 2003); reviewed by (Baroiller & D'Cotta, 2016; Budd et al., 2015; Devlin & Nagahama, 2002; Lee et al., 2017), but in some systems this is only successful when steroids are administered during the "sex determination" period, when the gonadal fate remains labile (Brion et al., 2004). Application of non-aromatizable androgens down-regulates the aromatase pathway in female fish leading to sex change (Bhandari et al., 2006; Govoroun et al., 2001; Li, Liu, Zhang, & Lin, 2006; Ohta et al., 2012). Aromatase inhibitors (AI) disrupt ovarian E2 production in protogynous (Higa et al., 2003; Nozu et al., 2009), protandrous (Lee et al., 2001; Nakamura, Miura, Nozu, & Kobayashi, 2015) and bidirectional hermaphrodites (Kroon, Munday, Westcott, Hobbs, & Liley, 2005). Although AI treatment leads to complete sex change in protogynous species, rescue is possible through the co-administration of E2 (Higa et al., 2003). That sex change is typically not sustained following withdrawal of hormonal treatments (e.g., Wu, Li, Luo, Chen, & Chang, 2015) indicates that, while sex steroids clearly regulate gonadal fate, a molecular switch is required to sustain the shift in hormone production and maintain sex change.

6. Neuroendocrine regulation of sex change

The neuroendocrine regulation of sex change has been most intensively studied in protogynous wrasses that display socially-controlled sex change (Lamm et al., 2015). In these species, behavioral changes frequently precede gonadal sex change and appear to be independent of the gonad (Godwin, Crews, & Warner, 1996; Nakamura et al., 1989; Warner & Swearer, 1991). The established view is that changes in the brain drive both behavioral and gonadal sex change, although through different neuroendocrine pathways (Godwin et al., 1996; Lamm et al., 2015) (Fig. 4). Cross-talk between these pathways and interactions among the neurochemical signals that mediate them likely coordinate the behavioral and physiological responses during sex change.

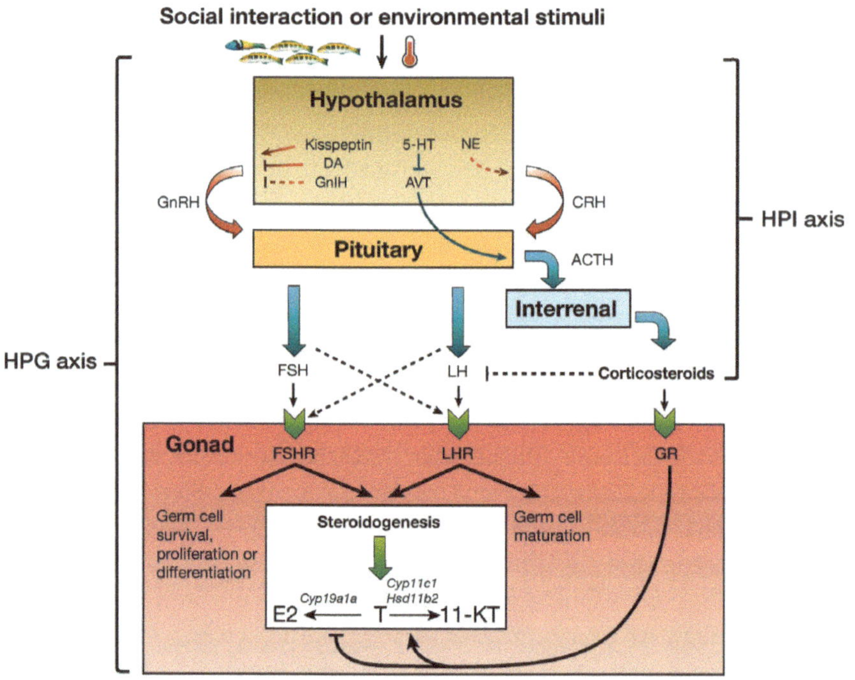

Fig. 4 Schematic illustration of neuroendocrine cross-talk between the HPG and HPI axes regulating steroidogenesis and behavior in teleosts. Solid lines indicate interactions with support from fish models; dashed lines indicate interactions with support from non-teleost systems. Abbreviations: ACTH, adrenocorticotropic hormone; AVT, arginine vasotocin; CRH, corticotropin-releasing hormone; DA, dopamine; E2, 17beta-estradiol; GnIH, gonadotropin inhibitory hormone; GnRH, gonadotropin releasing hormone; FSH, follicle-stimulating hormone; FSHR, FSH receptor; LH, luteinizing hormone; LHR, LH receptor; NE, norepinephrine; T, testosterone; 5-HT, serotonin; 11-KT, 11-ketotestosterone. *Adapted with permission from Todd, E. V., Liu, H., Muncaster, S., & Gemmell, N. J. (2016). Bending genders: The biology of natural sex change in fish.* Sexual Development, 10, *223–241.*

The oestrogen-androgen balance, and thus gonadal fate, is ultimately regulated through the Hypothalamic-Pituitary-Gonadal (HPG) axis and its interaction with the neighboring stress axis (Fig. 4). Gonadotropin releasing hormone (GnRH) is released in pulses from the hypothalamus, stimulating the pituitary to produce and release gonadotropins (GtHs), follicle stimulating hormone (FSH) and luteinizing hormone (LH), into the circulation. GtHs directly regulate gonadal steroidogenesis via receptor-mediated stimulation of ovarian follicle cells or somatic Leydig cells in the testis (reviewed by Devlin & Nagahama, 2002; Weltzien, Andersson, Andersen, Shalchian-Tabrizi, & Norberg, 2004).

The expression of GtH subunits and their receptors (LHR, FSHR) alter during sex change in protogynous (Kobayashi, Alam, Horiguchi, Shimizu, & Nakamura, 2010), protandrous (An, Lee, & Choi, 2010; An, Lee, Yun, & Choi, 2009) and bidirectional (Kobayashi et al., 2009) species. Direct manipulation of GnRH or GtH signaling induces partial or complete sex change in protogynous (Kobayashi, Alam, et al., 2010; Reinboth & Brusle-Sicard, 1997) and protandrous hermaphrodites (see Lee et al., 2001). However, the patterns of GnRH and GtH signaling across sex change are inconsistent even among closely-related species (e.g., contradictory patterns of expression for GtH receptors are observed in protogynous groupers (Alam, Kobayashi, Hirai, & Nakamura, 2010; Hu et al., 2011) making it hard to understand the precise roles of GnRH and GtH signaling in controlling sex change. One possibility is that this variability reflects species-specific gonadotropin functioning in teleosts (see Levavi-Sivan, Bogerd, Mananos, Gomez, & Lareyre, 2010).

The perception and processing of external cues into the coordinated physiological response that characterizes sex change remain poorly understood (reviewed by Lamm et al., 2015; Liu et al., 2017). In bluehead wrasse, rapid neurochemical changes in the brain drive behavioral sex change (Godwin, 2010; Lamm et al., 2015), which precedes by several days the gonadal restructuring coordinated via the HPG axis (Larson, Norris, Gordon Grau, & Summers, 2003; Larson, Norris, & Summers, 2003; Semsar & Godwin, 2003); reviewed by (Godwin & Thompson, 2012; Lamm et al., 2015). It seems probable that cross-talk among distinct, but likely overlapping, neuroendocrine pathways mediates the effects of the social environment on gonadal state to coordinate sex change at the whole-body level (Lamm et al., 2015).

Removing a TP male from its social group triggers dramatic neuroendocrine changes in the brain of large females that stimulates behavioral sex change (Godwin, 2010; Lamm et al., 2015). These neuroendocrine changes include fluctuations in neurochemicals known to modulate social rank and sexually-dimorphic reproductive behaviors (Godwin, Sawby, Warner, Crews, & Grober, 2000; Godwin & Thompson, 2012; Lamm et al., 2015; Semsar, Perreault, & Godwin, 2004) which are known to elicit rapid behavioral and reproductive changes in response to environmental and/or physiological signals (Godwin & Thompson, 2012; Lamm et al., 2015; Liu et al., 2017; Maruska & Fernald, 2013; Zohar, Muñoz-Cueto, Elizur, & Kah, 2010).

Norepinephrine (NE) and arginine vasotocin (AVT) stimulate behavioral sex change, while dopamine (DA) and serotonin (5-HT) have inhibitory effects (Kramer, Caddell, & Bubenheimer-Livolsi, 1993; Larson, Norris, Gordon Grau, et al., 2003; Larson, Norris, & Summers, 2003; Perreault, Semsar, & Godwin, 2003; Semsar & Godwin, 2003, 2004; Semsar, Kandel, & Godwin, 2001). AVT appears especially important in promoting male-typical behaviors, such as courtship and aggression, in protogynous wrasses (see Godwin & Thompson, 2012; Semsar et al., 2001, 2004). However, methodological differences across studies and species-specific variation highlight the need for further research in a variety of species exhibiting different sex change strategies.

Kisspeptins and isotocin (IT) are also likely regulators of early sex change. Kisspeptins regulate vertebrate reproduction by stimulating GnRH release (Elizur, 2009; Espigares, Zanuy, & Gómez, 2015; Mechaly, Vinas, & Piferrer, 2013), but may also mediate transitions in social status (see Maruska & Fernald, 2011) via their receptors on AVT and IT neurons (Kanda, Akazome, Mitani, Okubo, & Oka, 2013). Isotocin, the teleost homolog of mammalian oxytocin, is strongly associated with social status and sex-specific reproductive behaviors across a range of species (Goodson & Bass, 2000; Lema, Sanders, & Walti, 2015). These hormones are of interest, as social rank may serve as a primer for sex change in haremic breeders (Lamm et al., 2015; Liu et al., 2017; Todd, Liu, et al., 2016). In a recent transcriptomic analysis of bluehead wrasse forebrain, increased expression of the gene encoding IT was one of the few statistically significant changes detected across sex change (Liu et al., 2017). Further work exploring the transcriptomic differences between TP males, IP males and females found that IT expression was characteristic of the TP male phenotype and expressed specifically in the preoptic area of the hypothalamus (POA), a primary site for socio-sexual behavioral integration in vertebrates (Todd et al., 2018). However, bidirectionally sex changing gobies show the opposite pattern; males and late-stage sex-changers show lower IT activity in the POA compared with females (Black, Reavis, & Grober, 2004). To date, fluctuations in the expression of kisspeptin and its receptors (kiss2, kiss1r) across sex change are known only for orange spotted grouper, *Epinephelus coioides* (Shi et al., 2010). Recent transcriptome analyses have, somewhat unexpectedly, not uncovered significant expression changes for these and other candidate neuroregulatory genes during sex change (Casas et al., 2016; Liu et al., 2017; Pauletto et al., 2018). Such studies have sampled at either a whole brain or forebrain level, and so may fail to pick up subtle and/or localized expression changes. The social-rank hypothesis, and a potential regulatory role for kisspeptins and IT in socially controlled sex change, warrants further research. Future work should employ more fine-scale sampling of specific brain regions, of which the POA should be a primary target.

Gonadotropin-inhibitory-hormone (GnIH) a key regulator of the HPG axis, may also have a role in sex change, but as yet is poorly studied. GnIH is known to inhibit GnRH and LH release in birds (Ubuka et al., 2014) and has been reported to reduce serum LH levels in goldfish, although the effects on GtH synthesis and release are stage specific and complicated (Moussavi, Wlasichuk, Chang, & Habibi, 2013; Qi et al., 2013). In protogynous groupers, opposite expression patterns of *gnih* (Wang et al., 2015) and *kiss2* (Shi et al., 2010) were observed in the hypothalamus following MT-induced sex change.

Neurochemical regulation of sex change in protandrous and bidirectional species, and species where sex change is not under social control, remains largely uninvestigated. Targeted manipulations of these neurohormones in the brain are required to clarify their roles in sex change.

7. The stress response may mediate sex change

The stress response modulates many processes central to major life-history transitions, including changes in behavior, metabolism and growth (Goikoetxea, Todd, & Gemmell, 2017; Solomon-Lane, Crespi, & Grober, 2013; Todd, Liu, et al., 2016). The hypothalamic-pituitary-interrenal (HPI) axis (Fig. 4), a major part of the neuroendocrine system, controls reactions to stress through the actions of corticotropic releasing hormone (CRH) and glucocorticoid steroids (GCs) (Goikoetxea et al., 2017; Liu et al., 2017; Todd, Liu, et al., 2016). Cortisol, the main glucocorticoid in fish and the hormone most directly associated with stress, may act as a key factor linking social environmental stimuli and the onset of sex change by initiating a shift in steroidogenesis from estrogens to androgens (Goikoetxea et al., 2017). That cortisol apparently also mediates agonistic behavior and reflects social status, strongly implicates the HPI axis in socially-induced sex change (Solomon-Lane et al., 2013).

In a variety of gonochoristic fishes, such as pejerrey or Japanese flounder, elevated temperatures and other environmental stressors result in increased cortisol levels and the down-regulation of aromatase, which leads to the activation of androgen pathways and gonadal masculinization (Fernandino, Hattori, Moreno Acosta, Strussmann, & Somoza, 2013; Goikoetxea et al., 2017; Hattori et al., 2009; Hayashi et al., 2010; Yamaguchi, Yoshinaga, Yazawa, Gen, & Kitano, 2010). Cortisol likely stimulates 11β-HSD expression, resulting in the production of both 11-KT and cortisone, the deactivated metabolite of cortisol (Fernandino et al., 2013; Fernandino, Hattori, Kishii, Strussmann, & Somoza, 2012). Thus, through its influence on steroidogenic gene expression and by communicating environmental and social status information along the HPI axis, cortisol may regulate natural sex change in sequential hermaphrodites (Goikoetxea et al., 2017; Solomon-Lane et al., 2013). Studies in several sequentially hermaphroditic species suggest cortisol production fluctuates across sex change. A transient spike in serum cortisol was observed in protandrous cinnamon clownfish undergoing sex change (Godwin & Thomas, 1993). Bluebanded gobies, bidirectional hermaphrodites, experiencing a "permissive" social environment showed a spike in serum cortisol as they transitioned from female to male (Solomon-Lane et al., 2013). Sustained cortisol administration promoted protogynous sex change in three-spot wrasse (Nozu & Nakamura, 2015).

The exact ways in which cortisol may act as a mediator in the sex change process remains unclear. Goikoetxea et al. (2017) considered three non-exclusive mechanisms through which cortisol might trigger female to male sex change: (i) cross-talk between the corticosteroid and androgen pathways, (ii) the inhibition of aromatase expression through the binding of cortisol to the glucocorticoid response elements (GRE) in the *cyp19a1a* promoter, and (iii) the depletion of primordial germ cells (PGCs) via upregulation of *amh* (the gene encoding anti-Müllerian hormone) (Fernandino et al., 2013; Goikoetxea et al., 2017; Pfennig, Standke, & Gutzeit, 2015; Todd, Liu, et al., 2016).

Transcriptome sequencing in bluehead wrasse strongly supports a role for cortisol in sex change. Liu (2016) observed early upregulation of *cyp11c1*, responsible for cortisol production, concurrent with inhibition of *cyp19a1a* and activation of *amh*, inhibition of *cyp19a1a* and activation of *amh*, followed by rapid upregulation during mid-sex change of *hsd11b2*, which encodes the enzyme responsible for metabolizing cortisol to inactive cortisone and downregulation of *hsd11b1a* which converts cortisone to cortisol.

Recent evidence in other vertebrates suggests stress may play a broader role in sexual fate transitions. In the bearded dragon, sex-reversed females could be distinguished transcriptionally from normal chromosomal females and males by upregulation of the proopiomelanocortin (POMC)-mediated environmental stress response and the preferential retention of introns from each of two *Jumonji* family genes, JARID2 and JMJD3, that are important epigenetic regulators. The working hypothesis is that high-temperature induced stress, presumably mediated by cortisol, results in the retention of intron sequences containing premature stop codons that alter or abolish epigenetic regulation, thereby over-riding chromosomally determined sexual fates (Deveson et al., 2017). This work, and other studies linking thermal stress with sex reversal (Navarro-Martín et al., 2011; Ribas et al., 2017; Shao et al., 2014), suggest that cortisol may be a long-sought-after mediator translating environmental signals into downstream genetic and physiological responses resulting in sex change (Goikoetxea et al., 2017). The possible role of cortisol as a proximate regulator of natural sex change warrants further attention (Goikoetxea et al., 2017).

8. The molecular regulation of sex change

Sexual fate appears increasing less an irreversible developmental commitment made during early embryo development, and more an ongoing tug-of-war for primacy between competing male and female developmental trajectories (Fig. 5). Nowhere among the vertebrates is this battle for primacy and its effect on sexual phenotype more obvious that in fish (Capel, 2017; Liu et al., 2017; Todd, Liu, et al., 2016).

Multiple genes have been shown to act as master sex-determining genes in fish and other vertebrates (Capel, 2017). While these sex-determining switches vary, a common set of downstream effectors act antagonistically in feminizing (e.g., *cyp19a1a*, *foxl2*, *wnt4*) and masculinizing (e.g., *dmrt1*, *amh*, *sox9*) networks to, respectively, promote ovarian or testicular development (Munger & Capel, 2012). Consequently, sexual fate is not simply dependent on the activation of one or other sex-specific network, but also contingent on the suppression of the opposing network to maintain that fate throughout life (Herpin & Schartl, 2011a). This dynamic antagonism

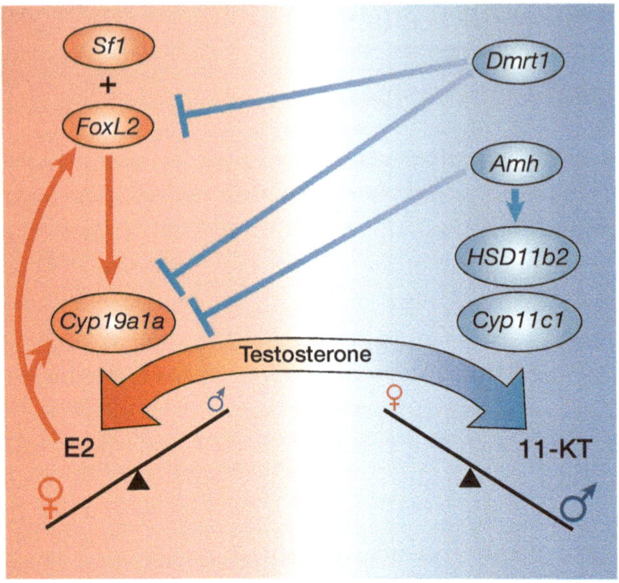

Fig. 5 Antagonistic sex-specific gene networks maintain sexual fate in fishes by promoting either an oestrogenic (left) or androgenic (right) environment. In females, *cyp19a1a* expression produces aromatase, which converts testosterone to estradiol (E2) and maintains the auto-regulatory feed-forward loop sustaining the high oestrogen levels that support ovarian function (Guiguen et al., 2010). Within this loop, transcription factors Foxl2 and Sf1 interact to up-regulate *cyp19a1a* expression (Wang et al., 2007), which also is controlled by gonadotropins like FSH through the synthesis of cAMP (Guiguen et al., 2010). An oestrogenic environment reinforces female-specific gene expression while suppressing male-promoting genes. In males, *cyp19a1a* expression and aromatase production is suppressed and *hsd11b2* and *cyp11c1* promote production of 11KT; such that androgenesis prevails supporting testicular function and male-specific gene expression. Inhibitory and activating effects of androgens on *Amh* are reported (+/−) (Pfennig et al., 2015). Dmrt1 suppresses *cyp19a1a* promoter activity directly (Wang et al., 2010) and indirectly via its antagonistic relationship with *foxl2* (see text). *Adapted with permission from Todd, E. V., Liu, H., Muncaster, S., & Gemmell, N. J. (2016). Bending genders: The biology of natural sex change in fish.* Sexual Development, 10, *223–241.*

makes sexual fate highly susceptible to manipulation by factors that interrupt the supremacy of the prevailing sexual network, enabling that of the opposing sex to gain dominance. The factors that might interrupt sex-specific networks are predominantly undetermined, but the actions of cortisol on epigenetic regulators, may be central to their regulation by environmental and physiological cues in sex-changing fish (Lamm et al., 2015; Liu et al., 2017; Todd, Liu, et al., 2016).

Sex change has evolved repeatedly in teleosts (Avise & Mank, 2009), but the molecular mechanism that triggers sex change remains unknown for any fish species (Todd, Liu, et al., 2016). In fishes, two widely conserved components of the molecular machinery essential for vertebrate sexual development, *cyp19a1a/b* (coding for gonadal/brain aromatase that catalyses conversion of androgens to oestrogens) and *dmrt1* (doublesex and mab-3 related transcription factor 1), are each vital to female- and male-promoting networks, respectively (Fig. 5). In females, *cyp19a1a* expression sustains an auto-regulatory loop that maintains the high oestrogen environment critical for ovarian function (Guiguen et al., 2010). In males, *dmrt1*, a critical transcriptional regulator, activates male-promoting genes (e.g., *sox9*, *sox8*) and suppresses ovarian pathways (e.g., *foxl2* and *rspo1/wnt/β-catenin*) (Herpin & Schartl, 2011a, 2011b).

In mice, under- or over-expression of either *dmrt1* or *foxl2* induces reprogramming of sexual cell fate and gonadal sex reversal (Lindeman et al., 2015; Matson et al., 2011; Uhlenhaut et al., 2009). Functional analyses, employing natural mutations, knockdown or genome editing of *dmrt1* have demonstrated its critical role in male development of various fish including medaka (Nanda et al., 2002), black porgy (Wu et al., 2012) sole (Cui et al., 2017). Impairing *dmrt1* expression alters *foxl2* or *cyp19a1* expression and leads the male gonad to develop as an ovotestis (Cui et al., 2017; Wu et al., 2012). Mutation of either *foxl2* or *cyp19a1* in Nile Tilapia (*Oreochromis niloticus*) resulted in female to male sex reversal, demonstrating that *foxl2* promotes ovarian development by upregulating *cyp19a1a* expression and repressing male pathway gene expression (Zhang et al., 2017). These, and other data, suggest that d*mrt1* interacts antagonistically with the female-specific transcription factor *foxl2* to influence *cyp19a1a* expression to control oestrogen production and the sexual fate of the gonad (Cui et al., 2017; Guiguen et al., 2010; Herpin & Schartl, 2011a; Kobayashi et al., 2013; Li et al., 2013; Zhang et al., 2017).

Thus, regulation of the androgen/oestrogen balance that determines sexual fate in fish seems to be governed predominantly by *dmrt1, foxl2* and *cyp19a1a* (Fig. 5). Expression of these genes is consistently sex-specific depending on which gonadal phenotype is developing, in both gonochoristic and sex-changing fishes (Alam, Kobayashi, Horiguchi, Hirai, & Nakamura, 2008; Liu et al., 2015; Wu, Tomy, Nakamura, & Chang, 2008; Xia, Zhou, Yao, Li, & Gui, 2007). The expression pattern of these genes also changes predictably in response to hormonal manipulations: *cyp19a1a/foxl2* is downregulated and *dmrt1* upregulated in fish treated with androgens, oestrogen antagonists or aromatase blockers, while treatment with oestrogen shows the opposite effect (Guiguen et al., 2010; Herpin & Schartl, 2011b).

Evidence that the expression patterns of a variety of genes linked to sexual fate (*cyp19a1a*, *dmrt1* and *amh*) alter in response to external environmental fluctuations, such as temperature and density (Fernandino et al., 2008; Guiguen et al., 2010; Pfennig et al., 2015; Ribas et al., 2017), raises the possibility that these and other key sex genes may be similarly sensitive to environmental cues that initiate sex change in sequential hermaphrodites (Todd, Liu, et al., 2016).

Investigations into the molecular control of sex change in sequential hermaphrodites, have predominantly focused on candidate gene approaches (Liu et al., 2017; Todd, Liu, et al., 2016). Extensive investigations into the patterns of expression for a core suite of genes in the vertebrate sex determination and differentiation cascade (e.g., *amh*, *foxl2*, *dmrt1*, *cyp19a1a*) and their likely proximate regulators have been undertaken in protogynous wrasse and grouper (Alam et al., 2008; Horiguchi et al., 2013; Kobayashi, Horiguchi, Nozu, & Nakamura, 2010; Li et al., 2006; Liu, Guiguen, & Liu, 2008; Miyake, Sakai, & Kuniyoshi, 2012; Nozu, Horiguchi, Kobayashi, & Nakamura, 2015; Xia et al., 2007), and protandrous porgy and clownfish (Nakamura et al., 2015; Wu et al., 2012, 2015; Wu & Chang, 2009).

From prior work in other vertebrates, there are a variety of promising candidates through which sex change could conceivably be initiated and controlled (Capel, 2017). However, to date the expression patterns observed for genes proposed as key initiators (e.g., *foxl2* and *dmrt1*) do not readily fit the pattern of natural sex change observed in those species studied. In protogynous groupers, gonadal *foxl2* expression decreases while *dmrt1* expression increases during natural sex change (Alam et al., 2008). However, the shifts in *foxl2* and *dmrt1* expression emerge only during the late transition stage (Alam et al., 2008), well downstream of the decline in E2 levels (Bhandari et al., 2003). More surprisingly, there was no sex-bias in gonadal expression of *foxl2* in protogynous three-spot wrasses; *foxl2* expression actually increased during sex change induced using aromatase inhibitor (Kobayashi, Horiguchi, et al., 2010). In bluehead wrasses, *foxl2* expression was much higher in ovaries than testes (Liu et al., 2015), but its expression peaked during mid-late sex change leading us to suggest it is not an initiator of sex change (Liu et al., 2017). Similarly, *dmrt1* expression during protogynous sex change occurs too late for this to be a key initiator of the process; rather its expression pattern seems to be driven by decreasing

E2 levels (Kobayashi, Nozu, Horiguchi, & Nakamura, 2014; Liu et al., 2007). Taken together, *foxl2* and *dmrt1* may be more important for progressing, as opposed to initiating gonadal sex change in protogynous fishes (Liu et al., 2017).

More recently, whole-transcriptome expression analysis has been used to understand sex differences in protogynous (Liu et al., 2015; Todd et al., 2018; Tsakogiannis et al., 2018) and protandrous systems (Casas et al., 2016). Two studies have also begun to explore how the transcriptome-wide expression landscape alters across natural sex change, in the brains and gonads of protogynous bluehead wrasse (Liu et al., 2017) and in clownfish (Casas et al., 2016). Collectively, these studies have identified several genes and regulatory factors that may be important orchestrators, and possible initiators, of natural sex change in sequential hermaphrodites. These studies are also shedding new light on how the balance between masculinizing and feminizing gene networks is altered to enable sex change in fishes (Todd, Liu, et al., 2016).

Whole-transcriptome expression profiling in bluehead wrasse (Liu et al., 2017) shows female-related gene expression gradually declines in the gonad (e.g., *dax1*, *figla*, *sox3*, *gdf9*, *hsd7b1*, *hsd11b3*) before the expression profiles become increasingly masculinized (e.g., *dmrt1*, *gsdf*, *cyp11c1*, *sox9*) (Fig. 3A). Candidate gene studies in protandrous black porgy (Wu & Chang, 2013) show the opposite pattern: male-related gene expression (e.g., *dmrt1*, *amh*, *amh2r*) declines coincidentally with testis volume, before expression profiles are increasingly feminized as the bisexual gonad becomes purely ovarian (e.g., *cyp19a1a*, *foxl2*, *wnt4*). Transcriptomic analyses in clownfish across sex change (Casas et al., 2016) also identified a steady decline in male-related genes (*dmrt1*, *amh*, *sox8*) and upregulation of female-pathway genes (*cyp19a1a*, *foxl2*, *wnt4*) as the gonad transitioned from an ovotestis to an ovary. In bluehead wrasse, the greatest shift in sex-specific gene expression occurs during mid-to-late sex change (Fig. 3C). Thus, it appears that the prevailing sexual network must first be shut down, and that sex change progresses as the suppression of the opposing network is lifted (Todd, Liu, et al., 2016). Genes that show changes in expression immediately prior to this shift are particularly interesting as potential elements of the switch that initiates sex change. As of now, it is not known whether expression changes in wrasses are due to reprogramming of existing cellular populations, or reflect broad scale replacement of the cells of one sex, with latent progenitor cells of the other. Future single cell genomic studies may resolve this issue.

9. Aromatase appears to trigger gonadal sex change in protogynous species

The gonadal aromatase gene, *cyp19a1a*, stands out in transcriptome studies as the only gene among those commonly known to influence sexual differentiation that is rapidly and completely shut-down in early protogynous sex change (Liu et al., 2017; Todd, Liu, et al., 2016). In bluehead wrasse, *cyp19a1a* expression plummets from the first sign of ovarian atresia, prior to the collapse of the feminizing expression landscape (Liu et al., 2017) (Fig. 3A). *Wnt4a* and *sf1* expression drop off concurrently, but are not completely arrested. This sharp decline in *cyp19a1a* expression has been reported in prior candidate gene studies of multiple protogynous species (Li et al., 2006; Zhang, Zhang, Liu, Zhang, & Zhang, 2013) and is completely consistent with the halt in E2 production observed at this stage (Fig. 3A). It therefore seems likely that *cyp19a1a* downregulation is the switch needed to initiate the gonadal transformation observed during protogynous sex change. However, there must also be upstream factors that act to negatively regulate its expression and so are potential triggers of sex change.

The promoter regions of teleost *cyp19a1a/b* genes contain putative DNA-binding motifs for multiple transcription and endocrine factors that have known roles in vertebrate sex differentiation (Todd, Liu, et al., 2016) including: FOXL2, steroidogenic factor 1 (SF-1), potential SRY-box (SOX9), Wilms tumor 1 protein (Wt1), GATA binding proteins, cAMP responsive elements (CRE), and response elements for glucocorticoids (GRE), oestrogens (ERE), progesterones (PRE) and androgens (ARE) (Gardner, Anderson, Place, Dixon, & Elizur, 2005; Guiguen et al., 2010). Thus, *cyp19a1a* expression appears to be regulated, or influenced, by numerous factors. Evidence is accumulating for (and against) a role of several of these during sex change, with epigenetic modification to *cyp19a1a* promoters one mechanism through which its responsiveness to these factors can be altered (see below).

As discussed earlier, candidate gene studies in protogynous fish show that *dmrt1* and *foxl2* expression alter only after serum E2 decreases, presumably in response to changes in *cyp19a1a* expression (Alam et al., 2008; Bhandari et al., 2003; Nozu et al., 2015). Further support that alterations in the expression of *dmrt1*, *dax1* and *foxl2* all occur well-downstream of *cyp19a1a* shut down come from transcriptomic studies in bluehead wrasse (Liu et al., 2017).

One possible regulator of *cyp19a1a* is steroidogenic factor 1 (*sf1* or *nr5a1*) (Liu et al., 2017), a transcription factor that has multiple roles in sex determination and differentiation networks in mammals (Bashamboo & McElreavey, 2015). Gonadal expression of *sf1* declined sharply during early female-to-male sex change in the bidirectional sex changing goby *Trimma okinawae* (Kobayashi et al., 2005). In bluehead wrasse gonads transitioning from ovary to testis, *sf1* expression also declined abruptly and concurrently with *cyp19a1a* expression, but recovered thereafter (Liu et al., 2017). The association of *sf1* with *cyp19a1a* in the initiating stages of female-to-male sex change is exciting in the wake of new evidence linking it to ovotesticular disorders of sexual development (Baetens et al., 2017). The role of *sf1* in sex changing fish needs further investigation.

10. Protandrous sex change and the role of dmrt1, amh and sox8

Studies of protandrous sex change have thus far focused strongly on single-gene studies, most of which have been undertaken in black porgy. A sharp decrease in *dmrt1* expression occurs at the onset of testicular degeneration in black porgy (Wu et al., 2012) and gilthead seabream (Liarte et al., 2007; Pauletto et al., 2018). Casas et al. (2016) also report a decrease in *dmrt1* across sex change in clownfish. It is possible that *dmrt1* may be as important in initiating protandrous sex change as *cyp19a1a* appears to be for protogynous species. Knocking down *dmrt1* in black porgy resulted in a loss of testicular germ cells and induced ovarian development in some experimental fish (Wu et al., 2012). In the same study, testicular *dmrt1* expression was significantly lower in fish that, several months later, underwent sex change to female, relative to those that remained male. However, in an earlier study, reduced *amh* but not *dmrt1* expression was reportedly predictive of sex change (Wu et al., 2010). Clearly, downregulation of both *dmrt1* and *amh* is important for protandrous sex change, but which acts earlier in the process is unclear. Transcriptome analysis of clownfish suggests that the *dmrt1* expression changes may precede those of *amh* (Casas et al., 2016), although the factors that affect *dmrt1* expression in early protandrous sex change remain to be determined. *Dmrt1* expression in the male testis appears to be maintained by androgens and gonadotropin signaling (e.g., LH) via the HPG axis (Herpin & Schartl, 2011b; Wu & Chang, 2013; Wu et al., 2012).

Further transcriptomic studies of the brain and gonad during sex change, particularly the early stages, are needed to identify the likely upstream regulators of *dmrt1,* followed by tests of regulatory interactions to determine just how the negative regulation of *dmrt1* is achieved in early protandrous sex change.

Anti-Müllerian hormone (Amh) is a multifunctional member of the Tgfβ superfamily of growth factors (Massagué, Seoane, & Wotton, 2005) that plays a critical role in vertebrate gonadal development for both males and females (reviewed by Pfennig et al., 2015). Amh is best known in mammals for its role in the regression of the Müllerian ducts in male embryos, from which the uterus and other female organs would otherwise develop (Behringer, 1994). Teleost fish have no Müllerian ducts, but possess orthologs of *amh* that are thought to have roles in early testicular differentiation, inhibition of germ cell proliferation and steroidogenesis to promote maleness (Pfennig et al., 2015).

Many of the physiological roles of *amh* in the ovary, testis and transitioning gonad during sex change remain speculative and need further investigation (Pfennig et al., 2015). Elevated levels of *amh* in male gonads are common during sexual differentiation in fish; and, as is observed in mammals, *amh* appears to directly inhibit *cyp19a1a* expression leading to an inverse relationship between *cyp19a1a* and *amh* expression (Pfennig et al., 2015). This inhibition is typical, but not universal for teleosts, as is inhibition of *amh* expression by oestrogens and FSH (Pfennig et al., 2015).

Transcriptomic analysis of protandrous clownfish (Casas et al., 2016) showed heightened expression of two different transcripts of *amh* in males compared to females, and expression of *amh* declined during the transition from male to female. Amh is also linked to sex change in protandrous black porgy, with expression of *amh* in cells bordering ectopic oocytes thought to suppress ovarian development in ovotestes (Wu et al., 2015).

In the protogynous hermaphroditic ricefield eel, a*mh* was also found to be upregulated early in the transitioning gonad (Hu, Guo, Gao, Tang, & Li, 2015), although the timing in relation to other male pathway genes or *cyp19a1a* was not investigated. However, whole-transcriptome data from bluehead wrasse showed that *amh* and its receptor *amhr2* are the first male-pathway genes upregulated in the gonad, and their initial surge is concurrent with the interruption of *cyp19a1a* expression (Liu et al., 2017). Whether *amh* contributes to the suppression of *cyp19a1a* at this time, or is simply responding to falling oestrogen levels and is a proximate effector of male-specific pathways, is not yet clear. More experimental evidence is needed to elucidate the exact functions of *amh* signaling in early protandrous and protogynous sex change. In the gonochoristic tilapia, *amh* and *amhr2* expression in the brain and pituitary appears to modulate FSH and LH release and occurs prior to *amh* upregulation in the gonad (Poonlaphdecha et al., 2013), suggesting a potential role in regulating brain-gonad communication via the HPG axis.

Sox8 is an important factor in the maintenance of cellular identify in mouse testis (Matson et al., 2011), and also an important regulator of adult Sertoli cell function and thus male fertility (O'Bryan et al., 2008). The role of *sox8* in sex change has not been widely considered, but it is known to be both a target of *dmrt1* and a regulator of *amh* expression (Salmon, Handyside, & Joyce, 2005). In the protandrous clownfish, *sox8* was strongly upregulated in males versus females alongside *dmrt1* and *amh* (Casas et al., 2016).

11. The role of epigenetics in sex change

Epigenetics provides a mechanism for cells to stably alter their gene expression patterns without requiring any underlying change in DNA sequence. Most modern epigenetic studies focus on modifications of DNA (e.g., methylation of cytosine bases) and histones (e.g., acetylation) which can regulate gene expression by reversibly altering the availability of genes, or specific exons, to transcription, and typically inhibit and promote transcription, respectively (Duncan, Gluckman, & Dearden, 2014; West-Eberhard, 2003). However, when the term epigenetics was first coined by Waddington (Goldberg, Allis, & Bernstein, 2007; Waddington, 1957) in explaining the process of cellular differentiation, whereby a multitude of cell- and tissue-types are created from essentially the same genetic code (Fig. 6). Essentially, Waddington imagined "epigenetics" representing a landscape within which cells progressively commit to a dedicated function, much like a marble rolling down a valley.

Natural sex change in fish almost certainly occurs without any underlying change in genetic content or DNA sequence. Thus, it seems likely that epigenetic mechanisms will play a central role in orchestrating sex change (Todd, Liu, et al., 2016). In this way, Waddington's epigenetic landscape may be helpful in conceptualizing sexual canalization and how it is reprogrammed during sex change. For example, in standard gonochoristic species, the bipotential gonad during early development could be imagined as a marble approaching the junction between two steep ravines, where entry into one makes transition to the other highly unlikely. When using this

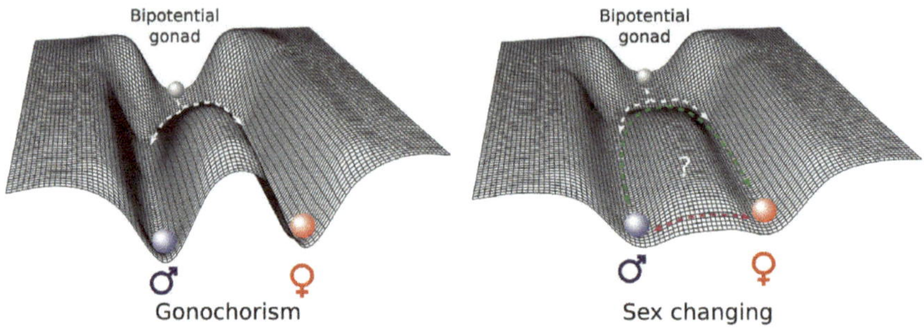

Fig. 6 Waddington's epigenetic landscapes and sexual phenotype. In gonochoristic and sex changing fish, the bipotential gonad undergoes sex determination and it is canalized through either a female or male pathway. For gonochoristic species, a strong genetic and epigenetic barrier precludes sex reversal during adult life. In sex changing species this barrier is reduced, sexual phenotype is more plastic and can be reprogrammed by environmental and social cues, resulting in sex change. The mechanistic process by which these barriers are overcome may involve transdifferentiation (Pink line), where cells of definitively one sex transition directly into cells of the other. Alternatively, a more complex reprogramming path, perhaps involving some form of dedifferentiation or developmental regression (Green line) may instead be involved.

analogy to consider sequential hermaphroditism, the critical question that emerges is what molecular processes form the barriers that normally separate males from females, and how are they routinely overcome?

11.1 Epigenetic barriers separate the sexes

DNA methylation at CG dinucleotides is unique compared to other epigenetic modifications because it has a well described mechanism by which it is propagated (Goll & Bestor, 2005). Essentially, the maintenance methyltransferase DNMT1 recognises methylated cytosines on the template strand, and copies this mark to the cognate cytosine on the newly synthesized strand. This elegant form of epigenetic memory allows stable transmission of (potentially lifelong) biological information. Sexually dimorphic DNA methylation patterns could represent a primary mechanism used to establish and maintain sexually divergent gene expression in species that are sexually isogenic. Alternatively, DNA methylation remodeling could be used to alter the expression of the prevailing sex pathway in favor of the other, possibly in an environmentally-sensitive manner, thus driving sex reversal.

For example, although the half-smooth tongue sole (*Cynoglossus semilaevis*) primarily has a ZW chromosomal sex determination system, analysis of gonads from ZZ and ZW fish revealed that sex-pathway genes differentially expressed between ovary and testis are major targets of methylation

(Shao et al., 2014). Thus, in addition to the genetic barriers separating male and female tongue sole, additional epigenetic barriers also exist. Interestingly, when ZW females become sex-reversed to "pseudomales" by temperature exposure, their DNA methylation signatures are indistinguishable from ZZ males and expression of female-specific W chromosomal genes is suppressed. ZW offspring in the F1 generation (produced by a cross between a ZW pseudomale and a normal ZW female) retain male epigenetic marks in their gonadal cells, and ~90% spontaneously sex-reverse in the absence of thermal influence (Shao et al., 2014). These results indicate that environmentally-induced sex reversal can reset heritable epigenetic marks, and that this reprogramming of the genome can override the opposing sexual pathway (Capel, 2017; Shao et al., 2014; Todd, Liu, et al., 2016).

It is likely that this is not an isolated example. The Nile Tilapia has a chromosomal sex determination system and with similar temperature-dependent sex reversal. It has recently been shown that *cyp19a1a* methylation levels are significantly higher in phenotypic males compared to females, irrespective of genotypic sex-type (Wang et al., 2017). Similarly, genetically female European sea bass (*Dicentrarchus labrax*) show down-regulation of *cyp19a1a* expression that is associated with the accumulation of DNA methylation and masculinization (Navarro-Martín et al., 2011).

In addition to these associations, bulk alteration of DNA methylation patterns has been shown to influence natural sex change. For example, in the protogynous ricefield eel, *Cyp19a1a* promoter regions were found to be hypermethylated and deacetylated in testis compared with ovary (Zhang et al., 2013). DNA methylation was concentrated within putative binding sites for cAMP response elements and sf1, and blocked gonadotropin-induced cAMP activation of *cyp19a1a* expression in vitro. Importantly, *cyp19a1a* promoter methylation increased as *cyp19a1a* expression decreased during sex change in ricefield eel, and implantation with DNA methylation inhibitors (5-aza-2′-deoxycytidine) could prevent or reverse gonadal sex change toward male. In a similar fashion, gonadal explants from Nile Tilapia treated with 5-aza-2′-deoxycytidine show increased expression of *cyp19a1a*, and subsequent feminization. Therefore, it seems likely that methylation of *cyp19a1a* DNA is an essential component of the epigenetic barrier separating males from females in these fish, and its remodeling is also likely essential for maintaining secondary male gonadal fate in protogynous hermaphrodites.

While histone modification does not have the same well-described memory system as DNA methylation, it has been linked to environmental sex determination in bearded dragons (Holleley et al., 2015), American alligator (Yatsu et al., 2016) and the red-eared slider turtle (Czerwinski, Natarajan, Barske, Looger, & Capel, 2016). KDM6B, a lysine-specific demethylase with a central role in epigenetic regulation of gene expression, is an early responder to male incubation temperature in both turtles and alligators (Capel, 2017) and a key regulator of sex in red-eared slider turtles (Ge et al., 2018). Suppressing *kdm6b* expression reduces demethylation of its target, trimethylated lysine 27 on histone 3 (H3K27), repressing *dmrt1* promoter activity. Thus, high amounts of KDM6B activate *dmrt1* expression and determine male sex, whereas reduced amounts of KDM6B repress *dmrt1* expression (Ge et al., 2018).

It is as yet unknown if *kdm6b*, a Jumonji gene family member, responds directly to temperature or is regulated by upstream temperature-sensitive elements yet to be discovered, such as CIRBP (cold-inducible RNA binding protein) (Schroeder, Metzger, Miller, & Rhen, 2016). However, in the red-eared slider turtle, American alligator, and the bearded dragon an intron is alternatively retained or excised during transcription of *kdm6b* and another Jumonji family member, *Jarid2* (Jumonji and AT-rich interaction domain containing 2) in response to temperature (Deveson et al., 2017). Alternative splicing of Jumonji genes is hypothesized to alter the targets of gene silencing, gene activation, and the recruitment of chromatin remodeling complexes [for example, Polycomb repressive complex 2 (PRC2)] in ways that are not yet fully understood to determine sex (Georges & Holleley, 2018).

Epigenetic modifications, such as those driven by *kdm6b*, which alter the expression of key sex-pathway genes is clearly a plausible, and indeed likely, mechanism through which sex change can be both initiated and maintained in sequential hermaphrodites. However, to date this area has received little attention. In whole-transcriptome data from bluehead wrasse, several genes encoding DNA methyltransferases (e.g., *dnmt1*, *dnmt3ab*, *dnmt3bb*) and histone acetyltransferases (e.g., *kat8*, *hat1*) or deacetylases (e.g., *hdac2*, *hdac10*) showed significant expression changes at the extremities of gonadal sex change (Liu et al., 2017). Specific functions for these genes during sex change cannot be inferred until more is known regarding epigenetic regulation of vertebrate sexual differentiation pathways. Moreover, much more attention should be focussed upon the nature of the intermediate states in these fish. Do the gonads of sequential hermaphrodites regress to some earlier developmental time point and re-differentiate into the other sex? Or perhaps a latent (germline?) stem cell population constantly present within the gonad reactivates at sex change to replace sexually divergent cells that are undergoing apoptosis? Or is there a more straightforward reprogramming of ovarian cells to a testicular fate (Fig. 6)? It is likely that novel genome-wide techniques such as single-cell RNA-seq (Stévant & Nef, 2018) will assist with answering these questions.

12. A new model of sex change

We have previously proposed a mechanistic model, featuring cross-talk between the HPG and HPI axes, through which a normal female reproductive cycle may be interrupted to initiate sex change in socially protogynous fishes (Goikoetxea et al., 2017; Liu et al., 2017). Briefly, following loss of a dominant male, rapid neurochemical changes in the hypothalamus of large females promote behavioral sex change (see previous section). Elevated AVT and NE levels specifically may then perturb GnRH and LH dynamics via the HPG axis to induce follicle apoptosis in the ovary, while driving up serum cortisol via the HPA/I axis (Liu et al., 2017). Elevated cortisol, together with epigenetic regulators, could block *cyp19a1a* transcription directly, and via activation of specific transcription factors (e.g., Amh), instigate the chain-reaction of falling E2 levels that accelerate ovarian degeneration and interrupt female-specific gene expression. Thereafter, elevated cortisol may stimulate 11-KT production to activate male-pathway genes and support testicular differentiation (Goikoetxea et al., 2017; Liu et al., 2017). There is much yet to test, but this current hypothesis is the first to link neurochemical changes occurring in the brain that are associated with behavioral sex change and stress, with the changes in gene expression and endocrine production during gonadal sex change in protogynous sequential hermaphrodites.

13. Conclusions

Studies from divergent systems are beginning to clarify how natural sex change may initiate and progress. Sex change clearly begins in the brain, but where this trigger lies, what this trigger is and how this then initiates gonadal sex change remain unknown (Lamm et al., 2015). Across many systems, re-direction of gonadal fate is initiated when expression of critical sex-maintenance genes (e.g., *cyp19a1a* in protogynous, and *drmt1* in protandrous species) is interrupted causing a cascading collapse of the prevailing expression landscape, endocrine environment, and gonadal anatomy. Once antagonistic suppression of the opposing sexual network lifts, establishment of a new sex-specific expression and endocrine environment drives gonadal development toward the secondary sex (Todd, Liu, et al., 2016). A tightly regulated, multi-component, trigger mechanism appears to sit at the top of this cascade to ensure sex change proceeds only under specific circumstances (Todd, Liu, et al., 2016). *Cyp19a1a* appears to initiate gonadal sex change in protogynous systems and we know many factors that appear to reduce the expression of this critical sexual pivot (e.g., cortisol, DNA methylation, *amh*). However, most of the data we have obtained to date is simple association, linking changes in gene expression levels and timing to key phenotypic events. We need more sophisticated experimental validation, as has been seen in mouse and other model systems, if we are to disentangle cause from effect. A fuller understanding of how complex neurological systems in the brain cross-talk with the stress (HPI) and reproductive (HPG) axes to translate external environmental signals into internal physiological responses is also needed.

Such work takes time—even in mouse our understanding of sex determination continues to evolve (Capel, 2017; Harris et al., 2018). In fish we know this process is also controlled by antagonistic interactions. In mice, the key players are the SRY-SOX9-FGF9 network that promotes testis determination partly by opposing pro-ovarian pathways, while RSPO1/WNT-β-catenin/FOXL2 signals control ovary development by inhibiting SRY-SOX9-FGF9 (Harris et al., 2018). The molecular basis of this mutual antagonism is unclear, but recent work suggests that ZNRF3, a WNT signaling antagonist and direct target of RSPO1-mediated inhibition, is involved (Harris et al., 2018). Male mice lacking ZNRF3 exhibit complete or partial gonadal sex reversal (Harris et al., 2018). Perhaps ZNRF3 plays a role in the dynamic antagonism we observe in sex changing fish?

Current insights into sex change derive from investigations into a handful of protogynous systems and studies on the protandrous black porgy and clownfish (Liu et al., 2017; Todd, Liu, et al., 2016). Contrasting across diverse representatives of each sex change strategy will be vital to determining how the switch between sex-specific developmental networks arises during sex change, and whether common mechanisms trigger and regulate the processes of female to male and male to female sex change.

The establishment of model systems for sex change remains a challenge, particularly where we seek to understand molecular genetic mechanisms. However, ongoing refinements in genomic sequencing and manipulative approaches are making the investigation and establishment of novel systems easier. RNA-sequencing (RNA-seq) (Todd, Black, & Gemmell, 2016;

Wang, Gerstein, & Snyder, 2009), especially, enables us to rapidly describe the co-ordinated pattern of expression, including that of uncharacterized genes and isoforms, across a time-series without the need for prior genetic resources. The value of this approach for identifying candidate triggers of sex change and genes with novel and unexpected roles in vertebrate sexual development has already been established for sex changing fish (Casas et al., 2016; Liu et al., 2015; Todd et al., 2018). Further refinement through the latest single-cell RNA-seq technologies are especially exciting with regards to characterizing genetic cascades involved in cellular reprogramming and cell fate determination (Farrell et al., 2018; Stévant & Nef, 2018; Stévant et al., 2018). This precision approach to understanding cell specific expression patterns is likely to herald significant advances in our understanding of sex change in both the brain and gonad. Single cell approaches will be especially valuable in determining the cellular origins of the secondary sex in protogynous species undergoing complete sex reversal, and in answering questions regarding the potential reprogramming or retained bipotency of gonia and somatic cells in teleost gonads. Further refinements through simultaneous genome-wide analysis of transcriptomes (RNA-seq) and methylomes (whole-genome bisulfite sequencing, BS-seq) from single cells (Angermueller et al., 2016) and chromatin accessibility (single-cell nucleosome, methylation and transcription sequencing, scNMT-seq) (Clark et al., 2018) will generate new testable hypotheses as we continue to search for components of the trigger mechanism used by sequential hermaphrodites to alter sexual fate in adulthood. However, manipulative studies will always be the acid test and an integrative approach will be necessary to answer how a common genetic toolkit can be flexibly adapted to achieve sexual plasticity both within the lifetime of an individual fish, and across the various teleost lineages that have independently evolved sequential hermaphroditism. Such in vitro manipulations have historically been confined to a handful of model systems, but once husbandry challenges are circumvented, gene editing approaches (Cong et al., 2013) will enable us to undertake sophisticated experiments quickly and affordably, to precisely document the pathway of events governing the transformation that is natural sex change.

Acknowledgments

We thank Blanche Capel for inviting us to contribute this article, and for editorial comments made that have improved our article. We are also grateful to Mark Lokman, Hui Lui, Jodi Thomas and Simon Muncaster for discussions that helped shaped this latest review. We thank

Robbie McPhee for his valuable assistance in preparing figures. This research was supported by Grant UOO1308 awarded to N.J.G. by the Marsden Fund, Royal Society of New Zealand and grants to N.J.G. and T.A.H. from the department of Anatomy, University of Otago. E.V.T. was supported by a University of Otago Health Sciences Career Postdoctoral Fellowship.

References

Ah-King, M., & Nylin, S. (2010). Sex in an evolutionary perspective: Just another reaction norm. *Evolutionary Biology*, *37*, 234–246.

Alam, M. A., Bhandari, R. K., Kobayashi, Y., Nakamura, S., Soyano, K., & Nakamura, M. (2006). Changes in androgen-producing cell size and circulating 11-ketotestosterone level during female-male sex change of honeycomb grouper Epinephelus merra. *Molecular Reproduction and Development*, *73*, 206–214.

Alam, M. A., Kobayashi, Y., Hirai, T., & Nakamura, M. (2010). Isolation, characterization and expression analyses of FSH receptor in protogynous grouper. *Comparative Biochemistry and Physiology. Part A, Molecular & Integrative Physiology*, *156*, 364–371.

Alam, M. A., Kobayashi, Y., Horiguchi, R., Hirai, T., & Nakamura, M. (2008). Molecular cloning and quantitative expression of sexually dimorphic markers Dmrt1 and Foxl2 during female-to-male sex change in Epinephelus merra. *General and Comparative Endocrinology*, *157*, 75–85.

An, K. W., Lee, J., & Choi, C. Y. (2010). Expression of three gonadotropin subunits and gonadotropin receptor mRNA during male-to-female sex change in the cinnamon clownfish, Amphiprion melanopus. *Comparative Biochemistry and Physiology. Part A, Molecular & Integrative Physiology*, *156*, 407–415.

An, K. W., Lee, K. Y., Yun, S. G., & Choi, C. Y. (2009). Molecular characterization of gonadotropin subunits and gonadotropin receptors in black porgy, Acanthopagrus schlegeli: Effects of estradiol-17beta on mRNA expression profiles. *Comparative Biochemistry and Physiology. Part B, Biochemistry & Molecular Biology*, *152*, 177–188.

Angermueller, C., Clark, S. J., Lee, H. J., Macaulay, I. C., Teng, M. J., Hu, T. X., et al. (2016). Parallel single-cell sequencing links transcriptional and epigenetic heterogeneity. *Nature Methods*, *13*, 229–232.

Aubin-Horth, N., & Renn, S. C. P. (2009). Genomic reaction norms: Using integrative biology to understand molecular mechanisms of phenotypic plasticity. *Molecular Ecology*, *18*, 3763–3780.

Avise, J. C., & Mank, J. E. (2009). Evolutionary perspectives on hermaphroditism in fishes. *Sexual Development*, *3*, 152–163.

Baetens, D., Stoop, H., Peelman, F., Todeschini, A.-L., Rosseel, T., Coppieters, F., et al. (2017). NR5A1 is a novel disease gene for 46,XX testicular and ovotesticular disorders of sex development. *Genetics in Medicine*, *19*, 367–376.

Baroiller, J. F., & D'Cotta, H. (2001). Environment and sex determination in farmed fish. *Comparative Biochemistry and Physiology, Part C: Toxicology & Pharmacology*, *130*, 399–409.

Baroiller, J.-F., & D'Cotta, H. (2016). The reversible sex of gonochoristic fish: Insights and consequences. *Sexual Development*, *10*, 242–266.

Barske, L. A., & Capel, B. (2008). Blurring the edges in vertebrate sex determination. *Current Opinion in Genetics & Development*, *18*, 499–505.

Bashamboo, A., & McElreavey, K. (2015). Human sex-determination and disorders of sex-development (DSD). *Seminars in Cell & Developmental Biology*, *45*, 77–83.

Behringer, R. R. (1994). The in vivo roles of Müllerian-inhibiting substance. *Current Topics in Developmental Biology*, *29*, 171–187.

Benvenuto, C., Coscia, I., Chopelet, J., Sala-Bozano, M., & Mariani, S. (2017). Ecological and evolutionary consequences of alternative sex-change pathways in fish. *Scientific Reports*, *7*, 9084.

Bhandari, R. K., Higa, M., Nakamura, S., & Nakamura, M. (2004). Aromatase inhibitor induces complete sex change in the protogynous honeycomb grouper (Epinephelus merra). *Molecular Reproduction and Development*, *67*, 303–307.

Bhandari, R. K., Komuro, H., Nakamura, S., Higa, M., & Nakamura, M. (2003). Gonadal restructuring and correlative steroid hormone profiles during natural sex change in protogynous honeycomb grouper (Epinephelus merra). *Zoological Science*, *20*, 1399–1404.

Bhandari, R. K., Nakamura, M., Kobayashi, T., & Nagahama, Y. (2006). Suppression of steroidogenic enzyme expression during androgen-induced sex reversal in Nile tilapia (Oreochromis niloticus). *General and Comparative Endocrinology*, *145*, 20–24.

Black, M. P., Reavis, R. H., & Grober, M. S. (2004). Socially induced sex change regulates forebrain isotocin in Lythrypnus dalli. *Neuroreport*, *15*, 185–189.

Brion, F., Tyler, C. R., Palazzi, X., Laillet, B., Porcher, J. M., Garric, J., et al. (2004). Impacts of 17β-estradiol, including environmentally relevant concentrations, on reproduction after exposure during embryo-larval-, juvenile- and adult-life stages in zebrafish (Danio rerio). *Aquatic Toxicology*, *68*, 193–217.

Budd, A., Banh, Q., Domingos, J., & Jerry, D. (2015). Sex control in fish: Approaches, challenges and opportunities for aquaculture. *Journal of Marine Science and Engineering*, *3*, 329–355.

Capel, B. (2017). Vertebrate sex determination: Evolutionary plasticity of a fundamental switch. *Nature Reviews. Genetics*, *18*, 675–689.

Casas, L., Saborido-Rey, F., Ryu, T., Michell, C., Ravasi, T., & Irigoien, X. (2016). Sex change in clownfish: Molecular insights from transcriptome analysis. *Scientific Reports*, *6*, 35461.

Chang, C. F., Lau, E. L., & Lin, B. Y. (1995). Estradiol-17β suppresses testicular development and stimulates sex reversal in protandrous black porgy, Acanthopagrus schlegeli. *Fish Physiology and Biochemistry*, *14*, 481–488.

Clark, S. J., Argelaguet, R., Kapourani, C.-A., Stubbs, T. M., Lee, H. J., Alda-Catalinas, C., et al. (2018). scNMT-seq enables joint profiling of chromatin accessibility DNA methylation and transcription in single cells. *Nature Communications*, *9*, 781.

Cole, K. S. (2010). Gonad morphology in hermaphroditic gobies. In K. S. Cole (Ed.), *Reproduction and sexuality in marine fishes, patterns and processes* (pp. 117–162). Berkeley, CA: University of California Press.

Cole, K. S. (2011). Patterns of reproductive morphology in the genus Gobiodon (Teleostei: Gobiidae). *Environmental Biology of Fishes*, *92*, 323.

Cong, L., Ran, F. A., Cox, D., Lin, S., Barretto, R., Habib, N., et al. (2013). Multiplex genome engineering using CRISPR/Cas systems. *Science*, *339*, 819–823.

Cui, Z., Liu, Y., Wang, W., Wang, Q., Zhang, N., Lin, F., et al. (2017). Genome editing reveals dmrt1 as an essential male sex-determining gene in Chinese tongue sole (Cynoglossus semilaevis). *Scientific Reports*, *7*, 42213.

Czerwinski, M., Natarajan, A., Barske, L., Looger, L. L., & Capel, B. (2016). A timecourse analysis of systemic and gonadal effects of temperature on sexual development of the red-eared slider turtle Trachemys scripta elegans. *Developmental Biology*, *420*, 166–177.

DeFalco, T., & Capel, B. (2009). Gonad morphogenesis in vertebrates: Divergent means to a convergent end. *Annual Review of Cell and Developmental Biology*, *25*, 457–482.

Deveson, I. W., Holleley, C. E., Blackburn, J., Marshall Graves, J. A., Mattick, J. S., Waters, P. D., et al. (2017). Differential intron retention in Jumonji chromatin modifier genes is implicated in reptile temperature-dependent sex determination. *Science Advances*, *3*, e1700731.

Devlin, R. H., & Nagahama, Y. (2002). Sex determination and sex differentiation in fish: An overview of genetic, physiological, and environmental influences. *Aquaculture, 208*, 191–364.

Duncan, E. J., Gluckman, P. D., & Dearden, P. K. (2014). Epigenetics, plasticity, and evolution: How do we link epigenetic change to phenotype? *Journal of Experimental Zoology. Part B, Molecular and Developmental Evolution, 322*, 208–220.

Elizur, A. (2009). The KiSS1/GPR54 system in fish. *Peptides, 30*, 164–170.

Erisman, B. E., Petersen, C. W., Hastings, P. A., & Warner, R. R. (2013). Phylogenetic perspectives on the evolution of functional hermaphroditism in teleost fishes. *Integrative and Comparative Biology, 53*, 736–754.

Espigares, F., Zanuy, S., & Gómez, A. (2015). Kiss2 as a regulator of LH and FSH secretion via paracrine/autocrine signaling in the teleost fish European Sea Bass (Dicentrarchus labrax). *Biology of Reproduction, 93*, 1–12.

Farrell, J. A., Wang, Y., Riesenfeld, S. J., Shekhar, K., Regev, A., & Schier, A. F. (2018). Single-cell reconstruction of developmental trajectories during zebrafish embryogenesis. *Science, 360*(6392), eaar3131. https://doi.org/10.1126/science.aar3131.

Feinberg, A. P. (2007). Phenotypic plasticity and the epigenetics of human disease. *Nature, 447*, 433–440.

Fernandino, J. I., Hattori, R. S., Kishii, A., Strussmann, C. A., & Somoza, G. M. (2012). The cortisol and androgen pathways cross talk in high temperature-induced masculinization: The 11beta-hydroxysteroid dehydrogenase as a key enzyme. *Endocrinology, 153*, 6003–6011.

Fernandino, J. I., Hattori, R. S., Moreno Acosta, O. D., Strussmann, C. A., & Somoza, G. M. (2013). Environmental stress-induced testis differentiation: Androgen as a by-product of cortisol inactivation. *General and Comparative Endocrinology, 192*, 36–44.

Fernandino, J. I., Hattori, R. S., Shinoda, T., Kimura, H., Strobl-Mazzulla, P. H., Strüssmann, C. A., et al. (2008). Dimorphic expression of *dmrt1* and *cyp19a1* (Ovarian Aromatase) during early gonadal development in Pejerrey, *Odontesthes bonariensis*. *Sexual Development, 2*, 316–324.

Fricke, H., & Fricke, S. (1977). Monogamy and sex change by aggressive dominance in coral reef fish. *Nature, 266*, 830–832.

Gamble, T., & Zarkower, D. (2012). Sex determination. *Current Biology, 22*, R257–R262.

Gardner, L., Anderson, T., Place, A. R., Dixon, B., & Elizur, A. (2005). Sex change strategy and the aromatase genes. *The Journal of Steroid Biochemistry and Molecular Biology, 94*, 395–404.

Ge, C., Ye, J., Weber, C., Sun, W., Zhang, H., Zhou, Y., et al. (2018). The histone demethylase KDM6B regulates temperature-dependent sex determination in a turtle species. *Science, 360*, 645–648.

Georges, A., & Holleley, C. E. (2018). How does temperature determine sex? *Science, 360*, 601–602.

Ghiselin, M. T. (1969). The evolution of hermaphroditism among animals. *The Quarterly Review of Biology, 44*, 189–208.

Godwin, J. (1994). Historical aspects of protandrous sex change in the anemonefish Amphiprion melanopus (Pomacentridae, Teleostei). *Journal of Zoology, 232*, 199–213.

Godwin, J. (2009). Social determination of sex in reef fishes. *Seminars in Cell & Developmental Biology, 20*, 264–270.

Godwin, J. (2010). Neuroendocrinology of sexual plasticity in teleost fishes. *Frontiers in Neuroendocrinology, 31*, 203–216.

Godwin, J., Crews, D., & Warner, R. R. (1996). Behavioural sex change in the absence of gonads in a coral reef fish. *Proceedings of the Biological Sciences, 263*, 1683–1688.

Godwin, J., Sawby, R., Warner, R. R., Crews, D., & Grober, M. S. (2000). Hypothalamic arginine vasotocin mRNA abundance variation across sexes and with sex change in a coral reef fish. *Brain, Behavior and Evolution*, *55*, 77–84.

Godwin, J. R., & Thomas, P. (1993). Sex change and steroid profiles in the protandrous anemonefish Amphiprion melanopus (Pomacentridae, Teleostei). *General and Comparative Endocrinology*, *91*, 144–157.

Godwin, J., & Thompson, R. (2012). Nonapeptides and social behavior in fishes. *Hormones and Behavior*, *61*, 230–238.

Goikoetxea, A., Todd, E. V., & Gemmell, N. J. (2017). Stress and sex: Does cortisol mediate sex change in fish? *Reproduction*, *154*, R149–R160.

Goldberg, A. D., Allis, C. D., & Bernstein, E. (2007). Epigenetics: A landscape takes shape. *Cell*, *128*, 635–638.

Goll, M. G., & Bestor, T. H. (2005). Eukaryotic cytosine methyltransferases. *Annual Review of Biochemistry*, *74*, 481–514.

Goodson, J. L., & Bass, A. H. (2000). Forebrain peptides modulate sexually polymorphic vocal circuitry. *Nature*, *403*, 769–772.

Govoroun, M., McMeel, O. M., D'Cotta, H., Ricordel, M. J., Smith, T., Fostier, A., et al. (2001). Steroid enzyme gene expressions during natural and androgen-induced gonadal differentiation in the rainbow trout, Oncorhynchus mykiss. *The Journal of Experimental Zoology*, *290*, 558–566.

Guiguen, Y., Cauty, C., Fostier, A., Fuchs, J., & Jalabert, B. (1994). Reproductive cycle and sex inversion of the seabass, Lates calcarifer, reared in sea cages in French Polynesia: Histological and morphometric description. *Environmental Biology of Fishes*, *39*, 231–247.

Guiguen, Y., Fostier, A., Piferrer, F., & Chang, C.-F. (2010). Ovarian aromatase and estrogens: A pivotal role for gonadal sex differentiation and sex change in fish. *General and Comparative Endocrinology*, *165*, 352–366.

Harris, A., Siggers, P., Corrochano, S., Warr, N., Sagar, D., Grimes, D. T., et al. (2018). ZNRF3 functions in mammalian sex determination by inhibiting canonical WNT signaling. *Proceedings of the National Academy of Sciences of the United States of America*, *115*, 5474–5479.

Hattori, A. (2012). Determinants of body size composition in limited shelter space: Why are anemonefishes protandrous? *Behavioral Ecology*, *23*, 512–520.

Hattori, R. S., Fernandino, J. I., Kishii, A., Kimura, H., Kinno, T., Oura, M., et al. (2009). Cortisol-induced masculinization: Does thermal stress affect gonadal fate in pejerrey, a teleost fish with temperature-dependent sex determination? *PLoS One*, *4*, e6548.

Hayashi, Y., Kobira, H., Yamaguchi, T., Shiraishi, E., Yazawa, T., Hirai, T., et al. (2010). High temperature causes masculinization of genetically female medaka by elevation of cortisol. *Molecular Reproduction and Development*, *77*, 679–686.

Herpin, A., Adolfi, M. C., Nicol, B., Hinzmann, M., Schmidt, C., Klughammer, J., et al. (2013). Divergent expression regulation of gonad development genes in medaka shows incomplete conservation of the downstream regulatory network of vertebrate sex determination. *Molecular Biology and Evolution*, *30*, 2328–2346.

Herpin, A., & Schartl, M. (2011a). Dmrt1 genes at the crossroads: A widespread and central class of sexual development factors in fish. *The FEBS Journal*, *278*, 1010–1019.

Herpin, A., & Schartl, M. (2011b). Sex determination: Switch and suppress. *Current Biology*, *21*, R656–R659.

Herpin, A., & Schartl, M. (2015). Plasticity of gene-regulatory networks controlling sex determination: Of masters, slaves, usual suspects, newcomers, and usurpators. *EMBO Reports*, *16*, 1260–1274.

Higa, M., Ogasawara, K., Sakaguchi, A., Nagahama, Y., & Nakamura, M. (2003). Role of steroid hormones in sex change of protogynous wrasse. *Fish Physiology and Biochemistry*, *28*, 149–150.

Holleley, C. E., O'Meally, D., Sarre, S. D., Marshall Graves, J. A., Ezaz, T., Matsubara, K., et al. (2015). Sex reversal triggers the rapid transition from genetic to temperature-dependent sex. *Nature*, *523*, 79–82.

Holmberg, J., & Perlmann, T. (2012). Maintaining differentiated cellular identity. *Nature Reviews. Genetics*, *13*, 429–439.

Horiguchi, R., Nozu, R., Hirai, T., Kobayashi, Y., Nagahama, Y., & Nakamura, M. (2013). Characterization of gonadal soma-derived factor expression during sex change in the protogynous wrasse, Halichoeres trimaculatus. *Developmental Dynamics*, *242*, 388–399.

Hu, Q., Guo, W., Gao, Y., Tang, R., & Li, D. (2015). Molecular cloning and characterization of amh and dax1 genes and their expression during sex inversion in rice-field eel Monopterus albus. *Scientific Reports*, *5*, 16667.

Hu, X., Liu, X., Zhang, H., Zhang, Y., Li, S., Sang, Q., et al. (2011). Expression profiles of gonadotropins and their receptors during 17alpha-methyltestosterone implantation-induced sex change in the orange-spotted grouper (Epinephelus coioides). *Molecular Reproduction and Development*, *78*, 376–390.

Jaenisch, R., & Bird, A. (2003). Epigenetic regulation of gene expression: How the genome integrates intrinsic and environmental signals. *Nature Genetics*, *33*(Suppl), 245–254.

Jones, G. P. (1980). Growth and reproduction in the protogynous hermaphrodite Pseudolabrus celidotus (Pisces: Labridae) in New Zealand. *Copeia*, *1980*, 660–675.

Kanda, S., Akazome, Y., Mitani, Y., Okubo, K., & Oka, Y. (2013). Neuroanatomical evidence that kisspeptin directly regulates isotocin and vasotocin neurons. *PLoS One*, *8*, e62776.

Kobayashi, Y., Alam, M. A., Horiguchi, R., Shimizu, A., & Nakamura, M. (2010). Sexually dimorphic expression of gonadotropin subunits in the pituitary of protogynous honeycomb grouper (Epinephelus merra): Evidence that follicle-stimulating hormone (FSH) induces gonadal sex change. *Biology of Reproduction*, *82*, 1030–1036.

Kobayashi, Y., Horiguchi, R., Nozu, R., & Nakamura, M. (2010). Expression and localization of forkhead transcriptional factor 2 (Foxl2) in the gonads of protogynous wrasse, Halichoeres trimaculatus. *Biology of Sex Differences*, *1*, 3.

Kobayashi, H., & Iwamatsu, T. (2005). Sex reversal in the medaka Oryzias latipes by brief exposure of early embryos to estradiol-17β. *Zoological Science*, *22*, 1163–1167.

Kobayashi, Y., Nagahama, Y., & Nakamura, M. (2013). Diversity and plasticity of sex determination and differentiation in fishes. *Sexual Development*, 7, 115–125.

Kobayashi, Y., Nakamura, M., Sunobe, T., Usami, T., Kobayashi, T., Manabe, H., et al. (2009). Sex change in the Gobiid fish is mediated through rapid switching of gonadotropin receptors from ovarian to testicular portion or vice versa. *Endocrinology*, *150*, 1503–1511.

Kobayashi, Y., Nozu, R., Horiguchi, R., & Nakamura, M. (2014). Histological observation of doublesex-mab 3-related transcription factor 1 (DMRT1) localization in the adult testis of three-spot wrasse. *International Aquatic Research*, *6*, 68.

Kobayashi, Y., Sunobe, T., Kobayashi, T., Nakamura, M., Suzuki, N., & Nagahama, Y. (2005). Molecular cloning and expression of Ad4BP/SF-1 in the serial sex changing gobiid fish, Trimma okinawae. *Biochemical and Biophysical Research Communications*, *332*, 1073–1080.

Koopman, P. (2008). Sex-determining cascades in gonadal development: Insights from ovotestes. *Biology of Reproduction*, *78*, 278.

Koulish, S., Kramer, C. R., & Grier, H. J. (2002). Organization of the male gonad in a protogynous fish, Thalassoma bifasciatum (Teleostei: Labridae). *Journal of Morphology*, *254*, 292–311.

Kramer, C. R., Caddell, M. T., & Bubenheimer-Livolsi, L. (1993). sGnRH-A [(d-Arg6, Pro9, NEt-)LHRH] in combination with domperidone induces gonad reversal in a protogynous fish, the bluehead wrasse, Thalassoma bifasciatum. *Journal of Fish Biology*, *42*, 185–195.

Kroon, F. J., Munday, P. L., & Pankhurst, N. W. (2003). Steroid hormone levels and bi-directional sex change in Gobiodon histrio. *Journal of Fish Biology*, *62*, 153–167.

Kroon, F. J., Munday, P. L., & Westcott, D. A. (2009). Equivalent whole-body concentrations of 11-ketotestosterone in female and male coral goby Gobiodon erythrospilus, a bidirectional sex-changing fish. *Journal of Fish Biology*, *75*, 685–692.

Kroon, F. J., Munday, P. L., Westcott, D. A., Hobbs, J.-P. A., & Liley, N. R. (2005). Aromatase pathway mediates sex change in each direction. *Proceedings of the Royal Society B: Biological Sciences*, *272*, 1399–1405.

Kuwamura, T., Kadota, T., & Suzuki, S. (2015). Bidirectional sex change in the magenta dottyback Pictichromis porphyrea: First evidence from the field in Pseudochromidae. *Environmental Biology of Fishes*, *98*, 201–207.

Lamm, M. S., Liu, H., Gemmell, N. J., & Godwin, J. R. (2015). The need for speed: Neuroendocrine regulation of socially-controlled sex change. *Integrative and Comparative Biology*, *55*, 307–322.

Larson, E. T., Norris, D. O., Gordon Grau, E., & Summers, C. H. (2003). Monoamines stimulate sex reversal in the saddleback wrasse. *General and Comparative Endocrinology*, *130*, 289–298.

Larson, E. T., Norris, D. O., & Summers, C. H. (2003). Monoaminergic changes associated with socially induced sex reversal in the saddleback wrasse. *Neuroscience*, *119*, 251–263.

Lee, Y. H., Du, J. L., Yueh, W. S., Lin, B. Y., Huang, J. D., Lee, C. Y., et al. (2001). Sex change in the protandrous black porgy, Acanthopagrus schlegeli: A review in gonadal development, estradiol, estrogen receptor, aromatase activity and gonadotropin. *Journal of Experimental Zoology*, *290*, 715–726.

Lee, S. L. J., Horsfield, J. A., Black, M. A., Rutherford, K., Fisher, A., & Gemmell, N. J. (2017). Histological and transcriptomic effects of 17α-methyltestosterone on zebrafish gonad development. *BMC Genomics*, *18*, 557.

Lee, M. F., Huang, J. D., & Chang, C. F. (2008). Development of ovarian tissue and female germ cells in the protandrous black porgy, Acanthopagrus schlegeli (Perciformes, Sparidae). *Zoological Studies*, *47*, 302–316.

Lema, S. C., Sanders, K. E., & Walti, K. A. (2015). Arginine vasotocin, isotocin and nonapeptide receptor gene expression link to social status and aggression in sex-dependent patterns. *Journal of Neuroendocrinology*, *27*, 142–157.

Levavi-Sivan, B., Bogerd, J., Mananos, E. L., Gomez, A., & Lareyre, J. J. (2010). Perspectives on fish gonadotropins and their receptors. *General and Comparative Endocrinology*, *165*, 412–437.

Li, G.-L., Liu, X.-C., Zhang, Y., & Lin, H.-R. (2006). Gonadal development, aromatase activity and P450 aromatase gene expression during sex inversion of protogynous red-spotted grouper Epinephelus akaara (Temminck and Schlegel) after implantation of the aromatase inhibitor, fadrozole. *Aquaculture Research*, *37*, 484–491.

Li, M.-H., Yang, H.-H., Li, M.-R., Sun, Y.-L., Jiang, X.-L., Xie, Q.-P., et al. (2013). Antagonistic roles of Dmrt1 and Foxl2 in sex differentiation via estrogen production in tilapia as demonstrated by TALENs. *Endocrinology*, *154*, 4814–4825.

Liarte, S., Chaves-Pozo, E., García-Alcazar, A., Mulero, V., Meseguer, J., & García-Ayala, A. (2007). Testicular involution prior to sex change in gilthead seabream is characterized by a decrease in DMRT1 gene expression and by massive leukocyte infiltration. *Reproductive Biology and Endocrinology*, *5*, 20.

Lindeman, R. E., Gearhart, M. D., Minkina, A., Krentz, A. D., Bardwell, V. J., & Zarkower, D. (2015). Sexual cell-fate reprogramming in the ovary by DMRT1. *Current Biology*, *25*, 764–771.

Liu, H. (2016). *Genomic basis of sex change in fish*. University of Otago. (Thesis, Doctor of Philosophy). Retrieved from http://hdl.handle.net/10523/6549.

Liu, J.-F., Guiguen, Y., & Liu, S.-J. (2008). Aromatase (P450arom) and 11β-hydroxylase (P45011β) genes are differentially expressed during the sex change process of the protogynous rice field eel, monopterus albus. *Fish Physiology and Biochemistry*, *35*, 511–518.

Liu, H., Lamm, M. S., Rutherford, K., Black, M. A., Godwin, J. R., & Gemmell, N. J. (2015). Large-scale transcriptome sequencing reveals novel expression patterns for key sex-related genes in a sex-changing fish. *Biology of Sex Differences*, *6*, 26.

Liu, H., Todd, E. V., Lokman, P. M., Lamm, M. S., Godwin, J. R., & Gemmell, N. J. (2017). Sexual plasticity: A fishy tale. *Molecular Reproduction and Development*, *84*, 171–194.

Liu, Z., Wu, F., Jiao, B., Zhang, X., Hu, C., Huang, B., et al. (2007). Molecular cloning of doublesex and mab-3-related transcription factor 1, forkhead transcription factor gene 2, and two types of cytochrome P450 aromatase in southern catfish and their possible roles in sex differentiation. *The Journal of Endocrinology*, *194*, 223–241.

Lo Nostro, F., Grier, H., Andreone, L., & Guerrero, G. A. (2003). Involvement of the gonadal germinal epithelium during sex reversal and seasonal testicular cycling in the protogynous swamp eel, Synbranchus marmoratus Bloch 1795 (Teleostei, Synbranchidae). *Journal of Morphology*, *257*, 107–126.

Lorenzi, V., Earley, R. L., & Grober, M. S. (2012). Differential responses of brain, gonad and muscle steroid levels to changes in social status and sex in a sequential and bidirectional hermaphroditic fish. *PLoS One*, 7, e51158.

Manabe, H., Toyoda, K., Nagamoto, K., Dewa, S.-I., Sakurai, M., Hagiwara, K., et al. (2013). Bidirectional sex change in seven species of Priolepis (Actinopterygii: Gobiidae). *Bulletin of Marine Science*, *89*, 635–642.

Maruska, K. P., & Fernald, R. D. (2011). Social regulation of gene expression in the hypothalamic-pituitary-gonadal axis. *Physiology*, *26*, 412–423.

Maruska, K. P., & Fernald, R. D. (2013). Social regulation of male reproductive plasticity in an African cichlid fish. *Integrative and Comparative Biology*, *53*, 938–950.

Massagué, J., Seoane, J., & Wotton, D. (2005). Smad transcription factors. *Genes & Development*, *19*, 2783–2810.

Matson, C. K., Murphy, M. W., Sarver, A. L., Griswold, M. D., Bardwell, V. J., & Zarkower, D. (2011). DMRT1 prevents female reprogramming in the postnatal mammalian testis. *Nature*, *476*, 101–104.

Matson, C. K., & Zarkower, D. (2012). Sex and the singular DM domain: Insights into sexual regulation, evolution and plasticity. *Nature Reviews. Genetics*, *13*, 163–174.

Mechaly, A. S., Vinas, J., & Piferrer, F. (2013). The kisspeptin system genes in teleost fish, their structure and regulation, with particular attention to the situation in Pleuronectiformes. *General and Comparative Endocrinology*, *188*, 258–268.

Miyake, Y., Sakai, Y., & Kuniyoshi, H. (2012). Molecular cloning and expression profile of sex-specific genes, Figla and Dmrt1, in the protogynous hermaphroditic fish, Halichoeres poecilopterus. *Zoological Science*, *29*, 690–701.

Moczek, A. P. (2015). Developmental plasticity and evolution–quo vadis? *Heredity*, *115*, 302–305.

Moussavi, M., Wlasichuk, M., Chang, J. P., & Habibi, H. R. (2013). Seasonal effect of gonadotrophin inhibitory hormone on gonadotrophin-releasing hormone-induced gonadotroph functions in the goldfish pituitary. *Journal of Neuroendocrinology*, *25*, 506–516.

Muncaster, S., Norberg, B., & Andersson, E. (2013). Natural sex change in the temperate protogynous Ballan wrasse Labrus bergylta. *Journal of Fish Biology*, *82*, 1858–1870.

Munday, P. L., Buston, P. M., & Warner, R. R. (2006). Diversity and flexibility of sex-change strategies in animals. *Trends in Ecology & Evolution*, *21*, 89–95.
Munday, P. L., Caley, M. J., & Jones, G. P. (1998). Bi-directional sex change in a coral-dwelling goby. *Behavioral Ecology and Sociobiology*, *43*, 371–377.
Munday, P. L., Kuwamura, T., & Kroon, F. J. (2010). Bidirectional sex change in marine fishes. In *Reproduction and sexuality in marine fishes*: University of California Press.
Munday, P. L., Wilson White, J., & Warner, R. R. (2006). A social basis for the development of primary males in a sex-changing fish. *Proceedings of the Biological Sciences*, *273*, 2845–2851.
Munger, S. C., & Capel, B. (2012). Sex and the circuitry: Progress toward a systems-level understanding of vertebrate sex determination. *Wiley Interdisciplinary Reviews. Systems Biology and Medicine*, *4*, 401–412.
Nakamura, M., Hourigan, T. F., Yamauchi, K., Nagahama, Y., & Grau, E. G. (1989). Histological and ultrastructural evidence for the role of gonadal steroid hormones in sex change in the protogynous wrasse Thalassoma duperrey. *Environmental Biology of Fishes*, *24*, 117–136.
Nakamura, M., Kobayashi, Y., Miura, S., Alam, M. A., & Bhandari, R. K. (2005). Sex change in coral reef fish. *Fish Physiology and Biochemistry*, *31*, 117–122.
Nakamura, M., Miura, S., Nozu, R., & Kobayashi, Y. (2015). Opposite-directional sex change in functional female protandrous anemonefish, Amphiprion clarkii: Effect of aromatase inhibitor on the ovarian tissue. *Zoological Letters*, *1*, 30.
Nakashima, Y., Kuwamura, T., & Yogo, Y. (2010). Why be a both-ways sex changer? *Ethology*, *101*, 301–307.
Nanda, I., Kondo, M., Hornung, U., Asakawa, S., Winkler, C., Shimizu, A., et al. (2002). A duplicated copy of DMRT1 in the sex-determining region of the Y chromosome of the medaka, Oryzias latipes. *Proceedings of the National Academy of Sciences of the United States of America*, *99*, 11778–11783.
Navarro-Martín, L., Viñas, J., Ribas, L., Díaz, N., Gutiérrez, A., Di Croce, L., et al. (2011). DNA methylation of the gonadal aromatase (cyp19a) promoter is involved in temperature-dependent sex ratio shifts in the European sea bass. *PLoS Genetics*, *7*, e1002447.
Nozu, R., Horiguchi, R., Kobayashi, Y., & Nakamura, M. (2015). Expression profile of doublesex/male abnormal-3-related transcription factor-1 during gonadal sex change in the protogynous wrasse, Halichoeres trimaculatus. *Molecular Reproduction and Development*, *82*, 859–866.
Nozu, R., Horiguchi, R., Murata, R., Kobayashi, Y., & Nakamura, M. (2013). Survival of ovarian somatic cells during sex change in the protogynous wrasse, Halichoeres trimaculatus. *Fish Physiology and Biochemistry*, *39*, 47–51.
Nozu, R., Kojima, Y., & Nakamura, M. (2009). Short term treatment with aromatase inhibitor induces sex change in the protogynous wrasse, Halichoeres trimaculatus. *General and Comparative Endocrinology*, *161*, 360–364.
Nozu, R., & Nakamura, M. (2015). Cortisol administration induces sex change from ovary to testis in the protogynous Wrasse, *Halichoeres trimaculatus*. *Sexual Development*, *9*, 118–124.
O'Bryan, M. K., Takada, S., Kennedy, C. L., Scott, G., Harada, S.-I., Ray, M. K., et al. (2008). Sox8 is a critical regulator of adult sertoli cell function and male fertility. *Developmental Biology*, *316*, 359–370.
Ohta, K., Mine, T., Yamaguchi, A., & Matsuyama, M. (2008). Sexually dimorphic expression of pituitary glycoprotein hormones in a sex-changing fish (Pseudolabrus sieboldi). *Journal of Experimental Zoology. Part A, Ecological Genetics and Physiology*, *309*, 534–541.

Ohta, K., Sakai, M., Sundaray, J. K., Kitano, T., Takeda, T., Yamaguchi, A., et al. (2012). Bidirectional sex change induced by sex steroid implantation in the hermaphrodite fish, Pseudolabrus sieboldi. *Journal of Experimental Zoology. Part A, Ecological Genetics and Physiology, 317*, 552–560.

Okutsu, T., Suzuki, K., Takeuchi, Y., Takeuchi, T., & Yoshizaki, G. (2006). Testicular germ cells can colonize sexually undifferentiated embryonic gonad and produce functional eggs in fish. *Proceedings of the National Academy of Sciences of the United States of America, 103*, 2725–2729.

Orban, L., Sreenivasan, R., & Olsson, P.-E. (2009). Long and winding roads: Testis differentiation in zebrafish. *Molecular and Cellular Endocrinology, 312*, 35–41.

Pauletto, M., Manousaki, T., Ferraresso, S., Babbucci, M., Tsakogiannis, A., Louro, B., et al. (2018). Genomic analysis of Sparus aurata reveals the evolutionary dynamics of sex-biased genes in a sequential hermaphrodite fish. *Communications Biology, 1*, 119.

Paul-Prasanth, B., Bhandari, R. K., Kobayashi, T., Horiguchi, R., Kobayashi, Y., Nakamoto, M., et al. (2013). Estrogen oversees the maintenance of the female genetic program in terminally differentiated gonochorists. *Scientific Reports, 3*, 2862.

Perreault, H. A., Semsar, K., & Godwin, J. (2003). Fluoxetine treatment decreases territorial aggression in a coral reef fish. *Physiology & Behavior, 79*, 719–724.

Pfennig, F., Standke, A., & Gutzeit, H. O. (2015). The role of Amh signaling in teleost fish—Multiple functions not restricted to the gonads. *General and Comparative Endocrinology, 223*, 87–107.

Piferrer, F., & Guiguen, Y. (2008). Fish gonadogenesis. Part II: Molecular biology and genomics of sex differentiation. *Reviews in Fisheries Science, 16*, 35–55.

Policansky, D. (1982). Sex change in plants and animals. *Annual Review of Ecology and Systematics, 13*, 471–495.

Poonlaphdecha, S., Pepey, E., Canonne, M., de Verdal, H., Baroiller, J.-F., & D'Cotta, H. (2013). Temperature induced-masculinisation in the Nile tilapia causes rapid up-regulation of both dmrt1 and amh expressions. *General and Comparative Endocrinology, 193*, 234–242.

Qi, X., Zhou, W., Lu, D., Wang, Q., Zhang, H., Li, S., et al. (2013). Sexual dimorphism of steroidogenesis regulated by GnIH in the goldfish, Carassius auratus. *Biology of Reproduction, 88*, 89.

Reinboth, R., & Brusle-Sicard, S. (1997). Histological and ultrastructural studies on the effects of hCG on sex inversion in the protogynous teleost Coris julis. *Journal of Fish Biology, 51*, 738–749.

Ribas, L., Liew, W. C., Díaz, N., Sreenivasan, R., Orbán, L., & Piferrer, F. (2017). Heat-induced masculinization in domesticated zebrafish is family-specific and yields a set of different gonadal transcriptomes. *Proceedings of the National Academy of Sciences of the United States of America, 114*, E941–E950.

Robertson, D. R. (1972). Social control of sex reversal in a coral-reef fish. *Science, 177*, 1007–1009.

Rodríguez-Marí, A., Cañestro, C., Bremiller, R. A., Nguyen-Johnson, A., Asakawa, K., Kawakami, K., et al. (2010). Sex reversal in zebrafish fancl mutants is caused by Tp53-mediated germ cell apoptosis. *PLoS Genetics, 6*, e1001034.

Sadovy de Mitcheson, Y., & Liu, M. (2008). Functional hermaphroditism in teleosts. *Fish and Fisheries, 9*, 1–43.

Salmon, N. A., Handyside, A. H., & Joyce, I. M. (2005). Expression of Sox8, Sf1, Gata4, Wt1, Dax1, and Fog2 in the mouse ovarian follicle: Implications for the regulation of Amh expression. *Molecular Reproduction and Development, 70*, 271–277.

Sato, T., Endo, T., Yamahira, K., Hamaguchi, S., & Sakaizumi, M. (2005). Induction of female-to-male sex reversal by high temperature treatment in Medaka, Oryzias latipes. *Zoological Science, 22*, 985–988.

Schroeder, A. L., Metzger, K. J., Miller, A., & Rhen, T. (2016). A novel candidate gene for temperature-dependent sex determination in the common snapping turtle. *Genetics*, *203*, 557–571.

Semsar, K., & Godwin, J. (2003). Social influences on the arginine vasotocin system are independent of gonads in a sex-changing fish. *The Journal of Neuroscience*, *23*, 4386–4393.

Semsar, K., & Godwin, J. (2004). Multiple mechanisms of phenotype development in the bluehead wrasse. *Hormones and Behavior*, *45*, 345–353.

Semsar, K., Kandel, F. L., & Godwin, J. (2001). Manipulations of the AVT system shift social status and related courtship and aggressive behavior in the bluehead wrasse. *Hormones and Behavior*, *40*, 21–31.

Semsar, K., Perreault, H. A. N., & Godwin, J. (2004). Fluoxetine-treated male wrasses exhibit low AVT expression. *Brain Research*, *1029*, 141–147.

Shao, C. W., Li, Q. Y., Chen, S. L., Zhang, P., Lian, J. M., Hu, Q. M., et al. (2014). Epigenetic modification and inheritance in sexual reversal of fish. *Genome Research*, *24*, 604–615.

Shapiro, D. Y. (1992). Plasticity of gonadal development and protandry in fishes. *The Journal of Experimental Zoology*, *261*, 194–203.

Shapiro, D. Y., Sadovy, Y., & McGehee, A. M. (1993). Periodicity of sex change and reproduction in the red hind, Epinephelus guttatus, a protogynous grouper. *Bulletin of Marine Science*, *53*, 1151–1162.

Shi, Y., Zhang, Y., Li, S., Liu, Q., Lu, D., Liu, M., et al. (2010). Molecular identification of the Kiss2/Kiss1ra system and its potential function during 17alpha-methyltestosterone-induced sex reversal in the orange-spotted grouper, Epinephelus coioides. *Biology of Reproduction*, *83*, 63–74.

Shibata, Y., Paul-Prasanth, B., Suzuki, A., Usami, T., Nakamoto, M., Matsuda, M., et al. (2010). Expression of gonadal soma derived factor (GSDF) is spatially and temporally correlated with early testicular differentiation in medaka. *Gene Expression Patterns*, *10*, 283–289.

Solomon-Lane, T. K., Crespi, E. J., & Grober, M. S. (2013). Stress and serial adult metamorphosis: Multiple roles for the stress axis in socially regulated sex change. *Frontiers in Neuroscience*, *7*, 210.

Stévant, I., & Nef, S. (2018). Single cell transcriptome sequencing: A new approach for the study of mammalian sex determination. *Molecular and Cellular Endocrinology*, *468*, 11–18.

Stévant, I., Neirijnck, Y., Borel, C., Escoffier, J., Smith, L. B., Antonarakis, S. E., et al. (2018). Deciphering cell lineage specification during male sex determination with single-cell RNA sequencing. *Cell Reports*, *22*, 1589–1599.

Sunobe, T., Nakamura, M., Kobayashi, Y., Kobayashi, T., & Nagahama, Y. (2005). Gonadal structure and P450scc and 3β-HSD immunoreactivity in the gobiid fish Trimma okinawae during bidirectional sex change. *Ichthyological Research*, *52*, 27–32.

Thomas, J. T., Liu, H., Todd, E. V., & Gemmell, N. J. (2018). Sex change in fish. In M. K. Skinner (Ed.), *Encyclopedia of reproduction* (2nd ed., pp. 192–197). Oxford: Academic Press.

Todd, E. V., Black, M. A., & Gemmell, N. J. (2016). The power and promise of RNA-seq in ecology and evolution. *Molecular Ecology*, *25*, 1224–1241.

Todd, E. V., Liu, H., Lamm, M. S., Thomas, J. T., Rutherford, K., Thompson, K. C., et al. (2018). Female mimicry by sneaker males has a transcriptomic signature in both the brain and the gonad in a sex-changing fish. *Molecular Biology and Evolution*, *35*, 225–241.

Todd, E. V., Liu, H., Muncaster, S., & Gemmell, N. J. (2016). Bending genders: The biology of natural sex change in fish. *Sexual Development*, *10*, 223–241.

Tsakogiannis, A., Manousaki, T., Lagnel, J., Sterioti, A., Pavlidis, M., Papandroulakis, N., et al. (2018). The transcriptomic signature of different sexes in two protogynous hermaphrodites: Insights into the molecular network underlying sex phenotype in fish. *Scientific Reports*, *8*, 3564.

Ubuka, T., Son, Y. L., Tobari, Y., Narihiro, M., Bentley, G. E., Kriegsfeld, L. J., et al. (2014). Central and direct regulation of testicular activity by gonadotropin-inhibitory hormone and its receptor. *Frontiers in Endocrinology*, *5*, 8.

Uhlenhaut, N. H., Jakob, S., Anlag, K., Eisenberger, T., Sekido, R., Kress, J., et al. (2009). Somatic sex reprogramming of adult ovaries to testes by FOXL2 ablation. *Cell*, *139*, 1130–1142.

Ungewitter, E. K., & Yao, H. H.-C. (2013). How to make a gonad: Cellular mechanisms governing formation of the testes and ovaries. *Sexual Development*, *7*, 7–20.

Vega-Frutis, R., Macías-Ordóñez, R., Guevara, R., & Fromhage, L. (2014). Sex change in plants and animals: A unified perspective. *Journal of Evolutionary Biology*, *27*, 667–675.

Waddington, C. H. (1957). *The strategy of the genes*. London: George Allen & Unwin, Ltd.

Wang, D. S., Zhou, L. Y., Kobayashi, T., Matsuda, M., Shibata, Y., Sakai, F., et al. (2010). Doublesex- and Mab-3-related transcription factor-1 repression of aromatase transcription, a possible mechanism favoring the male pathway in Tilapia. *Endocrinology*, *151*(3), 1331–1340.

Wang, Z., Gerstein, M., & Snyder, M. (2009). RNA-Seq: A revolutionary tool for transcriptomics. *Nature Reviews. Genetics*, *10*, 57–63.

Wang, D. S., Kobayashi, T., Zhou, L. Y., Paul-Prasanth, B., Ijiri, S., Sakai, F., et al. (2007). Foxl2 up-regulates aromatase gene transcription in a female-specific manner by binding to the promoter as well as interacting with Ad4 binding protein/steroidogenic factor 1. *Molecular Endocrinology*, *21*, 712–725.

Wang, Q., Qi, X., Guo, Y., Li, S., Zhang, Y., Liu, X., et al. (2015). Molecular identification of GnIH/GnIHR signal and its reproductive function in protogynous hermaphroditic orange-spotted grouper (Epinephelus coioides). *General and Comparative Endocrinology*, *216*, 9–23.

Wang, Y. Y., Sun, L. X., Zhu, J. J., Zhao, Y., Wang, H., Liu, H. J., et al. (2017). Epigenetic control of cyp19a1a expression is critical for high temperature induced Nile tilapia masculinization. *Journal of Thermal Biology*, *69*, 76–84.

Warner, R. R. (1975). The adaptive significance of sequential hermaphroditism in animals. *The American Naturalist*, *109*, 61–82.

Warner, R. R. (1984). Mating behavior and hermaphroditism in coral reef fishes. *American Scientist*, *72*, 128–136.

Warner, R. R., & Robertson, R. (1978). Sexual patterns in the labroid fishes of the western Caribbean, I the wrasses (Labridae). *Smithsonian Contributions to Zoology*, *254*, 1–27.

Warner, R. R., & Swearer, S. E. (1991). Social control of sex change in the Bluehead wrasse, Thalassoma bifasciatum (Pisces: Labridae). *The Biological Bulletin*, *181*, 199–204.

Weltzien, F.-A., Andersson, E., Andersen, Ø., Shalchian-Tabrizi, K., & Norberg, B. (2004). The brain-pituitary-gonad axis in male teleosts, with special emphasis on flatfish (Pleuronectiformes). *Comparative Biochemistry and Physiology. Part A, Molecular & Integrative Physiology*, *137*, 447–477.

West-Eberhard, M. J. (2003). *Developmental plasticity and evolution*. Oxford University Press.

Wilhelm, D., & Koopman, P. (2006). The makings of maleness: Towards an integrated view of male sexual development. *Nature Reviews. Genetics*, *7*, 620–631.

Windley, S. P., & Wilhelm, D. (2015). Signaling pathways involved in mammalian sex determination and gonad development. *Sexual Development*, *9*, 297–315.

Wu, G. C., & Chang, C. F. (2009). wnt4 is associated with the development of ovarian tissue in the protandrous black porgy, Acanthopagrus schlegeli. *Biology of Reproduction*, *81*, 1073–1082.

Wu, G.-C., & Chang, C.-F. (2013). The switch of secondary sex determination in protandrous black porgy, Acanthopagrus schlegeli. *Fish Physiology and Biochemistry*, *39*, 33–38.

Wu, G. C., Chiu, P. C., Lin, C. J., Lyu, Y. S., Lan, D. S., & Chang, C. F. (2012). Testicular dmrt1 is involved in the sexual fate of the ovotestis in the protandrous black porgy. *Biology of Reproduction*, *86*, 41.

Wu, G. C., Li, H. W., Luo, J. W., Chen, C., & Chang, C. F. (2015). The potential role of AMH to prevent ectopic female development in testicular tissue of the protandrous black porgy, *Acanthopagrus schlegelii*. *Biology of Reproduction*, *92*, 158.

Wu, G.-C., Tomy, S., Lee, M.-F., Lee, Y.-H., Yueh, W.-S., Lin, C.-J., et al. (2010). Sex differentiation and sex change in the protandrous black porgy, Acanthopagrus schlegeli. *General and Comparative Endocrinology*, *167*, 417–421.

Wu, G. C., Tomy, S., Nakamura, M., & Chang, C. F. (2008). Dual roles of cyp19a1a in gonadal sex differentiation and development in the protandrous black porgy, Acanthopagrus schlegeli. *Biology of Reproduction*, *79*, 1111–1120.

Xia, W., Zhou, L., Yao, B., Li, C.-J., & Gui, J.-F. (2007). Differential and spermatogenic cell-specific expression of DMRT1 during sex reversal in protogynous hermaphroditic groupers. *Molecular and Cellular Endocrinology*, *263*, 156–172.

Yamaguchi, T., Yoshinaga, N., Yazawa, T., Gen, K., & Kitano, T. (2010). Cortisol is involved in temperature-dependent sex determination in the Japanese flounder. *Endocrinology*, *151*, 3900–3908.

Yatsu, R., Miyagawa, S., Kohno, S., Parrott, B. B., Yamaguchi, K., Ogino, Y., et al. (2016). RNA-seq analysis of the gonadal transcriptome during Alligator mississippiensis temperature-dependent sex determination and differentiation. *BMC Genomics*, *17*, 77.

Yeh, S.-L., Kuo, C.-M., Ting, Y.-Y., & Chang, C.-F. (2003). Androgens stimulate sex change in protogynous grouper, Epinephelus coioides: Spawning performance in sex-changed males. *Comparative Biochemistry and Physiology, Part C: Toxicology & Pharmacology*, *135C*, 375–382.

Yoshizaki, G., Ichikawa, M., Hayashi, M., Iwasaki, Y., Miwa, M., Shikina, S., et al. (2010). Sexual plasticity of ovarian germ cells in rainbow trout. *Development*, *137*, 1227–1230.

Zhang, X., & Ho, S.-M. (2011). Epigenetics meets endocrinology. *Journal of Molecular Endocrinology*, *46*, R11–R32.

Zhang, X., Li, M., Ma, H., Liu, X., Shi, H., Li, M., et al. (2017). Mutation of foxl2 or cyp19a1a results in female to male sex reversal in XX Nile Tilapia. *Endocrinology*, *158*, 2634–2647.

Zhang, Y., Zhang, S., Liu, Z., Zhang, L., & Zhang, W. (2013). Epigenetic modifications during sex change repress gonadotropin stimulation of Cyp19a1a in a teleost ricefield eel (Monopterus albus). *Endocrinology*, *154*, 2881–2890.

Zohar, Y., Muñoz-Cueto, J. A., Elizur, A., & Kah, O. (2010). Neuroendocrinology of reproduction in teleost fish. *General and Comparative Endocrinology*, *165*, 438–455.

CHAPTER FOUR

Genetic regulation of sex determination and maintenance in zebrafish (*Danio rerio*)

Michelle E. Kossack, Bruce W. Draper*
Molecular and Cellular Biology, University of California, Davis, CA, United States
*Corresponding author: e-mail address: bwdraper@ucdavis.edu

Contents

Current Topics in Developmental Biology, Volume 134
ISSN 0070-2153
https://doi.org/10.1016/bs.ctdb.2019.02.004

Abstract

Over the last several decades zebrafish (*Danio rerio*) has become a major model organism for the study of vertebrate development and physiology. Given this, it may be surprising how little is known about the mechanism that zebrafish use to determine sex. While zebrafish are a gonochoristic species (having two sexes) that do not switch sex as adults, it was appreciated early on that sex ratios obtained from breeding lab domesticated lines were not typically a 1:1 ratio of male and female, suggesting that sex was not determined by a strict chromosomal mechanism. Here we will review the recent progress toward defining the genetic mechanism for sex determination in both wild and domesticated zebrafish.

1. Introduction to zebrafish sexual phenotypes

Adult zebrafish males and females can be distinguished based on several sexually dimorphic phenotypes. The three most useful external dimorphisms for identifying the sex of an adult zebrafish are described here. First, because ovaries are significantly larger than testes, females, in general, have a larger abdomen (Fig. 1A and B). A second reliable indicator is the

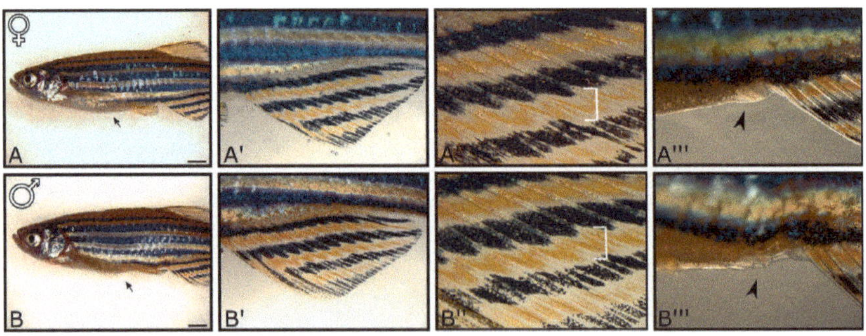

Fig. 1 Sexually dimorphic phenotypes that distinguish adult female and male zebrafish. 6-month-old female (A–A‴) and male (B–B‴) zebrafish. Low magnification view shows that females (A) have a larger abdomen (arrow) then males (B). Low magnification view of the anal fins (A′ and B′) shows that females have lighter yellow pigmentation then males. High magnification view of the anal fins (A" and B") shows that the yellow pigment stripes in females are narrower than those in males (brackets). The genital papilla protrudes from the ventral body surface in females (arrow in A‴), but not in males (arrow in B‴). Scale bar in A and B, 2 mM.

dimorphic color of the anal fin and abdomen. One of the main pigment cell types in zebrafish are the xanthaphores, which form a yellow strip of pigment that lies between the characteristic black pigment stripes formed by melanophores which give zebrafish their name. While xanthaphores are present in both males and females, males produce significantly more yellow pigment than females, and therefore appear more yellow when viewed under the appropriate light. Differences in color are most apparent on the anal fin (Fig. 1A′, B′ and A″, B″). A third and perhaps most reliable distinguishable feature between males and females is the appearance of the genital pore, which is located on the ventral midline of the abdomen just anterior to the anal fin and posterior to the anus (Menke, Spitsbergen, Wolterbeek, & Woutersen, 2011; Fig. 1A‴, B‴). As the name implies, the genital pore is the opening through which the gametes are released. When viewed with the aid of a dissecting microscope, the female genital pore protrudes from the body surface and has characteristic longitudinal ridges (Fig. 1A‴). By contrast, the genital pore of males does not protrude from the body surface and therefore can only be seen when viewed from the ventral surface (Fig. 1B ‴). Finally, there are subtler dimorphic structures, such as the presence of mating tubercles on the dorsal surface of the male pectoral fins, but these are more difficult to observe and therefore less commonly used to distinguish between sexes (Kang, Nachtrab, & Poss, 2013).

2. Wild zebrafish, but not domesticated lines, utilize a ZZ/ZW chromosomal sex determination mechanism

The major domesticated zebrafish lines that are widely used in the laboratory were all derived from fish that were obtained from pet stores. For example, the fish used by George Streisinger to produce the AB line were obtained from a pet store in Albany, Oregon, while those used to derive the TU (Tübingen) line originated from a pet store in Germany (Mullins, Hammerschmidt, Haffter, & Nüsslein-Volhard, 1994; Streisinger, Walker, Dower, Knauber, & Singer, 1981; Wilson et al., 2014). Importantly, because these lines were developed to be used for forward genetic screens, they were initially selected to be free of recessive lethal mutations (Mullins et al., 1994; Streisinger et al., 1981). It is possible that this selection inadvertently affected loci that regulated sex determination.

Early on it was recognized that the ratio of males to females in the domesticated lines could vary widely, suggesting an atypical sex determining mechanism (Streisinger et al., 1981). However, repeated mating of specific pairs of fish from the domesticated AB, TU and Toh lines produced progeny with reproducible sex ratios in a pair-specific manner, providing evidence that genetics plays an influential role in sex determination (Liew et al., 2012).

To further investigate the genetic component of sex determination in zebrafish, three independent groups used genome-wide association (GWA) methods to identify loci that contributed to sex determination in domesticated lines. While these studies confirmed that no major chromosomal sex-determining locus existed, they each found evidence for a more complex polygenic sex determining system where several unlinked loci each make measurable contributions to sex determination (Anderson et al., 2012; Bradley et al., 2011; Howe et al., 2013; Liew et al., 2012). Importantly, no two groups used the same genetic background for their analysis and as a result, each group identified different contributing loci. These studies therefore suggest that a polygenic system is regulating sex in domesticated lines, but that the particular genes involved vary between the independently selected lines.

While GWA studies of domesticated lines clearly pointed to a strain-specific polygenic system of sex determination, they conflicted with cytogenetic analysis by Sharma, Sharma, and Tripathi (1998) that found evidence that zebrafish collected from the wild possessed heteromorphic chromosomes consistent with a ZZ/ZW chromosomal sex determining system, where the heterogametic ZW animals were females and the homogametic ZZ animals were male. To further investigate this difference Wilson et al. (2014) performed a comparative GWA study using both domesticated lines as well as lines derived from fish collected more recently from the wild. This study included three types of lines: (1) the domesticated lines AB and TU that had been subjected to genetic selection, (2) The EKW (EkkWill) and WIK (Wild India Kolkata), lines that were derived from natural populations but cultured in the lab for many generations without genetic selection, and (3) two strains, NA (Nadia) and CB (Cooch Behar), that were derived from fish more recently collected from wild populations. In accordance to the previous work, this study did not identify any major sex-linked loci in the domesticated lines AB and TU. By contrast, all lines derived from wild-caught fish, regardless of how many generations in the lab, possessed the same significantly sex-liked locus that mapped to the end of the long arm of chromosome 4 (Ch 4q). While the molecular identity of this locus has yet to be determined, these data indicate that wild zebrafish utilize a ZW/ZZ sex determining system, in which female development was associated with the presence of the W chromosome. However, presence of the W chromosome did not guarantee female development as some ZW individuals develop as fertile males. This result suggested that female development requires either the presence of a female promoting factor that localizes to the W chromosome, or that female development can only occur if a single copy of a *Z*-linked male promoting factor(s) is present (Wilson et al., 2014).

The diversity of loci discovered to contribute to zebrafish sex determination suggests that the genetic mechanism is labile and potentially evolving either toward or away from a more rigid chromosomally-based system in the wild. Although the original analysis by Sharma and colleagues provided evidence for cytogenetically distinct Z and W chromosomes, similar to the mammalian X and Y analysis (Sharma et al., 1998), the molecular analysis of Wilson et al. (2014) did not find significant sequence differences between the Z and W chromosomes, suggesting that any molecular differences between these chromosomes is minor. However, the NA-, WIK- and CB-derived strains originated from fish collected in regions that were 2000 km from those used by Sharman et al. (1998) (Mansar Lake, Jammu, India). It is therefore possible that differences between the results of these studies are due to region-specific differences in the karyotypes of zebrafish sex chromosomes (Sharma et al., 1998; Wilson et al., 2014).

3. Overview of zebrafish gonad development and timing of sex determination and differentiation

The primary function of any sex determining mechanism is to specify the sexual characteristic of the somatic gonad cells, which are the major producers of the circulating sex hormones 11-ketotestosterone, the primary male androgen in teleost, and estradiol in females (see, Devlin & Nagahama, 2002 for review). These hormones then instruct the rest of the somatic cells in the body to adopt the sex appropriate appearance and also promote sex appropriate behaviors. Understanding sex determination in zebrafish can therefore be reduced to understanding how cells of the somatic gonad determine their sexual phenotype. While sexual differentiation is likely initiated between 20 and 25 days post-fertilization (dpf) in zebrafish (see below), sexual maturation, which is defined as the ability to mate, is not complete until zebrafish are 2.5–3 months old. Once determined, zebrafish maintain their sex throughout life. Thus, gonad development can be separated into three steps: sex determination, differentiation and maintenance.

Initial studies looking at the timing of sex differentiation in the gonad showed that all zebrafish gonads go through a transient female-like phase during early larval life (approx. 13–25 dpf), where all larvae produce some early-stage oocytes (Takahashi, 1977). It is therefore likely that sex determination occurs during this period. For animals that were determined as females, the early-stage oocytes they produced continue to mature. By contrast, for animals that were determined as males, their early-stage oocytes undergo apoptosis as the gonads transition from a proto-ovary into a functional testis. Because gonads during this transition period can contain both female- and male-type germ cells, zebrafish have been referred to as juvenile hermaphrodites (Takahashi, 1977). It should be noted that the times listed below are approximate and can vary between studies due to differences in rearing conditions.

3.1 Embryogenesis to 8 dpf

Primordial germ cells (PGC) in zebrafish are the first cell type to be specified in the developing embryo (Yoon, Kawakami, & Hopkins, 1997). PGCs can first be identified based on the localization of several maternally-supplied transcripts that encode germ cell-specific factors, such as Vasa and Nanos3 (Köprunner, Thisse, Thisse, & Raz, 2001; Yoon et al., 1997). During gastrulation the PGCs begin to migrate to the site of the future gonad, reaching this target before the end of the first day of development. One-day old embryos have an average of 30 PGCs that are distributed into two bilateral domains flanking the midline and adjacent to somite 5 (Weidinger et al., 2002; Yoon et al., 1997). While it is not known exactly when somatic gonad precursor cells first arise, somatic cells in close association with germ cells were identified in histological sections as early as 5 dpf (Braat et al., 1999). Between 1 and 8 dpf, germ cell numbers did not appear to increase suggesting that PGCs are quiescent during early larval development (Leerberg, Sano, & Draper, 2017; Tzung et al., 2015). Expression of genes involved in sex determination or differentiation of the gonads prior to 8 dpf have not been reported.

3.2 The larval bipotential/undifferentiated gonad (8–20 dpf)

The quiescent phase of early zebrafish gonad development appears to end around 8–10 dpf when, based on cell counts, EdU incorporation and detection of phospho-histone 3, somatic gonad cells, followed by germ cells, begin to proliferate (Leerberg et al., 2017; Tzung et al., 2015). As in most vertebrates, there is evidence that the zebrafish larval gonad is initially bipotential. The expression of many genes known or predicted to be involved in male and female sex determination or differentiation in other vertebrates were detected in zebrafish somatic gonad cells starting around 8 dpf (Leerberg et al., 2017; Rodríguez-Marí et al., 2005). For example, expression of *cyp19a1a*, which encodes aromatase that converts androgens to estrogen in females was detected in all gonads, even though approximately half of these will eventually develop as testes (Leerberg et al., 2017; Rodríguez-Marí et al., 2005). Similarly, cells expressing the gene encoding the TGF-ß ligand anti-Müllerian hormone, *amh*, were also detected in all gonads during this time period (Leerberg et al., 2017; Rodríguez-Marí et al., 2005). Importantly, at this stage the cells with male characteristic gene expression were distinct from those with female characteristic gene expression, indicating that early zebrafish gonads are composed of an apparent mosaic of male- and female-type cells (Leerberg et al., 2017).

During the bipotential period, early-stage oocytes (Stage 1a, also called perinuclear-stage oocytes; Selman, Wallace, Sarka, & Qi, 1993) were detected in all gonads starting around 13–14 days post-fertilization, indicating that germ cells initially follow a female-type mode of development regardless of the eventual sex of the animal (Takahashi, 1977; Tzung et al., 2015; Uchida, Yamashita, Kitano, & Iguchi, 2002). Between 13 to ~20 dpf the gonad increases in size due to continued proliferation of both germ cells and somatic gonad cells, though not all gonads appeared to increase in size at the same rate (Leerberg et al., 2017; Tzung et al., 2015).

3.3 Sex-specific differentiation (20–25 dpf)

The first signs of overt sexual differentiation begin around 20–25 dpf when the oocytes that are present in presumptive males begin to undergo apoptosis (Uchida et al., 2002). This period is often referred to as "transitioning" because these gonads are transitioning from an initial ovary-like development to definitive testis development. By contrast, oocytes in animals that will become females continue to mature (Uchida et al., 2002). The transitioning phase of testis development appeared to be over by ~30 dpf, when the majority of oocytes had been cleared from gonads that will differentiate as testes (Uchida et al., 2002). While sex of a fish cannot be reliably assigned based on the external criteria until ~2.5–3 months of age, histological analysis of 30 dpf gonads can reveal the eventual fate of the gonad, and therefore sex of the fish, as ovaries are significantly larger than testes at this time (Takahashi, 1977). Because of this difference in size, it is therefore possible to select for fish of a particular sex at 30 dpf based on the expression levels of various germ cell-expressed transgeneic reporter lines, such as Tg(*vas:egfp*) and Tg(*ziwi:egfp*) (Dranow et al., 2016; Krøvel & Olsen, 2002; Wang, Bartfai, Sleptsova-Freidrich, & Orban, 2007).

4. Oocytes are required for primary female sex determination

While the genetic determinants of primary sex determination have yet to be discovered in domesticated or wild zebrafish strains, it has become apparent over the last decade that germ cells, and in particular, oocytes, are a key regulator of gonadal sex determination in domesticated zebrafish. The first key discovery was that ablation of all germ cells during the first day of development resulted in all-male development (Slanchev, Stebler, de la Cueva-Méndez, & Raz, 2005). It was later shown that while these fish were sterile, the somatic cells of their testes expressed the appropriate male-specific genes, suggesting that formation of ovarian somatic cell types, but not testicular cell types, was dependent on germ cells (Siegfried & Nüsslein-Volhard, 2008). Consistent with these findings, mutations that caused germ cell loss during the first several days of development, such as *ziwi/piwil1* and *nanos3*, resulted in animals that developed as phenotypic, sterile males (e.g., Draper, McCallum, & Moens, 2007; Houwing et al., 2007). These studies argued that germ cells were required for female development, but the mechanism by which they promoted female sexual development was not identified.

A critical clue to how germ cells promote female development came from studies of *fancl* mutants (Rodríguez-Marí et al., 2010). *fancl* encodes a component of the Fanconi Anemia/BRACA DNA repair pathway and is expressed in early meiotic oocytes, including those that form during the bipotential stage of gonad development. It is likely that Fancl functions to repair Spo11-induced double-strand DNA breaks that are required for the initiation of meiotic recombination. *fancl* mutants had normal numbers of germ cells at the early larval stage and, like wild-type, all *fancl* mutants initially produce early-stage oocytes during the bipotential stage. However, unlike wild-type, all *fancl* mutant oocytes underwent apoptosis during the late larval stage, presumably due to their inability to repair Spo11-induced double strand DNA breaks. By ~30 dpf, all mutant gonads resembled wild-type testes and as adults, all mutants were male. The oocyte loss and sex reversal phenotypes could be rescued by producing *fancl;tp53* double mutants, which were unable to activate the Tp53-dependent apoptosis pathway, indicating that oocyte apoptosis was the primary reason why all gonads differentiated as testes. Thus, while previous studies showed the importance of germ cells for female development, this study identified that meiotic oocytes are the key germ cell-stage that promotes ovarian, and thus female, development in domesticated zebrafish. Since this landmark study, other mutants have been described that disrupt oocyte meiosis and/or cause early germ cell loss, and like *fancl*, result in an all-male phenotype (e.g., *zilli*, *tdrd1*, *vasa*, *mps1*; Hartung, Forbes, & Marlow, 2014; Houwing, Berezikov, & Ketting, 2008; Huang et al., 2011; Poss, Nechiporuk, Stringer, Lee, & Keating, 2004).

From these studies the general hypothesis emerged that oocytes produce a signal that acts on somatic gonad cells to stabilize the expression of ovary development-promoting genes. Further, it was hypothesized that a threshold amount of this signal was required to stabilize ovarian development and absent this, the somatic gonad upregulates testis-promoting genes. If each oocyte can produce a set amount of signal, then the total amount of signal will be directly correlated to the number of oocytes produced. In turn, the number of oocytes produced by each larva is likely to be proportional to the number of germ cells each animal has at the beginning of the bipotential stage of gonad development. As such, animals with more germ cells are more likely to develop as female. If true, the number of germ cells each animal has at the start of the bipotential stage should not be uniform within a population.

To begin to test this, Tzung and colleagues quantified total germ cell numbers in animals at 7 and 14 dpf, a time before and at the start of the bipotential stage, respectively (Tzung et al., 2015). At 7 dpf, they found that animals had a normal distribution of germ cells, with the average being 26–30 PGCs per animal. By contrast, at 14 dpf, they found that germ cell numbers followed a bimodal distribution, where nearly half of the animals had an average of ~40 PGCs while the other half had an average of ~90 PGCs. As a more direct test of the hypothesis, they experimentally manipulated the numbers of PGCs by injecting dilute amounts of antisense morpholino oligos (MO) targeting the germ cell-localized *dead-end* (dnd) mRNA into fertilized eggs, as reduction of *dnd* function leads to loss of

PGCs (Weidinger et al., 2003). Consistent with the hypothesis, *dnd* MO injected animals, but not control animals, had an increase in the number of animals that developed as male. Taken together these results strongly argue that the numbers of germ cells an animal has during the bipotential stage has a strong influence whether an ovary or a testis forms, and thus the eventual sexual fate of the animal. Given the strong evidence that a polygenetic sex determining mechanism influences sex ratios in domesticated zebrafish, it is tempting to speculate that the genes involved function to directly or indirectly regulate PGC proliferation during the early larval stage. It will be important to repeat these experiments in zebrafish strains that have maintained the ancestral ZZ/ZW chromosomal sex determining system to determine if oocyte signaling is also required for female development in these lines.

5. Maintenance of the sexual phenotype

There is growing evidence that sexual determination and differentiation in vertebrates are not irreversible once established, but instead require constant expression of sex-maintenance genes, such as the male- and female-specific transcription factors Dmrt1 and FoxL2, respectively. While mammals do not normally switch sex as adults, it is estimated that 2% of teleost species switch sex at some point during their life and in some species can switch sex multiple times (Avise & Mank, 2009). However, zebrafish do not belong to the 2% of species that switch sex as part of their normal life cycle. It was therefore a surprise that when oocytes were depleted in an adult zebrafish female, either by genetic mutation or chemical ablation, that the fish readily sex reversed to males that, in some cases, were fertile (Dranow, Tucker, & Draper, 2013; Dranow et al., 2016). This was first discovered through analysis of *nanos3* mutants, which showed loss of oocytes in adults. Nanos3 is required for germline stem cell maintenance in zebrafish ovaries but is not required for the initial wave of oocytes that are produced during the bipotential stage. This led to a situation in which germline stem cells were normally specified in mutant ovaries such that *nanos3* mutants were able to become females (Beer & Draper, 2013; Draper et al., 2007). However, the germline stem cell population was not maintained and all premeiotic germ cells entered meiosis by 40 dpf (Beer & Draper, 2013). Once all eggs were depleted by either spawning or atresia, which occurred between 4 and 5 months post fertilization (mpf), all mutant females sex-reversed to males, as assayed by their appearance and their ability to induce wild-type females to spawn. Not surprisingly, these sex-reversed males were sterile given that all germ cells were lost prior to sex reversal. By contrast, when oocytes were ablated but premeiotic germ cells, including germline stem cells, were left unaffected, the sex reversed males were fertile, likely a result of the germline stem cells switching from producing oocytes to producing functional sperm (Dranow et al., 2016, 2013). These results clearly showed that oocytes play an essential role in both primary sex determinations, as detailed above, and throughout life to maintain the differentiated state of the ovarian somatic cells, which in turn assures continued production of the female sex hormone estradiol.

6. Zebrafish sex determination can be influenced by environmental factors

While it is now clear that zebrafish in the wild utilize a genetic sex determination mechanism, it is also clear that sex ratios in domesticate lines can be greatly influenced by certain environmental factors, such as temperature, pH, oxygen concentration and rearing density. However, because these environmental factors cannot fully explain normal sex ratios, sex of domesticated zebrafish is likely not regulated by a strict environmental sex determining system. Instead, given that all of these factors likely induce stress (e.g., high temperature or high rearing density), and that these factors always increase the proportions of males in the population, it is possible that all of these factors alter sex ratios by indirectly affecting the proliferation of germ cells during the bipotential phase. Regardless, it is useful to review the environmental parameters that are known to influence sex ratios in zebrafish.

6.1 Temperature

Increased temperature during larval development significantly increases the proportion of males. The standard laboratory rearing temperature for zebrafish is 28.5 °C (Westerfield, 2007). Uchida et al. (2004) demonstrated that when fish were reared during the critical period of gonad determination (15–20 dpf) at 35 °C, 68.8% of the expected females developed as male. When the rearing temperature was increased to 37 °C, 100% of the expected females developed as male. Indicating that higher temperature can cause female-to-male sex reversal. Although increased temperature caused sex reversal, it also significantly impacted survival. At 37 °C only 54.7% of the individuals survived and no individuals survived at 39 °C. Thus, although temperature may influence gonad development, decreases in survivability at these temperatures suggest that this may be primarily a response to stress.

6.2 Dissolved oxygen content

Similarly, reduced dissolved oxygen content in the water increases the proportion of males. Zebrafish are able to develop normally within a wide range of oxygen concentrations (Padilla & Roth, 2001). Normal water oxygen content at 28.5 °C is 5.8 mg/L and rearing individuals at this temperature will cause about 61.9% of fish to develop as male (Shang, Yu, & Wu, 2006). If zebrafish were reared in hypoxic conditions, 0.8 mg/L, 74.4% of individuals would develop as male (Shang et al., 2006).

6.3 Food availability

It has often been anecdotally observed that rearing density affects sex ratios in laboratory zebrafish, where high rearing density leads to increased proportions of males. It is likely that food availability is what drives this phenomenon (Lawrence, 2007). Lawrence, Ebersole, and Kesseli (2008) observed that when food availability doubles the proportion of individuals that developed as female significantly increased from 51% to 76% females. Interestingly, a significant interaction between food and the proportion of females was observed in both selected domesticated strain (TU) and Non-selected domesticated strains (WIK) that have a ZZ/ZW karyotype. While not addressed in this study, it is likely in the WIK strain that the increase in females was due to an increased proportion of ZW individuals developing as female. It will be interesting to determine how food availability influences germ cell numbers during the bipotential gonad stage.

7. Genetic analysis of zebrafish sex determining pathway

Studies to identify genes involved in zebrafish sex determination and differentiation have, until recently, been hampered by lack of methods to obtain genetic mutations. Few mutants affecting sex determination or differentiation were identified in any of the major forward genetic screens because these screens were largely focused on identifying recessive lethal mutations that had phenotypes that manifest during the first 5 dpf

(Driever et al., 1996; Haffter & Nusslein-Volhard, 1996). In addition, screens for mutations that affect sex determination would be complicated by the highly variable sex ratios obtained in crosses of domesticated zebrafish. However, with the recent advances in genome editing reverse genetic technologies, such as TALENs and CRISPR/Cas9, it is now possible to directly assay the function of candidate genes using loss-of-function mutants. Below we review the current progress that has been made in this area. It should be noted that in all cases reviewed below, the mutants have been produced using one of the domesticated zebrafish strains, not a strain that contains the wild sex chromosomes. In addition, only genes for which there are data available from genetic mutations are present (i.e., genes that have support only from Morpholino-based knockdown experiments are not presented). We will first review the genes that function to promote male sex determination and/or differentiation, and then those that promote female.

8. Male-promoting genes

8.1 Double-sex and Mab-3 related transcription factor 1 (Dmrt1)

The first genes involved in sex determination and differentiation that were found to be widely conserved across animals were orthologs of the *Drosophila double-sex* and *C. elegans mab-3* transcription factors (Raymond et al., 1998). In vertebrates the orthologous genes are referred to as Dmrt's, for *double-sex mab-3* related transcription factors. In mammals *Dmrt1* is located on an autosome where it functions downstream of SRY, and as such is not involved in the initiation of sex determination, but instead in sex differentiation. By contrast, in several other species Dmrt orthologs have been identified as the major chromosomal sex determinant. For example, in the Japanese Medaka fish, which utilizes an XX/XY sex determination system, a paralog of *dmrt*, called *dmY*, is located on the Y chromosome where it functions to initiate male sex determination, analogous to *Sry* in mammals (Matsuda et al., 2002). Interestingly, while *Dmrt1* is not the male sex determinant in mammals, forced expression of *Dmrt1* in XX animals caused female-to-male sex reversal, indicating that it has the ability to drive male sex determination (Zhao, Svingen, Ng, & Koopman, 2015).

The zebrafish genome contains a single ortholog of *dmrt1*, and its role in zebrafish sexual development was directly tested using genetic point mutations generated by ENU mutagenesis as well as indel mutations generated using either TALEN or CRISPR/Cas9 genome editing (Lin et al., 2017; Webster et al., 2017). It was found that the majority of *dmrt1* mutants developed as females, consistent with the role of Dmrt1 orthologs in promoting male development. Interestingly, a small percentage of *dmrt1* mutants developed as phenotypic, though sterile, males, indicating that Dmrt1 is not absolutely required for the development of male secondary sexual characteristics. It is possible that even in the absence of Dmrt1, a threshold level of the female-promoting oocyte signal is still required for female development. If so, then the *dmrt1* mutants that developed as males likely produced too few oocytes during the bipotential stage to stabilize female development. That these males were sterile is consistent with the known germ cell-autonomous role of *Dmrt1* in mice (Matson et al., 2010; Zhang, Oatley, Bardwell, & Zarkower, 2016). However, it has yet to be determined in zebrafish if *dmrt1* is also cell autonomously required for germ cell survival in males.

8.2 Anti-Müllerian Hormone (Amh)

Amh is a member of the Tgf-ß superfamily of growth factors (Cate et al., 1986; Picard, Goulut, Bourrillon, & Josso, 1986). Mammalian early embryos initially form precursors for both the male and female-specific reproductive tracts, called Wolffian and Müllerian ducts, respectively. In mammals, AMH is expressed in fetal Sertoli cells in males and granulosa cells in adult females (Josso et al., 1993; Munsterberg & Lovell-Badge, 1991). Although *Amh* mutant females developed normally, loss of *Amh* function in males resulted in persistence of the Müllerian ducts, indicating that AMH functions to inhibit Müllerian duct development (Behringer, Finegold, & Cate, 1994; Belville, Josso, & Picard, 1999). In addition, although *Amh* mutant testes had Leydig cell hyperplasia, they were able to produce functional sperm (Behringer et al., 1994). Thus, in mammals, AMH is required for aspects of sexual differentiation, but not for sex determination.

By contrast to mammals, in some teleost fish, Amh, or its bone morphogenetic protein (BMP) type 2 receptor, Amhr, has evolved to be the master sex determining gene. For example, in two species of *pejerrey* a duplicate copy of *amh* located on the Y chromosome (*amhy*) appears to be the male sex determinant (Hattori et al., 2012; Yamamoto, Zhang, Sarida, Hattori, & Strüssmann, 2014). In zebrafish, as in mammals, *amh* is expressed in Sertoli cells in males, and to a lesser extent in granulosa cells that surround stage II and older oocytes in females (Rodríguez-Marí et al., 2005). The role of Amh in zebrafish was investigated using CRISPR/Cas9-induced mutations, and it was found that loss of *amh* caused a female-biased sex ratio, but that males could develop in the absence of *amh* function (Lin et al., 2017). Though mutant males had normal fertility at 2.5 mpf, they gradually lost fertility and were sterile by 6 mpf. As mutant animals aged, their abdomens became distended due to gonadal hyperplasia. Based on histology and marker gene analysis, 6 mpf testes contained mostly early stage proliferative germ cells but few meiotic spermatocytes or mature sperm. Though *amh* is commonly referred to as a male-specific factor, *amh* is expressed in granulosa cells in the ovary and *amh* mutant females had similar hyperplasic gonads to those of mutant males. Together these results suggest that in both adult males and females Amh plays a role in limiting germ cell proliferation while promoting gamete maturation (Lin et al., 2017). The phenotypes observed for zebrafish *amh* mutants appeared similar to those observed in medaka (*Oryzias latipes*) *amhr2* mutants, called *hotei* (Morinaga et al., 2007), suggesting that the functions of Amh in sex determination and differentiation are conserved in teleost.

Central to defining the role of *amh* in zebrafish and the degree to which its functions are conserved, will be the identification of its receptors. In mammals Amh signals through the Bmp receptors, composed of a complex between two type 1 and two type 2 receptors (Gouédard et al., 2000; Jamin, Arango, Mishina, Hanks, & Behringer, 2002). Unique to BMP ligands, AMH uses a dedicated Bmp type 2 receptor, called AMHR2, which is conserved in most vertebrates (Visser, 2003). While *AMHR2* orthologs have been identified in many teleost species, one has not been identified in the zebrafish genome raising the possibility that zebrafish Amh signaling utilizes a different Bmp type 2 receptor (Pfennig, Standke, & Gutzeit, 2015). In mammals, mutational analysis of the Bmp type 1 receptor BMPR1A provided strong evidence that this is the main Bmp type 1 receptor used by AMH for inhibition of Müllerian duct development, but unlike AMHR2, this receptor was not specific for the AMH ligand (Jamin et al., 2002). Interestingly, a mutation in the zebrafish Bmp type 1 receptor encoded by *bmpr1bb*, a zebrafish ortholog of mammalian BMPR1B, had a phenotype that was nearly identical to that of *amh* mutants, raising the possibility that this receptor serves as the Bmp type 1 receptor for Amh in zebrafish (Neumann, Dovey, Chandler, Carbajal, & Amatruda, 2009).

8.3 Gonadal soma derived factor (Gdsf)

Gsdf is a new member of the Tgf-ß ligand superfamily that was first identified in the trout testis and that appears to be teleost-specific (Mazurais, Montfort, Delalande, & Le Gac, 2005; Sawatari, Shikina, Takeuchi, & Yoshizaki, 2007). *gsdf* is expressed in Sertoli cells in the testis and granulosa cells in the ovary (Gautier, Sohm, Joly, Le Gac, & Lareyre, 2011; Sawatari et al., 2007; Shibata et al., 2010). In the Luzon rice fish (*Oryzias luzoneseis*), Gsdf was identified as the male sex determining gene located on the Y chromosome. While alleles of *gsdf* were found on both the X and Y chromosome, the Y allele, called $gsdf^{Y}$ was expressed at an earlier timepoint during development then $gsdf^{X}$, and early expression of $gsdf^{Y}$ in an XX fish led to male development (Myosho et al., 2012). In addition to *O. luzoneseis*, Gsdf was linked to sex determination in sablefish (Rondeau et al., 2013). In zebrafish, *gsdf* is expressed in somatic gonadal cells during the bipotential stage and in adults is expressed in granulosa and Sertoli cells (Gautier et al., 2011). *gsdf* is located on zebrafish chromosome 21, and loss of *gsdf* function had no apparent effect on sex ratios (Yan et al., 2017). Instead, mutant females, while initially fertile, developed hyperplastic ovaries that had a progressive loss of late-stage oocytes and increase of early stage oocytes, a phenotype that was strikingly similar to *amh* mutant females. However, unlike *amh* mutants, which had defects in testis development and spermatogenesis (Lin et al., 2017), *gsdf* mutant males had enlarged testes relative to wild-type males, but were otherwise fertile (Yan et al., 2017). Thus, while *gsdf* plays a role in sex determination in some teleost, it does not appear to influence sex determination in domesticated zebrafish, but instead appears to play a similar role to that of Amh in regulating germ cell proliferation and oocyte maturation in the ovary.

8.4 Androgen receptor (Ar)

Androgens are the male-promoting steroid sex hormones that function by binding to the androgen receptor (AR), a member of the nuclear hormone receptor superfamily (Gao, Bohl, & Dalton, 2005). Upon ligand binding, AR regulates sex-specific gene expression. Mutational analysis in mice showed that *AR* mutant mice formed a testis but were sterile due to a spermatogenesis defect. In addition, AR mutant male mice do not form normal reproductive tracts and had ambiguous or feminized external genitalia (De Gendt et al., 2004; Yeh et al., 2002). Mutant analysis in female mice showed that AR is also required for oocyte maturation, likely through regulating folliculogeneisis (Gill, Jamnongjit, & Hammes, 2004). In adult male zebrafish, AR is expressed in Sertoli cells (de Waal, Wang, Nijenhuis, Schulz, & Bogerd, 2008). While *ar* is also expressed in the zebrafish ovary, the particular cell type has not been determined (Hossain, Larsson, Scherbak, Olsson, & Orban, 2008). *ar* mutant zebrafish had a female biased sex ratio relative to wild-type controls, arguing that androgen signaling plays a role in promoting male sexual development (Crowder, Lassiter, & Gorelick, 2018; Yu et al., 2018). In addition, *ar* mutant males and females had reduced fertility as adults. *ar* mutant testes had disorganized tubules and did not release sperm during natural mating. However, mature sperm that were capable of fertilizing wild-type eggs were obtained from dissected mutant testes. Mutant ovaries had an increase in early stage oocytes relative to wild-type. In contrast to males, however mutant females were able to release mature eggs during normal mating, though the numbers released were significantly lower than wild-type females (Crowder et al., 2018). Similar to mice, *ar* mutant females displayed defects in oocyte maturation (Crowder et al., 2018; Yu et al., 2018).

9. Female-promoting genes

9.1 Nr0b1

The zebrafish ortholog of Human *DAX1/NR0B1,* called *nr0b1,* encodes an atypical member of the orphan nuclear hormone receptor transcription factor family that is involved in the development of steroidogenic organs (Zhao et al., 2006). In mice, *Nr0b1* is expressed in all tissues that constitute the hypothalamus-pituitary-gonad axis (HPG). In mammals *NR0B1* is X-linked and duplication of *NR0B1* in human males caused male-to-female sex reversal, while mutations in *NR0B1* resulted in adrenal hypoplasia congenita syndrome with associated hypogonadotropic hypogonadism (Bardoni et al., 1994; Battistin et al., 2012). In zebrafish, *nr0b1* is expressed in the brain and interrenal gland, the fish equivalent of the adrenal gland, during early development (Zhao et al., 2006). Beginning around 13 dpf, *nr0b1* is expressed in somatic cells of the early bipotential gonad, and expression persists into adulthood in both the ovaries and testes. Indel mutations were produced using TALENS, and homozygous *nr0b1* mutants developed predominantly as males, though some females were identified (Chen, Zhang, Wang, Zhang, & Peng, 2016). Regardless of sex, all mutants were fertile as adults. To investigate why most mutants developed as males, germ cell numbers were quantified during the bipotential phase and it was found that *nr0b1* mutants had, on average, 35% to 45% fewer germ cells than wild-type at 14 dpf and 18 dpf, respectively. It was unlikely that the reduction of germ cells in mutants vs. wild-type was due to germ cell apoptosis, as double *nr0b1;tp53* mutants, which cannot activate the Tp53-dependent apoptotic pathway, had similar sex ratios and germ cell numbers to those of *nrob1* single mutants. Given the known role of germ cells in promoting female sex determination (see above), this likely explains the skewed sex ratios. Finally, examination of gene expression using RT-qPCR showed that at 13 dpf, mutant larvae had significantly decreased expression of the female-specific aromatase-encoding gene *cyp19a1a* when compared to wild-type larvae, suggesting that *nr0b1* functions during the early bipotential stage to promote female development (Chen et al., 2016). Given that loss of *cyp19a1a* alone did not appear to alter early germ cell proliferation (Dranow et al., 2016) it is possible that *nr0b1* independently regulates *cyp19a1a* expression and germ cell proliferation. Together these results suggest that *nr0b1* influences sex determination by regulating both germ cell proliferation and expression of the pro-female *cyp19a1a* gene during the bipotential stage of gonad development.

9.2 Wnt4a

In mammals, Wnt4 is required for primary female sex determination where it antagonizes the action of the male-promoting FGF9 ligand during the bipotential stage of gonad development (Vainio et al., 1999). Although *Wnt4* orthologs have been identified in teleost, including zebrafish, Fgf9 is the only Fgf ortholog not present in teleost genomes. It has therefore remained an open question whether Wnt4 signaling is necessary for sex determination in teleost. As a first test of the role of Wnt signaling in zebrafish sex determination, Sreenivasan et al. (2014) ectopically expressed the Wnt antagonist *dikkof1* (*dkk1*) using a heat shock promoter and found that the majority of zebrafish developed as males. This result strongly argued that canonical Wnt signaling is required for female sex determination in zebrafish (Sreenivasan et al., 2014).

The zebrafish genome contains two Wnt4 orthologs, called *wnt4a* and *wnt4b* (Liu, Majumdar, Schauerte, Haffter, & Drummond, 2000; Ungar, Kelly, & Moon, 1995). Though many of the gene duplicates that are present in zebrafish resulted from the teleost-specific whole genome duplication that occurred after this lineage diverged from that of tetrapods, phylogenetic analysis performed by Kossack et al. (2019) argues instead that the duplication event that resulted in *wnt4a* and *wnt4b* occurred prior to the divergence of teleost and tetrapods. First, while tetrapod genomes contain a single *Wnt4* gene, two *Wnt4* genes are present in more basal tetrapods, such as turtles and Coelacanths. Sequence comparisons show that the Wnt4 orthologs group into two clear clades, called Wnt4a and Wnt4b, arguing for a common and more ancient origin. The single mammalian WNT4 groups within the WNT4a clade, arguing that the WNT4b ortholog was lost prior to the emergence of mammals. In zebrafish, expression of *wnt4a*, but not *wnt4b*, was detected in gonads during the bipotential phase, after which its expression becomes restricted to somatic cells in the ovary. Thus, the expression pattern of *wnt4a* in zebrafish is similar to the expression of *Wnt4* in mammals (Vainio et al., 1999). Kossack et al. (2019) tested the role of Wnt4a in sex determination using *wnt4a* point mutants and CRISPR/Cas9-induced indel mutants. They found that while wild-type siblings had normal ratios of males and females (near 50:50), ~95% of *wnt4a* mutant animals developed as males. The few *wnt4a* mutants that developed as females appeared to have normal ovarian development. It was proposed that Wnt4a sensitizes the somatic gonad cells to the proposed female promoting oocyte signal that is required during the bipotential stage for stabilizing female development. If so, then only animals that produce the most oocytes during the bipotential stage can develop as females. Thus, Wnt4a appears to promote female sex determination, but is not absolutely required for female development.

In addition to its role in promoting female sex determination, in mammals WNT4 is also required for development of the Müllerian ducts into the fallopian tubes and uterus in females. Though the similarities between reproductive duct development in fish and mammals were unknown, both male and female *wnt4a* mutants failed to develop functional reproductive ducts (Kossack et al., 2019). This result argues that Wnt4 has an ancient role in promoting reproductive duct development and that the teleost reproductive ducts and the mammalian Müllerian ducts may share a common evolutionary origin.

9.3 Cyp19a1a, Aromatase

Regardless of what genes functions as upstream regulators of sex determination in zebrafish, one of the key downstream steps must be to activate that production of the sex-appropriate hormone: the androgen 11-ketotestosterone in males, and estrogen in females. Estrogen is the key female sex hormone that regulates secondary sex differentiation in mammals, and pharmacological manipulation of estrogen production in zebrafish indicated it plays a key role in sex determination and/or differentiation (Fenske, Maack, Schäfers, & Segner, 2005; Guiguen, Fostier, Piferrer, & Chang, 2010; Uchida, Yamashita, Kitano, & Iguchi, 2004). The final step in estrogen production is catalyzed by *Cyp19a1* aromatase in mammals, which is expressed in ovarian granulosa cells. The zebrafish genome contains two ohnologs of *Cyp19a1*, called *cyp19a1a* and *cyp19a1b*, which are expressed in the ovary and brain, respectively (Chiang, Yan, et al., 2001). Two cell types are required for estrogen production in mammals: androgens produced by theca cells are converted to estrogen by CYP19A1/Aromatase, which is expressed in granulosa cells. By contrast, in zebrafish, *cyp19a1a* is expressed in both theca cells and granulosa cells, but its expression in granulosa cells is restricted to those that surround oocytes that have advanced past mid-Stage II (Dranow et al., 2016; Rodríguez-Marí et al., 2005). Similar expression had been reported in medaka suggesting that, in contrast to mammals, both theca and granulosa cells are capable of estrogen production in teleost (Nakamura, Kurokawa, Asakawa, Shimizu, & Tanaka, 2009).

Loss-of-function mutations in both *cyp19a1a* and *cyp19a1b* have been produced via TALEN and CRISPR/Cas9 (Dranow et al., 2016; Yin et al., 2017). While *cyp19a1b* mutants had no discernable defects in the brain or gonad, all *cyp19a1a* mutants developed as fertile males. Analysis of the developing mutant gonads showed that *cyp19a1a* mutants produced oocytes during the bipotential stage of gonad development, like wild-type animals, but that these oocytes underwent apoptosis as animals transitioned to spermatogenesis. By contrast to mammals, where estrogen is required for maturation of sperm in males (Robertson et al., 1999), zebrafish *cyp19a1a* mutant males appeared fully fertile (Dranow et al., 2016; Yin et al., 2017).

Interestingly, while mature sperm were not evident in wild-type zebrafish gonads until ~60 dpf, mature sperm were found in some *cyp19a1a* mutant gonads as early as 35 dpf (Dranow et al., 2016). This later result suggests that during normal development, *cyp19a1a* activity during the bipotential phase delays the time when overt testis differentiation can occur in individuals that will become male. To date, *cyp19a1a* mutants have the most penetrant early sex reversal phenotype, with 100% of prospective female animals being fully sex-reversed by 40 dpf. This result suggests that *cyp19a1a* plays an early role in female gonad function.

10. Genes required for maintenance of female sex differentiation

10.1 FoxL2a and FoxL2b

In mammals, the transcription factor *FoxL2* is expressed in granulosa cells and loss of *FoxL2* in adults results in female-to-male sex reversal of the somatic gonad (Uhlenhaut et al., 2009). FOXL2 is proposed to form a mutually antagonistic interaction with the male promoting transcription factor DMRT1. Both FOXL2 and DMRT1 are continually expressed in adult gonads and if the function of either gene is lost postnatally, the other gene is upregulated in somatic gonad cells resulting in partial sex reversal of the gonad (Matson et al., 2011; Uhlenhaut et al., 2009). Thus, the sexual phenotype of the somatic gonad must be actively maintained by the function of these factors.

Zebrafish have two ohnologs of *Foxl2*, called *foxl2a* and *foxl2b*, which were derived during the teleost-specific whole genome duplication (Yang, Wang, Li, & Gui, 2017). As in mice, *foxl2a* and *foxl2b* appeared to be expressed only in ovarian granulosa cells. Granulosa cells that surrounded early stage oocytes expressed the highest levels of *foxl2a* (stage I–II; Selman et al., 1993), while *foxl2b* was expressed in follicle cells around all oocytes (Stage I–IV). The functions of *foxl2a* and *foxl2b* were tested using TALEN-induced Indel (Yang et al., 2017). Single mutants for *foxl2a* had normal sex ratios. *foxl2a* mutant ovaries were indistinguishable from wild-type controls up to 150 dpf. However, by 180 dpf, mutant ovaries showed signs of premature ovarian failure (POF) as they contained no oocytes more mature than stage I. Thus, *foxl2a* is not required for initial oogenesis, but instead is required for maintaining oogenesis. *foxl2b* mutant ovaries followed a similar developmental trajectory to that of *foxl2a* mutants, undergoing POF by 150 dpf. However, in contrast to *foxl2a* mutants, following POF, the ovaries of *foxl2b* mutant underwent a partial sex reversal around 270 dpf, and contained regions of germ cells that were undergoing spermatogenesis. Finally, *foxl2a;foxl2b* double mutants had a more severe defect in maintaining female sex determination then *foxl2b* single mutants. At 35 dpf, *foxl2a;foxl2b* double mutants appeared to have normal sex ratios based on the morphology of the gonads. However, between 40 and 60 dpf, the oocytes in double mutants degraded and were replaced by spermatogenetic germ cells, indicating that these ovaries had undergone a more complete ovary-to-testis conversion. Analysis of gene expression during this time showed a correlation between sex reversal and upregulation of the male-specific factors *dmrt1* and *sox9a.* Together these results argue that, similar to the role of FOXL2 in mammals, zebrafish Foxl2a and Foxl2b function to maintain the female sex differentiated fate of somatic ovarian cells after primary sex determination (Yang et al., 2017).

10.2 Bone morphogenetic protein 15 (Bmp15)

During the juvenile and adult stages, oocytes continue to play a key role by producing a factor(s) that is necessary to maintain ovary fate. Bmp15 is a TGF-ß ligand that is expressed in all oocytes beginning at Stage Ib (Clelland et al., 2006; Dranow et al., 2016). In mice, *Bmp15* mutants had only slightly reduced fertility but were otherwise normal, while mutations in the close ortholog *growth and differentiation factor 9* (*Gdf9*) resulted in arrested development of pre-antral follicles (Dong et al., 1996; Yan et al., 2001). When *Gdf9* heterozygous mutations were introduced in a *Bmp15* background, female mice were sterile (Yan et al., 2001). This suggests that *Gdf9* and *Bmp15* synergize to promote oocyte maturation. It has been proposed that metabolic failure causes the phenotype observed in *Bmp15* and *Gdf9* mutants (Su et al., 2007). For example, in normal development, oocytes down regulated the expression of enzymes such as those required for glycolysis. Instead, the oocyte-produced Bmp15 and Gdf9 ligands instruct the overlying granulosa cells to upregulate expression of genes encoding glycolytic enzymes. The granulosa cells in turn provide pyruvate to the oocyte via gap junctions that form between these cells. It was proposed that this assures coordinated growth between oocytes and granulosa cells (Su et al., 2007).

In zebrafish, *bmp15* and *gdf9* are also expressed in oocytes. In contrast to mammals, however, mutational analysis has showed that loss of *bmp15*, but not *gdf9*, resulted in a failure of oocytes to progress past early stage II of oogenesis. Once *bmp15* mutant oocytes had reached this stage (~50 dpf), they underwent apoptosis and were replaced by spermatogenic germ cells. By 90 dpf, the mutant ovaries had fully sex reversed to fertile testes. As such, Bmp15 mutants had a phenotype that was identical to ablation of oocytes using pharmacogenetic methods (Dranow et al., 2013), which identifies Bmp15 as a probable oocyte-produced factor that maintains female sex differentiation in adults. Because loss of *bmp15* did not appear to affect premeiotic germ cells, the sex-reversed males were fertile. In contrast to *bmp15* mutants, *gdf9* mutants had no apparent defect in gonad development (Dranow et al., 2016).

11. Loose ends

The studies so far reviewed make it clear that many of the genes involved in the regulatory network that controls sex determination and differentiation in mammals are conserved in zebrafish. However, in some important cases, the role of a conserved gene in zebrafish has yet to be established. For example, there is ample evidence that SOX9 is a critical factor in mammalian sex determination, and one of the two zebrafish ohnologs of *Sox9*, called *sox9a*, is expressed in somatic gonad cells during the bipotential phase and in adult Sertoli cells of the testes, consistent with a role in male sex determination (Chiang, Pai, et al., 2001; Rodríguez-Marí et al., 2005). However, *sox9a* mutants are embryonic lethal due to defects in cartilage development (Yan et al., 2002). By contrast, *sox9b* is not expressed in the bipotential gonad or the adult testis, but instead is expressed in oocytes in the adult ovary (Rodríguez-Marí et al., 2005). Mutations in *sox9b* were viable but had defects in the development of the hepatopancreatic duct (Delous et al., 2012; Manfroid et al., 2012). Although *sox9b* mutant fish can survive to adulthood, these studies did not report the sex of mutant adults. Thus, it remains to be determined if *sox9a* and/or *sox9b* have roles in sex determination in zebrafish.

Another key player in mammalian sex determination is the FGF9 ligand. During the bipotential phase of mammalian gonad development, FGF9 is expressed in somatic gonadal cells in both males and females. If SRY is present and activates SOX9 expression, FGF9 expression is potentiated by SOX9 and positive feedback with FGFR2, leading to male development. Loss of *Fgf9* function caused male-to-female sex reversal (Colvin, Green, Schmahl, Capel, & Ornitz, 2001; Schmahl, Kim, Colvin, Ornitz, & Capel, 2004). While zebrafish have orthologs of *Sox9* and *Fgfr2*, *Fgf9* is the only Fgf ortholog that appears to be absent from all teleost genomes. It remains unclear if another Fgf ligand in teleost plays an analogous role to FGF9 during sex determination.

12. Concluding remarks

The development of genome editing tools over the past decade have led to rapid advancements in our understanding of the genetic regulation of sex determination, differentiation and maintenance in teleost. Studies in zebrafish, described here, as well as studies in other model and non-model teleost have firmly established that vertebrates share a core genetic regulatory module for sex determination and differentiation and that specific genes or gene duplicates within this module can evolve to become the major genetic determinant of sex (e.g., *dmy* in Medaka or *amhrY* in *pejerrey*; Hattori et al., 2012; Matsuda et al., 2002; Yamamoto et al., 2014). While the majority of zebrafish studies described above were conducted in derived domesticated lines that have lost the natural sex determining locus identified in wild zebrafish, this has not prevented the identification of many of the major genes that likely function downstream of the natural sex determinant. Once the natural determinant is identified, it will then be possible to build a unified genetic model of sex determination in zebrafish. Regardless, the progress to date has established zebrafish as a significant model for understanding the genetic regulation of sex determination and differentiation. In addition, the discovery that zebrafish can undergo complete functional sex reversion as adults has established zebrafish as a model for investigating genes that function to maintain a stable sexual phenotype during adult life.

Acknowledgments

This work was supported by funding from the National Institute of Environmental Health Sciences (T32 ES-0070599 to M.E.K.) and the National Science Foundation (IOS-1456737 to B.W.D.).

References

Anderson, J. L., Marí, A., Braasch, I., Amores, A., Hohenlohe, P., Batzel, P., et al. (2012). Multiple sex-associated regions and a putative sex chromosome in zebrafish revealed by RAD mapping and population genomics. *PLoS One*, *7*, e40701. https://doi.org/10.1371/journal.pone.0040701.

Avise, J. C., & Mank, J. E. (2009). Evolutionary perspectives on hermaphroditism in fishes. *Sexual Development*, *3*, 152–163. https://doi.org/10.1159/000223079.

Bardoni, B., Zanaria, E., Guioli, S., Floridia, G., Worley, K. C., Tonini, G., et al. (1994). A dosage sensitive locus at chromosome Xp21 is involved in male to female sex reversal. *Nature Genetics*, *7*, 497–501.

Battistin, C., Menezes Filho, H. C., Domenice, S., Nishi, M. Y., Della Manna, T., Kuperman, H., et al. (2012). A novel DAX1/NR0B1 mutation in a patient with adrenal hypoplasia congenita and hypogonadotropic hypogonadism. *Arquivos Brasileiros de Endocrinologia e Metabologia*, *56*, 496–500.

Beer, R. L., & Draper, B. W. (2013). nanos3 maintains germline stem cells and expression of the conserved germline stem cell gene nanos2 in the zebrafish ovary. *Developmental Biology*, *374*, 308–318.

Behringer, R. R., Finegold, M. J., & Cate, R. L. (1994). Müllerian-inhibiting substance function during mammalian sexual development. *Cell*, *79*, 415–425. https://doi.org/10.1016/0092-8674(94)90251-8.

Belville, C., Josso, N., & Picard, J. Y. (1999). Persistence of mullerian derivatives in males. *American Journal of Medical Genetics Part C Seminars in Medical Genetics*, *89*, 218–223. https://doi.org/10.1002/(SICI)1096-8628(19991229)89:4<218::AID-AJMG6>3.0.CO;2-E.

Braat, A. K., Zandbergen, T., Van De Water, S., Goos, H. J. T. H., & Zivkovic, D. (1999). Characterization of zebrafish primordial germ cells: Morphology and early distribution of vasa RNA. *Developmental Dynamics*, *216*, 153–167. https://doi.org/10.1002/(SICI)1097-0177(199910)216:2<153::AID-DVDY6>3.0.CO;2-1.

Bradley, K. M., Breyer, J. P., Melville, D. B., Broman, K. W., Knapik, E. W., & Smith, J. R. (2011). An SNP-based linkage map for zebrafish reveals sex determination loci. *G3 Bethesda*, *1*, 3–9. https://doi.org/10.1534/g3.111.000190.

Cate, R. L., Mattaliano, R. J., Hession, C., Tizard, R., Farber, N. M., Cheung, A., et al. (1986). Isolation of the bovine and human genes for müllerian inhibiting substance and expression of the human gene in animal cells. *Cell*, *45*, 685–698. https://doi.org/10.1016/0092-8674(86)90783-X.

Chen, S., Zhang, H., Wang, F., Zhang, W., & Peng, G. (2016). nr0b1 (DAX1) mutation in zebrafish causes female-to-male sex reversal through abnormal gonadal proliferation and differentiation. *Molecular and Cellular Endocrinology*, *433*, 105–116. https://doi.org/10.1016/j.mce.2016.06.005.

Chiang, E. F., Pai, C. I., Wyatt, M., Yan, Y. L., Postlethwait, J., & Chung, B. (2001). Two sox9 genes on duplicated zebrafish chromosomes: Expression of similar transcription activators in distinct sites. *Developmental Biology*, *231*, 149–163. https://doi.org/10.1006/dbio.2000.0129.

Chiang, E. F., Yan, Y. L., Tong, S. K., Hsiao, P. H., Guiguen, Y., Postlethwait, J., et al. (2001). Characterization of duplicated zebrafish cyp19 genes. *The Journal of Experimental Zoology*, *290*, 709–714.

Clelland, E., Kohli, G., Campbell, R. K., Sharma, S., Shimasaki, S., & Peng, C. (2006). Bone morphogenetic protein-15 in the zebrafish ovary: Complementary deoxyribonucleic acid cloning, genomic organization, tissue distribution, and role in oocyte maturation. *Endocrinology*, *147*, 201–209. https://doi.org/10.1210/en.2005-1017.

Colvin, J. S., Green, R. P., Schmahl, J., Capel, B., & Ornitz, D. M. (2001). Male-to-female sex reversal in mice lacking fibroblast growth factor 9. *Cell*, *104*, 875–889. https://doi.org/10.1016/S0092-8674(01)00284-7.

Crowder, C. M., Lassiter, C. S., & Gorelick, D. A. (2018). Nuclear androgen receptor regulates testes organization and oocyte maturation in zebrafish. *Endocrinology*, *159*, 980–993. https://doi.org/10.1210/en.2017-00617.

De Gendt, K., Swinnen, J. V., Saunders, P. T. K., Schoonjans, L., Dewerchin, M., Devos, A., et al. (2004). A Sertoli cell-selective knockout of the androgen receptor causes spermatogenic arrest in meiosis. *Proceedings of the National Academy of Sciences*, *101*, 1327–1332. https://doi.org/10.1073/pnas.0308114100.

de Waal, P. P., Wang, D. S., Nijenhuis, W. A., Schulz, R. W., & Bogerd, J. (2008). Functional characterization and expression analysis of the androgen receptor in zebrafish (*Danio rerio*) testis. *Reproduction*, *136*, 225–234. https://doi.org/10.1530/REP-08-0055.

Delous, M., Yin, C., Shin, D., Ninov, N., Debrito Carten, J., Pan, L., et al. (2012). Sox9B is a key regulator of Pancreaticobiliary ductal system development. *PLoS Genetics*, *8*, 1–16. https://doi.org/10.1371/journal.pgen.1002754.

Devlin, R. H., & Nagahama, Y. (2002). Sex determination and sex differentiation in fish: An overview of genetic, physiological, and environmental influences. *Aquaculture*, *208*, 191–364. https://doi.org/10.1016/S0044-8486(02)00057-1.

Dong, J., Albertini, D. F., Nishimori, K., Kumar, T. R., Lu, N., & Matzuk, M. M. (1996). Growth differentiation factor-9 is required during early ovarian folliculogenesis. *Nature*, *383*, 531–535. https://doi.org/10.1038/383531a0.

Dranow, D. B., Hu, K., Bird, A. M., Lawry, S. T., Adams, M. T., Sanchez, A., et al. (2016). Bmp15 is an oocyte-produced signal required for maintenance of the adult female sexual phenotype in zebrafish. *PLoS Genetics*, *12*, e1006323.

Dranow, D. B., Tucker, R. P., & Draper, B. W. (2013). Germ cells are required to maintain a stable sexual phenotype in adult zebrafish. *Developmental Biology*, *376*, 43–50.

Draper, B. W., McCallum, C. M., & Moens, C. B. (2007). nanos1 is required to maintain oocyte production in adult zebrafish. *Developmental Biology*, *305*, 589–598.

Driever, W., Schier, A. F., Neuhauss, S. C. F., Malicki, J., Stemple, D. L., Stainier, D. Y. R., et al. (1996). A genetic screen for mutations affecting embryogenesis in zebrafish. *Development*, *123*, 37–46.

Fenske, M., Maack, G., Schäfers, C., & Segner, H. (2005). An environmentally relevant concentration of estrogen induces arrest of male gonad development in zebrafish, *Danio rerio*. *Environmental Toxicology and Chemistry*, *24*, 1088–1098. https://doi.org/10.1897/04-096R1.1.

Gao, W., Bohl, C. E., & Dalton, J. T. (2005). Chemistry and structural biology of androgen receptor. *Chemical Reviews*, *105*, 3352–3370. https://doi.org/10.1021/cr020456u.

Gautier, A., Sohm, F., Joly, J.-S., Le Gac, F., & Lareyre, J.-J. (2011). The proximal promoter region of the zebrafish gsdf gene is sufficient to mimic the spatio-temporal expression pattern of the endogenous gene in Sertoli and granulosa cells. *Biology of Reproduction*, *85*, 1240–1251. https://doi.org/10.1095/biolreprod.111.091892.

Gill, A., Jamnongjit, M., & Hammes, S. R. (2004). Androgens promote maturation and signaling in mouse oocytes independent of transcription: A release of inhibition model for mammalian oocyte meiosis. *Molecular Endocrinology*, *18*, 97–104. https://doi.org/10.1210/me.2003-0326.

Gouédard, L., Chen, Y. G., Thevenet, L., Racine, C., Borie, S., Lamarre, I., et al. (2000). Engagement of bone morphogenetic protein type IB receptor and Smad1 signaling by anti-Müllerian hormone and its type II receptor. *The Journal of Biological Chemistry*, *275*, 27973–27978.

Guiguen, Y., Fostier, A., Piferrer, F., & Chang, C.-F. F. (2010). Ovarian aromatase and estrogens: A pivotal role for gonadal sex differentiation and sex change in fish. *General and Comparative Endocrinology*, *165*, 352–366. https://doi.org/10.1016/j.ygcen.2009.03.002.

Haffter, P., & Nusslein-Volhard, C. (1996). Large scale genetics in a small vertebrate, the zebrafish. *The International Journal of Developmental Biology*, *40*, 221–227.

Hartung, O., Forbes, M. M., & Marlow, F. L. (2014). Zebrafish vasa is required for germ-cell differentiation and maintenance. *Molecular Reproduction and Development*, *81*, 946–961. https://doi.org/10.1002/mrd.22414.

Hattori, R. S., Murai, Y., Oura, M., Masuda, S., Majhi, S. K., Sakamoto, T., et al. (2012). A Y-linked anti-Mullerian hormone duplication takes over a critical role in sex determination. *Proceedings of the National Academy of Sciences of the United States of America*, *109*, 2955–2959. https://doi.org/10.1073/pnas.1018392109.

Hossain, M. S., Larsson, A., Scherbak, N., Olsson, P.-E., & Orban, L. (2008). Zebrafish androgen receptor: Isolation, molecular, and biochemical characterization1. *Biology of Reproduction*, *78*, 361–369. https://doi.org/10.1095/biolreprod.107.062018.

Houwing, S., Berezikov, E., & Ketting, R. F. (2008). Zili is required for germ cell differentiation and meiosis in zebrafish. *The EMBO Journal*, *27*, 2702–2711. https://doi.org/10.1038/emboj.2008.204.

Houwing, S., Kamminga, L. M., Berezikov, E., Cronembold, D., Girard, A., Van Den Elst, H., et al. (2007). A role for Piwi and piRNAs in germ cell maintenance and transposon silencing in zebrafish. *Cell*, *129*, 69–82.

Howe, K., Clark, M. D., Torroja, C. F., Torrance, J., Berthelot, C., Muffato, M., et al. (2013). The zebrafish reference genome sequence and its relationship to the human genome. *Nature*, *496*, 498–503. https://doi.org/10.1038/nature12111.

Huang, H. Y., Houwing, S., Kaaij, L. J. T., Meppelink, A., Redl, S., Gauci, S., et al. (2011). Tdrd1 acts as a molecular scaffold for Piwi proteins and piRNA targets in zebrafish. *The EMBO Journal*, *30*, 3298–3308. https://doi.org/10.1038/emboj.2011.228.

Jamin, S. P., Arango, N. A., Mishina, Y., Hanks, M. C., & Behringer, R. R. (2002). Requirement of Bmpr1a of Mullerian duct regression during male sexual development. *Nature Genetics*, *32*, 408–410. https://doi.org/https://www.nature.com/articles/ng1003z.

Josso, N., Lamarre, I., Picard, J. Y., Berta, P., Davies, N., Morichon, N., et al. (1993). Anti-Müllerian hormone in early human development. *Early Human Development*, *33*, 91–99. https://doi.org/10.1016/0378-3782(93)90204-8.

Kang, J., Nachtrab, G., & Poss, K. D. (2013). Local Dkk1 crosstalk from breeding ornaments impedes regeneration of injured male zebrafish fins. *Developmental Cell*, *27*, 19–31. https://doi.org/10.1016/j.devcel.2013.08.015.

Köprunner, M., Thisse, C., Thisse, B., & Raz, E. (2001). A zebrafish nanos-related gene is essential for the development of primordial germ cells. *Genes & Development*, *15*, 2877–2885. https://doi.org/10.1101/gad.212401.

Kossack, M. E., High, S. K., Hopton, R. E., Yan, Y.-L., Postlethwait, J. H., & Draper, B. W. (2019). Female sex development and reproductive duct formation depend on Wnt4a in zebrafish. *Genetics*, *211*, 219–233.

Krøvel, A. V., & Olsen, L. C. (2002). Expression of a vas::EGFP transgene in primordial germ cells of the zebrafish. *Mechanisms of Development*, *116*, 141–150. https://doi.org/10.1016/S0925-4773(02)00154-5.

Lawrence, C. (2007). The husbandry of zebrafish (*Danio rerio*): A review. *Aquaculture*, *269*, 1–20. https://doi.org/10.1016/j.aquaculture.2007.04.077.

Lawrence, C., Ebersole, J. P., & Kesseli, R. V. (2008). Rapid growth and out-crossing promote female development in zebrafish (*Danio rerio*). *Environmental Biology of Fishes*, *81*, 239–246. https://doi.org/10.1007/s10641-007-9195-8.

Leerberg, D. M., Sano, K., & Draper, B. W. (2017). Fibroblast growth factor signaling is required for early somatic gonad development in zebrafish. *PLoS Genetics*, *13*, 1–28. https://doi.org/10.1371/journal.pgen.1006993.

Liew, W. C., Bartfai, R., Lim, Z., Sreenivasan, R., Siegfried, K. R., & Orban, L. (2012). Polygenic sex determination system in zebrafish. *PLoS One*, *7*, e343971. https://doi.org/10.1371/journal.pone.0034397.

Lin, Q., Mei, J., Li, Z., Zhang, X., Zhou, L., & Gui, J. (2017). Distinct and cooperative roles of amh and dmrt1 in self-renewal and differentiation of male germ cells in zebrafish. *Genetics*, *207*, 1007–1022. https://doi.org/10.1534/genetics.117.300274/-/DC1.1.

Liu, A., Majumdar, A., Schauerte, H. E., Haffter, P., & Drummond, I. A. (2000). Zebrafish wnt4b expression in the floor plate is altered in sonic hedgehog and gli-2 mutants. *Mechanisms of Development*, *91*, 409–413. https://doi.org/10.1016/S0925-4773(99)00308-1.

Manfroid, I., Ghaye, A., Naye, F., Detry, N., Palm, S., Pan, L., et al. (2012). Zebrafish sox9b is crucial for hepatopancreatic duct development and pancreatic endocrine cell regeneration. *Developmental Biology*, *366*, 268–278. https://doi.org/10.1016/j.ydbio.2012.04.002.

Matson, C. K., Murphy, M. W., Griswold, M. D., Yoshida, S., Bardwell, V. J., & Zarkower, D. (2010). The mammalian doublesex homolog DMRT1 is a transcriptional gatekeeper that controls the mitosis versus meiosis decision in male germ cells. *Developmental Cell*, *19*, 612–624. https://doi.org/10.1016/j.devcel.2010.09.010.

Matson, C. K., Murphy, M. W., Sarver, A. L., Griswold, M. D., Bardwell, V. J., & Zarkower, D. (2011). DMRT1 prevents female reprogramming in the postnatal mammalian testis. *Nature*, *476*, 101–105. https://doi.org/10.1038/nature10239.

Matsuda, M., Nagahama, Y., Shinomiya, A., Sato, T., Matsuda, C., Kobayashi, T., et al. (2002). DMY is a Y-specific DM-domain gene required for male development in the medaka fish. *Nature*, *417*, 559–563. https://doi.org/10.1038/nature751.

Mazurais, D., Montfort, J., Delalande, C., & Le Gac, F. (2005). Transcriptional analysis of testis maturation using trout cDNA macroarrays. *General and Comparative Endocrinology, 142*, 143–154. https://doi.org/10.1016/j.ygcen.2005.02.018.

Menke, A. L., Spitsbergen, J. M., Wolterbeek, A. P. M., & Woutersen, R. A. (2011). Normal anatomy and histology of the adult zebrafish. *Toxicologic Pathology, 39*, 759–775. https://doi.org/10.1177/0192623311409597.

Morinaga, C., Saito, D., Nakamura, S., Sasaki, T., Asakawa, S., Shimizu, N., et al. (2007). The hotei mutation of medaka in the anti-Mullerian hormone receptor causes the dysregulation of germ cell and sexual development. *Proceedings of the National Academy of Sciences, 104*, 9691–9696. https://doi.org/10.1073/pnas.0611379104.

Mullins, M. C., Hammerschmidt, M., Haffter, P., & Nüsslein-Volhard, C. (1994). Large-scale mutagenesis in the zebrafish: In search of genes controlling development in a vertebrate. *Current Biology, 4*, 189–202. https://doi.org/10.1016/S0960-9822(00)00048-8.

Munsterberg, A., & Lovell-Badge, R. (1991). Expression of the mouse anti-mullerian hormone gene suggests a role in both male and female sexual differentiation. *Development, 113*, 613–624. https://doi.org/10.1177/0146167211402216.

Myosho, T., Otake, H., Masuyama, H., Matsuda, M., Kuroki, Y., Fujiyama, A., et al. (2012). Tracing the emergence of a novel sex-determining gene in medaka, Oryzias luzonensis. *Genetics, 191*, 163–170. https://doi.org/10.1534/genetics.111.137497.

Nakamura, S., Kurokawa, H., Asakawa, S., Shimizu, N., & Tanaka, M. (2009). Two distinct types of theca cells in the medaka gonad: Germ cell-dependent maintenance of cyp19a1-expressing theca cells. *Developmental Dynamics, 238*, 2652–2657.

Neumann, J. C., Dovey, J. S., Chandler, G. L., Carbajal, L., & Amatruda, J. F. (2009). Identification of a heritable model of testicular germ cell tumor in the zebrafish. *Zebrafish, 6*, 319–327. https://doi.org/10.1089/zeb.2009.0613.

Padilla, P. A., & Roth, M. B. (2001). Oxygen deprivation causes suspended animation in the zebrafish embryo. *Proceedings of the National Academy of Sciences of the United States of America, 98*. 7331 LP–7335.

Pfennig, F., Standke, A., & Gutzeit, H. O. (2015). The role of Amh signaling in teleost fish—Multiple functions not restricted to the gonads. *General and Comparative Endocrinology, 223*, 87–107. https://doi.org/10.1016/j.ygcen.2015.09.025.

Picard, J. Y., Goulut, C., Bourrillon, R., & Josso, N. (1986). Biochemical analysis of bovine testicular anti-Mullerian hormone. *FEBS Letters, 195*, 73–76.

Poss, K. D., Nechiporuk, A., Stringer, K. F., Lee, C., & Keating, M. T. (2004). Germ cell aneuploidy in zebrafish with mutations in the mitotic checkpoint gene mps1. *Genes & Development, 18*, 1527–1532. https://doi.org/10.1101/gad.1182604.

Raymond, C. S., Shamu, C. E., Shen, M. M., Seifert, K. J., Hirsch, B., Hodgkin, J., et al. (1998). Evidence for evolutionary conservation of sex-determining genes. *Nature, 391*, 691–695. https://doi.org/10.1111/lre.12210.

Robertson, K. M., O'Donnell, L., Jones, M. E., Meachem, S. J., Boon, W. C., Fisher, C. R., et al. (1999). Impairment of spermatogenesis in mice lacking a functional aromatase (cyp 19) gene. *Proceedings of the National Academy of Sciences of the United States of America, 96*, 7986–7991. https://doi.org/10.1073/pnas.96.14.7986.

Rodríguez-Marí, A., Cañestro, C., Bre Miller, R. A., Nguyen-Johnson, A., Asakawa, K., Kawakami, K., et al. (2010). Sex reversal in zebrafish fancl mutants is caused by Tp53-mediated germ cell apoptosis. *PLoS Genetics, 6*, 1–14. https://doi.org/10.1371/journal.pgen.1001034.

Rodríguez-Marí, A., Yan, Y.-L., BreMiller, R. A., Wilson, C., Cañestro, C., & Postlethwait, J. H. (2005). Characterization and expression pattern of zebrafish anti-Müllerian hormone (amh) relative to sox9a, sox9b, and cyp19a1a, during gonad development. *Gene Expression Patterns, 5*, 655–667. https://doi.org/10.1016/j.modgep.2005.02.008.

Rondeau, E. B., Messmer, A. M., Sanderson, D. S., Jantzen, S. G., von Schalburg, K. R., Minkley, D. R., et al. (2013). Genomics of sablefish (*Anoplopoma fimbria*): Expressed genes, mitochondrial phylogeny, linkage map and identification of genetic sex markers. *BMC Genomics*, *14*, 452.

Sawatari, E., Shikina, S., Takeuchi, T., & Yoshizaki, G. (2007). A novel transforming growth factor-β superfamily member expressed in gonadal somatic cells enhances primordial germ cell and spermatogonial proliferation in rainbow trout (*Oncorhynchus mykiss*). *Developmental Biology*, *301*, 266–275. https://doi.org/10.1016/j.ydbio.2006.10.001.

Schmahl, J., Kim, Y., Colvin, J. S., Ornitz, D. M., & Capel, B. (2004). Fgf9 induces proliferation and nuclear localization of FGFR2 in Sertoli precursors during male sex determination. *Development*, *131*, 3627–3636. https://doi.org/10.1242/dev.01239.

Selman, K., Wallace, R. A., Sarka, A., & Qi, X. (1993). Stages of oocyte development in the zebrafish, Brachydanio rerio. *Journal of Morphology*, *218*, 203–224. https://doi.org/10.1002/jmor.1052180209.

Shang, E. H. H., Yu, R. M. K., & Wu, R. S. S. (2006). Hypoxia affects sex differentiation and development leading to a male-dominated population in zebrafish (*Danio rerio*). *Environmental Science & Technology*, *40*, 3118–3122. https://doi.org/10.1021/es0522579.

Sharma, K. K., Sharma, O. P., & Tripathi, N. K. (1998). Female hetero-gamety in *Danio rerio* (*Cypriniformes: Cyprinidae*). *Proceedings of the National Academy of Sciences India Section B, Biological Sciences*, *68*(B), 123–126.

Shibata, Y., Paul-Prasanth, B., Suzuki, A., Usami, T., Nakamoto, M., Matsuda, M., et al. (2010). Expression of gonadal soma derived factor (Gsdf) is spatially and temporally correlated with early testicular differentiation in medaka. *Gene Expression Patterns*, *10*, 283–289. https://doi.org/10.1016/j.gep.2010.06.005.

Siegfried, K. R., & Nüsslein-Volhard, C. (2008). Germ line control of female sex determination in zebrafish. *Developmental Biology*, *324*, 277–287. https://doi.org/10.1016/j.ydbio.2008.09.025.

Slanchev, K., Stebler, J., de la Cueva-Méndez, G., & Raz, E. (2005). Development without germ cells: The role of the germ line in zebrafish sex differentiation. *Proceedings of the National Academy of Sciences of the United States of America*, *102*, 4074–4079. https://doi.org/10.1073/pnas.0407475102.

Sreenivasan, R., Jiang, J., Wang, X., Bártfai, R., Kwan, H. Y., Christoffels, A., et al. (2014). Gonad differentiation in zebrafish is regulated by the canonical Wnt signaling pathway. *Biology of Reproduction*, *90*(1–10), 45. https://doi.org/10.1095/biolreprod.113.110874.

Streisinger, G., Walker, C., Dower, N., Knauber, D., & Singer, F. (1981). Production of clones of homozygous diploid zebra fish (Brachydanio rerio). *Nature*, *291*, 293–296.

Su, Y.-Q., Sugiura, K., Wigglesworth, K., O'Brien, M. J., Affourtit, J. P., Pangas, S. A., et al. (2007). Oocyte regulation of metabolic cooperativity between mouse cumulus cells and oocytes: BMP15 and GDF9 control cholesterol biosynthesis in cumulus cells. *Development*, *135*, 111–121. https://doi.org/10.1242/dev.009068.

Takahashi, H. (1977). Juvenile hermaphroditism in the zebrafish *Brachydanio rerio*. *Bulletin of the Faculty of Fisheries Hokkaido University*, *28*, 57–65.

Tzung, K.-W., Goto, R., Saju, J. M., Sreenivasan, R., Saito, T., Arai, K., et al. (2015). Early depletion of primordial germ cells in zebrafish promotes testis formation. *Stem Cell Reports*, *4*, 61–73.

Uchida, D., Yamashita, M., Kitano, T., & Iguchi, T. (2002). Oocyte apoptosis during the transition from ovary-like tissue to testes during sex differentiation of juvenile zebrafish. *The Journal of Experimental Biology*, *205*, 711–718.

Uchida, D., Yamashita, M., Kitano, T., & Iguchi, T. (2004). An aromatase inhibitor or high water temperature induce oocyte apoptosis and depletion of P450 aromatase activity in the gonads of genetic female zebrafish during sex-reversal. *Comparative Biochemistry and Physiology, Part A: Molecular & Integrative Physiology*, *137*, 11–20. https://doi.org/10.1016/S1095-6433(03)00178-8.

Uhlenhaut, N. H., Jakob, S., Anlag, K., Eisenberger, T., Sekido, R., Kress, J., et al. (2009). Somatic sex reprogramming of adult ovaries to testes by FOXL2 ablation. *Cell, 139*, 1130–1142.

Ungar, A. R., Kelly, G. M., & Moon, R. T. (1995). Wnt4 affects morphogenesis when misexpressed in the zebrafish embryo. *Mechanisms of Development, 52*, 153–164. https://doi.org/10.1016/0925-4773(95)00386-F.

Vainio, S., Heikkila, M., Kispert, A., Chin, N., McMahon, A. P., Heikkilä, M., et al. (1999). Female development in mammals is regulated by Wnt-4 signalling. *Nature, 397*, 405–409.

Visser, J. A. (2003). AMH signaling: From receptor to target gene. *Molecular and Cellular Endocrinology, 211*, 65–73. https://doi.org/10.1016/j.mce.2003.09.012.

Wang, X. G., Bartfai, R., Sleptsova-Freidrich, I., & Orban, L. (2007). The timing and extent of "juvenile ovary" phase are highly variable during zebrafish testis differentiation. *Journal of Fish Biology, 70*, 33–44. https://doi.org/10.1111/j.1095-8649.2007.01363.x.

Webster, K. A., Schach, U., Ordaz, A., Steinfeld, J. S., Draper, B. W., & Siegfried, K. R. (2017). Dmrt1 is necessary for male sexual development in zebrafish. *Developmental Biology, 422*, 33–46. https://doi.org/10.1016/j.ydbio.2016.12.008.

Weidinger, G., Stebler, J., Slanchev, K., Dumstrei, K., Wise, C., Lovell-Badge, R., et al. (2003). Dead end, a novel vertebrate germ plasm component, is required for zebrafish primordial germ cell migration and survival. *Current Biology, 13*, 1429–1434. https://doi.org/10.1016/S0960-9822(03)00537-2.

Weidinger, G., Wolke, U., Köprunner, M., Thisse, C., Thisse, B., & Raz, E. (2002). Regulation of zebrafish primordial germ cell migration by attraction towards an intermediate target. *Development, 129*, 25–36.

Westerfield, M. (2007). In M. Westerfield (Ed.), *The zebrafish book: A guide for the laboratory use of zebrafish Danio (Brachydanio) rerio* (5th ed). Eugene, OR: University of Oregon Press.

Wilson, C. A., High, S. K., McCluskey, B. M., Amores, A., Yan, Y. L., Titus, T. A., et al. (2014). Wild sex in zebrafish: Loss of the natural sex determinant in domesticated strains. *Genetics, 198*, 1291–1308. https://doi.org/10.1534/genetics.114.169284.

Yamamoto, Y., Zhang, Y., Sarida, M., Hattori, R. S., & Strüssmann, C. A. (2014). Coexistence of genotypic and temperature-dependent sex determination in pejerrey Odontesthes bonariensis. *PLoS One, 9*, 1–8. https://doi.org/10.1371/journal.pone.0102574.

Yan, Y.-L., Desvignes, T., Bremiller, R., Wilson, C., Dillon, D., High, S., et al. (2017). Gonadal soma controls ovarian follicle proliferation through Gsdf in zebrafish. *Developmental Dynamics, 246*, 925–945. https://doi.org/10.1002/dvdy.24579.

Yan, Y.-L., Miller, C. T., Nissen, R. M., Singer, A., Liu, D., Kirn, A., et al. (2002). A zebrafish sox9 gene required for cartilage morphogenesis. *Development, 129*, 5065–5079.

Yan, C., Wang, P., DeMayo, J., DeMayo, F. J., Elvin, J. A., Carino, C., et al. (2001). Synergistic roles of bone morphogenetic protein 15 and growth differentiation factor 9 in ovarian function. *Molecular Endocrinology, 15*, 854–866. https://doi.org/10.1210/mend.15.6.0662.

Yang, Y.-J., Wang, Y., Li, Z., & Gui, J.-F. (2017). Sequential, divergent and cooperative requirements of Foxl2a and Foxl2b in ovary development and maintenance of zebrafish. *Genetics, 205*, 1551–1572. https://doi.org/10.1534/genetics.116.199133.

Yeh, S., Tsai, M.-Y., Xu, Q., Mu, X.-M., Lardy, H., Huang, K.-E., et al. (2002). Generation and characterization of androgen receptor knockout (ARKO) mice: An in vivo model for the study of androgen functions in selective tissues. *PNAS, 99*, 13498–13503.

Yin, Y., Tang, H., Liu, Y., Chen, Y., Li, G., Liu, X., et al. (2017). Targeted disruption of aromatase reveals dual functions of cyp19a1a during sex differentiation in zebrafish. *Endocrinology, 158*, 3030–3041. https://doi.org/10.1210/en.2016-1865.

Yoon, C., Kawakami, K., & Hopkins, N. (1997). Zebrafish vasa homologue RNA is localized to the cleavage planes of 2- and 4-cell-stage embryos and is expressed in the primordial germ cells. *Development*, *124*, 3157–3165.

Yu, G., Zhang, D., Liu, W., Wang, J., Liu, X., & Zhou, C. (2018). Zebrafish androgen receptor is required for spermatogenesis and maintenance of ovarian function. *Oncotarget*, *9*, 24320–24334. https://doi.org/10.18632/oncotarget.24407.

Zhang, T., Oatley, J., Bardwell, V. J., & Zarkower, D. (2016). DMRT1 is required for mouse Spermatogonial stem cell maintenance and replenishment. *PLoS Genetics*, *12*, 1–18. https://doi.org/10.1371/journal.pgen.1006293.

Zhao, L., Svingen, T., Ng, E. T., & Koopman, P. (2015). Female-to-male sex reversal in mice caused by transgenic overexpression of Dmrt1. *Development*, *142*, 1083–1088. https://doi.org/10.1242/dev.122184.

Zhao, Y., Yang, Z., Phelan, J. K., Wheeler, D. A., Lin, S., & McCabe, E. R. B. (2006). Zebrafish *dax 1* is required for development of the Interrenal organ, the adrenal cortex equivalent. *Molecular Endocrinology*, *20*, 2630–2640. https://doi.org/10.1210/me.2005-0445.

CHAPTER FIVE

Regulation of germ cell sex identity in medaka

Minoru Tanaka*
Division of Biological Science, Nagoya University, Nagoya, Japan
*Corresponding author: e-mail address: mtanaka@bio.nagoya-u.ac.jp

Contents

Abstract

Germline stem cells are sexually indifferent or flexible even in the mature ovary and testis. Acquiring sex identity consistent with the sex of the body is a critical issue in germline stem cells for producing eggs or sperm. However, the molecular mechanism of the sexual fate decision in germ cells is unclear. Medaka is the first vertebrate in which germline stem cells were found in the mature ovary (Nakamura, Kobayashi, Nishimura, Higashijima, & Tanaka, 2010), and a germ cell autonomous switch gene involved in the sexual fate decision, *foxl3*, was identified (Nishimura et al., 2015) in vertebrates. Here, the mechanism underlying the sex identity of germ cells is described based on the current understanding of germ cell behavior during the sexual fate decision. The control of *foxl3* expression in germ cells and components acting downstream of *foxl3* are also described.

Current Topics in Developmental Biology, Volume 134
ISSN 0070-2153
https://doi.org/10.1016/bs.ctdb.2019.01.010

1. Introduction

One of the primary advantages of using medaka (*Oryzias* species) for studying reproduction is the large body of literature available for this organism (Kinoshita, Murata, Naruse, & Tanaka, 2009). This enables analysis from many different perspectives. One such example is sex reversal. The sex of medaka (male development) is initiated by expression of *dmy/dmrt1bY* on the Y chromosome (Matsuda et al., 2002; Nanda et al., 2002). *Dmy/dmrt1bY* was the second sex determination gene identified 10 years after discovery of the mammalian sex determination gene *SRY*. However, functional sex reversal can be induced completely, even after the sex is determined genetically (e.g., Kinoshita et al., 2009; Paul-Prasanth et al., 2013). This raises a fundamental question regarding the acquisition of sex identity in both the somatic and germ cells.

Medaka is also the only vertebrate in which a switch gene for the sexual fate decision of germ cells has been identified (Nishimura et al., 2015). The sex identity of germ cells, reflected as the decision to become eggs or sperm, is controlled by the autonomous function of *foxl3* in germ cells. The discovery of the switch gene has led to studies of the molecular mechanism of germ cell sex determination.

In this chapter, I first describe sexually different behaviors of germ cells during sex differentiation in medaka. This will provide a foundation for how germ cell sex determination proceeds. Next, I will focus on the function of *foxl3*. Phenotypically, in the *foxl3* mutant, the timing of sex determination is not restricted to the course of sex determination by *dmy/dmrt1bY*. Finally, I discuss the regulation of *foxl3*.

2. Timing of acquisition of sex identity in germ cells

During embryogenesis, germ cells are first established as primordial germ cells (PGCs) near the stage of gastrulation. Blastodermal cells containing germ granules, which are intracellular structures, can develop into PGCs. PGCs migrate posteriorly along the embryonic axis on the lateral plate mesoderm and enter the gonadal primordium, in synchrony with gonad formation (Kurokawa et al., 2006; Nakamura et al., 2006). PGC development progresses because of the expression of *dazl*, resulting in the formation of gonocytes in the gonadal primordium (Nishimura et al., 2018). Gonocytes and PGCs show differences in gene expression. Medaka contains several types of gonocytes that can be discriminated based on their size and appearance, suggesting that gonocytes are heterologous populations.

Dmy/dmrt1bY is expressed in the somatic cells surrounding the gonocytes soon after gonocytes appear in the gonadal primodium (5–6 days post-hatching; dph) (Hornung, Herpin, & Schartl, 2007; Kobayashi et al., 2004). The sex identity of germ cells may be acquired after *dmy/dmrt1bY* is expressed. This is true from a conventional perspective, and genetic studies indicated that germ cells obey sexual cue(s) directed by somatic cells. However, here, I introduce an interesting example in terms of the sex characteristics of germ cells before expression of *dmy/dmrt1bY*.

3. *sdgc*

Comparative transcriptome analyses revealed that in PGCs, one gene, *sdgc* (sex determination in germ cells), was expressed at different levels between female and male embryos (Nishimura et al., 2014). Interestingly, the male-biased expression is normalized after expression of *dmy/dmrt1bY*. Knockdown and overexpression of this gene by morpholino and RNA injection, respectively, caused no phenotype differences in vivo.

However, once the PGCs are isolated and cultured ex vivo, the proliferation rate was affected by *sdgc* in culture dishes. Isolated PGCs with XY chromosomes exhibit more frequent proliferation than those with XX. XX PGCs into which more *sdgc* transcripts are microinjected, show increased EdU incorporation, and knockdown of *sdgc* by morpholino causes lower proliferation of XY germ cells. Interestingly, *sdgc* was mapped very close to the *dmy/dmrt1bY* locus but in the region common to both the X and Y chromosomes.

The medaka sex chromosome is evolutionarily new, and the X and Y chromosomes are nearly identical except for a very small region in the *dmy/dmrt1bY* locus (Matsuda & Sakaizumi, 2016; Takehana, 2011). The authors detected single-nucleotide polymorphisms in the region of *sdgc* between the X and Y chromosomes (Nishimura et al., 2014). This gene is an example of a locus in close proximity to the small sex determining region of the Y that may be in the process of evolving to provide a reproductive advantage to the male sex. Although this change does not affect the sex identity of germ cells in vivo at present, subtle nucleotide changes may cause *sdgc* to undergo evolution to affect the sex identity of germ cells in the future.

4. Conventional view of sex differentiation of gonads in medaka

Gonocytes are characterized by their characteristics in isolation. Initially, just after arriving in the gonadal primordium, gonocytes are enclosed with supporting cells (Nakamura et al., 2008). In XY medaka, *dmy/dmrt1bY* is expressed in Sertoli cells, the supporting cell lineage.

The mitotic activity of gonocytes does not differ between XX females and XY males during an early stage of sex differentiation. This early type of mitotic activity is often referred to as type I division, and is characterized by the production of two equal daughter cells. The daughter cells are completely enclosed within the supporting cells. Germ cells (gonocytes) undergoing type I division do not express markers for germline stem cells at the earliest time points. However, because germline stem cells in the adult ovary and testis also exhibit this type of division, type I divisions may be a characteristic of cells differentiating into germline stem cells (Saito et al., 2007) (Fig. 1).

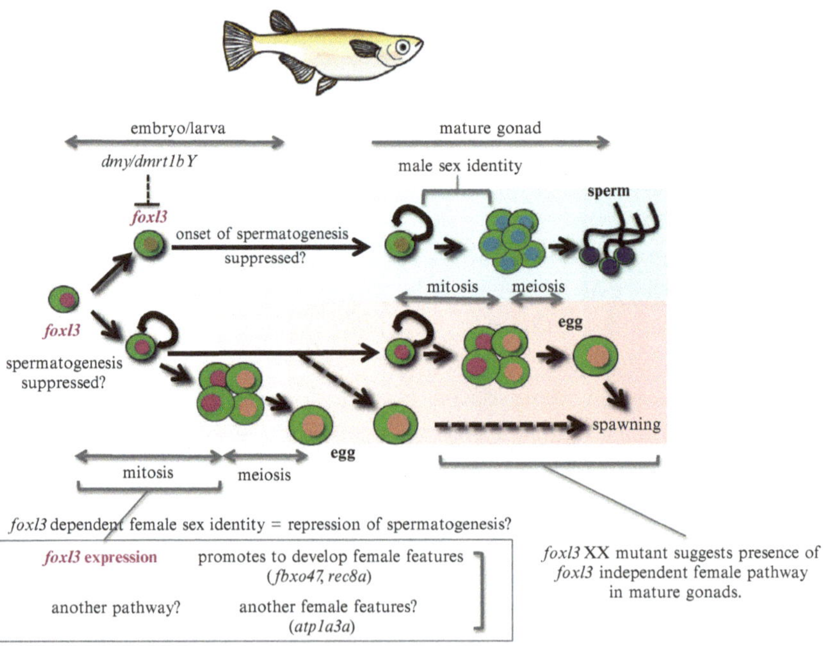

Fig. 1 Summary of early gametogenesis and germ cell sex determination. *Red circles* indicate nuclei expressing *foxl3*. *Foxl3* is kept expressed in mitotic germ cells in female gonads. In male, *foxl3* is repressed in accordance with expression of *dmrt1* in somatic cells.

By approximately day 5 dph (an early stage of sex differentiation), some populations of germ cells undergoing type I division in the female gonad begin to display a unique type of division known as type II division. The germ cells divide synchronously and rapidly to form germ cell-cysts consisting of 2–64 germ cells. Unlike type I division, a cluster of daughter cells is enclosed by supporting cells as a unit of cysts (Saito et al., 2007) (Fig. 1). These germ cells are connected to each other by intercellular bridges. After successive mitotic divisions, the germ cells enter meiosis to form oocytes. Thus, germ cells committed to gametogenesis may undergo this type of division, in contrast to the type I division characteristic of a stem type of germ cell. Importantly, germ cells in male gonads appear as gonocytes and continue type I division. Spermatogenesis does not begin until the onset of the puberty (approximately 1 month after hatching). Thus, sex differences in germ cell behavior can be observed by 5 dph at the latest.

The sex difference in germ cell behavior indicates that oogenesis initiates much earlier than spermatogenesis. Along with ongoing oogenesis, sex differentiation of the female gonad continues by the time of puberty with formation of the stromal compartment and ovarian cavity. Germinal cradles, which are niches serving as an area for very early oogenesis, become distinct as the stromal compartment and ovarian cavity forms (Nishimura & Tanaka, 2014). Germinal cradles are recognized as *sox9b*-expressing supporting cells harboring germline stem cells, cystic germ cells, and a very early stage of diplotene oocytes (Nakamura, Kobayashi, Nishimura, Higashijima, & Tanaka, 2010). These early stages of germ cells in the mature ovary have been described as germ cell clusters or as oogonia based on classical histology. Diplotene oocytes in the germinal cradles do not yet form follicles. Diplotene oocytes become follicles with granulosa cells (female supporting cells) as they exit from germinal cradles into a stromal compartment (Nakamura et al., 2010; Nishimura, Nakamura, & Tanaka, 2016).

In males, as the time of puberty approaches, several germ cells (gonocytes) form clusters with supporting cells. These clusters are origins of lobule structures (functionally equivalent to seminiferous tubules in mice), and stem cell markers such as *nanos2* begin to be detected in germ cells (Aoki, Nakamaura, Ishikawa, & Tanaka, 2009). Spermatogenic divisions proceed in a similar manner to oogenesis. Some populations of germ cells with type I division (germline stem cells or histologically called type A spermatogonia) produce germ cells through type II division (alternatively known as type B spermatogonia). Again, they undergo successive mitotic divisions prior to entry into meiosis (spermatocytes).

I have briefly described the process of sex differentiation from the perspective of germ cells. Collectively, sex differentiation typically occurs by 5 dph. However, germline stem cells remain sexually indifferent or undetermined even if they are present in the mature ovary or testis. The next critical question is how and when the sex identity of germ cells is acquired in medaka.

5. Switch gene for sexual fate decision in germ cells

An intensive search to understand the mechanism of acquisition of sex identity identified a switch gene, *foxl3* (Nishimura et al., 2015). *Foxl3* belongs to a family of transcription factors with a fork-head domain. Expression is initially detected in some populations of germ cells undergoing type I division at approximately 5 dpf in both female and male developing gonads. As oogenesis initiates, expression is observed in mitotic germ cells with both type I and type II divisions but not in meiotic germ cells. Expression persists in female developing gonads, while *foxl3* is no longer detected several days after hatching in male gonads. Germ cells in germinal cradles in the mature ovary continue to express this gene.

The phenotype of *foxl3* disruption was unexpected. The mutant female develops an ovary but produces sperm. Sperm accumulates in the germinal epithelium where the germinal cradles are located. Although they are not spawned, artificial insemination studies showed that the sperm is fertile. The mutant ovary appears normal histologically, consisting of a typical structure of an ovary with a stromal compartment and ovarian cavity. Cells expressing the gene for the estrogen-producing enzyme, P450 aromatase (*cyp19a*), are found on the ventral side of the mutant ovary, suggesting that the development of steroidogenic cells is normal and the internal endocrine environment is female. This is consistent with female secondary characteristics in female mutants. The function of foxl3 is germ cell-autonomous because the mutant germ cells, transplanted into normal host ovaries, continue producing sperm, while host germ cells differentiate into oocytes. Collectively, it was shown that foxl3 represses the initiation of spermatogenesis in a germ cell-autonomous manner. Thus, *foxl3* acts as a switch gene for the sexual fate decision in germ cells (Nishimura et al., 2015).

The expression of *foxl3* detected by in situ hybridization temporally coincides with the presence of foxl3 protein, based on antibody detection. This suggests that the sex identity of germ cells is determined by the time germ cells enter meiosis and foxl3 expression is lost (Fig. 1). Thus, mitotic germ cells with type II division are still acquiring their sex identity.

Sex is determined at some point between germline stem cells and the end of successive mitotic divisions, because the rounds of division following this stage differ between spermatogenesis and oogenesis. Germ cells with type II division in the ovary repeat mitosis successively for a maximum of six rounds. However, germ cells undergoing spermatogenesis divide for more than six rounds before producing spermatocytes (Nakamura et al., 2010).

This suggests that during prophase of meiosis germ cells already show sexual differences between oogenesis and spermatogenesis. Consistent with this, we found that the shapes of chromosomes in spreads can be sexually distinguished at the zygotene stage. The chromosomes show greater expansion and are thinner during oogenesis than during spermatogenesis (unpublished), indicating that the sexually dimorphic chromatin status is related to the mechanism conferring the sex identity of germ cells.

Thus, a question arises regarding the mechanistic relationship between the acquisition of sex identity and commitment to meiosis. The timing of meiotic commitment remains unclear. If meiotic commitment occurs at the halfway point before the end of successive type II division, the acquisition of sex identity, triggered by foxl3, may be distinct from the meiotic commitment mechanistically. Notably, meiosis occurs at the same developmental time during normal oogenesis in the wild-type ovary and during spermatogenesis in the *foxl3* mutant ovary. Both initiate by approximately 5 dph, whereas normal spermatogenesis begins much later. This indicates that meiosis is primed to occur by the time the identity of either sex is acquired in germ cells and is halted in normal spermatogenesis until puberty. Thus, the initiation of meiosis may not be the same as the mechanism triggering the acquisition of sex identity in germ cells.

In mouse and human, it has been reported that the sites of DNA breaks and frequency of crossing over during meiosis differ in germ cells between females and males (Brick et al., 2018; Halldorsson et al., 2016). The recombination rate also differs in the medaka chromosomes (Kasahara et al., 2007). This suggests a link between the two distinct mechanisms of meiosis and sex identity in germ cells. Thus, it will be interesting to examine the recombination rate between normal spermatogenesis and spermatogenesis that occurs in female *foxl3* mutants.

6. Mechanism of sexual fate switch in germ cells

To understand the switch mechanism of germ cell sex determination, it is important to identify the genes acting downstream from *foxl3*. The *cis*-elements to which foxl3 binds were revealed using medaka embryos, and are very similar to other foxl family members. Genes targeted by the transcription factor foxl3 have also been searched by comparison of whole transcriptomes between wild-type and mutant germ cells to evaluate female-specific expression during sex differentiation. The candidate genes were further screened based on their expression and epistatic analyses. Finally, two candidate genes fulfilled the strict criteria of screening, *rec8a* and *fbxO47* (Kikuchi et al., 2019). Foxl3 binding elements are found in the evolutionally conserved genomic regions near the two genes. ChIP analysis showed that foxl3 is enriched in these regions. Germ cells in female *foxl3* mutants do not express these genes. Based on these data, it is likely that the two genes function downstream of *foxl3* (Fig. 1).

Rec8a is one factor constituting the multisubunit complex known as cohesin. Cohesin is essential for the proper segregation of chromosomes during cell division and is involved in the cohesion of sister chromatids after DNA replication. The core components are SMC1 and SMC3 and two factors of the kleisin family. One of the kleisin factors that functions during mitosis, Rad21, is replaced by another kleisin factor, Rec8, during meiosis. Medaka has duplicated *rec8* genes, *rec8a* and *rec8b*. Interestingly, the transcripts of *rec8a* are detected in cystic germ cells with type II division in female medaka. This also indicates that a pathway associated with female meiosis begins to function before entering meiosis. Fbxo47 is a component of the SCF (Skp1/Cul1/F-box) ubiquitin ligases and belongs to a family of F-box-only proteins. The specific function of fbxo47 is not completely understood but is thought to selectively identify the target protein for polyubiquitination.

Preliminary results have shown that disruption of these genes explains a part of the phenotype in the *foxl3* mutant ovary (unpublished). Collectively, this indicates that establishment of female germ cells, also described as the commitment to oogenesis, involves the activation of different pathways leading to several features of oocytes and these pathways are primed during type II mitotic division. Interestingly, one gene, *atp1a3a*, was found to be expressed in female mitotic germ cells but is not regulated by *foxl3* (Kikuchi et al., 2019). This suggests that another pathway acts in parallel with *foxl3* and contributes to female sex identity.

Based on studies in mammals, the epigenetic status may be involved in sex identity in germ cells. Reprograming of epigenetic modification including DNA methylation occurs after PGCs reside in the gonad. The level of DNA methylation in germ cells is lowest when the PGCs reach a gonadal area and are higher in males than in females because of dnmt3a/b activity.

Derivation of female germ cells from cultured cells appears to require demethylation of the promoters of several meiotic genes such as Scp1, indicating a different role for DNA methylation in initiation of the sexual fate decision in germ cells (Miyauchi et al., 2017). However, it is unclear if sex-specific differences occur in species other than mammals.

7. Default state of germ cell sex

Studies on *foxl3* have revealed a germ cell autonomous mechanism of "commitment to oogenesis." This raises another question regarding whether the sexual switch, "not committed to oogenesis" is synonymous with "commitment to spermatogenesis." Specifically, it is currently unknown whether there is a spermatogenic mechanism repressed by *foxl3* (*rec8a* and *fbxo47* are the genes upregulated by *foxl3* for oogenesis). However, during the normal process of germ cell sex determination, germ cells in the male undergo less frequent type I division and show no signs of differentiation into sperm until surrounded by the lobule testicular structure at puberty. Germ cells in the male do not express *foxl3* during this period. As observed in *foxl3* female mutants, however, the germ cells readily enter spermatogenesis in an ovarian environment. This suggests that, in the normal male, germ cells become prepared for spermatogenesis at the time when foxl3 is downregulated but are suppressed by either *foxl3* and/or other signals in the male environment.

Currently, which sex is the default state in germ cells is unclear. In the earliest gonocytes, *foxl3* is expressed in both XX female and XY male gonads. Additionally, medaka germ cells possess an innate feature, a feminizing power in which large numbers of germ cells antagonize masculinization by *Dmy/dmrt1bY*, even if the fish contains a Y chromosome with intact *Dmy/dmrt1bY*. The XY male with many germ cells develops into a functional female (Kurokawa et al., 2007; Tanaka, 2016). Based on both *foxl3* expression and this feminizing power, it is expected that the default stage is female.

However, as discussed above, it is possible that the female identity of germ cells observed in embryonic stages is based on repression of the male pathway by *foxl3*. It is conceivable that the male pathway has been implemented by the time *foxl3* is functional. Thus, initially, the sex of germ cells show a biphasic status between female and male pathways. One further twist is that the feminizing power of germ cells does not depend on the female sex of germ cells. It must instead be an innate feature of germ cells, because germ cells committed to spermatogenesis can still induce feminization (Nishimura et al., 2018). The feminizing power of germ cells is distinct from the female sex identity.

8. Regulation of *foxl3*

Analyses of *foxl3* indicate that a mechanism exists to cause sex identity to be consistent with the body sex. Thus, when *dmy/dmrt1bY* is expressed, the expression of *foxl3* must be repressed. One of the earliest genes expressed downstream of *dmy/dmrt1bY* is *gsdf*, a member of the *TGFβ* gene family that is teleost-specific (Sawatari, Shikina, Takeuchi, & Yoshizaki, 2007; Zhang et al., 2016). Expression of this gene in male supporting cells (Sertoli cells) can be detected immediately following expression of *dmy/dmrt1bY*, at the beginning of sex determination (Shibata et al., 2010). In the absence of gsdf function, a high ratio of XY medaka develops into a female in the adult, suggesting that *gsdf* is required for male development (Imai, Saino, & Matsuda, 2015).

In the *gsdf* mutant, the mitotic activity of gonocytes increases in both the female and male, and some gonocytes enter an oogenetic pathway with the production of germ cells undergoing type II division in both medaka and zebrafish (Yan et al., 2017; Zhang et al., 2016). Type II division substantially contributes to an increase in the germ cell number during an early stage of male sex determination in the mutant (until several days posthatching). *Foxl3* expression is observed in both the female and male mutant during this early stage (Nishimura et al., 2018). Thus, *gsdf* appears to be involved in downregulation of *foxl3* in the male gonad. However, during this early stage, *foxl3* expression is often detected in wild-type male germ cells. Thus, it is difficult to conclude that a lack of gsdf promotes feminization of germ cells. Evaluation of receptor(s) expression may reveal the location of cells that respond to gsdf. However, the receptor(s) for gsdf have not been identified.

Another complication is that because XY *gsdf* mutants contain more germ cells than wild-type males, sex reversal may occur as a result of the feminizing power of germ cells. Once the number of germ cells crosses beyond some threshold, the gonad will develop into an ovary and male-specific genes such as *dmrt1* may be suppressed.

In mammals, the involvement of retinoid acid in oogenesis has been intensively examined (Spiller, Koopman, & Bowles, 2017). The factors affecting germ cell sex in relation to the effect of retinoic acid have also been analyzed. For example, nanos2 blocks the entry into meiosis by suppressing Stra8, an effector of retinoic acid in germ cells. Retinoic acid is reported to interact with TGFβ signaling (Suzuki & Saga, 2008; Wu et al., 2016). However, in medaka, retinoic acid does not appear to be critical during this early stage of germ cell sex determination because inhibition of retinoic acid function does not have an obvious effect on the sex of germ cells (Adolfi et al., 2016). Consistent with this observation, nanos2 is not detected during this early stage in the medaka gonad but is expressed after the establishment of germline stem cells at puberty (Aoki et al., 2009; Nakamura et al., 2010). Nanos2 expression in medaka germ cells is not a hallmark of germ cell maleness but rather of the establishment of germline stem cells.

What regulates the expression of *foxl3*? Foxl3 expression in male germ cells completely diminishes several days after hatching. This indicates that some signal(s) negatively regulating the expression of *foxl3* at this stage are important for determining germ cell sex. The signal(s) must be transmitted from somatic cells, but they remain unknown.

In rice eels, expression of *foxl3* differs from that in medaka. The piRNA *miR-9* was shown to contribute to posttranscriptional regulation of foxl3 and was reported to be required for the female to male sex reversal of rice eels (Gao, Jia, Qing, & Li, 2016). However, whether this piRNA is also functional in medaka has not been investigated.

9. Another pathway for determining germ cell sex?

The phenotype of the female *foxl3* mutant suggests that there is another mechanism that causes germline stem cells to become committed to oogenesis. For the first 1–2 months after hatching, the XX female mutants continue producing only sperm in the ovary. Subsequently however, oocytes appear in the sperm-filled mature ovary (Fig. 1). The oocytes are fertile and are even spawned (Nishimura et al., 2015). The female mutant behaves as a female, which results in the fertilization of spawned eggs from another male. The resulting embryos develop normally to adulthood.

To explain this sudden onset of oogenesis in the mature gonads of XX mutants, two explanations are possible in terms of stem cell types. (1) Germline stem cells that have been producing sperm change their fate and begin producing eggs. Some signal(s) associated with changes in the internal environment, which we do not address here, may emerge at this stage (1–2 months after hatching), and cause the sexual fate to change from spermatogenesis to oogenesis. In this scenario, identical germ cells alter their fate. (2) A second possibility is that dormant germline stem cells that did not contribute to spermatogenesis remain in the adult ovary. Our previous study indicated the presence of mitotically quiescent germ cells in embryonic and adult ovaries (Nakamura et al., 2010, 2012). It may be that these germline stem cells remain dormant for several months and then begin to produce eggs without the function of foxl3.

In either scenario (fate change or de novo oogenesis), the signal(s) causing germ cells to acquire their female sex identity remain unclear. The normal female body becomes an estrogenic environment by this adult stage. Therefore, it may be predicted that the female endocrinological environment causes this change. Indeed, long-term treatment with estrogen inhibitors causes the adult female to become a fertile male (Paul-Prasanth et al., 2013). Estrogen, however, does not appear to be involved in this abrupt onset of oogenesis in the *foxl3* mutant (Nishimura et al., 2015). Inhibition of estrogenic action did not repress the appearance of oocytes. Administration of estrogen before the appearance of oocytes did not promote the precocious oogenesis. Thus, the female endocrinological environment is unlikely to directly cause this oogenesis.

Notably, we recently observed that *foxl3* is transiently expressed in a very small number of spermatogonia (possibly germline stem cells) in the adult testis. This indicates that even in the mature testis, an internal environment affecting the sex identity of germ cells fluctuates or is unstable. In this situation, germ cells that begin to commit to oogenesis may not require *foxl3* function as a repressor of spermatogenesis but can be driven into oogenesis. It will be interesting to evaluate *fbxo47*, *rec8a* and *atp1a2* expression in the mature ovary.

Identification of a switch gene for sex identity in medaka germ cells would provide insight into how sex identity is acquired in germ cells. Ongoing studies of mammals have shown different results. This may be because different species developed different mechanisms of regulating sex identity during evolution, such as the presence of many different sex determination genes on sex chromosomes. Alternatively, the developmental processes evaluated by researchers when studying mammals and fish are different. Most analyses in mammals have used germ cells just before or during meiosis, while studies in fish have focused on mitotic germ cells before the sexual dimorphism of germ cells becomes apparent. The details of these mechanisms are under investigation. However, one emerging aspect is that there appears to be several mechanisms of conferring sex identity in germ cells.

Acknowledgments

I appreciate all the current and previous lab members who have contributed to understanding of the mechanism of sex in medaka. The works described above have been supported by a Grant-in-Aid for Research Activity Start-up (15H06838:TN), a Grant-in-Aid for Scientific Research on Innovative Areas (17H06430:MT), a Grant-in-Aid for Scientific Research (A) (16H02514:MT), NBRP medaka and NIBB Collaborative Research Program.

References

Adolfi, M. C., Herpin, A., Regensburger, M., Sacquegno, J., Waxman, J. S., & Schartl, M. (2016). Retinoic acid and meiosis induction in adult versus embryonic gonads of medaka. *Scientific Reports*, *6*, 34281.

Aoki, Y., Nakamaura, S., Ishikawa, Y., & Tanaka, M. (2009). Expresssion and systenic analysis of four *nanos* genes in medaka. *Zoological Science*, *26*, 112–118.

Brick, K., Thibault-Sennett, S., Smagulova, F., Lam, K. G., Pu, Y., Pratto, F., et al. (2018). Extensive sex differences at the initiation of genetic recombination. *Nature*, *561*, 338–342.

Gao, Y., Jia, D., Qing, H., & Li, D. (2016). Foxl3, a target of miR-9, stimulates spermatogenesis in spermatogonia during natural sex change in *Monopterus albus*. *Endocrinology*, *157*, 4388–4399.

Halldorsson, B. V., Hardarson, M. T., Kehr, B., Styrkarsdottir, U., Gylfason, A., Thorleifsson, G., et al. (2016). The rate of meiotic gene conversion varies by sex and age. *Nature Genetics*, *48*, 1377–1384.

Hornung, U., Herpin, A., & Schartl, M. (2007). Expression of the male determining gene dmrt1bY and its autosomal coorthologue dmrt1a in medaka. *Sexual Development*, *1*, 197–206.

Imai, T., Saino, K., & Matsuda, M. (2015). Mutation of gonadal soma-derived factor induces medaka XY gonads to undergo ovarian development. *Biochemical and Biophysical Research Communications*, *467*, 109–114.

Kasahara, M., Naruse, K., Sasaki, S., Nakatani, Y., Qu, W., Ahsan, B., et al. (2007). The medaka draft genome and insights into vertebrate genome evolution. *Nature*, *447*, 714–719.

Kikuchi, M., Nishimura, T., Saito, D., Shigenobu, S., Takada, R., Gutierrez-Triana, J. A., et al. (2019). Novel components of germline sex determination acting downstream of *foxl3* in medaka. *Developmental Biology*, *445*, 80–89.

Kinoshita, M., Murata, K., Naruse, K., & Tanaka, M. (2009). *Medaka—Biology, management and experimental protocols*. Iowa, USA: Wiley-Blackwell.

Kobayashi, T., Matsuda, M., Kajiura-Kobayashi, H., Suzuki, A., Saito, N., et al. (2004). Two DM domain genes, *DMY* and *DMRT1*, involved in testicular differentiation and development in the medaka, *Oryzias latipes*. *Developmental Dynamics*, *231*, 518–526.

Kurokawa, H., Aoki, Y., Nakamura, S., Ebe, Y., Kobayashi, D., & Tanaka, M. (2006). Time-lapse analysis reveals different modes of primordial germ cell migration in the medaka *Oryzias latipes*. *Development, Growth & Differentiation*, *48*, 209–221.

Kurokawa, H., Saito, D., Nakamura, S., Katoh-Fukui, Y., Ohta, K., Aoki, Y., et al. (2007). Germ cells are essential for sexual dimorphism in the medaka gonad. *Proceedings of the National Academy of Sciences of the United States of America*, *104*, 16958–16963.

Matsuda, M., Nagahama, Y., Shinomiya, A., Sato, T., Matsuda, C., Kobayashi, T., et al. (2002). DMY is a Y-specific DM-domain gene required for male development in the medaka fish. *Nature*, *417*, 559–563.

Matsuda, M., & Sakaizumi, M. (2016). Evolution of the sex-determining gene in the teleostean genus *Oryzias*. *General and Comparative Endocrinology*, *239*, 80–88.

Miyauchi, H., Miyauchi, H., Ohta, H., Nagaoka, S., Nakaki, F., Sasaki, K., et al. (2017). Bone morphogenetic protein and retinoic acid synergistically specify female germ-cell fate in mice. *The EMBO Journal*, *36*, 3100–3119.

Nakamura, S., Aoki, Y., Saito, D., Kuroki, Y., Fujiyama, A., Naruse, K., et al. (2008). Sox9b/sox9a2-EGFP transgenic medaka reveals the morphological reorganization of the gonads and a common presursor of both the female and male supporting cells. *Molecular Reproduction and Development*, *75*, 472–476.

Nakamura, S., Kobayashi, D., Aoki, Y., Yokoi, H., Ebe, Y., Wittbrodt, J., et al. (2006). Identification and lineage tracing of two populations of somatic gonadal precursors in medaka embryos. *Developmental Biology*, *295*, 678–688.

Nakamura, S., Kobayashi, K., Nishimura, T., Higashijima, S., & Tanaka, M. (2010). Identification of germline stem cells in the ovary of the teleost medaka. *Science*, *328*, 1561–1563.

Nakamura, S., Watakabe, I., Nishimura, T., Picard, J.-Y., Toyoda, A., Taniguchi, Y., et al. (2012). Hyperproliferation of mitotically active germ cells dues todefective anti-Müllerian hormone signaling mediates sex reversal in medaka. *Development*, *139*, 2283–2287.

Nanda, I., Kondo, M., Hornung, U., Asakawa, S., Winkler, C., Shimizu, A., et al. (2002). A duplicated copy of DMRT1 in the sex-determining region of the Y chromosome of the medaka, Oryzias latipes. *Proceedings of the National Academy of Sciences of the United States of America*, *99*, 11778–11783.

Nishimura, T., Herpin, A., Kimura, T., Hara, I., Kawasaki, T., Nakamura, S., et al. (2014). Analysis of a novel gene, *Sdgc*, reveals sex chromosome-dependent differences of medaka germ cells prior to gonad formation. *Development*, *141*, 3363–3369.

Nishimura, T., Nakamura, S., & Tanaka, M. (2016). A structurally and functionally common unit in testes and ovaries of medaka (*Oryzias latipes*), a teleost fish. *Sexual Development*, *10*, 159–165.

Nishimura, T., Sato, T., Yamamoto, Y., Watakabe, I., Ohkawa, Y., Suyama, M., et al. (2015). *foxl3* is a germ cell-intrinsic factor involved in sperm-egg fate decision in medaka. *Science*, *349*, 328–331.

Nishimura, T., & Tanaka, M. (2014). Gonadal development in fish. *Sexual Development*, *8*, 252–261.

Nishimura, T., Yamada, K., Fujimori, C., Kikuchi, M., Kawasaki, T., Siegfried, K. R., et al. (2018). Germ cells in the teleost fish medaka have an inherent feminizing effect. *PLoS Genetics*, *14*(3), e1007259.

Paul-Prasanth, B., Bhandari, R. K., Kobayashi, T., Horiguchi, R., Kobayashi, Y., Nakamoto, M., et al. (2013). Estrogen oversees the maintenance of the female genetic program in terminally differentiated gonochorists. *Scientific Reports*, *3*, 2862.

Saito, D., Morinaga, C., Aoki, Y., Nakamura, S., Mitani, H., Furutani-Seiki, M., et al. (2007). Proliferation of germ cells during gonadal sex differentiation in medaka: Insights from germ cell depleted mutant *zenzai*. *Developmental Biology*, *310*, 280–290.

Sawatari, E., Shikina, S., Takeuchi, T., & Yoshizaki, G. (2007). A novel transforming growth factor-beta superfamily member expressed in gonadal somatic cells enhances primordial germ cell and spermatogonial proliferation in rainbow trout (*Oncorhynchus mykiss*). *Developmental Biology*, *301*, 266–275.

Shibata, Y., Paul-Prasanth, B., Suzuki, A., Usami, T., Nakamoto, M., Matsuda, M., et al. (2010). Expression of gonadal soma derived factor (GSDF) is spatially and temporally correlated with early testicular differentiation in medaka. *Gene Expression Patterns*, *10*, 283–289.

Spiller, C., Koopman, P., & Bowles, J. (2017). Sex determination in the mammalian germline. *Annual Review of Genetics*, *51*, 265–285.

Suzuki, A., & Saga, Y. (2008). Nanos2 suppresses meiosis and promotes male germ cell differentiation. *Genes & Development*, *22*, 430–435.

Takehana, Y. (2011). Frequent turnover of sex chromosome in the medaka fishes. In K. Naruse, M. Tanaka, & H. Takeda (Eds.), *Medaka—A model organogenesis, human disease, and evolution* (pp. 229–240). Springer.

Tanaka, M. (2016). Germline stem cells are critical for sexual fate decision of germ cells. *BioEssays*, *38*, 1227–1233.

Wu, Q., Fukuda, K., Kato, Y., Zhou, Z., Deng, C. X., & Saga, Y. (2016). Sexual fate change of XX germ cells caused by the deletion of SMAD4 and STRA8 independent of somatic sex reprogramming. *PLoS Biology*, *14*(9), e1002553.

Yan, Y. L., Desvignes, T., Bremiller, R., Wilson, C., Dillon, D., High, S., et al. (2017). Gonadal soma controls ovarian follicle proliferation through Gsdf in zebrafish. *Developmental Dynamics*, *246*, 925–945.

Zhang, X., Guan, G., Li, M., Zhu, F., Liu, Q., Naruse, K., et al. (2016). Autosomal gsdf acts as a male sex initiator in the fish medaka. *Scientific Reports*, *6*, 19738.

Further reading

Cañestro, C., Catchen, J. M., Rodríguez-Marí, A., Yokoi, H., & Postlethwait, J. H. (2009). Consequences of lineage-specific gene loss on functional evolution of surviving paralogs: ALDH1A and retinoic acid signaling in vertebrate genomes. *PLoS Genetics*, *5*(5), e1000496.

Naruse, K., Tanaka, M., & Takeda, H. (Eds.), (2011). *Medaka a model for organogenesis, human disease, and evolution* (pp. 229–240). Tokyo: Springer.

Watari, E., Shikina, S., Takeuchi, T., & Yoshizaki, G. (2007). A novel transforming growth factor-beta superfamily member expressed in gonadal somatic cells enhances primordial germ cell and spermatogonial proliferation in rainbow trout (*Oncorhynchus mykiss*). *Developmental Biology*, *301*, 266–275.

CHAPTER SIX

Characterizing the bipotential mammalian gonad

Serge Nef[a,*], Isabelle Stévant[a], Andy Greenfield[b,*]
[a]Department of Genetic Medicine and Development, University of Geneva Medical School, Geneva, Switzerland
[b]Mammalian Genetics Unit, Medical Research Council Harwell Institute, Oxfordshire, United Kingdom
*Corresponding authors: e-mail address: serge.nef@unige.ch; a.greenfield@har.mrc.ac.uk

Contents

Abstract

Primary sex determination is the decision by which the bipotential embryonic gonad commits to either the testicular or ovarian fate. The developing gonad constitutes a unique paradigm for the study of lineage specification, cell fate commitment and the exploration of how distinct cell populations diverge from multipotent progenitors. After the separation of the adreno-gonadal primordium into two distinct primordia, somatic progenitor cells of the gonadal primordium undergo several cell fate decisions and sex-specific cell differentiation. The specification of the supporting and steroidogenic cell lineages into either Sertoli and Leydig cells in the testis, or granulosa and theca cells in the ovary is essential for germ cell development and endocrine function of the gonads. In this review, we focus on the early events leading to gonad formation, including the identity of gonadal progenitors, the genetic networks involved in cell lineage specification, the mutual antagonism of the pro-testis and pro-ovary networks and

Current Topics in Developmental Biology, Volume 134
ISSN 0070-2153
https://doi.org/10.1016/bs.ctdb.2019.01.002

the importance of timing of developmental events orchestrating testis and ovary development. We discuss, and put into perspective, a number of experiments performed in mice or humans that have shed light on sex-determining mechanisms and, where possible, the clinical significance and limitations of such model organism data.

1. Introduction

Sexual reproduction is an evolved strategy to generate biological diversity and novelty at each generation and is of fundamental importance to all vertebrates and many invertebrates. At the core of this strategy is the generation of distinct sexes, with distinct reproductive organs, producing distinct gametes. Mammals use a chromosomal sex-determining mechanism: XY individuals develop testes; XX individuals form ovaries. While the gonads are germ cell factories they are also endocrine glands, shaping the development of the reproductive organs and ensuring sexual dimorphism and fertility. Thus, the development of gonads is crucial to sex identity. The process by which the embryo develops either a testis or an ovary is known as gonadal sex determination or primary sex determination. Once differentiated, the gonads produce sex hormones that promote development of sexually dimorphic structures that characterize distinctive male and female anatomy, a process also known as secondary sex determination.

The study of gonadal sex determination is a paradigm of developmental biology, based as it is on the sexually dimorphic development of a gonadal primordium, the genital ridge. This primordium is commonly described as "bipotential": whether it is found in an XY or XX embryo it has the capacity to develop into either a testis or an ovary. It is something of a dogma in the field of mammalian sex determination that most of the male and female gonadal cell-types arise from precursor cell lineages that are homologous, i.e., the Sertoli cells of the testis and the granulosa cells of the ovary arise from a single precursor lineage in this bipotential primordium. Recent studies in the mouse suggest that the story is more complicated than this (see Sections 4 and 5). Nevertheless, this framework allows us to characterize sex determination as a process that fundamentally concerns cell lineage commitment or cell fate determination. Questions that still fascinate developmental biologists include how an embryonic cell lineage can become increasingly specialized throughout development, giving rise to multiple daughter lineages, and how the final differentiated cell-types achieve a stability, i.e., how they maintain their differentiated state. Both these questions will be discussed in this review. The first question is increasingly being answered by careful study of transcription factors and signaling molecules required for sexually dimorphic gonad development and an analysis of the transcriptomes and epigenomes that such molecules regulate. Techniques for performing such analyses are becoming increasingly sophisticated and we will discuss these.

Sex determination raises some distinct questions: How is sexually dimorphic development of a bipotential embryonic primordium achieved? What is the relationship between the testis- and ovary-determining gene regulatory networks? What is the consequence of disrupting these networks by genetic mutation? The answers to these questions, mostly discovered over the last 25 years, are fascinating and sometimes unexpected. In this review, we will discuss the mutual antagonism that exists between the pro-testis and pro-ovary developmental pathways, the importance of timing of developmental events and why gonadal sex reversal is a relatively common phenotype when key genes are disrupted.

The significance of sex determination, however, goes beyond the insights its study offers into the role of genes in organogenesis and development more broadly. In humans, mutations in key sex-determining genes can give rise to a group of developmental abnormalities known as disorders (or differences) of sex development (DSD). Diagnosing and managing such conditions requires an understanding of human sex determination and the molecular mechanisms underpinning it. While human fetal material and human cells are increasingly used as tools for experimentation, the mouse has been the primary model for understanding mammalian sex determination. Here, focusing on events prior to overt differentiation of the gonads, we will discuss a number of experiments performed in mice which have shed light on sex-determining mechanisms and, where possible, the clinical significance of such model organism data.

2. Beginnings: Formation of the adreno-genital primordium (AGP)

Both the gonads and the adrenal cortex share a common developmental origin, the adreno-gonadal primordium (AGP) (see Fig. 1). In the mouse,

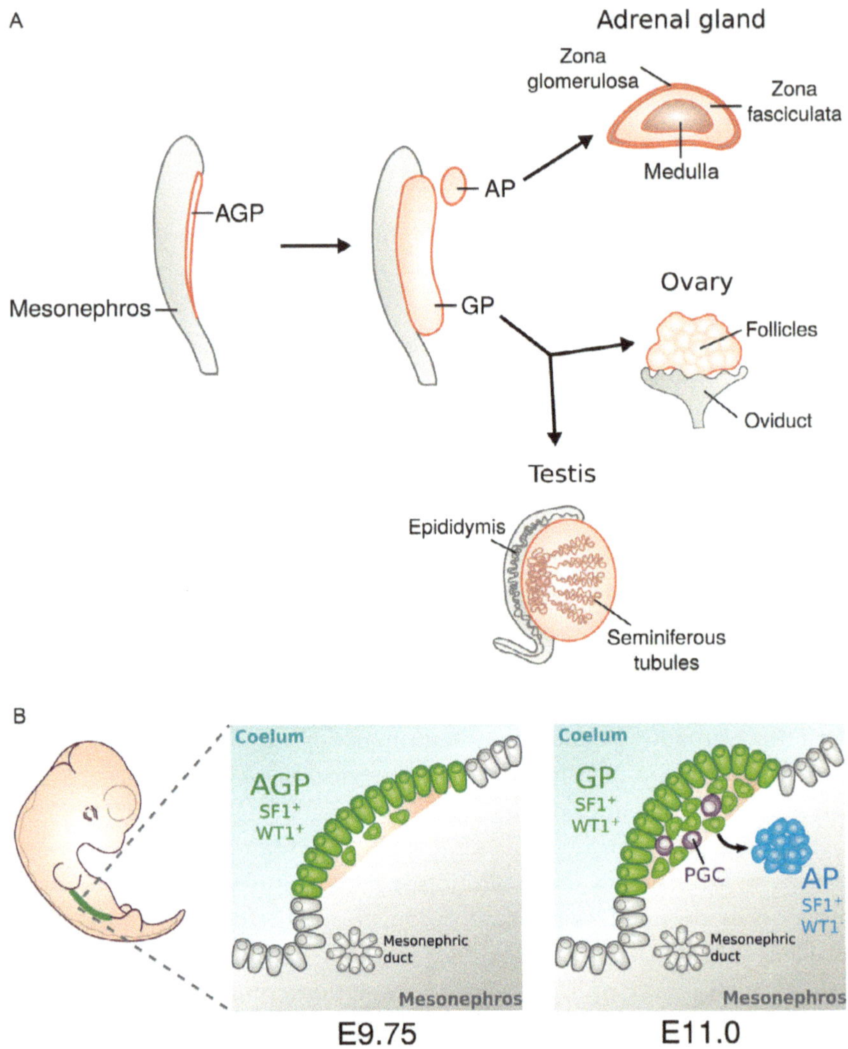

Fig. 1 Adrenogonadal primordium (AGP) development in the mouse. (A and B) Schematic representations of AGP differentiation into gonadal primordium (GP) and adrenal primordium (AP).

the AGP is visible at around 9.5 days *post coitum* (dpc) as a thickening of the coelomic epithelium on the ventral surface of the mesonephros (Hatano, Takakusu, Nomura, & Morohashi, 1996; Ikeda, Shen, Ingraham, & Parker, 1994). It is composed of a population of uncharacterized progenitor cells expressing the orphan nuclear receptor steroidogenic factor 1 (SF1, also named NR5A1 or AD4BP) (Hatano et al., 1996; Morohashi, 1997) and primordial germ cells (PGCs). Germ cells arise at approximately 6.25 dpc in a cluster of 45 epiblast cells in the proximal region of the pre-gastrulation mouse embryo (Seki et al., 2005). These PGCs proliferate, relocate to the endoderm and are incorporated into the hindgut invagination. PGCs migrate anteriorly along the hindgut, through the dorsal mesentery, and ultimately colonize the AGP at around 10.0 dpc (Ginsburg, Snow, & McLaren, 1990; Molyneaux, Stallock, Schaible, & Wylie, 2001). By 10.5 dpc, the AGP separates to give rise to the adrenal and gonadal primordia (Morohashi, 1997). Together with PGCs, ingressing SF1-positive ($SF1^+$) cells along the rostrocaudal axis contribute to the formation of the gonadal primordium. In contrast, the more medial $SF1^+$ cells, located in the rostral region of the AGP, migrate and condense to form the adrenal primordium at the cranial end of the mesonephros. AGP specification and its separation into two distinct primordia, the gonadal and adrenal, remain poorly understood. What is known is that the transcription factors Wilms' tumor 1 protein (WT1) and SF1 play central roles in AGP precursor specification (Kreidberg et al., 1993; Luo, Ikeda, & Parker, 1994). While both WT1 and SF1 are expressed in the AGP, WT1 is switched off within the adrenal primordium soon after it separates, but remains expressed in the gonadal primordium (Vidal & Schedl, 2000).

3. Genital ridge formation

The gonadal primordium (also known as the genital ridge) first appears on the coelomic surface of the mesonephros at around 10.0 dpc in the mouse or at gestational week 4 in the human embryo. The cells of the coelomic epithelium possess a unique molecular signature which includes the GATA zinc finger transcription factor GATA4 and WT1, as well as SF1. Formation of the genital ridge results from the proliferation of these coelomic epithelial cells, which then undergo epithelial-to-mesenchymal transition (EMT). By 11.5 dpc, the basement membrane beneath the coelomic epithelium is discontinuous, thereby permitting ingression or cell-oriented division of these progenitor cells into the gonad. Cell lineage tracing experiments using organ culture indicate that the coelomic epithelium is itself a source of supporting cell precursors for the male and female gonads (Karl & Capel, 1998). This developmental potential is, however, temporally regulated. Labeling of coelomic epithelial cells prior to 11.5 dpc in XX and XY gonads revealed subsequent ingression into the gonad; in XY gonads these cells contributed to the Sertoli cell population and interstitial lineages. Labeling after 11.5 dpc resulted in the incorporation of cells only into the interstitial compartment of the testis. This migratory potential ceased by 12.5 dpc. In the XX gonad, ingressing cells had no obvious fate restriction in the same developmental period. From these data, we can infer that at least some Sertoli cell progenitors arise in the coelomic epithelium. It is unclear whether there are other sources. Interestingly, deleting the Notch signaling inhibitors *Numb* and *Numbl* in the developing gonad from early stages disrupted the polarity of coelomic epithelial cells and eliminated or reduced cell numbers of multiple gonadal cell lineages, including supporting cells and Leydig cells, and led to clusters of undifferentiated cells within the gonad (Lin, Barske, DeFalco, & Capel, 2017). The paired-like homeobox protein EMX2 and the homeodomain proteins SIX1 and SIX4 also contribute to the regulation of EMT and subsequent ingression of gonadal progenitor cells. In the forming genital ridges of *Emx2*$^{-/-}$ mouse mutants, coelomic epithelial cells again lose their cell polarity and the number of migrating cells is reduced, resulting in impaired gonad formation (Kusaka et al., 2010). In the case of *Six1* and *Six4* double knockout mutants, genital ridges exhibit delayed/reduced EMT and subsequent ingression of progenitor cells, resulting in smaller gonads and adrenal glands (Fujimoto et al., 2013). Overall, these data suggest that asymmetrically-dividing coelomic epithelial cells are multipotent progenitors of gonadal somatic cell lineages.

SF1 is a critical factor for genital ridge formation and development. It acts in a dose-dependent manner by modulating the expression of target genes implicated in proliferation, survival and differentiation of gonadal progenitor cells. As a consequence, mouse embryos lacking one functional copy of *Sf1* exhibit underdeveloped adrenal glands and reduced cellular proliferation, while complete inactivation of *Sf1* leads to regression of the gonads by 12.5 dpc due to apoptosis of somatic cells and complete agenesis of the adrenal gland. In addition, impaired gonadal and adrenal formation is associated with downregulation or ectopic upregulation of *Sf1*, as is also the case for mice with targeted inactivation of *Gata4* (Hu, Okumura, & Page, 2013), *Wt1* (Kreidberg et al., 1993), the polycomb factor M33 (*Cbx2*) (Katoh-Fukui et al., 2005), the transcription co-factor *Cited2* (Val, Martinez-Barbera, & Swain, 2007), the homeodomain protein *Pbx1* (Schnabel, Selleri, & Cleary, 2003), the transcription factor Odd-skipped related 1 (*Odd1*) (Wang, Lan, Cho, Maltby, & Jiang, 2005), the LIM homeobox protein *Lhx9* (Birk et al., 2000) and the Insulin/IGF1R signaling pathway (Pitetti et al., 2013).

4. Reconstructing developmental lineage trajectories and cell fate in the developing gonad by single-cell RNA sequencing (scRNA-seq)

A major goal in the field of sex determination, and more broadly in developmental biology, is to understand cell fate potential and the specification of progenitor cells. Lineage tracing is an important tool for following a group of cells and observing their descendants. There are a variety of complementary approaches available for in vivo lineage tracing (Spanjaard & Junker, 2017). Currently, the most popular approach for lineage tracing relies on genetic labeling of cells using a combination of inducible recombinases and fluorescent reporters, which provides exquisite control by enabling cell-type and stage-specific labeling of cells. While powerful, there are several limitations and technical challenges associated with the use of genetic labeling by means of the Cre/*lox* system. In particular, it requires prior knowledge of the progenitor cells (i.e., identification of cell-specific marker genes) and the availability of specific, inducible Cre lines that recapitulate the endogenous expression of these key markers. In addition, while this method provides data on the cell fate potential of labeled cells, it gives no information about the intermediate cellular states, their molecular identity and the potential branch points along the route (Fletcher, Das, & Ngai, 2018).

Single-cell transcriptomics based on RNA sequencing (scRNA-seq) has emerged as a powerful technique for measuring gene expression in individual cells. Within a tissue or organ of interest, it allows the identification and classification of cell populations in a comprehensive and unbiased manner, a perspective that would be obscured in bulk level analysis. When performed in times-series, scRNA-seq data can be analyzed with lineage prediction tools to reconstruct developmental trajectories of cell lineages and characterize the transcriptional dynamics controlling their differentiation. This approach has been applied to the sex determination process and has provided new insights into the differentiation of the somatic progenitor cells during testis and ovarian differentiation (see Section 5).

5. Specification of the somatic lineages in the developing gonads

A population of coelomic epithelial cells that expresses *Gata4* and *Sf1* is thought to give rise to the somatic lineages of the genital ridge. However, little is known about these early somatic progenitors. In particular, it is unclear whether there are distinct subclasses of progenitors, with discrete characteristics, sex differences and cell fate. The classical model of gonadal sex determination states that the bipotential gonad contains at least two pre-established somatic cell lineages: the supporting and the steroidogenic cell lineages. The supporting and steroidogenic progenitors will differentiate, respectively, into Sertoli cells and Leydig cells in the testis and granulosa cells and theca cells in the ovary (Svingen & Koopman, 2013). This model has been recently challenged with the report of Leydig-to-Sertoli cell transdifferentiation (Zhang et al., 2015) and the observation that both lineages derive from $WT1^+$ cells present at E10.5 in the genital ridge (Liu, Rodriguez, & Yao, 2016).

Recent studies taking advantage of scRNA-seq have provided new insights into these somatic progenitor cells and the fate decisions that lead to their differentiation into the supporting and steroidogenic cell lineages (for a review, see Stevant & Nef, 2018). scRNA-seq has shown great effectiveness in identifying the different cell types composing the developing gonads. It also helps to reconstruct the cell lineages and the expression dynamics controlling cell differentiation during the process of sex determination. scRNA-seq performed on XY and XY mouse gonads during sex determination identified a single multipotent $Sf1^+$ progenitor cell population prior to the onset of sex determination (Stevant, Kuhne, et al., 2018; Stevant, Neirijnck, et al., 2018). These early somatic progenitors, present at 10.5 dpc, express genes related to epithelial stem cell identity and also negative regulators of cell differentiation, consistent with their coelomic epithelial origin. XX and XY early progenitors have a similar transcriptomic identity and do not exhibit obvious sexual dimorphism. Although the presence of additional rare populations of progenitor cells cannot be excluded, these findings support the "common progenitor identity" hypothesis and demonstrate the absence of a cell-lineage-specific fate prior to sex determination. Reconstruction of the fates and sex-specific differentiation of the gonadal $Sf1^+$ cell lineage, based on these scRNA-seq studies, is depicted in Fig. 2.

It has been shown that supporting cell commitment and sex-specific differentiation into either Sertoli or granulosa cells occurs sequentially. Initially, both XY and XX multipotent progenitors that adopt a supporting cell fate share a similar transcriptomic identity, before initiating the sex-dependent genetic program leading to their differentiation into Sertoli and granulosa cells, respectively. Both Sertoli and granulosa cell differentiation are

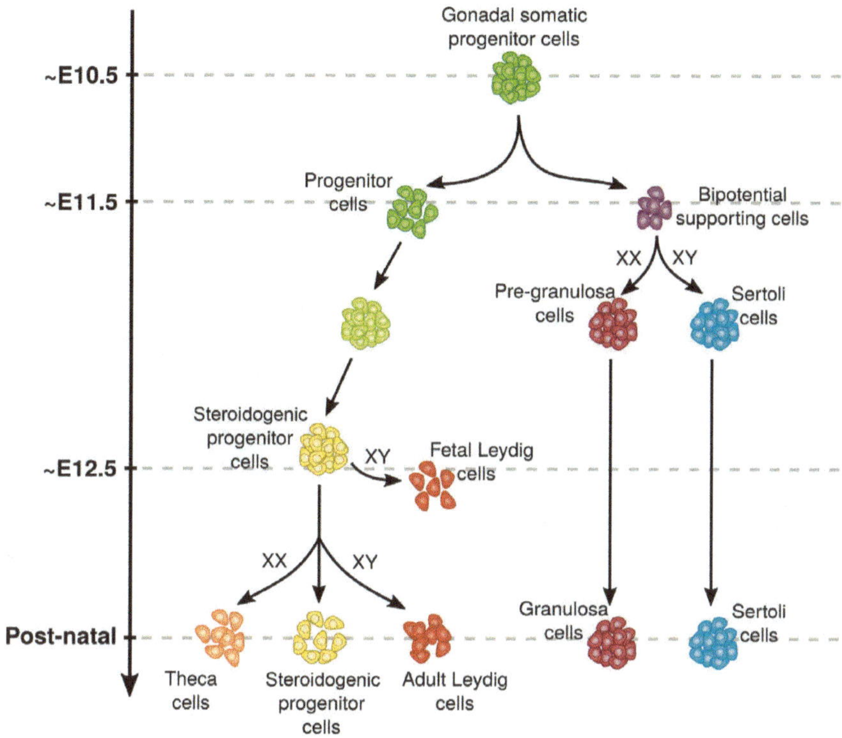

Fig. 2 Model of cell lineage specification during sex determination. Single-cell RNA sequencing data identified a unique *Sf1*$^{+}$ progenitor cell population at 10.5 dpc, at the origin of the supporting and steroidogenic lineages in XX and XY gonads. Around 11.5 dpc, a subset of these multipotent cells commit toward the supporting lineage. Rapidly, pre-Sertoli cells initiate the sex-specific genetic program leading toward Sertoli cell differentiation. In contrast, pre-granulosa cell differentiation into granulosa cells is delayed and progenitors remain in an early stage of differentiation. XX and XY supporting progenitors share similar transcriptomes prior to the sex-specific differentiation into either Sertoli or granulosa cells. Another subset of multipotent cells maintain their progenitor state and, by 12.5 dpc, undergo gradual transcriptional changes, acquiring a steroidogenic fate by progressively expressing *Pdgfra*, *Arx* or *Ptch1*. The transcriptomes of XX and XY steroidogenic precursors are similar and do not display strong sexual dimorphism.

mediated by dynamic and highly complex transcriptional programs, including thousands of genes differentially expressed in granulosa and Sertoli cells between 10.5 and 16.5 dpc. Both genetic programs are complex, with sequential waves of activation or inactivation of genes, reflecting the process of Sertoli or granulosa cell commitment and differentiation. The narrow, transient expression profile of numerous genes suggests that they are required during a precise stage of cellular differentiation (e.g., *Nr0b1*, *Sry*), while prolonged expression profiles might reveal a role in maintenance of cell identity (e.g., *Sox9*, *Dmrt1*). However, XX and XY supporting cells initially share similar transcriptomic profiles and then bifurcate toward either the Sertoli or granulosa cell branch (Stevant, Kuhne, et al., 2018). This indicates that supporting cell commitment is initially sex-independent and disconnected from sex-specific differentiation into either Sertoli and granulosa cells.

In addition, scRNA-seq data reveal that XY interstitial cells and XX stromal cells are transcriptionally similar and evolve to express a set of genes that specify their fate as steroidogenic cell precursors, at the origin of both fetal Leydig cells and theca cells (Stevant, Kuhne, et al., 2018). In contrast to the highly dynamic program mediating granulosa and Sertoli cell differentiation, the remaining XX and XY progenitor cells that do not commit toward the supporting lineage display much less variation in gene expression and undergo progressive transcriptomic changes through time, switching from a multipotent identity to a steroidogenic precursor identity. Initially, the XX and XY progenitor cells express multipotent markers such as *Wt1* and *Cbx2*. Then, starting at 11.5 dpc, the progenitor cells lose expression of these genes and gradually express markers of steroidogenic cell precursors such as *Wnt5a*, *Arx*, *Pdgfra*, *Gli2*, and *Tcf21*.

6. Molecular control of gonadal cell fate choice: Gene networks and signaling events

6.1 SRY and SOX9: Initiating testis determination

The Y-linked testis-determining gene, *SRY*, encodes an HMG box transcriptional regulator that initiates testis determination in the developing XY gonad by driving differentiation of supporting cell precursors into Sertoli cells. The control of expression of *Sry* has been studied in great detail in the mouse and it reveals a complex story of genetic and epigenetic regulation aimed at ensuring timely expression of *Sry*. The consequences of a delay to (or reduction in) *Sry* expression are discussed below.

6.1.1 Early events: Cell proliferation

It has been clear for many years that the number of somatic cells that have a Y chromosome, and so express *Sry*, in the developing XY gonad is an important determinant of gonadal fate. The injection of XY embryonic stem cells into an XX blastocyst can generate chimeric male mice, but only if sufficient numbers of XY cells are incorporated into the gonadal cell lineages (Burgoyne, Buehr, & McLaren, 1988; Patek et al., 1991). It is likely, therefore, that supporting cell precursors must be present in sufficient numbers to permit testis determination. What is known about the origin of the supporting cell lineage and the role played by cell proliferation in testis determination?

Transcriptome characterization by scRNA-Seq of $Sf1^+$ cells at 10.5 dpc confirms the presence of proliferating, multipotent progenitors possessing an epithelial identity as might be expected for coelomic epithelial cells (Stevant, Neirijnck, et al., 2018). In particular, these cells significantly overexpress genes related to cell division, as well as *Fox* genes (*Foxc1*, *Foxc2* and *Foxd2*) essential for patterning of the intermediate mesoderm (Kume, Deng, & Hogan, 2000) from which the testis develops, stem cell maintenance- and renewal-related genes (*Sall1*, *Sall4*) (Basta, Robbins, Kiefer, Dorsett, & Rauchman, 2014; Yang, Gao, Chai, & Ma, 2010), and *Cbx1* and *Cbx2*, necessary for *Sry* expression (Katoh-Fukui et al., 2012). Two genes known to repress miRNA activity and to be involved in stem cell proliferation are also detected: *Trim71* and *Lin28* (Rybak et al., 2009; Yang et al., 2015).

Given the role that the proliferative coelomic epithelium plays in the provision of supporting cell precursors (see Section 3), is there any direct evidence for a role for cell proliferation in sex determination? The use of cell proliferation inhibitors, again in organ culture models, revealed an 8-h window, between approximately 10.8 and 11.2 dpc, that was critical for establishing expression of pro-testis genes and testis cord formation (Schmahl & Capel, 2003). Inhibition of cell proliferation either side of this window did not block testis development. This window overlaps with the developmental potential of coelomic epithelial cells to contribute to Sertoli cells (see above) and also with the onset of *Sry* expression. However, unlike male-specific cell proliferation in the coelomic region after 11.5 dpc, initiated by SRY (Schmahl, Eicher, Washburn, & Capel, 2000), there is no indication that these early events are sex-specific. Indeed, this absence of sex-specificity is predicted by the fact that XX embryos transgenic for *Sry* are (often) capable of initiating and completing testis determination (Koopman, Gubbay, Vivian, Goodfellow, & Lovell-Badge, 1991); this suggests that events required for *Sry* expression and subsequent testis determination can occur in an XX embryo. Thus, molecular and cellular events "upstream" of the initiation of Sry expression occur in both sexes—whether they result in testis determination depends on whether *Sry* is present or not.

6.1.2 Regulation of *Sry* expression

A number of studies have demonstrated that mouse *Sry* is expressed in a narrow temporal window between 10.5 and 12.5 dpc. RNAse protection (Hacker, Capel, Goodfellow, & Lovell-Badge, 1995), wholemount in situ hybridization (Bullejos & Koopman, 2001; Warr et al., 2012), qRT-PCR (Jeske, Bowles, Greenfield, & Koopman, 1995; Koopman, Münsterberg, Capel, Vivian, & Lovell-Badge, 1990) and immunodetection of expression from an epitope-tagged *Sry* transgene (Sekido, Bar, Narvaez, Penny, & Lovell-Badge, 2004) all indicate an onset at around 10.5 dpc, with cells in the center of the gonad initiating expression. The basis of this initial spatial restriction is unclear, but it may relate to the source of supporting cell precursors and where they first enter the gonad. This early urogenital ridge *Sry* expression is germ cell-independent and is extinguished by 12.5 dpc.

Loss of *Sry* causes complete XY gonadal sex reversal of the fetal and adult gonad, i.e., XY female development (Lovell-Badge & Robertson, 1990). Thus, ensuring that *Sry* is expressed in the appropriate place and time is vital for the development of maleness. Loss of function genetic studies have revealed a role for a relatively large number of genes in the regulation of *Sry* expression (see Fig. 3). Not surprisingly, several transcription factors are implicated, along with signaling molecules and, more recently, chromatin modifiers. What have studies of these factors revealed about the regulation of *Sry*?

The insulin family of growth factors plays an essential role in mediating different aspects of adreno-gonadal development, such as adrenal specification, testicular differentiation, and ovarian development (Griffeth, Bianda, & Nef, 2014; Pitetti et al., 2013). The action of insulin, and the related growth factors insulin-like growth factor 1 (IGF1) and IGF2 is mediated by the insulin receptor (INSR) and the IGF type I receptor (IGF1R), two membrane-associated tyrosine kinase receptors. Mouse embryos lacking both *Insr* and *Igf1r* exhibit: (i) complete agenesis of the adrenal cortex; (ii) ovary development and a completely female phenotype in XY mutants and (iii) a delay in ovarian differentiation so that *Insr;Igf1r* mutant gonads, irrespective of genetic sex, remain in an extended undifferentiated state, before the ovarian differentiation program ultimately is initiated at around 16.5 dpc. The primary gonadal defect is due in part to the early failure in genital ridge expansion following reduced proliferation rates of somatic progenitor cells in both XX and XY gonads prior to sex determination. The absence of insulin/IGF signaling affects also the expression of transcription factors essential for adreno-gonadal development and function, such as *Wt1*,

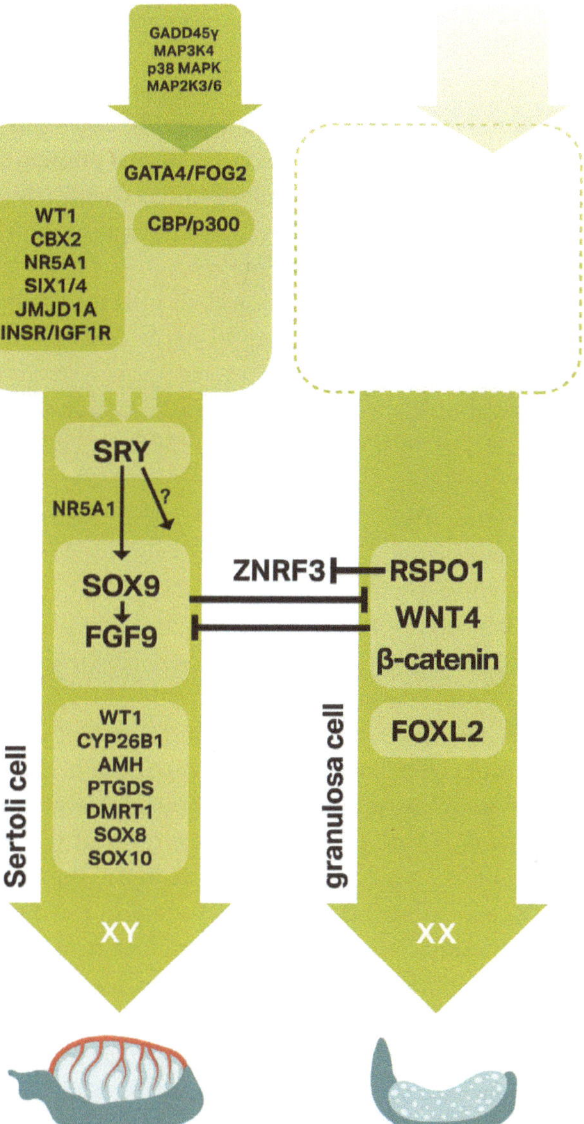

Fig. 3 Mutually antagonistic genetic networks regulating sex determination in mammals. A schema is shown that includes key (but not all) molecules that function in XY testis and XX ovary development. *Sry* expression is controlled by a number of transcription factors, signaling molecules and chromatin modifiers. These pro-*Sry* factors are also present in the XX gonad (dotted empty box), explaining the possibility of *Sry* transgene expression in XX gonads. SRY promotes *Sox9* expression (arrow), but it may have other targets. SOX9 drives Sertoli cell differentiation with the assistance of FGF9/FGFR2. The SRY-SOX9-FGF9 pro-testis axis inhibits (hammered lines) the pro-ovarian network of RSPO1-WNT4/CTNNB1-FOXL2, which itself inhibits the male pathway and drives granulosa cell differentiation. RSPO1 directly inhibits the anti-WNT factor, ZNRF3, which sequesters Frizzled receptor. Other potential regulatory relationships are shown, with question marks where these are unclear or uncertain. The specification of supporting cell fate results in the commitment of other lineages to testicular or ovarian fates.

Lhx9, *Bmp2*, *Dax1* and *Sf1*, but also 50 additional genes defined as lineage-primed by Jameson, Natarajan, et al. (2012) prior to and/or at the time of sex determination. Together, reduced proliferation and premature alteration of the testicular and ovarian genetic programs in these cells could explain both adrenal agenesis and the incapacity of mutant gonads to develop into either ovaries or testes at the time of sex determination. The reduction of *Sf1* expression (and *Gadd45g*—see below) and the failure of *Sry* upregulation support the idea that IGFs act upstream of SRY action and are important for regulating AGP development.

Recent investigations have revealed a signaling pathway that is required for normal expression of *Sry*: GADD45γ-MAP3K4-MAP2K3/6-p38 MAPK (Bogani et al., 2009; Gierl, Gruhn, von Seggern, Maltry, & Niehrs, 2012; Johnen et al., 2013; Warr et al., 2012, 2014; Warr, Siggers, Carre, Wells, & Greenfield, 2016). GADD45γ, a member of the GADD45 family that is implicated in active DNA demethylation, can activate MAP3K4 (Miyake, Takekawa, Ge, & Saito, 2007), a kinase at the top of a mitogen-activated protein kinase (MAPK) phospho-relay signal transduction system that can phosphorylate the transcription factor GATA4. Since GATA4 (and its cofactor FOG2) is known to be required for *Sry* expression, this phosphorylation is thought to be an important component in activating the onset of *Sry* transcription. One interesting observation about mutants lacking *Map3k4* or *Gadd45g* is that XY gonadal sex reversal is associated with a significant reduction in levels of *Sry* transcript at 11.5 dpc, but levels recover by around 12.0 dpc, indicating a delay in reaching peak expression. This delay is thought to result in sex reversal due to a failure to repress expression of pro-ovarian genes in the mutant XY gonadal somatic cells. Careful experimentation with an inducible *Sry* transgene indicates a critical window (between 11.0 and 11.25 dpc) for SRY to establish the male pathway during XY gonad development: any delay to *Sry* activation beyond 11.3 dpc caused a sex reversal phenotype (Hiramatsu et al., 2008). Below, we discuss the inhibitory interactions between the testis- and ovary-determining pathways in more detail.

The regulation of *Sry* transcription must ultimately be explained at the level of chromatin, by describing epigenomic events required at the locus. Such events create a chromatin context conducive to recruitment of transcription factors and co-factors. Interestingly, demethylation of several CpG dinucleotides in the vicinity of the proximal *Sry* promoter in XY gonadal somatic cells is one of the earliest events in testis determination, preceding *Sry* expression itself (Gierl et al., 2012). Again, mouse knockout studies have also shed light on chromatin regulators required for normal *Sry* expression. The balanced activity of the histone demethylase KDM3A (JMJD1A), and the histone methyltransferase GLP/G9A, regulates levels of the inhibitory heterochromatin mark H3K9me2 at the *Sry* promoter, permitting accumulation of the positive mark, H3K4me3, and activation of *Sry* expression (Kuroki et al., 2013, 2017). Control of histone acetylation has also been shown to be required for normal *Sry* expression: loss of three out of four functional copies of the genes encoding the histone acetyltransferases CBP and p300 in gonadal somatic cells causes XY gonadal sex reversal, associated with loss of *Sry* expression and reduced levels of acetylated H3K27 at the *Sry* promoter (Carre et al., 2018). It is unclear why such commonly employed chromatin regulators as KDM3A and CBP/p300 are required for testis determination, but are apparently dispensable for fetal ovary development.

The upshot of *Sry*/SRY accumulation in supporting cell precursors of the XY gonad is the rapid up-regulation of another key testis-determining gene, *Sox9*, which also encodes an HMG box transcription factor. This up-regulation is mediated by the cooperation between SF1 and SRY and their occupancy of gonadal enhancers 5′ to the *Sox9* locus (Gonen et al., 2018; Gonen, Quinn, O'Neill, Koopman, & Lovell-Badge, 2017; Sekido & Lovell-Badge, 2008). SOX9 then drives Sertoli cell differentiation by its regulation of numerous target genes (Rahmoun et al., 2017). Loss of *Sox9* also results in complete gonadal sex reversal (Barrionuevo et al., 2006). We now turn to the phenomenon of gonadal sex reversal and the relationship between the pro-testis and pro-ovary genetic networks.

6.2 Testis determination: Opposing the ovarian genetic program

When genes functioning in testis determination, such as *Sry* and *Sox9*, are disrupted, gonadal sex reversal can occur. This might be complete, resulting in XY ovary development, or partial, causing ovotestis development (Wilhelm et al., 2009). But why? The explanation consists in the mutually antagonistic relationship between the testis- and ovary-determining gene networks: when one is disrupted, it fails to appropriately inhibit the other and that alternate pathway becomes dominant. In numerous cases of testis-determining gene knockouts, XY gonadal sex reversal occurs due to the ectopic activation of pro-ovary pathways of gene expression. One example of this phenomenon is offered by the analysis of mouse knockouts for the fibroblast growth factor gene, *Fgf9*, and the canonical WNT signaling molecule, *Wnt4*. XY fetuses lacking FGF9 undergo complete gonadal sex reversal, when on the sensitized C57BL/6J background, and this is associated with activation of canonical WNT signals (Colvin, Green, Schmahl, Capel, & Ornitz, 2001; Kim et al., 2006). The corollary of this is that loss of *Wnt4* in the XX gonad results in partial masculinization and up-regulation of *Fgf9* (Kim et al., 2006). (In the case of knockouts for other pro-ovary genes, such as *Foxl2*, *Ctnnb1* (which encodes β-catenin) or *Rspo1*, XX gonadal sex reversal is also only partial and occurs postnatally: possible explanations for this will be addressed below.) Thus, FGF9 and WNT4 act in a mutually antagonistic fashion. Indeed, analysis of XY fetuses lacking both *Fgf9* and *Wnt4* reveals that testis development can occur in the absence of both, suggesting that the primary role of FGF9 is to oppose the anti-testis activity of WNT4 (Jameson, Lin, & Capel, 2012). Other pairwise deletions of testis- and ovary-determining genes, such as *Sox9/Rspo1* (Lavery et al., 2012) and *Sox9/Ctnnb1* (Nicol & Yao, 2015), also result in XY testicular differentiation, suggesting that testis determination can occur when anti-testis genes such as *Rspo1* and *Ctnnb1* are absent, even if key testis-determining genes are also missing. These data reveal the importance of the anti-ovary roles of testis-determining genes like *Fgf9* and *Sox9*. They also suggest that other pro-testis factors, such as other SOX proteins, can positively drive testis development in the absence of SRY and SOX9.

Mechanistic understanding of these mutually antagonistic interactions is sketchy. For example, it is not immediately obvious how two secreted signaling molecules, FGF9 and WNT4, might interact at the molecular level. One recent study reveals a key antagonistic molecular interaction in the sex-determining mechanism (Harris et al., 2018). As mentioned above, canonical WNT signals are an important determinant of ovarian fate. WNT4 activity results in the accumulation of nuclear β-catenin and the activation/inhibition of target genes (Bernard et al., 2012). RSPO1, another key ovarian determinant, is a secreted molecule known to potentiate canonical WNT signaling, and it does so through its negative impact on two widespread transmembrane E3 ubiquitin ligases, ZNRF3 and RNF43 (Hao et al., 2012; Koo et al., 2012). ZNRF3/RNF43, in turn, inhibit WNT signals by increasing membrane turnover of the WNT receptor, Frizzled (FZ). A study of XY fetuses lacking ZNRF3 reveals variable testis determination defects, ranging from complete gonadal sex reversal to morphologically abnormal testis development, often within a single fetus (Harris et al., 2018). These defects are associated with ectopic canonical WNT signals at 11.5–12.5 dpc in the mutant XY gonad and disruption to *Sox9* expression. These data suggest a model in which high levels of RSPO1 in the XX gonad result in inhibition of ZNRF3, itself a negative regulator of ovary development. In the XY gonad, *Rspo1* expression is negligible (in contrast to the XX gonad) and, as a consequence, ZNRF3 can actively inhibit WNT4/FZ-mediated signals.

6.3 Ovary development: Granulosa cell heterogeneity, germ cells and XX gonadal sex reversal

In addition to WNT signals, the transcription factor FOXL2 plays an important role in the specification of ovarian granulosa cell fate (Auguste et al., 2011; Boulanger et al., 2014; Pannetier, Chassot, Chaboissier, & Pailhoux, 2016). However, recent studies suggest a cellular heterogeneity in precursors of the granulosa cell lineage (Mork et al., 2012; Rastetter et al., 2014; Zheng et al., 2014). Lineage tracing experiments indicate that FOXL2-positive somatic cells in early fetal gonads give rise to granulosa cells of ovarian medullary follicles, which are activated before puberty and contribute to early fertility. These cells are thought to be homologous to SRY-positive cells of the XY gonad and originate in the coelomic epithelium. LGR5, which is a receptor of RSPO1, marks another population of fetal ovarian somatic cells; these lack FOXL2 and give rise to cortical granulosa cells, which form the primordial follicles of the definitive ovarian reserve. Although distinct granulosa cell populations have been observed based on the mutually exclusive expression of FOXL2 and LGR5 (Rastetter et al., 2014), recent studies suggest that these two populations represent granulosa cells at different stages of differentiation. Both ovarian granulosa cells of medullary and cortical follicles are derived from a common precursor cell that expresses GATA4 (Gustin et al., 2016). These GATA4-positive precursors may undergo one of two differentiation processes: proliferation or entry into mitotic arrest. The first set of precursor cells directly contributes to medullary granulosa cells by differentiating at the onset of ovary development via maintained expression of the cell cycle inhibitor p27 and activation of FOXL2 expression. Around 15.5 dpc and postnatal day 4 (PND4), a second population of precursors contributes indirectly by initially establishing a proliferative pregranulosa cell pool regulated by WNT4-LGR5 and β-catenin (Mork et al., 2012). Subsequently, these cells will differentiate in response to unknown signals that initiate the expression of p27 and FOXL2, thus contributing to the life-long cortical follicular reserve. It is unclear what are the factors and signals regulating the balance between WNT/β-catenin signaling, which promotes proliferation of supporting cells in the developing ovary and initiation of p27, and FOXL2 expression required for differentiation of granulosa cell precursors.

Nevertheless, it is remarkable to compare the temporal discordances between, on the one hand, early differentiation of Sertoli cells and the rapid formation of cords in the testis, with, on the other, the delay in formation of ovarian follicles and the late differentiation of granulosa cells of cortical follicles. Such temporal delays in specifying ovarian cell fate have been also observed when comparing the transcriptional changes occurring during supporting cell lineage differentiation in XX and XY gonads. Differentiation of both granulosa and Sertoli cells is mediated by a complex, highly dynamic program. Initially, XX and XY supporting cells share a common transcriptomic profile, which then bifurcates toward either the Sertoli or the granulosa branch. Differentially expressed genes between XX and XY supporting cells at each embryonic stage reveal that sexual dimorphism in XX supporting cells appears delayed compared to XY supporting cells (Stevant, Kuhne, et al., 2018).

For many years, it had been a dogma in the field of sex determination that, while germ cells are not required for testis determination, they do play a role in development of the ovary during folliculogenesis. Germ cells are critical for morphological development of the ovary at birth: clusters of germ cells break down and primordial follicles form. In addition, postnatal XX gonadal sex reversal in mutants such as *Wnt4* and *Rspo1*, involving somatic cell transdifferentiation to a SOX9-positive Sertoli cell fate, occurs after germ cell loss, suggesting a link between germ cells and inhibition of testicular somatic cell fate (Guigon & Magre, 2006). Determining whether germ cells play a role in the specification of somatic cell fate in the fetal ovary requires separating the effects of germ cell depletion from any independent effects of somatic cell disruption. Two recent studies have reported such an approach. Targeted ablation of germ cells in the developing XX gonad, using Cre-regulated diphtheria toxin expression, revealed that specification of granulosa cell fate, indicated by FOXL2 expression, was not affected by their absence (Uhlenhaut et al., 2009). In addition, early germ cell depletion using the chemotherapy drug busulfan, or genetic disruption using the Kit^{Wv} allele that disrupts germ cell migration, did not disrupt the establishment or maintenance of granulosa cells based on a variety of molecular markers (Maatouk et al., 2012).

However, other data suggest that meiotic germ cells may still play a role in XX gonadal sex reversal. In XX gonads lacking *Rspo1* and *Wnt4*, in which perinatal partial sex reversal occurs, somatic cell transdifferentiation occurs when germ cells survive the pre-meiotic stage of development and die only after meiotic entry. This may indicate that granulosa cells acquire the competence to transdifferentiate only following contact with meiotic germ cells (Maatouk et al., 2012). Analysis of *Wnt4*- and *Rspo1*-deficient gonads indicates that prior to transdifferentiation, mutant pregranulosa cells exit mitotic arrest and undergo precocious differentiation (Maatouk, Mork, Chassot, Chaboissier, & Capel, 2013). Since depletion of germ cells in these mutants prior to meiosis prevents this exit from quiescence, it appears that germ cell-somatic cell interactions around the time of meiosis are normally important for maintenance of quiescence and these are disrupted in the absence of WNT4/RSPO1, leading to precocious differentiation and transdifferentiation. It has been suggested that *Sox9* up-regulation cannot occur in quiescent pregranulosa cells, even in the absence of WNT4 or RSPO1, and that this may account for the absence of fetal XX gonadal sex reversal in such cases. This could be tested by attempting an early disruption to mitotic arrest of pregranulosa cells in conjunction with deletion of *Wnt4*/*Rspo1*/*Foxl2* and assaying for fetal female-to-male gonadal sex reversal.

6.4 Comparisons between human and mouse sex determination

At the outset of any comparison of mouse and human sex determination it should be made clear that several key sex-determining genes have conserved roles in these two species. A number of *SRY* and *SOX9* mutations have been reported to disrupt human testis determination, resulting in 46,XY gonadal dysgenesis and female presentation (Berta et al., 1990; Wagner et al., 1994). *RSPO1* mutations have been described in very rare cases of 46,XX testicular disorder of sex development (DSD) (Parma et al., 2006). Even this modest description of gene conservation already suggests that human sex determination in the fetal gonad, at around 6 weeks, consists of SOX protein-driven testis determination and canonical WNT signal-driven ovary development, each opposing the other. These observations suggest that the mouse is an excellent model for studying potential mechanisms of human sex determination.

There is genetic evidence supporting a role for a number of human genes in sex determination. However, DSDs that disrupt human sex determination are rare and there are often limited numbers of mutations available for functional studies. Moreover, such functional studies rely on heterologous cell line assays or mouse mutant phenotypes. Nevertheless, mutations in the following human genes have been reported in DSDs and also support the case for the similarity between mouse and human testis determination: *FOG2* (*ZFPM2*) (Bashamboo et al., 2014), *GATA4* (Lourenco et al., 2011), *CBX2* (Biason-Lauber, Konrad, Meyer, DeBeaufort, & Schoenle, 2009), *SF1* (*NR5A1*) (Allali et al., 2011; Bashamboo et al., 2016), *DMRT1* (Murphy et al., 2015), *FGFR2* (Bagheri-Fam et al., 2015), *WT1* (Barbaux et al., 1997), *DHH* (Canto, Soderlund, Reyes, & Mendez, 2004; Clark, Garland, & Russell, 2000), *HHAT* (Callier et al., 2014) and *ZNRF3* (Harris et al., 2018). A homozygous *WNT4* mutation is associated with 46,XX testicular DSD (Mandel et al., 2008), as are structural variants associated with the mis-regulation of *SOX9*, *SOX10* and *SOX3* (Benko et al., 2011; Polanco, Wilhelm, Davidson, Knight, & Koopman, 2010; Sutton et al., 2011).

Differences between mouse and human are also sometimes apparent, as one might expect. One interesting example involves the role of MAPK signals in testis determination. As discussed earlier, MAP3K4 is required for mouse testis determination (Bogani et al., 2009). However, mutations in *MAP3K1* are associated with 46,XY complete gonadal dysgenesis in humans (Granados et al., 2017; Pearlman et al., 2010), but loss of *Map3k1* does not disrupt mouse testis determination (Warr et al., 2011). The simplest explanation for this apparent discrepancy is that reported human *MAP3K1* mutations have gain-of-function effects, perhaps disrupting protein interactions required for testis determination (Loke et al., 2014). But inferring causality in an individual case of mutation is not without risk.

One other significant difference consists in the frequency of human sex reversal reported in individuals heterozygous for a mutation. In contrast, homozygosity for a mutant allele tends to be required for disruption to sex determination in mice. However, how these observations are interpreted requires some care. There are cases where mouse XY sex reversal occurs in the absence of one copy of a gene, such as in the case of *Fog2* and *Gata4* (Bouma, Washburn, Albrecht, & Eicher, 2007), *Map3k4* (Warr et al., 2014) and *Znrf3* (Harris et al., 2018). In all these cases, sex reversal requires the highly sensitized B6.Y^{AKR} genetic background, suggesting that genetic background may be similarly important in cases of human disorders of sex development. Indeed, a growing body of evidence suggests an oligogenic basis for some DSDs (Camats, Fernandez-Cancio, Audi, Schaller, & Fluck, 2018; Mazen et al., 2016; Robevska et al., 2018).

7. Future research perspectives

Sex determination is a complex developmental process by which the bipotential gonads develop as either testes or ovaries. Despite several decades of research revealing the delicate balance of pro-male and pro-female genes mediating sex determination, the majority of DSD cases related to gonadal development and differentiation remain undiagnosed at the genetic level (Bashamboo, Eozenou, Rojo, & McElreavey, 2017). The success rate of identifying causal variants through exome sequencing in patients with DSD is rather low compared to other genetic disorders, suggesting that a significant proportion of pathogenic variants and in/dels may be localized in intergenic regions, perhaps regulating the balance of expression of key sex-determining genes. In particular, regions such as enhancers, silencers and insulators are potentially attractive candidates, but the lack of knowledge regarding the locations of these crucial gene-regulatory elements limits our ability to study the regulatory mechanisms that control gene expression. We might reasonably expect that in the near future, with the advent of new, more sensitive genome-wide approaches, a precise characterization of the dynamics of gene expression during the process of sex determination will be available for each of the major cell-lineages of the gonad, both in the mouse model and human (Stevant & Nef, 2018; Stevant, Neirijnck, et al., 2018). In parallel, we are also starting to witness the emergence of genome-wide data describing the chromatin landscape in the major cell types of the gonad (Garcia-Moreno, Plebanek, & Capel, 2018; Maatouk et al., 2017). These genome-wide assays include DNaseI hypersensitive site sequencing (DNaseI-seq), ATAC-Seq, chromatin immunoprecipitation followed by sequencing (ChIP-seq) of histone modifications that mark promoters of active or silenced genes (H3K4me2/3, H3K27ac, H3K9ac, H4K16ac, H3K27me3, H3K9me2/3) and chromosome conformation capture methods such as Hi-C, all of which will be instrumental in identifying active enhancer regions regulating the complex genetic network that underlies sex determination. Indeed, the observation that open regions of chromatin in developing Sertoli cells are detectable at gene loci expressed specifically in granulosa cells (Maatouk et al., 2017) suggests that such regions may have a dual use, being occupied by repressive or activating protein complexes depending on the sex of the gonad. Such sites may therefore mediate the mutual antagonism of the testis- and ovary-determining networks on a genome-wide scale.

Another major limitation in the field is the lack of an in vitro system to test whether a putative candidate gene or potential pathogenic variant is important for the process of sex determination. So far, the classic method of validation relies on in vivo loss- or gain-of-function experiments using mouse functional genomics. The generation of knock-out mice using traditional methods, or CRISPR/Cas9-mediated genome editing, is still expensive and time consuming. Unfortunately, there are no reliable somatic gonadal cell lines, such as gonadal somatic precursor cell lines, Sertoli or granulosa cell lines, although attempts have been made (Buganim et al., 2012). Here also, significant resources are currently being invested to develop culture protocols aimed at differentiating human or mouse iPS/ES cells toward the Sertoli or granulosa cell lineage, in order to derive stably reprogrammed cell lines. The development of such in vitro systems may be extremely important in testing and validating the potential pathogenic impact of DSD variants and also in establishing genome-wide or proteome-wide screens for novel genes and pathways that regulate gonadal lineage specification.

Acknowledgments

We would like to thank Gareth Clarke for assistance with the production of Fig. 3. A.G. acknowledges support from the UK Medical Research Council for his laboratory's research on sex determination at the Harwell Institute (MC_U142684167). S.N. acknowledges support of grants from the Swiss National Science Foundation (grants 31003A_173070 and 51PHI0-141994) and from the Département de l'Instruction Publique of the State of Geneva.

References

Allali, S., Muller, J. B., Brauner, R., Lourenco, D., Boudjenah, R., Karageorgou, V., et al. (2011). Mutation analysis of NR5A1 encoding steroidogenic factor 1 in 77 patients with 46, XY disorders of sex development (DSD) including hypospadias. *PLoS One*, *6*, e24117.

Auguste, A., Chassot, A. A., Gregoire, E. P., Renault, L., Pannetier, M., Treier, M., et al. (2011). Loss of R-spondin1 and Foxl2 amplifies female-to-male sex reversal in XX mice. *Sexual Development*, *5*, 304–317.

Bagheri-Fam, S., Ono, M., Li, L., Zhao, L., Ryan, J., Lai, R., et al. (2015). FGFR2 mutation in 46,XY sex reversal with craniosynostosis. *Human Molecular Genetics*, *24*, 6699–6710.

Barbaux, S., Niaudet, P., Gubler, M. C., Grunfeld, J. P., Jaubert, F., Kuttenn, F., et al. (1997). Donor splice-site mutations in WT1 are responsible for Frasier syndrome. *Nature Genetics*, *17*, 467–470.

Barrionuevo, F., Bagheri-Fam, S., Klattig, J., Kist, R., Taketo, M. M., Englert, C., et al. (2006). Homozygous inactivation of Sox9 causes complete XY sex reversal in mice. *Biology of Reproduction*, *74*, 195–201.

Bashamboo, A., Brauner, R., Bignon-Topalovic, J., Lortat-Jacob, S., Karageorgou, V., Lourenco, D., et al. (2014). Mutations in the FOG2/ZFPM2 gene are associated with anomalies of human testis determination. *Human Molecular Genetics*, *23*, 3657–3665.

Bashamboo, A., Donohoue, P. A., Vilain, E., Rojo, S., Calvel, P., Seneviratne, S. N., et al. (2016). A recurrent p.Arg92Trp variant in steroidogenic factor-1 (NR5A1) can act as a molecular switch in human sex development. *Human Molecular Genetics*, *25*(16), 3446–3453.

Bashamboo, A., Eozenou, C., Rojo, S., & McElreavey, K. (2017). Anomalies in human sex determination provide unique insights into the complex genetic interactions of early gonad development. *Clinical Genetics*, *91*, 143–156.

Basta, J. M., Robbins, L., Kiefer, S. M., Dorsett, D., & Rauchman, M. (2014). Sall1 balances self-renewal and differentiation of renal progenitor cells. *Development*, *141*, 1047–1058.

Benko, S., Gordon, C. T., Mallet, D., Sreenivasan, R., Thauvin-Robinet, C., Brendehaug, A., et al. (2011). Disruption of a long distance regulatory region upstream of SOX9 in isolated disorders of sex development. *Journal of Medical Genetics*, *48*, 825–830.

Bernard, P., Ryan, J., Sim, H., Czech, D. P., Sinclair, A. H., Koopman, P., et al. (2012). Wnt signaling in ovarian development inhibits Sf1 activation of Sox9 via the Tesco enhancer. *Endocrinology*, *153*, 901–912.

Berta, P., Hawkins, J. R., Sinclair, A. H., Taylor, A., Griffiths, B. L., Goodfellow, P. N., et al. (1990). Genetic evidence equating SRY and the testis-determining factor. *Nature*, *348*, 448–450.

Biason-Lauber, A., Konrad, D., Meyer, M., DeBeaufort, C., & Schoenle, E. J. (2009). Ovaries and female phenotype in a girl with 46,XY karyotype and mutations in the CBX2 gene. *American Journal of Human Genetics*, *84*, 658–663.

Birk, O. S., Casiano, D. E., Wassif, C. A., Cogliati, T., Zhao, L., Zhao, Y., et al. (2000). The LIM homeobox gene Lhx9 is essential for mouse gonad formation. *Nature*, *403*, 909–913.

Bogani, D., Siggers, P., Brixey, R., Warr, N., Beddow, S., Edwards, J., et al. (2009). Loss of mitogen-activated protein kinase kinase kinase 4 (MAP3K4) reveals a requirement for MAPK signalling in mouse sex determination. *PLoS Biology*, 7, e1000196.

Boulanger, L., Pannetier, M., Gall, L., Allais-Bonnet, A., Elzaiat, M., Le Bourhis, D., et al. (2014). FOXL2 is a female sex-determining gene in the goat. *Current Biology*, *24*, 404–408.

Bouma, G. J., Washburn, L. L., Albrecht, K. H., & Eicher, E. M. (2007). Correct dosage of Fog2 and Gata4 transcription factors is critical for fetal testis development in mice. *Proceedings of the National Academy of Sciences of the United States of America*, *104*, 14994–14999.

Buganim, Y., Itskovich, E., Hu, Y. C., Cheng, A. W., Ganz, K., Sarkar, S., et al. (2012). Direct reprogramming of fibroblasts into embryonic Sertoli-like cells by defined factors. *Cell Stem Cell*, *11*, 373–386.

Bullejos, M., & Koopman, P. (2001). Spatially dynamic expression of Sry in mouse genital ridges. *Developmental Dynamics*, *221*, 201–205.

Burgoyne, P. S., Buehr, M., & McLaren, A. (1988). XY follicle cells in ovaries of XX—XY female mouse chimaeras. *Development*, *104*, 683–688.

Callier, P., Calvel, P., Matevossian, A., Makrythanasis, P., Bernard, P., Kurosaka, H., et al. (2014). Loss of function mutation in the palmitoyl-transferase HHAT leads to syndromic 46,XY disorder of sex development by impeding Hedgehog protein palmitoylation and signaling. *PLoS Genetics*, *10*, e1004340.

Camats, N., Fernandez-Cancio, M., Audi, L., Schaller, A., & Fluck, C. E. (2018). Broad phenotypes in heterozygous NR5A1 46,XY patients with a disorder of sex development: An oligogenic origin? *European Journal of Human Genetics*, *26*(9), 1329–1338.

Canto, P., Soderlund, D., Reyes, E., & Mendez, J. P. (2004). Mutations in the desert hedgehog (DHH) gene in patients with 46,XY complete pure gonadal dysgenesis. *The Journal of Clinical Endocrinology and Metabolism*, *89*, 4480–4483.

Carre, G. A., Siggers, P., Xipolita, M., Brindle, P., Lutz, B., Wells, S., et al. (2018). Loss of p300 and CBP disrupts histone acetylation at the mouse Sry promoter and causes XY gonadal sex reversal. *Human Molecular Genetics*, *27*, 190–198.

Clark, A. M., Garland, K. K., & Russell, L. D. (2000). Desert hedgehog (Dhh) gene is required in the mouse testis for formation of adult-type Leydig cells and normal development of peritubular cells and seminiferous tubules. *Biology of Reproduction*, *63*, 1825–1838.

Colvin, J. S., Green, R. P., Schmahl, J., Capel, B., & Ornitz, D. M. (2001). Male-to-female sex reversal in mice lacking fibroblast growth factor 9. *Cell*, *104*, 875–889.

Fletcher, R. B., Das, D., & Ngai, J. (2018). Creating lineage trajectory maps via integration of single-cell RNA-sequencing and lineage tracing: Integrating transgenic lineage tracing and single-cell RNA-sequencing is a robust approach for mapping developmental lineage trajectories and cell fate changes. *Bioessays*, *40*(8), e1800056.

Fujimoto, Y., Tanaka, S. S., Yamaguchi, Y. L., Kobayashi, H., Kuroki, S., Tachibana, M., et al. (2013). Homeoproteins Six1 and Six4 regulate male sex determination and mouse gonadal development. *Developmental Cell*, *26*, 416–430.

Garcia-Moreno, S. A., Plebanek, M. P., & Capel, B. (2018). Epigenetic regulation of male fate commitment from an initially bipotential system. *Molecular and Cellular Endocrinology*, *468*, 19–30.

Gierl, M. S., Gruhn, W. H., von Seggern, A., Maltry, N., & Niehrs, C. (2012). GADD45G functions in male sex determination by promoting p38 signaling and Sry expression. *Developmental Cell*, *23*, 1032–1042.

Ginsburg, M., Snow, M. H., & McLaren, A. (1990). Primordial germ cells in the mouse embryo during gastrulation. *Development*, *110*, 521–528.

Gonen, N., Futtner, C. R., Wood, S., Garcia-Moreno, S. A., Salamone, I. M., Samson, S. C., et al. (2018). Sex reversal following deletion of a single distal enhancer of *Sox9*. *Science*, *360*(6396), 1469–1473.

Gonen, N., Quinn, A., O'Neill, H. C., Koopman, P., & Lovell-Badge, R. (2017). Normal levels of Sox9 expression in the developing mouse testis depend on the TES/TESCO enhancer, but this does not act alone. *PLoS Genetics*, *13*, e1006520.

Granados, A., Alaniz, V. I., Mohnach, L., Barseghyan, H., Vilain, E., Ostrer, H., et al. (2017). MAP3K1-related gonadal dysgenesis: Six new cases and review of the literature. *American Journal of Medical Genetics Part C, Seminars in Medical Genetics*, *175*, 253–259.

Griffeth, R. J., Bianda, V., & Nef, S. (2014). The emerging role of insulin-like growth factors in testis development and function. *Basic and Clinical Andrology*, *24*, 12.

Guigon, C. J., & Magre, S. (2006). Contribution of germ cells to the differentiation and maturation of the ovary: Insights from models of germ cell depletion. *Biology of Reproduction*, *74*, 450–458.

Gustin, S. E., Hogg, K., Stringer, J. M., Rastetter, R. H., Pelosi, E., Miles, D. C., et al. (2016). WNT/beta-catenin and p27/FOXL2 differentially regulate supporting cell proliferation in the developing ovary. *Developmental Biology*, *412*, 250–260.

Hacker, A., Capel, B., Goodfellow, P., & Lovell-Badge, R. (1995). Expression of *Sry*, the mouse testis-determining gene. *Development*, *121*, 1603–1614.

Hao, H. X., Xie, Y., Zhang, Y., Charlat, O., Oster, E., Avello, M., et al. (2012). ZNRF3 promotes Wnt receptor turnover in an R-spondin-sensitive manner. *Nature*, *485*, 195–200.

Harris, A., Siggers, P., Corrochano, S., Warr, N., Sagar, D., Grimes, D. T., et al. (2018). ZNRF3 functions in mammalian sex determination by inhibiting canonical WNT signaling. *Proceedings of the National Academy of Sciences of the United States of America*, *115*, 5474–5479.

Hatano, O., Takakusu, A., Nomura, M., & Morohashi, K. (1996). Identical origin of adrenal cortex and gonad revealed by expression profiles of Ad4BP/SF-1. *Genes to Cells*, *1*, 663–671.
Hiramatsu, R., Matoba, S., Kanai-Azuma, M., Tsunekawa, N., Katoh-Fukui, Y., Kurohmaru, M., et al. (2008). A critical time window of Sry action in gonadal sex determination in mice. *Development*, *136*, 129–138.
Hu, Y. C., Okumura, L. M., & Page, D. C. (2013). Gata4 is required for formation of the genital ridge in mice. *PLoS Genetics*, *9*, e1003629.
Ikeda, Y., Shen, W. H., Ingraham, H. A., & Parker, K. L. (1994). Developmental expression of mouse steroidogenic factor-1, an essential regulator of the steroid hydroxylases. *Molecular Endocrinology*, *8*, 654–662.
Jameson, S. A., Lin, Y. T., & Capel, B. (2012). Testis development requires the repression of Wnt4 by Fgf signaling. *Developmental Biology*, *370*, 24–32.
Jameson, S. A., Natarajan, A., Cool, J., Defalco, T., Maatouk, D. M., Mork, L., et al. (2012). Temporal transcriptional profiling of somatic and germ cells reveals biased lineage priming of sexual fate in the fetal mouse gonad. *PLoS Genetics*, *8*, e1002575.
Jeske, Y. W., Bowles, J., Greenfield, A., & Koopman, P. (1995). Expression of a linear Sry transcript in the mouse genital ridge. *Nature Genetics*, *10*, 480–482.
Johnen, H., Gonzalez-Silva, L., Carramolino, L., Flores, J. M., Torres, M., & Salvador, J. M. (2013). Gadd45g is essential for primary sex determination, male fertility and testis development. *PLoS One*, *8*, e58751.
Karl, J., & Capel, B. (1998). Sertoli cells of the mouse testis originate from the coelomic epithelium. *Developmental Biology*, *203*, 323–333.
Katoh-Fukui, Y., Miyabayashi, K., Komatsu, T., Owaki, A., Baba, T., Shima, Y., et al. (2012). Cbx2, a polycomb group gene, is required for Sry gene expression in mice. *Endocrinology*, *153*, 913–924.
Katoh-Fukui, Y., Owaki, A., Toyama, Y., Kusaka, M., Shinohara, Y., Maekawa, M., et al. (2005). Mouse Polycomb M33 is required for splenic vascular and adrenal gland formation through regulating Ad4BP/SF1 expression. *Blood*, *106*, 1612–1620.
Kim, Y., Kobayashi, A., Sekido, R., DiNapoli, L., Brennan, J., Chaboissier, M. C., et al. (2006). Fgf9 and Wnt4 act as antagonistic signals to regulate mammalian sex determination. *PLoS Biology*, *4*, e187.
Koo, B. K., Spit, M., Jordens, I., Low, T. Y., Stange, D. E., van de Wetering, M., et al. (2012). Tumour suppressor RNF43 is a stem-cell E3 ligase that induces endocytosis of Wnt receptors. *Nature*, *488*, 665–669.
Koopman, P., Gubbay, J., Vivian, N., Goodfellow, P., & Lovell-Badge, R. (1991). Male development of chromosomally female mice transgenic for *Sry*. *Nature*, *351*, 117–121.
Koopman, P., Münsterberg, A., Capel, B., Vivian, N., & Lovell-Badge, R. (1990). Expression of a candidate sex-determining gene during mouse testis differentiation. *Nature*, *348*, 450–452.
Kreidberg, J. A., Sariola, H., Loring, J. M., Maeda, M., Pelletier, J., Housman, D., et al. (1993). WT-1 is required for early kidney development. *Cell*, *74*, 679–691.
Kume, T., Deng, K., & Hogan, B. L. (2000). Murine forkhead/winged helix genes Foxc1 (Mf1) and Foxc2 (Mfh1) are required for the early organogenesis of the kidney and urinary tract. *Development*, *127*, 1387–1395.
Kuroki, S., Matoba, S., Akiyoshi, M., Matsumura, Y., Miyachi, H., Mise, N., et al. (2013). Epigenetic regulation of mouse sex determination by the histone demethylase Jmjd1a. *Science*, *341*, 1106–1109.
Kuroki, S., Okashita, N., Baba, S., Maeda, R., Miyawaki, S., Yano, M., et al. (2017). Rescuing the aberrant sex development of H3K9 demethylase Jmjd1a-deficient mice by modulating H3K9 methylation balance. *PLoS Genetics*, *13*, e1007034.
Kusaka, M., Katoh-Fukui, Y., Ogawa, H., Miyabayashi, K., Baba, T., Shima, Y., et al. (2010). Abnormal epithelial cell polarity and ectopic epidermal growth factor receptor

(EGFR) expression induced in Emx2 KO embryonic gonads. *Endocrinology*, *151*, 5893–5904.

Lavery, R., Chassot, A. A., Pauper, E., Gregoire, E. P., Klopfenstein, M., de Rooij, D. G., et al. (2012). Testicular differentiation occurs in absence of R-spondin1 and Sox9 in mouse sex reversals. *PLoS Genetics*, *8*, e1003170.

Lin, Y. T., Barske, L., DeFalco, T., & Capel, B. (2017). Numb regulates somatic cell lineage commitment during early gonadogenesis in mice. *Development*, *144*, 1607–1618.

Liu, C., Rodriguez, K., & Yao, H. H. (2016). Mapping lineage progression of somatic progenitor cells in the mouse fetal testis. *Development*, *143*, 3700–3710.

Loke, J., Pearlman, A., Radi, O., Zuffardi, O., Giussani, U., Pallotta, R., et al. (2014). Mutations in MAP3K1 tilt the balance from SOX9/FGF9 to WNT/beta-catenin signaling. *Human Molecular Genetics*, *23*, 1073–1083.

Lourenco, D., Brauner, R., Rybczynska, M., Nihoul-Fekete, C., McElreavey, K., & Bashamboo, A. (2011). Loss-of-function mutation in GATA4 causes anomalies of human testicular development. *Proceedings of the National Academy of Sciences of the United States of America*, *108*, 1597–1602.

Lovell-Badge, R., & Robertson, E. (1990). XY female mice resulting from a heritable mutation in the murine primary testis determining gene, *Tdy*. *Development*, *109*, 635–646.

Luo, X., Ikeda, Y., & Parker, K. L. (1994). A cell-specific nuclear receptor is essential for adrenal and gonadal development and sexual differentiation. *Cell*, *77*, 481–490.

Maatouk, D. M., Mork, L., Chassot, A. A., Chaboissier, M. C., & Capel, B. (2013). Disruption of mitotic arrest precedes precocious differentiation and transdifferentiation of pregranulosa cells in the perinatal Wnt4 mutant ovary. *Developmental Biology*, *383*, 295–306.

Maatouk, D. M., Mork, L., Hinson, A., Kobayashi, A., McMahon, A. P., & Capel, B. (2012). Germ cells are not required to establish the female pathway in mouse fetal gonads. *PLoS One*, *7*, e47238.

Maatouk, D. M., Natarajan, A., Shibata, Y., Song, L., Crawford, G. E., Ohler, U., et al. (2017). Genome-wide identification of regulatory elements in Sertoli cells. *Development*, *144*, 720–730.

Mandel, H., Shemer, R., Borochowitz, Z. U., Okopnik, M., Knopf, C., Indelman, M., et al. (2008). SERKAL syndrome: An autosomal-recessive disorder caused by a loss-of-function mutation in WNT4. *American Journal of Human Genetics*, *82*, 39–47.

Mazen, I., Abdel-Hamid, M., Mekkawy, M., Bignon-Topalovic, J., Boudjenah, R., El Gammal, M., et al. (2016). Identification of NR5A1 mutations and possible digenic inheritance in 46,XY gonadal dysgenesis. *Sexual Development*, *10*, 147–151.

Miyake, Z., Takekawa, M., Ge, Q., & Saito, H. (2007). Activation of MTK1/MEKK4 by GADD45 through induced N-C dissociation and dimerization-mediated trans autophosphorylation of the MTK1 kinase domain. *Molecular and Cellular Biology*, *27*, 2765–2776.

Molyneaux, K. A., Stallock, J., Schaible, K., & Wylie, C. (2001). Time-lapse analysis of living mouse germ cell migration. *Developmental Biology*, *240*, 488–498.

Mork, L., Maatouk, D. M., McMahon, J. A., Guo, J. J., Zhang, P., McMahon, A. P., et al. (2012). Temporal differences in granulosa cell specification in the ovary reflect distinct follicle fates in mice. *Biology of Reproduction*, *86*, 37.

Morohashi, K. (1997). The ontogenesis of the steroidogenic tissues. *Genes to Cells*, *2*, 95–106.

Murphy, M. W., Lee, J. K., Rojo, S., Gearhart, M. D., Kurahashi, K., Banerjee, S., et al. (2015). An ancient protein-DNA interaction underlying metazoan sex determination. *Nature Structural & Molecular Biology*, *22*, 442–451.

Nicol, B., & Yao, H. H. (2015). Gonadal identity in the absence of pro-testis factor SOX9 and pro-ovary factor beta-catenin in mice. *Biology of Reproduction*, *93*, 35.

Pannetier, M., Chassot, A. A., Chaboissier, M. C., & Pailhoux, E. (2016). Involvement of FOXL2 and RSPO1 in ovarian determination, development, and maintenance in mammals. *Sexual Development, 10*, 167–184.

Parma, P., Radi, O., Vidal, V., Chaboissier, M. C., Dellambra, E., Valentini, S., et al. (2006). R-spondin1 is essential in sex determination, skin differentiation and malignancy. *Nature Genetics, 38*, 1304–1309.

Patek, C. E., Kerr, J. B., Gosden, R. G., Jones, K. W., Hardy, K., Muggleton-Harris, A. L., et al. (1991). Sex chimaerism, fertility and sex determination in the mouse. *Development, 113*, 311–325.

Pearlman, A., Loke, J., Le Caignec, C., White, S., Chin, L., Friedman, A., et al. (2010). Mutations in MAP3K1 cause 46,XY disorders of sex development and implicate a common signal transduction pathway in human testis determination. *American Journal of Human Genetics, 87*, 898–904.

Pitetti, J. L., Calvel, P., Romero, Y., Conne, B., Truong, V., Papaioannou, M. D., et al. (2013). Insulin and IGF1 receptors are essential for XX and XY gonadal differentiation and adrenal development in mice. *PLoS Genetics, 9*, e1003160.

Polanco, J. C., Wilhelm, D., Davidson, T. L., Knight, D., & Koopman, P. (2010). Sox10 gain-of-function causes XX sex reversal in mice: Implications for human 22q-linked disorders of sex development. *Human Molecular Genetics, 19*, 506–516.

Rahmoun, M., Lavery, R., Laurent-Chaballier, S., Bellora, N., Philip, G. K., Rossitto, M., et al. (2017). In mammalian foetal testes, SOX9 regulates expression of its target genes by binding to genomic regions with conserved signatures. *Nucleic Acids Research, 45*(12), 7191–7211.

Rastetter, R. H., Bernard, P., Palmer, J. S., Chassot, A. A., Chen, H., Western, P. S., et al. (2014). Marker genes identify three somatic cell types in the fetal mouse ovary. *Developmental Biology, 394*, 242–252.

Robevska, G., van den Bergen, J. A., Ohnesorg, T., Eggers, S., Hanna, C., Hersmus, R., et al. (2018). Functional characterization of novel NR5A1 variants reveals multiple complex roles in disorders of sex development. *Human Mutation, 39*, 124–139.

Rybak, A., Fuchs, H., Hadian, K., Smirnova, L., Wulczyn, E. A., Michel, G., et al. (2009). The let-7 target gene mouse lin-41 is a stem cell specific E3 ubiquitin ligase for the miRNA pathway protein Ago2. *Nature Cell Biology, 11*, 1411–1420.

Schmahl, J., & Capel, B. (2003). Cell proliferation is necessary for the determination of male fate in the gonad. *Developmental Biology, 258*, 264–276.

Schmahl, J., Eicher, E. M., Washburn, L. L., & Capel, B. (2000). Sry induces cell proliferation in the mouse gonad. *Development, 127*, 65–73.

Schnabel, C. A., Selleri, L., & Cleary, M. L. (2003). Pbx1 is essential for adrenal development and urogenital differentiation. *Genesis, 37*, 123–130.

Seki, Y., Hayashi, K., Itoh, K., Mizugaki, M., Saitou, M., & Matsui, Y. (2005). Extensive and orderly reprogramming of genome-wide chromatin modifications associated with specification and early development of germ cells in mice. *Developmental Biology, 278*, 440–458.

Sekido, R., Bar, I., Narvaez, V., Penny, G., & Lovell-Badge, R. (2004). SOX9 is up-regulated by the transient expression of SRY specifically in Sertoli cell precursors. *Developmental Biology, 274*, 271–279.

Sekido, R., & Lovell-Badge, R. (2008). Sex determination involves synergistic action of SRY and SF1 on a specific Sox9 enhancer. *Nature, 453*, 930–934.

Spanjaard, B., & Junker, J. P. (2017). Methods for lineage tracing on the organism-wide level. *Current Opinion in Cell Biology, 49*, 16–21.

Stevant, I., Kuhne, F., Greenfield, A., Chaboissier, M.-C., Dermitzakis, E. T., & Serge, N. (2018). Single-cell transcriptomics of the mouse gonadal soma reveals the establishment of sexual dimorphism in distinct cell lineages. *BioRxiv*. https://doi.org/10.1101/410407.

Stevant, I., & Nef, S. (2018). Single cell transcriptome sequencing: A new approach for the study of mammalian sex determination. *Molecular and Cellular Endocrinology*, *468*, 11–18.

Stevant, I., Neirijnck, Y., Borel, C., Escoffier, J., Smith, L. B., Antonarakis, S. E., et al. (2018). Deciphering cell lineage specification during male sex determination with single-cell RNA sequencing. *Cell Reports*, *22*, 1589–1599.

Sutton, E., Hughes, J., White, S., Sekido, R., Tan, J., Arboleda, V., et al. (2011). Identification of SOX3 as an XX male sex reversal gene in mice and humans. *The Journal of Clinical Investigation*, *121*, 328–341.

Svingen, T., & Koopman, P. (2013). Building the mammalian testis: Origins, differentiation, and assembly of the component cell populations. *Genes & Development*, *27*, 2409–2426.

Uhlenhaut, N. H., Jakob, S., Anlag, K., Eisenberger, T., Sekido, R., Kress, J., et al. (2009). Somatic sex reprogramming of adult ovaries to testes by FOXL2 ablation. *Cell*, *139*, 1130–1142.

Val, P., Martinez-Barbera, J. P., & Swain, A. (2007). Adrenal development is initiated by Cited2 and Wt1 through modulation of Sf-1 dosage. *Development*, *134*, 2349–2358.

Vidal, V., & Schedl, A. (2000). Requirement of WT1 for gonad and adrenal development: Insights from transgenic animals. *Endocrine Research*, *26*, 1075–1082.

Wagner, T., Wirth, J., Meyer, J., Zabel, B., Held, M., Zimmer, J., et al. (1994). Autosomal sex reversal and campomelic dysplasia are caused by mutations in and around the *SRY*-related gene *SOX9*. *Cell*, *79*, 1111–1120.

Wang, Q., Lan, Y., Cho, E. S., Maltby, K. M., & Jiang, R. (2005). Odd-skipped related 1 (Odd 1) is an essential regulator of heart and urogenital development. *Developmental Biology*, *288*, 582–594.

Warr, N., Bogani, D., Siggers, P., Brixey, R., Tateossian, H., Dopplapudi, A., et al. (2011). Minor abnormalities of testis development in mice lacking the gene encoding the MAPK signalling component, MAP3K1. *PLoS One*, *6*, e19572.

Warr, N., Carre, G. A., Siggers, P., Faleato, J. V., Brixey, R., Pope, M., et al. (2012). Gadd45gamma and Map3k4 interactions regulate mouse testis determination via p38 MAPK-mediated control of Sry expression. *Developmental Cell*, *23*, 1020–1031.

Warr, N., Siggers, P., Carre, G. A., Bogani, D., Brixey, R., Akiyoshi, M., et al. (2014). Transgenic expression of Map3k4 rescues T-associated sex reversal (Tas) in mice. *Human Molecular Genetics*, *23*, 3035–3044.

Warr, N., Siggers, P., Carre, G. A., Wells, S., & Greenfield, A. (2016). Genetic analyses reveal functions for MAP2K3 and MAP2K6 in mouse testis determination. *Biology of Reproduction*, *94*, 103.

Wilhelm, D., Washburn, L. L., Truong, V., Fellous, M., Eicher, E. M., & Koopman, P. (2009). Antagonism of the testis- and ovary-determining pathways during ovotestis development in mice. *Mechanisms of Development*, *126*, 324–336.

Yang, J., Gao, C., Chai, L., & Ma, Y. (2010). A novel SALL4/OCT4 transcriptional feedback network for pluripotency of embryonic stem cells. *PLoS One*, *5*, e10766.

Yang, M., Yang, S. L., Herrlinger, S., Liang, C., Dzieciatkowska, M., Hansen, K. C., et al. (2015). Lin28 promotes the proliferative capacity of neural progenitor cells in brain development. *Development*, *142*, 1616–1627.

Zhang, L., Chen, M., Wen, Q., Li, Y., Wang, Y., Wang, Y., et al. (2015). Reprogramming of Sertoli cells to fetal-like Leydig cells by Wt1 ablation. *Proceedings of the National Academy of Sciences of the United States of America*, *112*, 4003–4008.

Zheng, W., Zhang, H., Gorre, N., Risal, S., Shen, Y., & Liu, K. (2014). Two classes of ovarian primordial follicles exhibit distinct developmental dynamics and physiological functions. *Human Molecular Genetics*, *23*, 920–928.

CHAPTER SEVEN

Role of epigenetic regulation in mammalian sex determination

Shingo Miyawaki[a,b], **Makoto Tachibana**[a,b,*]
[a]Institute of Advanced Medical Sciences, Tokushima University, Tokushima, Japan
[b]Graduate School of Frontier Biosciences, Osaka University, Osaka, Japan
*Corresponding author: e-mail address: tachiban@fbs.osaka-u.ac.jp

Contents

Abstract

Mammalian sex determination is triggered by activation of the mammalian sex-determining gene, *Sry*, in a spatially and temporally controlled manner. Because reduced or delayed *Sry* expression results in male-to-female sex reversal, male development is highly dependent on the accurate transcription of *Sry*. *SRY* dysregulation is a potential cause of human disorders of sex development (DSD). In addition to changes in DNA sequences, gene expression is regulated by epigenetic mechanisms. Epigenetic regulation ensures spatial and temporal accuracy of the expression of developmentally regulated genes. Epigenetic regulation such as histone tail modification, DNA methylation, chromatin remodeling, and non-coding RNA regulation engages several biological processes in multicellular organisms. In recent years, it has been revealed that various types of epigenetic regulation are involved in accurate gonadal differentiation

Current Topics in Developmental Biology, Volume 134
ISSN 0070-2153
https://doi.org/10.1016/bs.ctdb.2019.01.008

in mammals. In particular, histone modification plays an integral part in sex determination, which is the first step of gonadal differentiation. Here, we focus on the findings on the epigenetic modifications that regulate *Sry* expression. Finally, we discuss the role of metabolism that potentially alters the epigenetic state in response to environmental cues.

1. Introduction

Many species on the earth are classified into two sexes, male and female. Sex determination involves the development of the gonads into testes or ovaries, which determines the phenotypic sex of the internal and external genitalia. Sex determination is usually categorized as either genotypic sex determination (GSD) or environmental sex determination (ESD), and the former is the most common type of mammalian sex determination system. In eutherian mammals, phenotypic sex is determined by the presence or absence of the sex-determining region of the chromosome Y (*Sry*) gene, which is located in the male-specific region of the Y chromosome (Gubbay et al., 1990; Koopman, Gubbay, Vivian, Goodfellow, & Lovell-Badge, 1991; Sinclair et al., 1990). The expression of *Sry* is only detectable during a short period from embryonic day (E) 10.5 to 12.5 in mouse (Hacker, Capel, Goodfellow, & Lovell-Badge, 1995; Hiramatsu et al., 2009; Jeske, Mishina, Cohen, Behringer, & Koopman, 1996; Koopman, Münsterberg, Capel, Vivian, & Lovell-Badge, 1990; Wilhelm et al., 2009). Such temporal accuracy of *Sry* expression is essential for male development (Hiramatsu et al., 2009; Wilhelm et al., 2009).

Accurate gene regulation generally depends on epigenetic modifications. Epigenetics is defined as the study of heritable changes in gene expression that cannot be explained by alterations in the DNA sequence (Berger, Kouzarides, Shiekhattar, & Shilatifard, 2009; Wolffe & Matzke, 1999). Epigenetic regulation can occur through post-translational histone tail modification, DNA modification, chromatin remodeling and changes in chromosomal organization, and long non-coding RNA regulating local chromatin structures. During mammalian development, structural changes in chromatin occur in order to activate or suppress gene expression in a stage-specific manner. Therefore, epigenetic regulatory mechanisms play indispensable roles in mammalian development. In recent years, a compelling body of evidence regarding the role of epigenetics in mammalian sex determination has been accumulated (Garcia-Moreno, Plebanek, & Capel, 2018; Kuroki & Tachibana, 2018). For example, it has been found that a histone demethylase, JmjC-domain-containing protein 1a (*Jmjd1a*), histone acetyltransferases, cAMP response element binding protein (CREB)-binding protein (CBP) and its paralog p300 are crucial for *Sry* activation; as such, multiple epigenetic regulatory mechanisms might be required to ensure the accurate expression of *Sry* (Carré et al., 2018; Kuroki et al., 2013). It is also shown that epigenetics is involved in ESD animals. For instance, DNA methylation controls *Cyp19a1* expression in European sea bass, and histone methylation regulates *Dmrt1* expression in the red-eared slider turtle (Ge et al., 2018; Navarro-Martín et al., 2011). In this review, we provide an overview of our current knowledge about the epigenetic regulation of mammalian sex development.

2. Transcriptional regulation of the *Sry* gene

Differentiation of Sertoli cells is the first step in male sex development in mammals (Albrecht & Eicher, 2001; Koopman et al., 1990). It is driven by the expression of *Sry* in the supporting cell lineage in the sexually undifferentiated gonad (Gubbay et al., 1990; Sinclair et al., 1990). Sertoli cells are essential for the coordination of testicular development. The sexually undifferentiated cells expressing *Sry* are called pre-Sertoli cells. Sertoli cells form the testis cords that encapsulate and direct germ cell differentiation, and orchestrate the differentiation of other somatic cells, including the steroidogenic cell lineage (Habert, Lejeune, & Saez, 2001).

Developmental gene expression is fine-tuned through cross-talk between the function of transcription factors and epigenetic modifications. Moreover, recent studies have revealed that transcription factors (TFs) and epigenetic modifications are involved in *Sry* regulation to ensure the spatial and temporal accuracy of its expression. In this section, we introduce TFs and kinases involved in the regulation of *Sry* expression. *Sry* regulators such as GATA-binding factor 4 (GATA4), nuclear receptor subfamily 5 group A member 1 (NR5A1), also referred to as Ad4BP/SF-1, Wilms tumor protein (WT1), chromobox protein homolog 2 (CBX2), mitogen-activated protein kinase kinase kinase 4 (MAP3K4), sine oculis-related homeobox (SIX1/4), the insulin-like growth factor 1 receptor (IGF1R), and growth arrest and DNA-damage-inducible protein 45G (GADD45G) have been identified (Ono & Harley, 2013). The binding sites of WT1 and GATA4 on the *Sry* locus have been identified (Gierl, Gruhn, von Seggern, Maltry, & Niehrs, 2012; Miyamoto, Taniguchi, Hamel, Silversides, & Viger, 2008). These sites are restricted to the proximal promoter of *Sry*, within just a few 100 base pairs of the transcription start site. NR5A1 binding sites in the pig and human *Sry* promoters have been experimentally determined by electrophoretic mobility shift assays (de Santa Barbara et al., 2001; Pilon et al., 2003). Because of low homology of the mouse 5′ flanking *Sry* sequence, the NR5A1-binding sites in pig and human are not found in the corresponding position in mouse. *Sry* regulation has been well explained by Larney et al. in terms of the "three-module model." Briefly, several *Sry* regulators described above converge on each of the three key TFs, GATA4, NR5A1, and WT1. It is anticipated that future work will reveal novel elements (e.g., enhancers or silencers) involved in *Sry* regulation and their binding proteins. Such findings would deepen our understanding of how *Sry* expression is triggered (Larney, Bailey, & Koopman, 2014; Quinn et al., 2014).

3. Epigenetic regulation of mammalian sex determination

Several studies have demonstrated that the *Sry* locus undergoes epigenetic modifications, which have been implicated in the regulation of *Sry* expression. In this section, we present an overview of the role of histone modifications in gene regulation, especially in *Sry* expression.

3.1 Histone modification in the regulation of gene transcription

The packaging of DNA and histones into chromatin is an essential aspect of gene regulation. In addition, *trans*-acting TFs and *cis*-acting chromatin structures (e.g., those that are accessible or inaccessible to TFs) influence gene expression. Epigenetic modifications make substantial contributions to gene regulation through changing the chromatin structure, especially in regulatory elements such as promoters and enhancers. Recently, as our understanding of histone modification-mediated changes of chromatin structures has increased, the complex interplay among different epigenetic modes is starting to be revealed (the histone code hypothesis). The epigenetic modes based on histone modification involve enzymes that add histone modifications to chromatin (writers), proteins that recognize and bind to chromatin through histone modification-recognition domains (readers), and enzymes that remove histone modifications from chromatin (erasers) (Fig. 1).

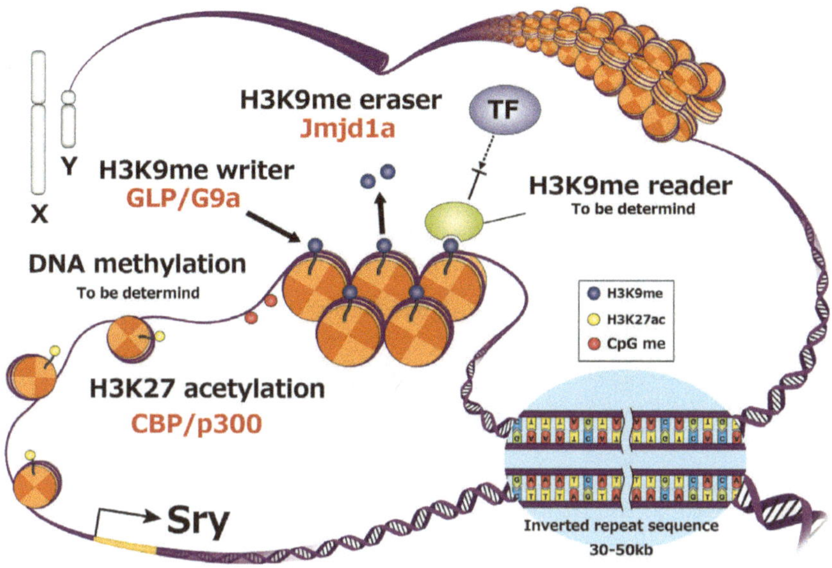

Fig. 1 The contributors for the epigenetic regulation of *Sry*. *Sry* gene is surrounded by large inverted repeats located in the male-specific region of the Y chromosome. Epigenetic modifications such as DNA methylation, histone methylation, and histone acetylation can alter the accessibility of DNA to transcription machinery and therefore influence gene expression. The epigenetic regulation is achieved by functional coupling between the enzymes that modify chromatin (writers), proteins that bind to specific modification and trigger the downstream pathway (readers), and the enzymes that remove modifications (erasers). In recent years, it has been revealed that multiple epigenetic regulatory mechanisms are involved in ensuring the accurate expression of *Sry*. Jmjd1a and GLP/G9a complex are the enzymes that demethylate and methylate H3K9 at the *Sry* locus. The reader protein that recognizes H3K9me at the *Sry* locus has not been identified. The histone acetyltransferases p300/CBP targets H3K27ac at the *Sry* locus. DNA demethylation concomitantly occurs with *Sry* activation. However, DNA demethylation pathway of the *Sry* locus should be determined. TF: Transcription factor.

Histone lysine methylation occurs on the H3 and H4 core histone molecules. Unlike histone acetylation, which is generally associated with transcriptional activation, histone lysine methylation contributes to transcriptional activation or repression, depending on the methylation site (Martin & Zhang, 2005). For example, methylation of histone H3 lysine 4 (H3K4) is seen in the promoter regions of actively transcribed genes. In contrast, methylation of H3K9 is enriched in transcriptionally silenced heterochromatin (Nakayama, Rice, Strahl, Allis, & Grewal, 2001; Peters et al., 2001). Histone modification is dynamically regulated by the opposing activities of writers and erasers. A protein complex that includes G9A and its related protein GLP regulates di-methylation of H3K9 (H3K9me2) (Tachibana, Sugimoto, Fukushima, & Shinkai, 2001; Tachibana et al., 2002, 2005). Depletion of either G9A or GLP in mouse leads to embryonic lethality at approximately E 9.5, indicating that H3K9me2 is crucial for post-implantation development. *Jmjd1a* (which is also called *Tsga/Jhdm2a/Kdm3a*) is an enzyme that demethylates H3K9 (Yamane et al., 2006). An analysis of *Jmjd1a*-deficient mice revealed that *Jmjd1a* plays vital roles during spermatogenesis (Liu et al., 2010; Okada, Scott, Ray, Mishina, & Zhang, 2007) and metabolic regulation (Inagaki et al., 2009; Tateishi, Okada, Kallin, & Zhang, 2009). Furthermore, *Jmjd1a*-deficient mice exhibit male-to-female sex reversal, as described below (Kuroki et al., 2013).

3.2 Role of histone methylation and demethylation in sex development

The expression of *Jmjd1a* in gonadal somatic cells increases from E 10.5 and reaches a plateau around E 11.5, namely, about the same time as *Sry* expression occurs. Single-cell RNA-sequence analysis confirmed that *Jmjd1a* and *Sry* show the same expression kinetics during gonadal somatic cell differentiation (Stévant et al., 2018). *Jmjd1a* deficiency did not induce a delay of *Sry* expression time, but weakened the peak of *Sry* expression in developing gonads. Notably, the expression levels of known regulators of *Sry* were unaffected. Furthermore, the expression levels of genes encoding TFs essential for gonadal development, such as LIM homeobox 9 (*Lhx9*), dosage-sensitive sex reversal, adrenal hypoplasia critical region, on chromosome X, gene 1 (*Dax-1*), aristaless-related homeobox (*Arx*), and doublesex and mab-3-related transcription factor 1 (*Dmrt1*), were also not changed in the *Jmjd1a*-deficient gonads. Exogenous expression of *Sry* rescued testicular development, strongly indicating that perturbed *Sry* expression is solely responsible for the sex reversal seen in XY *Jmjd1a*-deficient mice (Kuroki et al., 2013).

Interestingly, *Jmjd1a*-deficient XY mice showed varying degrees of sex reversal (Kuroki et al., 2013). A certain population of *Jmjd1a*-deficient XY mice develops to males carrying two testes, while the remainder undergoes sex reversal, either partially by exhibiting an ovary and a testis or completely by exhibiting two ovaries. The genetic background of the Y chromosome (Y^{CBA} or Y^{B6}) influences the frequencies of sex reversal in *Jmjd1a*-deficient mice; it occurs more often in Y^{CBA} than in Y^{B6}. For example, abnormal sex differentiation was found in 88% of *Jmjd1a*-deficient XY mice carrying Y^{CBA}, whereas the rate was only 14% in *Jmjd1a*-deficient

XY mice carrying Y^{B6}. Moreover, it has been shown that the expression of *Sry* is lower in Y^{CBA} than in Y^{B6}, in a B6 autosomal background (Kuroki et al., 2013). It is conceivable that the *Sry* expression levels in *Jmjd1a*-deficient gonads at E 11.5 might be near the threshold level for triggering the male pathway, and therefore differences in the genetic background may critically influence the subsequent sexual development (Kuroki et al., 2013).

ChIP experiments using isolated gonadal somatic cells at E 11.5 demonstrated that JMJD1A binds to the *Sry* locus. *Jmjd1a* deficiency resulted in an increase of H3K9me2 levels, whereas H3K4me2, a mark of actively transcribed genes, was decreased in the *Sry* locus. Taking these findings together, it was concluded that JMJD1A directly targets *Sry* and removes suppressive H3K9 methylation to activate expression (Kuroki et al., 2013) (Fig. 1). H3K9 methylation can be recognized by specific modules of the reader proteins, such as chromodomains (Bannister et al., 2001; Lachner, O'Carroll, Rea, Mechtler, & Jenuwein, 2001), ankyrin repeats (Collins et al., 2008), and tandem Tudor domains (Nady et al., 2011). It is conceivable that the H3K9me reader protein is involved in the negative regulation of *Sry* through mediating heterochromatinization. Temporal and cell-type-specific expression of *Jmjd1a* may play a role in loosening the heterochromatin structure at the *Sry* locus (discussed below). However, at present, it remains unresolved how *Jmjd1a* expression is regulated and how JMJD1A is recruited to the *Sry* promoter.

The increased levels of H3K9 methylation at the *Sry* locus in *Jmjd1a*-deficient gonads indicate the existence of an enzyme catalyzing H3K9 methylation, thereby inducing heterochromatinization at the *Sry* locus. Kuroki et al. identified the GLP/G9A H3K9 methyltransferase complex as the enzyme that counteracts *Jmjd1a*-mediated H3K9 demethylation at the *Sry* locus (Kuroki et al., 2017) (Fig. 1). Based on this finding, they attempted to rescue the sex-reversal phenotype of *Jmjd1a*-deficient mice by modulating GLP/G9A complex activity via the introduction of a heterozygous *GLP* mutation. This attempt successfully restored *Sry* expression and rescued the sex reversal (Kuroki et al., 2017). In addition, the administration of UNC0642, a chemical inhibitor of the GLP/G9A enzyme, into *Jmjd1a*-deficient embryos successfully rescued the sex-reversal phenotype. These findings revealed that the level of H3K9 methylation at the *Sry* locus is finely controlled by the balance of activity between the H3K9 demethylase *Jmjd1a* and the H3K9 methyltransferase GLP/G9A complex to ensure appropriate levels of *Sry* expression.

H3K4me3, a permissive histone-modification mark, is seen in the promoter regions of transcriptionally activated genes. It was shown that

H3K4me3 is abundant in the *Sry* promoter in pre-Sertoli cells, but not in adjacent mesonephric cells, where *Sry* is not expressed (Gierl et al., 2012). The enzymes responsible for H3K4me3 in the *Sry* locus have not yet been determined. The cause-and-effect relationship between H3K4me3 and *Sry* activation also remains to be elucidated.

Recently, it was discovered that histone demethylation is also involved in sex determination in reptilian species (Deveson et al., 2017; Ge et al., 2018). In many reptilian species, gonadal sex is determined by the environmental temperature during embryogenesis. Ge et al. showed that, in the red-eared slider turtle, an epigenetic modifier, the histone demethylase lysine demethylase 6B (KDM6B) (which is also called JMJD3), exhibits temperature-dependent sexually dimorphic expression in early embryos. KDM6B binds to the promoter of the male sex-determining gene *Dmrt1* and promotes its expression by eliminating H3K27me3 from gonadal cells (Ge et al., 2018). The knockdown of KDM6B under temperature conditions that would normally produce males resulted in female development. Thus, it was shown that the histone demethylase KDM6B plays an essential role in the temperature-dependent sex determination in this turtle species. Building on these findings, the next important step is to determine how temperature differences trigger the sexually dimorphic expression of *KDM6B*.

3.3 Histone acetylation in sex development

Histone acetylation is abundant in transcriptionally activated chromatin. It was shown that H3ac (detected by anti-pan-acetyl-histone H3 antibody), H3K9ac, and H3K27ac are enriched in the *Sry* promoter in pre-Sertoli cells (Gierl et al., 2012). Recently, the histone acetyltransferases CBP and its paralog p300 were reported to play roles in murine sex determination (Carré et al., 2018) (Fig. 1). Mutant embryos lacking either CBP or p300 demonstrated partial male-to-female sex reversal, and embryos lacking three of the four functional copies of *p300/Cbp* exhibited reduced expression of *Sry* and complete male-to-female sex reversal. The levels of H3K27ac were reduced in the mutant gonads at E 11.5, indicating that p300/CBP targets H3K27ac at the *Sry* locus. These findings suggest that this histone acetyltransferase complex is a key component for *Sry* activation during testicular development. However, normal H3K9ac levels were detected in the gonads of mice lacking p300/CBP, indicating that p300/CBP might not be the catalytic enzyme responsible for H3K9ac deposition at the *Sry* locus.

In addition to the acetylation of histone proteins, the SRY protein itself is known to be acetylated (Thevenet et al., 2004). The K136 residue of human SRY, which is embedded in the C-terminal nuclear localization signal of the high-mobility group domain of SRY, is acetylated by the histone acetyltransferase p300. It is postulated that SRY acetylation enhances its nuclear localization via an interaction with importin-β. Histone deacetylase 3 (HDAC3) can deacetylate the K136 residue of SRY (Thevenet et al., 2004). The biological significance of the nucleocytoplasmic shuttling of SRY for sex determination warrants further investigation.

3.4 Gene regulation by polycomb group proteins

Polycomb (Pc) was identified and genetically characterized by Pam and Ed Lewis in *Drosophila* (Lewis, 1978). The Pc group (PcG) proteins form macromolecular complexes composed of multiple protein subunits, in which the core subunits are highly conserved from *Drosophila* to mammals. The PcG proteins are classified into two distinct sets according to the constituting protein subunits, called polycomb-repressive complexes 1 and 2 (PRC1 and PRC2). It was shown that both PRC complexes exhibit enzymatic activity related to histone modification; specifically, PRC1 and PRC2 catalyze the ubiquitination of H2AK119 and methylation of H3K27, respectively (Simon & Kingston, 2013). It is proposed that PRC complexes act as transcriptional repressors and play roles in gene regulation during development (Margueron & Reinberg, 2011; Simon & Kingston, 2013). In addition, some reports have demonstrated that PRC complexes are involved in gene activation (Kondo et al., 2014; Maezawa et al., 2017).

The core components of the PRC1 complex are PcG RING finger protein (PCGF), RING finger protein, polyhomeotic-like protein (PHC), and CBX. CBX family proteins possess chromodomains that can recognize H3K27me3. Ring finger proteins catalyze the ubiquitination of H2AK119 (Kaustov et al., 2011). The loss of *Cbx2* (also referred to as *M33*) leads to complete male-to-female sex reversal and hypoplasia of gonads and adrenal glands in mice (Katoh-Fukui et al., 2005, 1998) and humans (Biason-Lauber, Konrad, Meyer, deBeaufort, & Schoenle, 2009). Expression analyses have demonstrated that *Sry* and *Sox9* transcripts were downregulated in *Cbx2*-deficient XY gonads in mice. The forced expression of *Sry* rescued the sex reversal in XY KO mice, but their testes were clearly smaller than those in control mice. These results suggest that *Cbx2* acts as an activator of *Sry* and governs the determination of gonadal size through regulating genes other than *Sry* (Katoh-Fukui et al., 2012). However, it remains unresolved how *Cbx2* positively regulates *Sry* expression.

3.5 DNA methylation

The enzymes that catalyze DNA methylation have been well characterized. DNA methylation occurs at cytosine bases, predominantly at cytosines located next to guanine (CpG sites) dinucleotide sequences, in the mammalian genome. DNA methylation is also regulated by the opposing activities of modifying and de-modifying enzymes. Maintenance DNA methyltransferase 1 (DNMT1) prefers DNA substrates known as hemi-methylated DNA, in which the CpG site is methylated in only one strand (Bestor & Ingram, 1983; Li, Bestor, & Jaenisch, 1992). DNMT3A and DNMT3B are de novo DNA methyltransferases that do not show any preference for hemi-methylated DNA (Okano, Bell, Haber, & Li, 1999; Okano, Xie, & Li, 1998).

Passive demethylation of newly synthesized DNA occurs automatically during replication if the maintenance DNA methylation pathway is inactive. The mechanisms for active DNA demethylation have long been disputed. Until recently, several reports had indicated that known DNA-modifying pathways are involved in active DNA demethylation, but further exploration was required to confirm this (Nabel, Manning, & Kohli, 2012). More recent findings have indicated that the oxidation of 5-methylcytosine by the ten-eleven translocation (TET) family enzymes plays an essential role in active DNA demethylation processes (Ito et al., 2010; Kriaucionis & Heintz, 2009; Tahiliani et al., 2009). The 5-hydroxymethylcytosine produced by TET enzymes is passively removed through DNA replication. Alternatively, the 5-hydroxymethylcytosine reverts to cytosine residues through oxidation and the subsequent base excision repair pathway, representing a new model for a complete cycle of dynamic cytosine modification (Kohli & Zhang, 2013). DNA methylation is generally associated with gene silencing and plays crucial roles in developmental processes, such as X-chromosome inactivation and genomic imprinting. DNA methylation can also act as a positive regulator of gene expression, particularly when methylation occurs in gene bodies as opposed to promoter or enhancer regions (Baubec et al., 2015). Moreover, there is much evidence to suggest that DNA methylation also contributes to the establishment of tissue-specific gene expression patterns during mammalian development.

Nishino et al. compared the DNA methylation profile of the *Sry* locus between several embryonic tissues (Nishino, Hattori, Tanaka, & Shiota, 2004). There are 16 CpG sites in the 4.5-kb 5′-flanking region of the murine *Sry* locus. It is known that there are two types of *Sry* transcript with different transcription start sites, linear and circular types (Capel, Rasberry, et al., 1993, Capel, Swain, et al., 1993; Dolci, Grimaldi, Geremia, Pesce, & Rossi, 1997; Jeske, Bowles, Greenfield, & Koopman, 1995). The linear transcript produces the functional SRY protein while the circular transcript is considered to be untranslated (Dolci et al., 1997). The CpG site is located upstream of the *Sry* transcription start site for both circular and linear transcripts. Bisulfite sequencing analysis revealed that both circular and linear promoter regions were hypermethylated in E 8.5 whole embryos, in which *Sry* expression had not yet been initiated. At the time of *Sry* activation (E 11.5), both circular and linear promoter regions exhibited low levels of methylation in the gonads, but high levels of methylation were observed in the liver. These sites were hypermethylated again by E 15.5, indicating that there is an inverse relationship between promoter CpG methylation and *Sry* expression in the gonads throughout sex determination (Nishino et al., 2004). An in vitro promoter activity assay demonstrated that a reporter construct containing the linear promoter region of *Sry* repressed reporter activity when the construct was methylated (Nishino et al., 2004). Non-CpG methylation of the internal cytosine in the CCTGG pentanucleotide sequence in the *Sry* gene has been reported (Nishino et al., 2011). An in vitro promoter activity assay suggested that CCTGG methylation at the *Sry* locus seems to act as a transcriptional activator. These findings suggest possible involvement of DNA methylation/demethylation in *Sry* regulation. Further studies are required to determine the cause-and-effect relationship between DNA methylation and *Sry* activation in the in vivo situation.

It has been shown that, upon the cloning of canines generated by somatic cell nuclear transfer (SCNT), male-to-female sex reversal occasionally occurs (Jeong et al., 2016). In the XY canine generated by SCNT, reduced *Sry* expression and hypermethylation of the *Sry* locus were found (Jeong et al., 2016). The molecular mechanism underlying SCNT-induced DNA hypermethylation at the *Sry* locus warrants further study. In addition, it is important to determine whether SCNT-induced male-to-female sex reversal is found in other animal species.

GADD45 has been considered to recruit DNA repair proteins to specific loci to trigger DNA demethylation (Barreto et al., 2007). It was shown that mice deficient in GADD45G, a member of the GADD45 family, exhibited male-to-female sex reversal (Gierl et al., 2012; Warr et al., 2012). It had been speculated that GADD45G might induce DNA demethylation at the *Sry* locus. However, loss of GADD45G did not alter CpG hypomethylation in the proximal *Sry* promoter, although *Sry* expression was reduced to 25% of that in the wild type. On the contrary, it was shown that GADD45G interacts with MAP3K4. In this model, MAP3K4 regulates *Sry* expression via the phosphorylation and activation of GATA4 (Gierl et al., 2012; Warr et al., 2012). Further study is required to reveal the molecular mechanism responsible for DNA hypomethylation at the *Sry* locus in developing gonads (Fig. 1).

3.6 ATP-dependent chromatin remodeling

Chromatin dynamics is crucial for gene regulation and chromosome function. Chromatin dynamics involves the action of specialized ATP-dependent chromatin remodeling complexes (Clapier, Iwasa, Cairns, & Peterson, 2017), of which there are four different families: switch/sucrose nonfermentable (SWI/SNF), ISWI, CHD, and INO80 families. Chromatin remodeling complexes include enzymes that ensure the appropriate density and spacing of nucleosomes. Chromatin remodeling complexes contribute to multiple developmental processes. For example, BAF60C, a subunit of the SWI/SNF-like BAF complexes, interacts with GATA4 and regulates the cardiac transcription factor NKX2.5, which is responsible for cardiac development (Lickert et al., 2004). hSWI/SNF complexes are also involved in muscle differentiation via *MyoD* induction (de la Serna, Carlson, & Imbalzano, 2001). The nucleosome remodeling and deacetylase complexes are also involved in cell fate decisions, such as B-cell differentiation and reprogramming to pluripotent stem cells (Fujita et al., 2004; Rais et al., 2013).

Developmental disorders characterized by neurological and genital abnormalities have been linked to mutations of chromatin remodelers. Alpha-thalassemia/mental retardation, X-linked (ATRX) syndrome is a human congenital disorder. The *ATRX* gene, which encodes the SWI/SNF-like chromatin remodeling protein, is responsible for this syndrome (Gibbons, 2006; Gibbons, Picketts, Villard, & Higgs, 1995). XY patients with deletions or mutations in the *ATRX* gene exhibit severe intellectual disabilities and male-to-female sex reversal, suggesting that ATRX is involved in testis development in humans. Sertoli cell lineage-specific *Atrx*-knockout mice (*Amh*-Cre; $Atrx^{flox/Y}$) were shown to develop small testes and discontinuous tubules due to prolongation of the G_2/M phase. In addition, proliferating Sertoli cells from these mice underwent apoptosis during the fetal stage. Unlike XY human patients with *ATRX* mutations, *Amh*-Cre; *Atrx*$^{flox/Y}$ mice did not show a phenotype of XY sex reversal (Bagheri-Fam et al., 2011). In *Amh*-Cre; *Atrx*$^{flox/Y}$ mice, it seems likely that the deletion of *Atrx* occurs after Sertoli cell differentiation (Hacker et al., 1995). Therefore, it is worth examining whether the deletion of *Atrx* prior to Sertoli cell differentiation may influence the subsequent sex differentiation.

CHARGE syndrome is characterized by a complex array of congenital malformations. A recent study identified a gene called *family with sequence similarity 172, member A* (*FAM172A*), as a novel CHARGE syndrome-associated gene. *Fam172a* mutant mice display male-to-female sex reversal, at least partly due to reduced *Sry* expression (Bélanger et al., 2018). CHARGE syndrome is mainly caused by mutations in the *chromodomain-helicase-DNA-binding protein 7* (*CHD7*) gene, encoding a chromatin remodeler. Bélanger et al. reported that not only *Fam172A* mutant mice but also *Chd7* mutant mice show XY sex reversal. It is considered that *Fam172A* and *Chd7* may interact and form a chromatin remodeler holo-complex. It is proposed that the chromatin remodelers can expose binding sites for transcription activators at a promoter or enhancer to promote gene expression (Boeger, Griesenbeck, Strattan, & Kornberg, 2004; Clapier et al., 2017). However, it is still unknown whether chromatin remodelers are involved in *Sry* regulation. Insights into these issues should improve our understanding of how male sex differentiation is regulated via chromatin remodeling.

3.7 Chromosomal structure

Large deletions of the *Rbm* locus on the Y chromosome, estimated to be 3–4Mb in size, cause sex reversal in mice (Mahadevaiah et al., 1998). It is considered that positional effects may account for this phenomenon; namely, large deletions of the *Rbm* locus bring the *Sry* locus closer to a heterochromatic region, thereby inducing perturbation of *Sry* expression (Capel, Rasberry, et al., 1993, Capel, Swain, et al., 1993; Laval et al., 1995). The murine *Sry* locus is surrounded by large inverted repeats (Gubbay et al., 1992). It is known that repeated DNA elements build transcriptionally inert heterochromatin that is rich in H3K9 methylation (Nishibuchi & Déjardin, 2017). The heterochromatin structure often spreads from the core regions along chromosomes, thereby influencing the expression of neighboring genes. According to this model, silencing of the murine *Sry* locus might be achieved by heterochromatin spreading out from large inverted repeats (Fig. 1). Further study is required to determine whether the large inverted repeats are actually heterochromatinized by H3K9 methylation. If this does occur, it would also be worth examining how the GLP/G9A complex contributes to this process. Alternatively, it has been reported that the ubiquitous expression of *Sry* induces embryonic lethality in mice, indicating an adverse effect of *Sry* expression on development (Yoshida, 2010). According to this scenario, *Sry* expression must be suppressed in most organs except the gonads in the sex-determining period. As the GLP/G9A complex is ubiquitously expressed in adult organs in mice (Tachibana et al., 2002, 2005), it is plausible that this complex contributes to *Sry* silencing not only in embryonic gonads but also in other organs.

3.8 Non-coding RNA and transcriptional gene silencing

Recent developments in high-throughput technologies have enabled genome-wide analyses of the non-coding genome. Among the various types of non-protein-coding transcripts, long non-coding RNAs (lncRNAs) have attracted attention in biomedical research. lncRNAs are defined as RNA transcripts of more than 200 nucleotides in length that do not encode proteins (Rinn & Chang, 2012). They include long intergenic ncRNA, transcripts partially overlapping with annotated genes, promoter-associated transcripts, enhancer RNA, and transcribed pseudogenes (Derrien et al., 2012; Kung, Colognori, & Lee, 2013; Rinn & Chang, 2012). ncRNA molecules can be classified into two types: *cis*-acting ncRNA, which influences the expression and/or chromatin state of nearby genes, and *trans*-acting ncRNA. It is particularly challenging to distinguish between the functions of the RNA molecule itself and those of the DNA from which it is transcribed. At least three potential mechanisms by which an lncRNA locus regulates gene expression have been postulated: First, the lncRNA transcript itself regulates the expression of neighboring genes through recruiting regulatory factors to its own locus. Activity of this type is sequence-dependent; Second, the process of transcription and/or splicing of the lncRNA confers a gene-regulating function-independent of the RNA sequence; Third, a functional DNA element within the lncRNA locus acts as a promoter or enhancer of neighboring genes. The activity of this type is independent of the production of lncRNA (Kopp & Mendell, 2018). Recent studies have revealed that lncRNAs play roles in multiple developmental processes

(Fatica & Bozzoni, 2014; Wapinski & Chang, 2011). Hox transcript antisense RNA (*Hotair*) controls skeletal patterning and gene repression. *Hotair* is transcribed from the *HoxC* locus and represses *HoxD* expression (Li et al., 2013; Rinn et al., 2007). Upperhand (*Uph*) is transcribed from the same promoter as *Hand2*, a transcription factor that is crucial for heart development. Premature termination of *Uph* transcription was shown to impair *Hand2* expression, resulting in cardiac defects (Anderson et al., 2016). These observations indicate that lncRNAs play a vital role in the temporal and spatial control of TFs with key functional roles in development.

The role of lncRNA in sex determination has been identified in the crustacean *Daphnia magna* (Kato et al., 2018). Specifically, an lncRNA named doublesex1 alpha promoter-associated long RNA (*DAPALR*) regulates the male-specific expression of the male-determining gene doublesex1 (*Dsx1*), which produces males in response to environmental stimuli. In mammals, sex determination-related lncRNAs have been found to be expressed during sex determination, but have not yet been shown to play a functional role (Rastetter, Smith, & Wilhelm, 2015). However, recent technical advances, especially genome-wide expression analysis such as global run-on sequences (GRO-seq) to detect nascent transcription, may provide the opportunity to identify sex determination-related lncRNAs (Core, Waterfall, & Lis, 2008; Meng et al., 2014; Wang et al., 2011).

4. Environmental cues for epigenetic regulation

4.1 Transcriptional links with intermediary metabolism

The discovery of the fundamental role of epigenetic regulation of the mammalian sex-determining gene *Sry* led to a new research area, namely, on the mechanistic link between environmental cues and sex determination not only in ESD system but also in GSD system (Tachibana, 2016). In recent years, the links between metabolism and cellular differentiation have become a focus of particular interest (Reid, Dai, & Locasale, 2017). The idea that intermediate metabolic products epigenetically influence gene expression was suggested several years ago. It is proposed that metabolic changes can alter gene expression through modulating the substrates or cofactors for chromatin-modifying enzymes (Fig. 2). It is also postulated that metabolite changes act as an environmental-sensing mechanism, thereby inducing chromatin modification. In this section, we discuss the possible involvement of metabolic pathways in epigenetic gene regulation by introducing some previous studies.

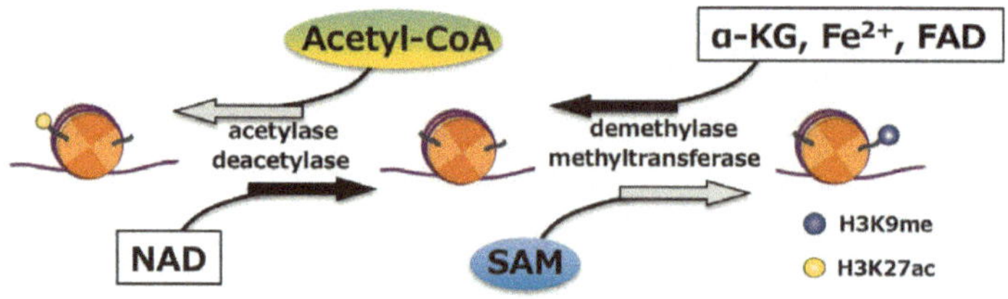

Fig. 2 Metabolites potentially influencing histone-modification status. Histone tails are modified by various enzymes, many of which require metabolites as substrates or cofactors. Histone lysine methylation is controlled by balanced activities of methyltransferase (e.g., GLP/G9a) and demethylase (e.g., Jmjd1a). Histone lysine acetylation is regulated by acetyltransferase (e.g., CBP/p300) and deacetylase. Histone methyltransferase and acetyltransferase use *S*-adenosyl methionine (SAM) and acetyl-CoA as methyl and acetyl donors, respectively. Histone demethylase family enzymes require either flavin adenine dinucleotide (FAD) or α-ketoglutarate (α-KG) and Fe^{2+} as cofactors for exerting demethylation activity. Class III histone deacetylases require nicotinamide adenine dinucleotide (NAD) as a cofactor for exerting deacetylation activity.

First, we provide evidence for possible involvement of amino acid metabolism in epigenetic gene regulation. In mammals, the agouti viable yellow (A^{vy}) locus, which determines hair color, is known to be affected epigenetically in a manner depending on the metabolic status. High and low expressions of A^{vy} result in the agouti (yellow) phenotype and the pseudo-agouti (brown) phenotype, respectively. The coat color of A^{vy} mice strictly depends on DNA methylation at this locus. Wolff et al. showed that supplementation of the maternal diet with methyl donor that affects *S*-adenosyl methionine (SAM) synthesis leads to promoter methylation of A^{vy} and pseudo-agouti coat color of offspring (Wolff, Kodell, Moore, & Cooney, 1998). As another example of nutritional state affecting epigenetics, pluripotent stem cells were shown to require high levels of α-ketoglutarate (α-KG) to promote the demethylation of H3K27me3, H3K9me3, H4K20me3, and TET-dependent DNA demethylation, and to contribute to the regulation of pluripotency-associated gene expression (Carey, Finley, Cross, Allis, & Thompson, 2015).

Second, we discuss possible involvement of sugar and fatty acid metabolism in epigenetic gene regulation. Acetyl coenzyme A (acetyl-CoA) is the universal donor for acetylating reactions in eukaryotes. During fasting, the organism switches to fatty acid oxidation. Acetyl-CoA produced by fatty acid oxidation is used to fuel the tricarboxylic acid cycle. In contrast, when sufficient amounts of nutrients are consumed, glycolysis becomes the primary pathway for acetyl-CoA generation. It was revealed that the excess amount of acetyl-CoA is exported from the mitochondria to the cytosol, serving as a carbon donor for anabolic reactions, including lipogenesis and the synthesis of cholesterol and amino acids. This acetyl-CoA transfer was catalyzed by ATP-citrate lyase (ACLY). It is an important issue whether the abundance of acetyl-CoA influences transcription through modulating histone acetylation. Recent studies have demonstrated that enzymes involved in acetyl-CoA synthesis, such as acyl-CoA synthetase short-chain family member 2 (ACSS2) (Mews et al., 2017), Acly (Wellen et al., 2009), and phosducin (PDC) (Sutendra et al., 2014), can translocate into the nucleus and regulate gene expression. During neurogenesis, ACSS2 becomes abundant in the nuclei of neuronal cells, and it then activates neuronal genes through histone acetylation (Mews et al., 2017). During T-cell differentiation, lactate dehydrogenase A (LDHA) is used to maintain high concentrations of acetyl-CoA, which enhances histone acetylation (Peng et al., 2016).

In summary, numerous metabolites influence epigenetic modification. Histone acetyltransferases and methyltransferases use acetyl-CoA and SAM as acetyl and methyl donors, respectively (Fig. 2). Class III histone deacetylases, such as sirtuins, require nicotinamide adenine dinucleotide (NAD^+) (Haigis & Guarente, 2006). Amine oxidase-type histone demethylases such as LSD1 require flavin adenine dinucleotide (FAD) (Shi et al., 2004), whereas JmjC-domain-containing histone demethylases rely on Fe(II) and α-KG for their enzymatic activity (Loenarz & Schofield, 2008; Tsukada et al., 2006). In the regulation of DNA methylation and demethylation, DNA methyltransferases and TET enzymes use SAM and a dioxygenation reaction that requires Fe^{2+}, O_2, and α-KG, respectively (Feil & Fraga, 2012; Kohli & Zhang, 2013). These findings suggest that cellular metabolism can affect gene expression through modulating epigenetic status and may also be involved in sex determination.

4.2 The role of metabolism in sex determination

NR5A1 (which is also referred to as AD4BP/SF-1) is known to play a number of roles in sex development, including in the establishment of the genital ridges (Luo, Ikeda, & Parker, 1994). *Nr5a1* mutations result in XY sex reversal in humans (Achermann, Ito, Ito, Hindmarsh, & Jameson, 1999), and *Nr5a1*-deficient mice exhibit gonadal agenesis (Luo et al., 1994). *Nr5a1* is expressed in most of the embryonic gonadal somatic cells, including *Sry*-expressing Sertoli cells and fetal Leydig cells (Karl & Capel, 1998; Schmahl, Eicher, Washburn, & Capel, 2000; Stévant et al., 2018). *Nr5a1* is also involved in glucose metabolism, as it regulates genes for the glycolytic pathway and the pentose phosphate pathway (Baba et al., 2014). Glycogen accumulation in pre-Sertoli cells is one of the earliest cellular events after *Sry* activation in the developing gonads (Matoba et al., 2005). Nef et al. also showed that *Sry* expression was severely reduced in XY mice with mutations in all three insulin receptor family members, suggesting that insulin signaling upstream of *Sry* might play a role in controlling *Sry* expression (Nef et al., 2003). These findings support the possible involvement of metabolism-based mechanisms in mammalian sex determination. It was shown that global metabolic change occurs during the developmental stage of E 10.5, which is the time at which *Sry* expression is initiated (Miyazawa et al., 2017). This is the chorioallantoic branching stage, when the maternal–fetal exchange of nutrients is promoted. As shown in the case of agouti viable yellow (A^{vy}) expression, embryonic development may be influenced by the maternal nutritional and metabolic state. It is an important issue whether this state can also influence embryonic sex differentiation. However, much about this issue remains to be revealed. Studies of the mechanistic link between metabolic pathways and epigenetic regulators will confer a more complete understanding of cellular and organismal functions during sex determination.

5. Summary

In this review, we have highlighted the epigenetic regulation of mammalian sex-determining gene *Sry*. *Sry* expression is quite strictly controlled in terms of space and time to induce the male pathway. Several TFs have been shown to be involved in *Sry* regulation. In contrast, the research on epigenetic regulation of *Sry* has just begun (Garcia-Moreno et al., 2018; Kuroki & Tachibana, 2018). To date, only two studies have demonstrated that mice lacking histone modifying enzymes exhibit XY sex reversal accompanied by altered levels of epigenetic modifications of *Sry*. First, it was shown that the removal of the suppressive mark H3K9me2 by JMJD1A is crucial for *Sry* activation (Kuroki et al., 2013). Second, the deposition of the active mark H3K27ac by p300/CBP was also shown to be crucial for *Sry* activation (Carré et al., 2018). In addition to these mutations, the active mark H3K4me3 has been reported to be correlated with *Sry* expression (Gierl et al., 2012). Moreover, another active mark, H3K9ac, has been reported to be abundant at the *Sry* locus in *Sry*-expressing gonadal somatic cells (Carré et al., 2018; Gierl et al., 2012). As mentioned above, the CpG methylation levels of the *Sry* promoter region have been found to be inversely correlated with *Sry* expression (Nishino et al., 2004). To evaluate the biological significance of these epigenetic marks for *Sry* regulation, identifying the enzymes or regulatory molecules controlling the levels of these marks appears to be critical. Elucidating such epigenetic regulation should help in the comprehensive understanding of the complex system of sex determination in mammals. In addition, it may be important to know how chromatin remodeling and lncRNA contribute to *Sry* regulation.

The dysregulation of epigenetic machinery has been found to be associated with several human diseases (Egger, Liang, Aparicio, & Jones, 2004; Jiang, Bressler, & Beaudet, 2004). DSDs are complex disorders, involving atypical chromosomal, gonadal, and/or anatomical sexual characteristics. More than half of human DSDs cannot be explained by alterations in the genes that are known to be required for sex determination or gonadogenesis (Munger & Capel, 2012). In this regard, it is possible that epigenetic alterations during embryogenesis are responsible for the onset of many DSDs. To address this issue, it is important to determine how environmental factors contribute to the operation of epigenetic machinery during mammalian development.

References

Achermann, J. C., Ito, M., Ito, M., Hindmarsh, P. C., & Jameson, J. L. (1999). A mutation in the gene encoding steroidogenic factor-1 causes XY sex reversal and adrenal failure in humans. *Nature Genetics*, *22*, 125–126. https://doi.org/10.1038/9629.

Albrecht, K. H., & Eicher, E. M. (2001). Evidence that Sry is expressed in pre-Sertoli cells and Sertoli and granulosa cells have a common precursor. *Developmental Biology*, *240*, 92–107. https://doi.org/10.1006/dbio.2001.0438.

Anderson, K. M., Anderson, D. M., McAnally, J. R., Shelton, J. M., Bassel-Duby, R., & Olson, E. N. (2016). Transcription of the non-coding RNA upperhand controls Hand2 expression and heart development. *Nature*, *539*, 433–436. https://doi.org/10.1038/nature20128.

Baba, T., Otake, H., Sato, T., Miyabayashi, K., Shishido, Y., Wang, C.-Y., et al. (2014). Glycolytic genes are targets of the nuclear receptor Ad4BP/SF-1. *Nature Communications*, *5*, 3634. https://doi.org/10.1038/ncomms4634.

Bagheri-Fam, S., Argentaro, A., Svingen, T., Combes, A. N., Sinclair, A. H., Koopman, P., et al. (2011). Defective survival of proliferating Sertoli cells and androgen receptor function in a mouse model of the ATR-X syndrome. *Human Molecular Genetics*, *20*, 2213–2224. https://doi.org/10.1093/hmg/ddr109.

Bannister, A. J., Zegerman, P., Partridge, J. F., Miska, E. A., Thomas, J. O., Allshire, R. C., et al. (2001). Selective recognition of methylated lysine 9 on histone H3 by the HP1 chromo domain. *Nature*, *410*, 120–124. https://doi.org/10.1038/35065138.

Barreto, G., Schäfer, A., Marhold, J., Stach, D., Swaminathan, S. K., Handa, V., et al. (2007). Gadd45a promotes epigenetic gene activation by repair-mediated DNA demethylation. *Nature*, *445*, 671–675. https://doi.org/10.1038/nature05515.

Baubec, T., Colombo, D. F., Wirbelauer, C., Schmidt, J., Burger, L., Krebs, A. R., et al. (2015). Genomic profiling of DNA methyltransferases reveals a role for DNMT3B in genic methylation. *Nature*, *520*, 243–247. https://doi.org/10.1038/nature14176.

Bélanger, C., Bérubé-Simard, F.-A., Leduc, E., Bernas, G., Campeau, P. M., Lalani, S. R., et al. (2018). Dysregulation of cotranscriptional alternative splicing underlies CHARGE syndrome. *Proceedings of the National Academy of Sciences of the United States of America*, *115*, E620–E629. 201715378. https://doi.org/10.1073/pnas.1715378115.

Berger, S. L., Kouzarides, T., Shiekhattar, R., & Shilatifard, A. (2009). An operational definition of epigenetics. *Genes & Development*, *23*, 781–783. https://doi.org/10.1101/gad.1787609.

Bestor, T. H., & Ingram, V. M. (1983). Two DNA methyltransferases from murine erythroleukemia cells: Purification, sequence specificity, and mode of interaction with DNA. *Proceedings of the National Academy of Sciences of the United States of America*, *80*, 5559–5563. https://doi.org/10.1073/pnas.80.18.5559.

Biason-Lauber, A., Konrad, D., Meyer, M., deBeaufort, C., & Schoenle, E. J. (2009). Ovaries and female phenotype in a girl with 46,XY karyotype and mutations in the CBX2 gene. *American Journal of Human Genetics*, *84*, 658–663. https://doi.org/10.1016/j.ajhg.2009.03.016.

Boeger, H., Griesenbeck, J., Strattan, J. S., & Kornberg, R. D. (2004). Removal of promoter nucleosomes by disassembly rather than sliding in vivo. *Molecular Cell*, *14*, 667–673. https://doi.org/10.1016/j.molcel.2004.05.013.

Capel, B., Rasberry, C., Dyson, J., Bishop, C. E., Simpson, E., Vivian, N., et al. (1993). Deletion of Y chromosome sequences located outside the testis determining region can cause XY female sex reversal. *Nature Genetics*, *5*, 301–307. https://doi.org/10.1038/ng1193-301.

Capel, B., Swain, A., Nicolis, S., Hacker, A., Walter, M., Koopman, P., et al. (1993). Circular transcripts of the testis-determining gene Sry in adult mouse testis. *Cell*, *73*, 1019–1030.

Carey, B. W., Finley, L. W. S., Cross, J. R., Allis, C. D., & Thompson, C. B. (2015). Intracellular α-ketoglutarate maintains the pluripotency of embryonic stem cells. *Nature*, *518*, 413–416. https://doi.org/10.1038/nature13981.

Carré, G. A., Siggers, P., Xipolita, M., Brindle, P., Lutz, B., Wells, S., et al. (2018). Loss of p300 and CBP disrupts histone acetylation at the mouse Sry promoter and causes XY gonadal sex reversal. *Human Molecular Genetics*, *27*, 190–198. https://doi.org/10.1093/hmg/ddx398.

Clapier, C. R., Iwasa, J., Cairns, B. R., & Peterson, C. L. (2017). Mechanisms of action and regulation of ATP-dependent chromatin-remodelling complexes. *Nature Reviews. Molecular Cell Biology*, *18*, 407–422. https://doi.org/10.1038/nrm.2017.26.

Collins, R. E., Northrop, J. P., Horton, J. R., Lee, D. Y., Zhang, X., Stallcup, M. R., et al. (2008). The ankyrin repeats of G9a and GLP histone methyltransferases are mono- and dimethyllysine binding modules. *Nature Structural & Molecular Biology*, *15*, 245–250. https://doi.org/10.1038/nsmb.1384.

Core, L. J., Waterfall, J. J., & Lis, J. T. (2008). Nascent RNA sequencing reveals widespread pausing and divergent initiation at human promoters. *Science*, *322*, 1845–1848. https://doi.org/10.1126/science.1162228.

de la Serna, I. L., Carlson, K. A., & Imbalzano, A. N. (2001). Mammalian SWI/SNF complexes promote MyoD-mediated muscle differentiation. *Nature Genetics*, *27*, 187–190. https://doi.org/10.1038/84826.

de Santa Barbara, P., Méjean, C., Moniot, B., Malclès, M. H., Berta, P., & Boizet-Bonhoure, B. (2001). Steroidogenic factor-1 contributes to the cyclic-adenosine monophosphate down-regulation of human SRY gene expression. *Biology of Reproduction*, *64*, 775–783. https://doi.org/10.1095/biolreprod64.3.775.

Derrien, T., Johnson, R., Bussotti, G., Tanzer, A., Djebali, S., Tilgner, H., et al. (2012). The GENCODE v7 catalog of human long noncoding RNAs: Analysis of their gene structure, evolution, and expression. *Genome Research*, *22*, 1775–1789. https://doi.org/10.1101/gr.132159.111.

Deveson, I. W., Holleley, C. E., Blackburn, J., Marshall Graves, J. A., Mattick, J. S., Waters, P. D., et al. (2017). Differential intron retention in Jumonji chromatin modifier genes is implicated in reptile temperature-dependent sex determination. *Science Advances*, *3*, 1–8. https://doi.org/10.1126/sciadv.1700731.

Dolci, S., Grimaldi, P., Geremia, R., Pesce, M., & Rossi, P. (1997). Identification of a promoter region generating Sry circular transcripts both in germ cells from male adult mice and in male mouse embryonal gonads. *Biology of Reproduction*, *57*, 1128–1135. https://doi.org/10.1095/biolreprod57.5.1128.

Egger, G., Liang, G., Aparicio, A., & Jones, P. A. (2004). Epigenetics in human disease and prospects for epigenetic therapy. *Nature*, *429*, 457–463. https://doi.org/10.1038/nature02625.

Fatica, A., & Bozzoni, I. (2014). Long non-coding RNAs: New players in cell differentiation and development. *Nature Reviews. Genetics*, *15*, 7–21. https://doi.org/10.1038/nrg3606.

Feil, R., & Fraga, M. F. (2012). Epigenetics and the environment: Emerging patterns and implications. *Nature Reviews. Genetics*, *13*, 97–109. https://doi.org/10.1038/nrg3142.

Fujita, N., Jaye, D. L., Geigerman, C., Akyildiz, A., Mooney, M. R., Boss, J. M., et al. (2004). MTA3 and the Mi-2/NuRD complex regulate cell fate during B lymphocyte differentiation. *Cell*, *119*, 75–86. https://doi.org/10.1016/j.cell.2004.09.014.

Garcia-Moreno, S. A., Plebanek, M. P., & Capel, B. (2018). Epigenetic regulation of male fate commitment from an initially bipotential system. *Molecular and Cellular Endocrinology*, *468*, 19–30. https://doi.org/10.1016/j.mce.2018.01.009.

Ge, C., Ye, J., Weber, C., Sun, W., Zhang, H., Zhou, Y., et al. (2018). The histone demethylase KDM6B regulates temperature-dependent sex determination in a turtle species. *Science*, *360*, 645–648. https://doi.org/10.1126/science.aap8328.

Gibbons, R. (2006). Alpha thalassaemia-mental retardation, X linked. *Orphanet Journal of Rare Diseases*, *1*, 15. https://doi.org/10.1186/1750-1172-1-15.

Gibbons, R. J., Picketts, D. J., Villard, L., & Higgs, D. R. (1995). Mutations in a putative global transcriptional regulator cause X-linked mental retardation with alpha-thalassemia (ATR-X syndrome). *Cell*, *80*, 837–845.

Gierl, M. S., Gruhn, W. H., von Seggern, A., Maltry, N., & Niehrs, C. (2012). GADD45G functions in male sex determination by promoting p38 signaling and sry expression. *Developmental Cell*, *23*, 1032–1042. https://doi.org/10.1016/j.devcel.2012.09.014.

Gubbay, J., Collignon, J., Koopman, P., Capel, B., Economou, A., Münsterberg, A., et al. (1990). A gene mapping to the sex-determining region of the mouse Y chromosome is a member of a novel family of embryonically expressed genes. *Nature*, *346*, 245–250. https://doi.org/10.1038/346245a0.

Gubbay, J., Vivian, N., Economou, A., Jackson, D., Goodfellow, P., & Lovell-Badge, R. (1992). Inverted repeat structure of the Sry locus in mice. *Proceedings of the National Academy of Sciences of the United States of America*, *89*, 7953–7957. https://doi.org/10.1073/pnas.89.17.7953.

Habert, R., Lejeune, H., & Saez, J. M. (2001). Origin, differentiation and regulation of fetal and adult Leydig cells. *Molecular and Cellular Endocrinology*, *179*, 47–74. https://doi.org/10.1016/S0303-7207(01)00461-0.

Hacker, A., Capel, B., Goodfellow, P., & Lovell-Badge, R. (1995). Expression of [i]Sry[/i], the mouse sex determining gene. *Development*, *121*, 1603–1614. https://doi.org/10.1006/geno.1993.1395.

Haigis, M. C., & Guarente, L. P. (2006). Mammalian sirtuins—Emerging roles in physiology, aging, and calorie restriction. *Genes & Development*, *20*, 2913–2921. https://doi.org/10.1101/gad.1467506.

Hiramatsu, R., Matoba, S., Kanai-Azuma, M., Tsunekawa, N., Katoh-Fukui, Y., Kurohmaru, M., et al. (2009). A critical time window of Sry action in gonadal sex determination in mice. *Development*, *136*, 129–138. https://doi.org/10.1242/dev.029587.

Inagaki, T., Tachibana, M., Magoori, K., Kudo, H., Tanaka, T., Okamura, M., et al. (2009). Obesity and metabolic syndrome in histone demethylase JHDM2a-deficient mice. *Genes to Cells*, *14*, 991–1001. https://doi.org/10.1111/j.1365-2443.2009.01326.x.

Ito, S., Dalessio, A. C., Taranova, O. V., Hong, K., Sowers, L. C., & Zhang, Y. (2010). Role of tet proteins in 5mC to 5hmC conversion, ES-cell self-renewal and inner cell mass specification. *Nature*, *466*, 1129–1133. https://doi.org/10.1038/nature09303.

Jeong, Y. H., Lu, H., Park, C. H., Li, M., Luo, H., Kim, J. J., et al. (2016). Stochastic anomaly of methylome but persistent SRY hypermethylation in disorder of sex development in canine somatic cell nuclear transfer. *Scientific Reports*, *6*, 1–11. https://doi.org/10.1038/srep31088.

Jeske, Y. W. A., Bowles, J., Greenfield, A., & Koopman, P. (1995). Expression of a linear Sry transcript in the mouse genital ridge. *Nature Genetics*, *10*, 480–482. https://doi.org/10.1038/ng0895-480.

Jeske, Y. W. A., Mishina, Y., Cohen, D. R., Behringer, R. R., & Koopman, P. (1996). Analysis of the role of Amh and Fra 1 in the Sry regulatory pathway. *Molecular Reproduction and Development*, *44*, 153–158. https://doi.org/10.1002/(SICI)1098-2795(199606)44:2<153::AID-MRD3>3.0.CO;2-M.

Jiang, Y.-H., Bressler, J., & Beaudet, A. L. (2004). Epigenetics and human disease. *Annual Review of Genomics and Human Genetics*, *5*, 479–510. https://doi.org/10.1146/annurev.genom.5.061903.180014.

Karl, J., & Capel, B. (1998). Sertoli cells of the mouse testis originate from the coelomic epithelium. *Developmental Biology*, *203*, 323–333. https://doi.org/10.1006/dbio.1998.9068.

Kato, Y., Perez, C. A. G., Mohamad Ishak, N. S., Nong, Q. D., Sudo, Y., Matsuura, T., et al. (2018). A 5′ UTR-overlapping LncRNA activates the male-determining gene doublesex1 in the Crustacean Daphnia magna. *Current Biology*, *28*. https://doi.org/10.1016/j.cub.2018.04.029. 1811.e4–1817.e4.

Katoh-Fukui, Y., Miyabayashi, K., Komatsu, T., Owaki, A., Baba, T., Shima, Y., et al. (2012). Cbx2, a polycomb group gene, is required for Sry gene expression in mice. *Endocrinology*, *153*, 913–924. https://doi.org/10.1210/en.2011-1055.

Katoh-Fukui, Y., Owaki, A., Toyama, Y., Kusaka, M., Shinohara, Y., Maekawa, M., et al. (2005). Mouse polycomb M33 is required for splenic vascular and adrenal gland formation through regulating Ad4BP/SF1 expression. *Blood*, *106*, 1612–1620. https://doi.org/10.1182/blood-2004-08-3367.

Katoh-Fukui, Y., Tsuchiya, R., Shiroishi, T., Nakahara, Y., Hashimoto, N., Noguchi, K., et al. (1998). Male-to-female sex reversal in M33 mutant mice. *Nature*, *393*, 688–692. https://doi.org/10.1038/31482.

Kaustov, L., Ouyang, H., Amaya, M., Lemak, A., Nady, N., Duan, S., et al. (2011). Recognition and specificity determinants of the human Cbx chromodomains. *The Journal of Biological Chemistry*, *286*, 521–529. https://doi.org/10.1074/jbc.M110.191411.

Kohli, R. M., & Zhang, Y. (2013). TET enzymes, TDG and the dynamics of DNA demethylation. *Nature*, *502*, 472–479. https://doi.org/10.1038/nature12750.

Kondo, T., Isono, K., Kondo, K., Endo, T. A., Itohara, S., Vidal, M., et al. (2014). Polycomb potentiates Meis2 activation in midbrain by mediating interaction of the promoter with a tissue-specific enhancer. *Developmental Cell*, *28*, 94–101. https://doi.org/10.1016/j.devcel.2013.11.021.

Koopman, P., Gubbay, J., Vivian, N., Goodfellow, P., & Lovell-Badge, R. (1991). Male development of chromosomally female mice transgenic for Sry. *Nature*, *351*, 117–121. https://doi.org/10.1038/351117a0.

Koopman, P., Münsterberg, A., Capel, B., Vivian, N., & Lovell-Badge, R. (1990). Expression of a candidate sex-determining gene during mouse testis differentiation. *Nature, 348*, 450–452. https://doi.org/10.1038/348450a0.

Kopp, F., & Mendell, J. T. (2018). Functional classification and experimental dissection of long noncoding RNAs. *Cell, 172*, 393–407. https://doi.org/10.1016/j.cell.2018.01.011.

Kriaucionis, S., & Heintz, N. (2009). The nuclear DNA base 5-hydroxymethylcytosine is present in purkinje neurons and the brain. *Science, 324*, 929–930. https://doi.org/10.1126/science.1169786.

Kung, J. T. Y., Colognori, D., & Lee, J. T. (2013). Long noncoding RNAs: Past, present, and future. *Genetics, 193*, 651–669. https://doi.org/10.1534/genetics.112.146704.

Kuroki, S., Matoba, S., Akiyoshi, M., Matsumura, Y., Miyachi, H., Mise, N., et al. (2013). Epigenetic regulation of mouse sex determination by the histone demethylase Jmjd1a. *Science, 341*, 1106–1109. https://doi.org/10.1126/science.1239864.

Kuroki, S., Okashita, N., Baba, S., Maeda, R., Miyawaki, S., Yano, M., et al. (2017). Rescuing the aberrant sex development of H3K9 demethylase Jmjd1a-deficient mice by modulating H3K9 methylation balance. *PLoS Genetics, 13,* e1007034. https://doi.org/10.1371/journal.pgen.1007034.

Kuroki, S., & Tachibana, M. (2018). Epigenetic regulation of mammalian sex determination. *Molecular and Cellular Endocrinology, 468*, 31–38. https://doi.org/10.1016/j.mce.2017.12.006.

Lachner, M., O'Carroll, D., Rea, S., Mechtler, K., & Jenuwein, T. (2001). Methylation of histone H3 lysine 9 creates a binding site for HP1 proteins. *Nature, 410*, 116–120. https://doi.org/10.1038/35065132.

Larney, C., Bailey, T. L., & Koopman, P. (2014). Switching on sex: Transcriptional regulation of the testis-determining gene Sry. *Development, 141*, 2195–2205. https://doi.org/10.1242/dev.107052.

Laval, S. H., Glenister, P. H., Rasberry, C., Thornton, C. E., Mahadevaiah, S. K., Cooke, H. J., et al. (1995). Y chromosome short arm-Sxr recombination in XSxr/Y males causes deletion of Rbm and XY female sex reversal. *Proceedings of the National Academy of Sciences of the United States of America, 92*, 10403–10407.

Lewis, E. B. (1978). A gene complex controlling segmentation in Drosophila. *Nature, 276*, 565–570.

Li, E., Bestor, T. H., & Jaenisch, R. (1992). Targeted mutation of the DNA methyltransferase gene results in embryonic lethality. *Cell, 69*, 915–926. https://doi.org/10.1016/0092-8674(92)90611-F.

Li, L., Liu, B., Wapinski, O. L., Tsai, M.-C., Qu, K., Zhang, J., et al. (2013). Targeted disruption of Hotair leads to homeotic transformation and gene derepression. *Cell Reports, 5*, 3–12. https://doi.org/10.1016/j.celrep.2013.09.003.

Lickert, H., Takeuchi, J. K., Von Both, I., Walls, J. R., McAuliffe, F., Adamson, S. L., et al. (2004). Baf60c is essential for function of BAF chromatin remodelling complexes in heart development. *Nature, 432*, 107–112. https://doi.org/10.1038/nature03071.

Liu, Z., Zhou, S., Liao, L., Chen, X., Meistrich, M., & Xu, J. (2010). Jmjd1a demethylase-regulated histone modification is essential for cAMP-response element modulator-regulated gene expression and spermatogenesis. *The Journal of Biological Chemistry, 285*, 2758–2770. https://doi.org/10.1074/jbc.M109.066845.

Loenarz, C., & Schofield, C. J. (2008). Expanding chemical biology of 2-oxoglutarate oxygenases. *Nature Chemical Biology, 4*, 152–156. https://doi.org/10.1038/nchembio0308-152.

Luo, X., Ikeda, Y., & Parker, K. L. (1994). A cell-specific nuclear receptor is essential for adrenal and gonadal development and sexual differentiation. *Cell*, 77, 481–490. https://doi.org/10.1016/0092-8674(94)90211-9.

Maezawa, S., Hasegawa, K., Yukawa, M., Sakashita, A., Alavattam, K. G., Andreassen, P. R., et al. (2017). Polycomb directs timely activation of germline genes in spermatogenesis. *Genes & Development*, *31*, 1693–1703. https://doi.org/10.1101/gad.302000.117.

Mahadevaiah, S. K., Odorisio, T., Elliott, D. J., Rattigan, A., Szot, M., Laval, S. H., et al. (1998). Mouse homologues of the human AZF candidate gene RBM are expressed in spermatogonia and spermatids, and map to a Y chromosome deletion interval associated with a high incidence of sperm abnormalities. *Human Molecular Genetics*, 7, 715–727.

Margueron, R., & Reinberg, D. (2011). The Polycomb complex PRC2 and its mark in life. *Nature*, *469*, 343–349. https://doi.org/10.1038/nature09784.

Martin, C., & Zhang, Y. (2005). The diverse functions of histone lysine methylation. *Nature Reviews. Molecular Cell Biology*, *6*, 838–849. https://doi.org/10.1038/nrm1761.

Matoba, S., Kanai, Y., Kidokoro, T., Kanai-Azuma, M., Kawakami, H., Hayashi, Y., et al. (2005). A novel Sry-downstream cellular event which preserves the readily available energy source of glycogen in mouse sex differentiation. *Journal of Cell Science*, *118*, 1449–1459. https://doi.org/10.1242/jcs.01738.

Meng, F.-L., Du, Z., Federation, A., Hu, J., Wang, Q., Kieffer-Kwon, K.-R., et al. (2014). Convergent transcription at intragenic super-enhancers targets AID-initiated genomic instability. *Cell*, *159*, 1538–1548. https://doi.org/10.1016/j.cell.2014.11.014.

Mews, P., Donahue, G., Drake, A. M., Luczak, V., Abel, T., & Berger, S. L. (2017). Acetyl-CoA synthetase regulates histone acetylation and hippocampal memory. *Nature*, *546*, 381–386. https://doi.org/10.1038/nature22405.

Miyamoto, Y., Taniguchi, H., Hamel, F., Silversides, D. W., & Viger, R. S. (2008). A GATA4/WT1 cooperation regulates transcription of genes required for mammalian sex determination and differentiation. *BMC Molecular Biology*, *9*, 1–18. https://doi.org/10.1186/1471-2199-9-44.

Miyazawa, H., Yamaguchi, Y., Sugiura, Y., Honda, K., Kondo, K., Matsuda, F., et al. (2017). Rewiring of embryonic glucose metabolism via suppression of PFK-1 and aldolase during mouse chorioallantoic branching. *Development*, *144*, 63–73. https://doi.org/10.1242/dev.138545.

Munger, S. C., & Capel, B. (2012). Sex and the circuitry: Progress toward a systems-level understanding of vertebrate sex determination. *Wiley Interdisciplinary Reviews. Systems Biology and Medicine*, *4*, 401–412. https://doi.org/10.1002/wsbm.1172.

Nabel, C. S., Manning, S. A., & Kohli, R. M. (2012). The curious chemical biology of cytosine: Deamination, methylation, and oxidation as modulators of genomic potential. *ACS Chemical Biology*, 7, 20–30. https://doi.org/10.1021/cb2002895.

Nady, N., Lemak, A., Walker, J. R., Avvakumov, G. V., Kareta, M. S., Achour, M., et al. (2011). Recognition of multivalent histone states associated with heterochromatin by UHRF1 protein. *The Journal of Biological Chemistry*, *286*, 24300–24311. https://doi.org/10.1074/jbc.M111.234104.

Nakayama, J., Rice, J. C., Strahl, B. D., Allis, C. D., & Grewal, S. I. S. (2001). Role of histone H3 lysine 9 methylation in epigenetic control of heterochromatin assembly. *Science*, *292*, 110–113. https://doi.org/10.1126/science.1060118.

Navarro-Martín, L., Viñas, J., Ribas, L., Díaz, N., Gutiérrez, A., Di Croce, L., et al. (2011). DNA methylation of the gonadal aromatase (cyp19a) promoter is involved in temperature-dependent sex ratio shifts in the European sea bass. *PLoS Genetics*, 7. https://doi.org/10.1371/journal.pgen.1002447.

Nef, S., Verma-Kurvari, S., Merenmies, J., Vassalli, J.-D., Efstratiadis, A., Accili, D., et al. (2003). Testis determination requires insulin receptor family function in mice. *Nature*, *426*, 291–295. https://doi.org/10.1038/nature02059.

Nishibuchi, G., & Déjardin, J. (2017). The molecular basis of the organization of repetitive DNA-containing constitutive heterochromatin in mammals. *Chromosome Research*, *25*, 77–87. https://doi.org/10.1007/s10577-016-9547-3.

Nishino, K., Hattori, N., Sato, S., Arai, Y., Tanaka, S., Nagy, A., et al. (2011). Non-CpG methylation occurs in the regulatory region of the Sry gene. *The Journal of Reproduction and Development*, *57*, 586–593. https://doi.org/10.1262/jrd.11-033A.

Nishino, K., Hattori, N., Tanaka, S., & Shiota, K. (2004). DNA methylation-mediated control of Sry gene expression in mouse gonadal development. *The Journal of Biological Chemistry*, *279*, 22306–22313. https://doi.org/10.1074/jbc.M309513200.

Okada, Y., Scott, G., Ray, M. K., Mishina, Y., & Zhang, Y. (2007). Histone demethylase JHDM2A is critical for Tnp1 and Prm1 transcription and spermatogenesis. *Nature*, *450*, 119–123. https://doi.org/10.1038/nature06236.

Okano, M., Bell, D. W., Haber, D. A., & Li, E. (1999). DNA methyltransferases Dnmt3a and Dnmt3b are essential for de novo methylation and mammalian development. *Cell*, *99*, 247–257. https://doi.org/10.1016/S0092-8674(00)81656-6.

Okano, M., Xie, S., & Li, E. (1998). Cloning and characterization of a family of novel mammalian DNA (cytosine-5) methyltransferases. *Nature Genetics*, *19*, 219–220. https://doi.org/10.1038/890.

Ono, M., & Harley, V. R. (2013). Disorders of sex development: New genes, new concepts. *Nature Reviews. Endocrinology*, *9*, 79–91. https://doi.org/10.1038/nrendo.2012.235.

Peng, M., Yin, N., Chhangawala, S., Xu, K., Leslie, C. S., & Li, M. O. (2016). Aerobic glycolysis promotes T helper 1 cell differentiation through an epigenetic mechanism. *Science*, *354*, 481–484. https://doi.org/10.1126/science.aaf6284.

Peters, A. H. F. M., O'Carroll, D., Scherthan, H., Mechtler, K., Sauer, S., Schöfer, C., et al. (2001). Loss of the Suv39h histone methyltransferases impairs mammalian heterochromatin and genome stability. *Cell*, *107*, 323–337. https://doi.org/10.1016/S0092-8674(01)00542-6.

Pilon, N., Daneau, I., Paradis, V., Hamel, F., Lussier, J. G., Viger, R. S., et al. (2003). Porcine SRY promoter is a target for steroidogenic factor 1. *Biology of Reproduction*, *68*, 1098–1106. https://doi.org/10.1095/biolreprod.102.010884.

Quinn, A., Kashimada, K., Davidson, T.-L., Ng, E. T., Chawengsaksophak, K., Bowles, J., et al. (2014). A site-specific, single-copy transgenesis strategy to identify 5′ regulatory sequences of the mouse testis-determining gene Sry. *PLoS One*, *9*, e94813. https://doi.org/10.1371/journal.pone.0094813.

Rais, Y., Zviran, A., Geula, S., Gafni, O., Chomsky, E., Viukov, S., et al. (2013). Deterministic direct reprogramming of somatic cells to pluripotency. *Nature*, *502*, 65–70. https://doi.org/10.1038/nature12587.

Rastetter, R. H., Smith, C. A., & Wilhelm, D. (2015). The role of non-coding RNAs in male sex determination and differentiation. *Reproduction*, *150*, R93–R107. https://doi.org/10.1530/REP-15-0106.

Reid, M. A., Dai, Z., & Locasale, J. W. (2017). The impact of cellular metabolism on chromatin dynamics and epigenetics. *Nature Cell Biology*, *19*, 1298–1306. https://doi.org/10.1038/ncb3629.

Rinn, J. L., & Chang, H. Y. (2012). Genome regulation by long noncoding RNAs. *Annual Review of Biochemistry*, *81*, 145–166. https://doi.org/10.1146/annurev-biochem-051410-092902.

Rinn, J. L., Kertesz, M., Wang, J. K., Squazzo, S. L., Xu, X., Brugmann, S. A., et al. (2007). Functional demarcation of active and silent chromatin domains in human HOX loci by noncoding RNAs. *Cell*, *129*, 1311–1323. https://doi.org/10.1016/j.cell.2007.05.022.

Schmahl, J., Eicher, E. M., Washburn, L. L., & Capel, B. (2000). Sry induces cell proliferation in the mouse gonad. *Development*, *127*, 65–73.

Shi, Y., Lan, F., Matson, C., Mulligan, P., Whetstine, J. R., Cole, P. A., et al. (2004). Histone demethylation mediated by the nuclear amine oxidase homolog LSD1. *Cell*, *119*, 941–953. https://doi.org/10.1016/j.cell.2004.12.012.

Simon, J. A., & Kingston, R. E. (2013). Occupying chromatin: Polycomb mechanisms for getting to genomic targets, stopping transcriptional traffic, and staying put. *Molecular Cell*, *49*, 808–824. https://doi.org/10.1016/j.molcel.2013.02.013.

Sinclair, A. H., Berta, P., Palmer, M. S., Hawkins, J. R., Griffiths, B. L., Smith, M. J., et al. (1990). A gene from the human sex-determining region encodes a protein with homology to a conserved DNA-binding motif. *Nature*, *346*, 240–244. https://doi.org/10.1038/346240a0.

Stévant, I., Neirijnck, Y., Borel, C., Escoffier, J., Smith, L. B., Antonarakis, S. E., et al. (2018). Deciphering cell lineage specification during male sex determination with single-cell RNA sequencing. *Cell Reports*, *22*, 1589–1599. https://doi.org/10.1016/j.celrep.2018.01.043.

Sutendra, G., Kinnaird, A., Dromparis, P., Paulin, R., Stenson, T. H., Haromy, A., et al. (2014). A nuclear pyruvate dehydrogenase complex is important for the generation of acetyl-CoA and histone acetylation. *Cell*, *158*, 84–97. https://doi.org/10.1016/j.cell.2014.04.046.

Tachibana, M. (2016). Epigenetics of sex determination in mammals. *Reproductive Medicine and Biology*, *15*, 59–67. https://doi.org/10.1007/s12522-015-0223-7.

Tachibana, M., Sugimoto, K., Fukushima, T., & Shinkai, Y. (2001). SET domain-containing protein, G9a, is a novel lysine-preferring mammalian histone methyltransferase with hyperactivity and specific selectivity to lysines 9 and 27 of histone H3. *The Journal of Biological Chemistry*, *276*, 25309–25317. https://doi.org/10.1074/jbc.M101914200.

Tachibana, M., Sugimoto, K., Nozaki, M., Ueda, J., Ohta, T., Ohki, M., et al. (2002). G9a histone methyltransferase plays a dominant role in euchromatic histone H3 lysine 9 methylation and is essential for early embryogenesis. *Genes & Development*, *16*, 1779–1791. https://doi.org/10.1101/gad.989402.

Tachibana, M., Ueda, J., Fukuda, M., Takeda, N., Ohta, T., Iwanari, H., et al. (2005). Histone methyltransferases G9a and GLP form heteromeric complexes and are both crucial for methylation of euchromatin at H3-K9. *Genes & Development*, *19*, 815–826. https://doi.org/10.1101/gad.1284005.

Tahiliani, M., Koh, K. P., Shen, Y., Pastor, W. A., Bandukwala, H., Brudno, Y., et al. (2009). Conversion of 5-methylcytosine to 5-hydroxymethylcytosine in mammalian DNA by MLL partner TET1. *Science*, *324*, 930–935. https://doi.org/10.1126/science.1170116. (80-).

Tateishi, K., Okada, Y., Kallin, E. M., & Zhang, Y. (2009). Role of Jhdm2a in regulating metabolic gene expression and obesity resistance. *Nature*, *458*, 757–761. https://doi.org/10.1038/nature07777.

Thevenet, L., Méjean, C., Moniot, B., Bonneaud, N., Galéotti, N., Aldrian-Herrada, G., et al. (2004). Regulation of human SRY subcellular distribution by its acetylation/deacetylation. *The EMBO Journal*, *23*, 3336–3345. https://doi.org/10.1038/sj.emboj.7600352.

Tsukada, Y. I., Fang, J., Erdjument-Bromage, H., Warren, M. E., Borchers, C. H., Tempst, P., et al. (2006). Histone demethylation by a family of JmjC domain-containing proteins. *Nature*, *439*, 811–816. https://doi.org/10.1038/nature04433.

Wang, D., Garcia-Bassets, I., Benner, C., Li, W., Su, X., Zhou, Y., et al. (2011). Reprogramming transcription by distinct classes of enhancers functionally defined by eRNA. *Nature*, *474*, 390–394. https://doi.org/10.1038/nature10006.

Wapinski, O., & Chang, H. Y. (2011). Long noncoding RNAs and human disease. *Trends in Cell Biology*, *21*, 354–361. https://doi.org/10.1016/j.tcb.2011.04.001.

Warr, N., Carre, G. A., Siggers, P., Faleato, J. V., Brixey, R., Pope, M., et al. (2012). Gadd45γ and Map 3k4 interactions regulate mouse testis determination via p38 MAPK-mediated control of Sry expression. *Developmental Cell*, *23*, 1020–1031. https://doi.org/10.1016/j.devcel.2012.09.016.

Wellen, K. E., Hatzivassiliou, G., Sachdeva, U. M., Bui, T. V., Cross, J. R., & Thompson, C. B. (2009). ATP-citrate lyase links cellular metabolism to histone acetylation. *Science, 324*, 1076–1080. https://doi.org/10.1126/science.1164097.

Wilhelm, D., Washburn, L. L., Truong, V., Fellous, M., Eicher, E. M., & Koopman, P. (2009). Antagonism of the testis- and ovary-determining pathways during ovotestis development in mice. *Mechanisms of Development, 126*, 324–336. https://doi.org/10.1016/j.mod.2009.02.006.

Wolff, G. L., Kodell, R. L., Moore, S. R., & Cooney, C. A. (1998). Maternal epigenetics and methyl supplements affect *agouti* gene expression in A^{vy}/a mice. *The FASEB Journal, 12*, 949–957. https://doi.org/10.1096/fasebj.12.11.949.

Wolffe, A. P., & Matzke, M. A. (1999). Epigenetics: Regulation through repression. *Science, 286*, 481–486. https://doi.org/10.1126/science.286.5439.481.

Yamane, K., Toumazou, C., Tsukada, Y. i., Erdjument-Bromage, H., Tempst, P., Wong, J., et al. (2006). JHDM2A, a JmjC-containing H3K9 demethylase, facilitates transcription activation by androgen receptor. *Cell, 125*, 483–495. https://doi.org/10.1016/j.cell.2006.03.027.

Yoshida, K. (2010). Ubiquitous expression of Sry induces embryonic lethality related to suppression of Tie2/Tek expression. *Advances in Bioscience and Biotechnology, 01*, 444–452. https://doi.org/10.4236/abb.2010.15058.

CHAPTER EIGHT

The regulation of *Sox9* expression in the gonad

Nitzan Gonen*, Robin Lovell-Badge*
The Francis Crick Institute, London, United Kingdom
*Corresponding authors: e-mail address: nitzan.gonen@crick.ac.uk; robin.lovell-badge@crick.ac.uk

Contents

Abstract

The bipotential nature of cell types in the early developing gonad and the process of sex determination leading to either testis or ovary differentiation makes this an interesting system in which to study transcriptional regulation of gene expression and cell fate decisions. SOX9 is a transcription factor with multiple roles during development, including being a key player in mediating testis differentiation and therefore subsequent male development. Loss of *Sox9* expression in both humans and mice results in XY female development, whereas its inappropriate activation in XX embryonic gonads can give male development. Multiple cases of Disorders of Sex Development in human patients or sex reversal in mice and other vertebrates can be explained by mutations affecting upstream regulators of *Sox9* expression, such as the product of the Y chromosome gene *Sry* that triggers testis differentiation. Other cases are due to mutations in the *Sox9* gene itself, including its own regulatory region. Indeed, rearrangements in and around the *Sox9* genomic locus indicate the presence of multiple critical enhancers and the complex nature of its regulation. Here we summarize what is known about the role of *Sox9* and its regulation during gonad development, including recently discovered critical enhancers. We also discuss higher order chromatin organization and how this might

Current Topics in Developmental Biology, Volume 134
ISSN 0070-2153
https://doi.org/10.1016/bs.ctdb.2019.01.004

be involved. We end with some interesting future directions that have the potential to further enrich our understanding on the complex, multi-layered regulation controlling *Sox9* expression in the gonads.

1. Introduction

The processes of sex determination and gonadal differentiation serve as beautiful systems to understand the role of genes during development and to study gene regulation and cell fate decisions. This is in part because cases of Disorders of Sex Development (DSD) in humans and cases of sex reversal in animals have led to the discovery of many of the genes controlling sex determination and differentiation (Wilhelm, Palmer, & Koopman, 2007). But this is also due to the bipotential nature of many cell types within the gonad and the fact that the early gonad (or genital ridge) is able to differentiate into either ovary or testis. Thus, the supporting cell precursors give rise to Sertoli cells or granulosa cells, the steroid cell precursors give rise to Leydig cells or theca cells, and the primordial germ cells (PGCs) give rise to spermatogonia or oogonia, in the testis and ovary respectively (Lin & Capel, 2015). It is now clear that in addition to the determination of gonadal sex that occurs in the embryo, these cell fate decisions need to be actively maintained in the adult, otherwise Sertoli cells transdifferentiate to granulosa cells or *vice versa*, and this can be followed by more general gonadal sex reversal (Huang, Ye, & Chen, 2017; Matson et al., 2011; Uhlenhaut et al., 2009).

The determination of gonadal sex in mammals, but also in many other vertebrates, relies on a delicate balance of the expression levels of several key pro-male and pro-female transcription factors. Of course, these both interact with and direct signaling pathways, many of which, such as Wnt signaling and steroid hormones, can play critical roles, but it seems that it is the balance of transcriptional controls that both initiates and maintains gonadal fate. Any shift from this delicate balance can result in sex reversal. Because the type of gonad that develops, and the hormones it produces, can affect other sex-specific characteristics, this can lead to many types of DSD (Carre & Greenfield, 2016; Lin & Capel, 2015). While some types of DSD result from the development of gonads of mixed sex, so-called ovotestes, there are also amplification or canalisation processes that usually work to ensure that only one type of gonad develops, even though the whole process is finely balanced at the beginning. In order to achieve this, multiple layers of regulation are expected to play a role, such as histone modifications, the activity of both enhancers and silencers, as well as higher order of chromatin organization, and these are all likely to operate in a system involving both positive and negative feedback mechanisms (Garcia-Moreno, Plebanek, & Capel, 2018).

Genes located at the top of regulatory cascades involved in sex determination have been proposed to have relatively unique roles, perhaps reflecting recent evolutionary origins (Wilkins, 1995), with some support for this in both invertebrates (Hediger et al., 2004) and vertebrates, including *Sry* (sex determining region Y), the Y chromosome linked testis-determining gene in mammals. While *Dmrt1* (doublesex and mab-3 related transcription factor 1) may have a more complex role, being involved in germ cells as well as playing a sometimes critical role in sex determination *via* its action in somatic cells, notably in birds where it maps to the Z chromosome, derivative homologs, such as *Dmy*, the male-determining gene located on the Y in some species of *Medaka*, and *Dmw*, which acts as a dominant female-determining gene mapping to the W in *Xenopus laevis*. Other genes lower down the pathway, leading to the development of ovaries or testes, have been co-opted for multiple roles during development (Capel, 2017). *Sox9*, which plays a pivotal role in mammalian sex determination, and is probably involved in testis development in all vertebrates, is one such gene.

1.1 The *Sox9* gene

SOX9 is a member of the SRY-related HMG box (SOX) family of transcription factors, which have roles in many tissues. Several types of DNA-binding proteins contain HMG boxes, including the "high mobility group" proteins HMG1 and HMG2, after which the domain is named, UBF, and the LEF/TCF family of transcription factors (Koopman, Sinclair, & Lovell-Badge, 2016; Lovell-Badge, 2010). The class of HMG box found in SOX9 was first identified in SRY (Gubbay et al., 1990; Sinclair et al., 1990). There are 20 members of the SOX family found in mammals, with a well-conserved HMG box of 79 amino acids, which can be divided into eight subgroups, A-H, based on the structural homology outside the box (Schepers, Teasdale, & Koopman, 2002; Wegner, 2010). SOX proteins usually form dimers while regulating target genes. SOX proteins can dimerize with other transcription factors or with homologous or heterologous SOX proteins. It is only when they bind DNA as a dimer that their gene regulation activity commences (Kamachi, Uchikawa, & Kondoh, 2000).

SOX9, along with SOX8 and SOX10, belong to the SOXE subfamily that shares high homology outside the HMG domain as well as two additional functional domains: a self-dimerization domain located at the N-terminal side and a transactivation domain located at the C-terminus (Gubbay et al., 1990; Jo et al., 2014; Kamachi & Kondoh, 2013; She & Yang, 2017).

SOX9 has been shown to act as both a transcriptional activator and a transcriptional repressor, depending on its binding partner, the target gene and the subsequent recruitment of either co-activators or co-repressors (Ikeda et al., 2004; Leung et al., 2011; Sekido & Lovell-Badge, 2008; Symon & Harley, 2017).

Sox9 is expressed in multiple tissues and cell types during embryonic development and through to adulthood, including cartilage, brain, pituitary, lung, heart, pancreas, testes, hair follicles, retina and several tissue-specific stem cell types (Jo et al., 2014). Genetic and molecular evidence suggests that the regulatory region of *Sox9* is spread over at least a 2 Mb gene desert 5′ to the coding sequence. This is predicted and partially known to contain multiple tissue-specific enhancers (Symon & Harley, 2017), as will be elaborated below.

1.2 The role of *Sox9* in gonadal development

In the embryonic gonad, *Sox9* is a direct downstream target gene, and possibly the only critical target, of SRY (Sekido, Bar, Narvaez, Penny, & Lovell-Badge, 2004; Sekido & Lovell-Badge, 2008). The gonads, which are colonized by PGCs, originate from the genital ridges, structures that develop from intermediate mesoderm associated with the mesonephros (a primitive kidney). The genital ridges can follow either a testicular or an ovarian path, depending on the expression (and repression) of key transcription factors (Jakob & Lovell-Badge, 2011; Lin & Capel, 2015; Maatouk & Capel, 2008). The supporting cell precursors that give rise to Sertoli cells in testes and granulosa cells in ovaries originate from a population of rapidly dividing cells in the coelomic epithelium that express several transcription factors, such as WT1 (Wilms tumor 1) and GATA4 (GATA binding protein 4) that are both quite widely expressed in the intermediate mesoderm, and notably, more specifically, the orphan nuclear receptor SF1 (steroidogenic factor-1), although this also has roles in the adrenal and pituitary (Karl & Capel, 1998; Kohler & Achermann, 2010; Schmahl, Eicher, Washburn, & Capel, 2000). After these cells leave the coelomic epithelium, they begin to express *Sox9* at a low, basal level irrespective of chromosomal sex (Morais da Silva et al., 1996; Zhao et al., 2018). It is SRY activity in the supporting cell precursors in XY gonads that dramatically increases *Sox9* expression and leads to the differentiation of Sertoli cells and the development of a testis (Jakob & Lovell-Badge, 2011; Kent, Wheatley, Andrews, Sinclair, & Koopman, 1996; Morais da Silva et al., 1996; Sekido et al., 2004). Unlike in humans, where SRY appears to be expressed in Sertoli cells throughout life, in the mouse SRY is only transiently expressed starting from embryonic day (E) 10.75, peaks at E11.5, and then disappears by E12.5. However, its expression can persist in XY gonads that have failed to initiate Sertoli cell and testis differentiation (Lee & Taketo, 1994; Sekido et al., 2004). SRY acts with SF1 to upregulate the expression of *Sox9* (Sekido, 2010). Once SOX9 reaches a certain threshold, it is involved in the repression of *Sry* expression (Sekido et al., 2004) as well as the activation of its own expression by interacting with SF1 (Sekido & Lovell-Badge, 2008). SOX9 is also able to further reinforce its expression by activating the FGF9 (Fibroblast Growth Factor 9) and PGD2 (Prostaglandin D2 Synthase) signaling pathways, directly or indirectly repressing ovary determining/anti-testis genes, and probably activating the expression of the related gene *Sox8* (Chaboissier et al., 2004; Schepers, Wilson, Wilhelm, & Koopman, 2003). SOX8 and SOX9 are then maintained in Sertoli cells throughout life, acting redundantly (probably always with SF1) to help maintain the expression of a range of genes required for the differentiation and function of Sertoli cells, including the production of signals instructing other cell lineages in the gonad to follow the testis pathway (Bagheri-Fam et al., 2008; Barrionuevo et al., 2009; Barrionuevo & Scherer, 2010; Chaboissier et al., 2004; Colvin, Green, Schmahl, Capel, & Ornitz, 2001; Kim et al., 2007; Moniot et al., 2009) (Fig. 1).

The key role of *Sox9* in testis determination has been shown by both gain and loss of function studies in mouse and human (Barrionuevo et al., 2006; Chaboissier et al., 2004; Foster et al., 1994; Huang, Wang, Ning, Lamb, & Bartley, 1999; Vidal, Chaboissier, de Rooij, & Schedl, 2001; Wagner et al., 1994). Notably, in humans, heterozygous loss-of-function mutations of *SOX9* lead to campomelic dysplasia (CD, OMIM 114290) (Houston et al., 1983), a severe skeletal malformation syndrome where 70% of XY patients show female development (Foster et al., 1994; Wagner et al., 1994). Duplication of *SOX9* has also been reported in a 46, XX mosaic male patient, suggesting that extra dose of *SOX9* is sufficient to initiate the male pathway in the absence of SRY (Huang et al., 1999).

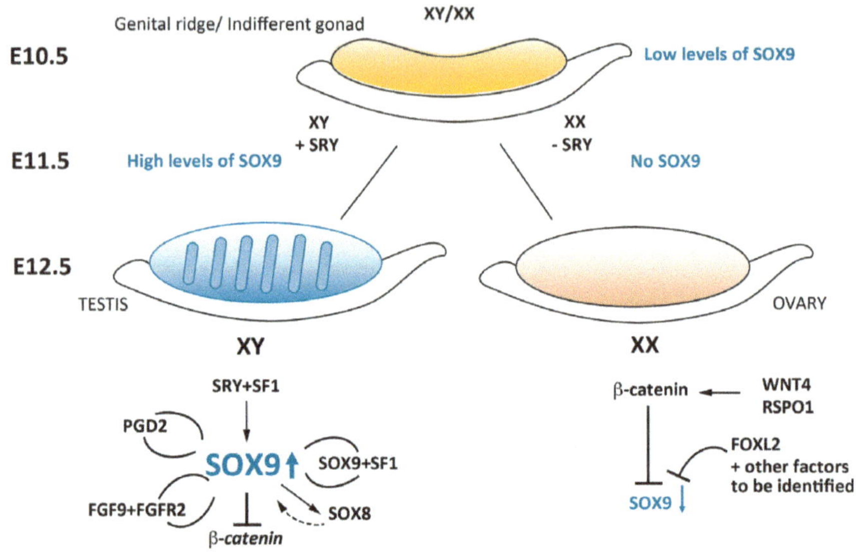

Fig. 1 The expression pattern of *Sox9* during mouse sex determination. SOX9 is expressed at low levels in the supporting cell precursors of both sexes in the bipotential gonad/genital ridge at E10.5. In XY gonads, SRY expression begins in the supporting cell precursors at E10.75, reaches a peak at E11.5 and disappears by E12.5. Once SRY is expressed, it acts along with SF1 to upregulate SOX9 expression, which initiates the differentiation of the supporting cell precursors to Sertoli cells. Once above a critical threshold, SOX9 binds to SF1 and acts in a feed-forward loop to activate its own expression. SOX9 also activates the expression of FGF9 and FGF2R, as well as PGD2 that in turn further increase SOX9 expression levels. *Sox8* expression begins at E12.5, as a direct or indirect target of SOX9, and acts to further enhance and reinforce SOX9 expression, as well as on the same targets as SOX9 in a redundant fashion (Barrionuevo et al., 2009). SOX9 is expressed in Sertoli cells throughout life. In addition to activating pro-males genes, SOX9 represses the expression of β-catenin, a pro-female gene. In XX gonads, SRY is not present and hence other genes and signaling pathways, including WNT4/RSPO1/β-catenin and *Foxl2* are expressed and act to repress the expression of *Sox9* in the ovary.

In the mouse, unlike in humans, heterozygous *Sox9* mutations do not cause XY sex reversal (Bi et al., 2001). Mice carrying a homozygous mutation in *Sox9* die at around E11.5, just at the onset of testis differentiation; however, *ex vivo* organ cultures showed no signs of testis cord formation after 3 days in culture (Chaboissier et al., 2004). To better understand the role of *Sox9* before and after the stage of sex determination, mice with a conditional null mutation of *Sox9* (*Sox9*$^{fl/fl}$) were used. Early, gonad-specific, loss of *Sox9*, triggered using an *Sf1-Cre* driver allele, demonstrated complete XY male-to-female sex reversal (Chaboissier et al., 2004; Lavery et al., 2011). In contrast, when *Sox9* was deleted using *Amh-Cre*, which resulted in complete elimination of *Sox9* expression in Sertoli cells, but not until E14, i.e., following sex determination and the onset of *Sox8* expression, testis development was normal and the mice were initially fertile. However, like *Sox8* homozygous mutants (O'Bryan et al., 2008; Sock, Schmidt, Hermanns-Borgmeyer, Bosl, & Wegner, 2001), they became sterile at about 5 months (Barrionuevo et al., 2009). Loss of both *Sox9* and *Sox8* does lead to a transdifferentiation of Sertoli cells into granulosa-like cells, proving their functional redundancy (Barrionuevo et al., 2016). Furthermore, using Wt1 regulatory sequences to drive overexpression of SOX9 in XX gonads leads to XX male development (Vidal et al., 2001) as does the chance insertion of a transgene about 1 Mb upstream of the *Sox9* coding region in "Odd Sex" mice, which also leads to XX male development (Bishop et al., 2000; Qin et al., 2004). When the latter is bred into XY mice lacking *Sry*, this leads to fully functional male testis and normal fertility, indicating that SOX9 is sufficient to trigger all the downstream events needed for the development of a fully fertile male (Qin & Bishop, 2005). All of the above clearly demonstrate that, in human as well as mouse, *Sox9* is both necessary and sufficient to induce testis differentiation and all subsequent aspects of male development.

Proper development of an ovary, on the other hand, requires the repression of *Sox9*, which involves activation of the WNT4 (wingless-type MMTV integration site family, member 4)/RSPO1 (R-spondin homolog 1) /β-catenin signaling pathway (Chassot et al., 2008; Kim et al., 2006; Maatouk et al., 2008; Vainio, Heikkila, Kispert, Chin, & McMahon, 1999) and expression of the transcription factor FOXL2 (forkhead box L2) (Ottolenghi et al., 2005; Schmidt et al., 2004; Uda et al., 2004). Other, as yet unknown factors are probably also involved, including factors that actively promote granulosa cell differentiation. In order to maintain ovarian cell fate, *Sox9* must be actively repressed throughout life in granulosa cells, otherwise they transdifferentiate into Sertoli cells, which then induce the rest of the gonad to become testis-like (Lindeman et al., 2015; Uhlenhaut et al., 2009; Zhao, Svingen, Ng, & Koopman, 2015) (Fig. 1).

In this review, we summarize and discuss the regulatory landscape controlling the gonadal expression of *Sox9*, in both humans and mice. We describe the gene-desert surrounding the *Sox9* locus, both upstream and downstream and how alterations in these non-coding regions such as deletions, duplications and inversions have been implicated in patients with DSD. We describe the previously known as well as newly discovered gonadal enhancers of *Sox9* and discuss the higher order of chromatin organization as an additional layer of regulation of gonadal *Sox9* expression. We conclude with some exciting future directions that can further accelerate our understanding on the multi-layered, complex, regulation of *Sox9* expression within the gonads.

2. The regulatory genomic locus of *Sox9*

The *Sox9* gene is located on the long arm (q) of chromosome 17 in humans and chromosome 11 in mice. It is embedded within a large gene desert with the neighboring genes located 2MB upstream and 0.5MB downstream in humans and 1.7MB upstream and 0.5MB downstream in mice (Fig. 2). Several lines of evidence support this region as being highly involved in the regulation of *Sox9* in the various tissues in which it is normally expressed. In this section, we summarize the literature regarding the regulatory, non-coding, regions implicated in the control of the gonadal expression of *Sox9*.

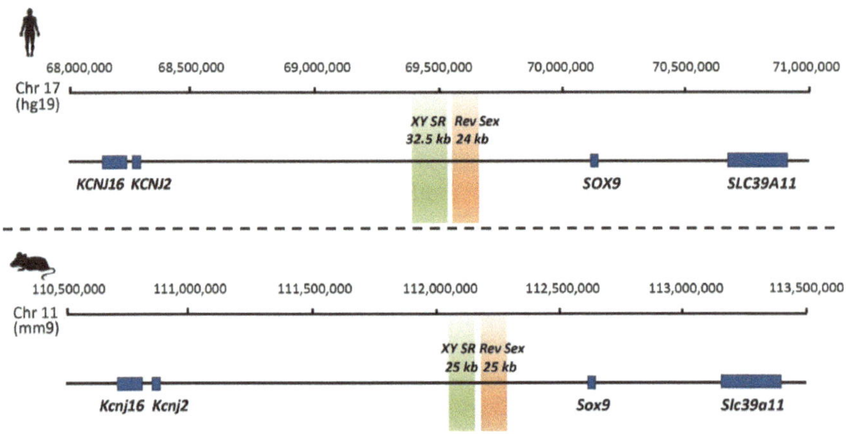

Fig. 2 Genomic view of the *SOX9/Sox9* locus in human and mouse. The *SOX9/Sox9* gene is located on chromosome 17 in human and chromosome 11 in the mouse. It is surrounded by a gene desert, with the neighboring genes located 2MB upstream in human and 1.7MB upstream in mouse, while in both species the next downstream gene is located 0.5MB away. Based on several DSD human patients, two regions have been associated with sex reversal. The first is called *RevSex* (peach box) and is a minimal region of 24kb located 559–583kb upstream of *SOX9,* duplications of which result in XX male development. The second is further upstream and termed XY SR (green box). This is a minimal 32.5kb region, located 607–639kb upstream of *SOX9* and when deleted in XY patients leads to female development. The regions in the mouse that show conserved synteny with the human *RevSex* and XY SR are both 25kb in size and are indicated by peach and green boxes, respectively.

2.1 Lessons from human patients

Much of what we know about the regulatory region of *SOX9,* and the fact that many tissue-specific enhancers lie in the upstream and downstream regions of *SOX9,* comes from human patients with deletions, duplications, translocations and inversions in and around the human *SOX9* gene. Altogether, data from patients indicate that the regulatory region of *SOX9* might span a 3 MB region located both upstream and downstream of the gene itself (Gordon et al., 2009). Heterozygous mutations in *SOX9* lead to campomelic dysplasia (CD, OMIM 114290), characterized by a severe skeletal malformation syndrome and male-to-female sex reversal in about 70% of the XY patients (Foster et al., 1994; Wagner et al., 1994). It is a severe syndrome and children with CD usually die within a year, often from breathing difficulties due to weakened cartilage in the upper respiratory tract (Foster et al., 1994; Gordon et al., 2009). The majority of mutations causing CD have been identified within the *SOX9* coding sequence and are predicted to result from haploinsufficiency, although dominant negative effects have not been ruled out in cases where an altered protein product is made. Additionally, many other patients have been characterized over the years with an unaffected *SOX9* coding region, but instead carry rearrangements upstream or downstream (Gordon et al., 2009; Symon & Harley, 2017). In CD patients, the severity of the campomelia (the bowing of the long bones) can vary depending on the distance of the translocation breakpoint from the *SOX9* gene. In the more distant lesions, the campomelic feature is missing and the disorder is described as acampomelic campomelic dysplasia (ACD) (Gordon et al., 2009). It is speculated that in these patients, a substantial portion of the *SOX9* regulatory region is present and hence some aspects of skeletal development can occur normally. Another disorder that has been implicated with long-distance alterations of the regulatory regions of *SOX9* is Pierre Robin Sequence (PRS; OMIM 261800), a disorder affecting the craniofacial skeleton (Benko et al., 2009; Jakobsen et al., 2007). Patients with isolated PRS have been identified with translocations and micro-deletions further than 1 Mb upstream of *SOX9* as well as 1.3–1.5 Mb downstream of *SOX9* (Benko et al., 2009; Velagaleti et al., 2005). PRS syndrome appears in most of the CD patients, but can also appear isolated without the campomelia syndrome. As mentioned above, 70% of XY patients with CD present with sex reversal and a female phenotype (Baetens, Mendonca, Verdin, Cools, & De Baere, 2017; Croft, Ohnesorg, & Sinclair, 2018; Symon & Harley, 2017). It is believed that the reason behind the finding that only 70% of XY patients are sex reversed is that the *SOX9* expression levels required in the gonads to direct male development are around 50% of normal levels. Hence, genetic variability between individuals that either alter the initial expression levels of *SOX9*, modulate how it acts, or affect how ovotestes develop and resolve into an ovary or a testis, can allow some to develop as XY males.

Looking specifically at the gonadal phenotypes, two regions have been identified in humans that affect *SOX9* expression in the testis (Fig. 2). One is a (minimal overlap) 24 kb region located 559–583 kb upstream of *SOX9* (hg19 chr17: 69,534,526–69,558,832), termed RevSex (or XX SR) that was found to be duplicated in XX male patients. The minimal overlap duplication in all known patients has been recently refined from 178 kb (Cox, Willatt, Homfray, & Woods, 2011) to 98 kb (Vetro et al., 2011), 78 kb (Benko et al., 2011), 67 kb (Xiao, Ji, Xing, Chen, & Tao, 2013), 41 kb (Hyon et al., 2015) and most recently to a 24 kb minimal region (Croft, Ohnesorg, & Sinclair, 2018; Ohnesorg et al., 2017). The other region has been defined by overlapping deletions associated with XY female development as a 32.5 kb region (termed XY SR) located 607–639 kb upstream of *SOX9* (hg19 chr17: 69,477,571–69,510,055) (Hyon et al., 2015; Kim et al., 2015) (Fig. 2). Most XY female patients that were originally discovered with deletions also covered the RevSex region, but it was only in Kim et al. that a patient was identified (termed DSD2 patient) with a deletion located 672–607 kb upstream of *SOX9* that did not contained the previously identified RevSex. This led to the separation of the patients into two groups, those with duplications in RevSex, resulting in XX males and those with deletions in XY SR, resulting in XY females (Baetens et al., 2017; Croft, Ohnesorg, & Sinclair, 2018; Kim et al., 2015; Symon & Harley, 2017). For a comprehensive review, which summarizes all identified patients, please see Baetens et al. (2017).

2.2 Lessons from mouse models

Several studies in mice have also strongly supported the existence of a highly complex regulatory landscape involved in *Sox9* gene regulation and pinpointed the location of multiple enhancers mediating tissue-specific expression of *Sox9* for the various tissues in which it is normally expressed, including chondrocytes, pancreas, liver, gut, neural crest and more (Barrionuevo et al., 2006; Mead et al., 2013; Symon & Harley, 2017; Wunderle, Critcher, Hastie, Goodfellow, & Schedl, 1998; Yao et al., 2015). Very recently, Mochizuki and colleagues used CRISPR/Cas9 based approaches to look for distant upstream enhancers governing cartilage-specific *Sox9* expression (Mochizuki et al., 2018). They used hemagglutinin (HA)-tagged catalytically dead Cas9 (dCas9) and a sgRNA directed against the *Sox9* promoter and using anti-HA ChIP, performed in primary chondrocytes, they found binding of the *Sox9* promoter to an enhancer located ~1 MB upstream of the *Sox9* TSS (transcription start site) which they termed as a rib cage-specific enhancer (RCSE). Deletion of this enhancer *in vivo* resulted in mice with shorter and narrower rib cages compared to wild-type mice. *Sox9* and *Col2a1* (Collagen Type II Alpha 1 Chain) expression levels were downregulated by ~50% in the costal cartilage of RCSE null mice. Since the reduction in *Sox9* expression levels as well as the phenotype observed in the RCSE null mice are weaker than those observed with *Sox9* KO mice, it is likely that other enhancers also participate in regulating *Sox9* expression in cartilage (Mochizuki et al., 2018).

In relation to *Sox9* regulation of expression in the gonads, two mouse models were described. The first is *Odd Sex* (*Ods*), which involves a 134 kb deletion located around 1 MB upstream of the *Sox9* gene associated with the insertion of a tyrosinase minigene under the control of the dopachrome tautomerase (*Dct*) promoter (Bishop et al., 2000). Heterozygous *Ods* mice exhibit XX male development due to constitutive expression of SOX9 in the gonads in the absence of SRY. It was initially hypothesized that the reason for the sex reversal was the loss of a critical gonad-specific repressor of *Sox9* located within the deleted 134 kb region (Bishop et al., 2000). However, a later study recapitulated the 134 kb deletion without the insertion of the *Dct* promoter and demonstrated that the deletion, in itself, was not sufficient to cause the sex reversal. This suggests that the gonad-specific expression of *Sox9* in *Ods* mice was due to the *Dct* promoter element interacting with gonad-specific enhancer elements located upstream of *Sox9*, and strongly inducing *Sox9* expression in XX gonads despite the absence of *Sry* (Qin et al., 2004).

The second model studied is provided by C57BL/6J-YPOS (B6-YPOS) mice. These mice carry autosomes from the B6 strain and a Y chromosome from the *Mus domesticus poschiavinus* (YPOS) wild-derived strain. They show a significant delay in *Sry* expression, and often present with complete XY female sex reversal (Bullejos & Koopman, 2005). The presence of a 55 MB congenic region on chromosome 11 was shown to protect B6-YPOS from sex reversal in a dose-dependent manner. This region, derived from a "semi-inbred" strain POSA, was further refined into a 1.62 MB genomic region that, when homozygous, confers full protection from XY B6-YPOS female sex reversal (Arboleda et al., 2014). This 1.62 MB region, which had been maintained despite 30 years of back-crossing onto C57BL/6J-YPOS, is located upstream of the *Sox9* gene, covering most of the 1.7 MB upstream mouse gene-desert, and is predicted to promote sufficient expression of *Sox9* to drive testis development despite the delayed *Sry* expression associated with the B6-YPOS background (Arboleda et al., 2014).

Taken altogether, these studies in human patients and mouse models strongly support the presence of several potential gonad-specific enhancers of *Sox9*. In the next section, we describe in detail the discovery of the first known gonad-specific enhancer of *Sox9*, termed TES (Testis-specific Enhancer of *Sox9*), and the very recent finding that several others, more upstream enhancers are able to drive gonadal *Sox9* expression, with one of them, called Enh13, having a critical role specifically at the stage of sex determination.

3. Gonadal enhancers of *Sox9*

An early attempt to explore the regulatory region of human *SOX9* involved making transgenic mice with a yeast artificial chromosome (YAC) vector containing up to 350 kb 5′ and 250 kb 3′ to the human SOX9 coding region, with a *lacZ* reporter gene engineered into the first exon (Wunderle et al., 1998). This gave reporter expression in several mouse tissues known to express *Sox9*, but not in testes. It was suspected that this might be due to lack of gonadal regulatory sequences or to differences in these between human and mouse. However, the analysis was rather minimal in terms of numbers of lines examined, and the latest stage examined was E11.5, which may have been too early. Bagheri-Fam et al. (2006) compared *Sox9* genomic sequences between human and pufferfish (Fugu) and identified eight highly conserved sequence elements between 290 kb 5′ and 450 kb 3′ to human *SOX9*, and then tested these for enhancer activity in transgenic mice, again using a *lacZ* reporter gene. They also failed to detect any gonadal expression, although several elements did give *lacZ* activity in other relevant tissues.

3.1 The TES enhancer

In order to avoid problems associated with potential lack of evolutionary conservation and the use of human genomic sequences in mice, another early attempt to look for a gonad-specific enhancer controlling *Sox9* expression in the testis was carried out by Sekido and Lovell-Badge (2008). They used a mouse *Sox9* bacterial artificial chromosome (BAC) clone (−70 to +50 kb relative to the *Sox9* transcriptional start site) carrying a minimal *heat-shock protein 68 (hsp68)* promoter and a *lacZ* reporter gene. Mice carrying this 120 kb construct mimicked several of the characteristics of the endogenous *Sox9* expression pattern, notably including expression in the fetal testis. LacZ gonadal expression started at E10.5 and increased by E11.5 in both sexes and then became restricted to the testis only by E12.5. Serial dissection of the initial 120 kb fragment led to the discovery of a 3.2 kb element termed TES (for Testis-specific Enhancer of *Sox9*) located −13 to −10 kb upstream of the *Sox9* TSS, which was able to fully recapitulate the gonadal expression pattern of *Sox9* (Fig. 3). A core element within TES of 1.4 kb was further defined and termed TESCO (for TES Core) (Sekido & Lovell-Badge, 2008). This enhancer can drive Sertoli cell-specific expression; moreover, by chromatin immunoprecipitation (ChIP) assays, both SRY and SF1 are found bound to TES *in vivo* at E11.5 in XY gonads. SOX9 also interacts with TESCO *in vivo*, suggesting a feed-forward regulatory loop by which SOX9 helps to maintain its own transcription once it reaches a critical threshold level (Sekido & Lovell-Badge, 2008). There has been much support for this

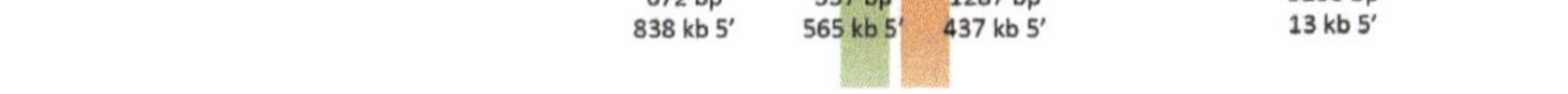

Fig. 3 Enhancers involved in mediating *Sox9* gonadal expression in the mouse. Several gonadal enhancers of *Sox9* have been identified. TES is a 3.2 kb enhancer with a core element of 1.4 kb, located 13 kb upstream of *Sox9*. Three other enhancers giving expression in the testis have been characterized and are labeled in turquoise and called Enh13, Enh14 and Enh32. The sizes and locations are indicated above or below each enhancer. Enh13 is located toward the 5′ side of the mouse homolog of the human XY SR region. Mice carrying homozygous deletions of Enh13 exhibit full XY female development indicating that Enh13 is the sole critical enhancer controlling *Sox9* expression at the stage of sex determination. No enhancer within the *RevSex* region was found in the mouse. Enh8 shows expression in the ovary and may serve either as a repressor, or as an enhancer that under normal conditions in inactive, but under cases of transdifferentiation of granulosa cells to Sertoli cells becomes active.

model and the discovery of TES and TESCO has allowed direct tests of the ability of both mutant SRY and SF1 proteins to act on *Sox9* transcription (Racca et al., 2014; Sreenivasan et al., 2018). For example, Sreenivasan and colleagues showed that the underlying mechanism explaining the variability of phenotypes observed in DSD patients carrying various mutations in SF1 is whether or not they can activate the TESCO enhancer.

In addition to TES being involved in the positive regulation of *Sox9* expression in the testis, it has also been implicated to be involved in the repression of *Sox9* expression in the ovary. It was shown that the WNT4 signaling pathway can inhibit *Sox9* expression at the transcriptional level by reducing the ability of SF1 to activate the TESCO enhancer (Bernard et al., 2012). Similarly, *Dax1* overexpression was shown to repress *Sox9* expression in a TES-dependent manner *via* competition with SF1 over the binding to TES. This study provides an explanation for 46, XY female DSD patients carrying a duplication of the DAX1 gene (Ludbrook et al., 2012). Furthermore, using ChIP techniques on adult ovaries, both FOXL2 and ESR1 (Estrogen Receptor 1) were found bound to TESCO *in vivo*. Transgenic mouse assays using TES sequences mutated for the FOXL2 sites suggested that the latter were also important. This explains the mechanism behind the ability of FOXL2 to repress *Sox9* expression in adult ovaries and hence to maintain granulosa cell fate. Conditional knock out of *Foxl2* in adult ovaries leads to transdifferentiation of the latter to testis, with onset of *Sox9* expression occurring within 2 days of the loss of FOXL2 protein (Jakob & Lovell-Badge, 2011; Uhlenhaut et al., 2009).

Despite having many of the properties expected for an enhancer with a critical role in sex determination, targeted deletion of either TES or TESCO did not result in XY female development (or in XX male sex reversal) (Gonen, Quinn, O'Neill, Koopman, & Lovell-Badge, 2017). Nevertheless, these studies did reveal that TES/TESCO accounts for about 55% of *Sox9* expression levels in the embryonic and postnatal mouse testis. Unlike in humans, where this is likely to correspond to a threshold below which ovary development occurs, in the mouse, 50% of *Sox9* expression levels are more than enough to form a functional testis, as indicated by mice heterozygous for a null mutation of *Sox9* itself (Barrionuevo et al., 2006; Gonen et al., 2017; Lavery et al., 2011). Using the TES deletion, it was possible to further reduce the expression levels of *Sox9*. Mouse embryos carrying one allele of the TES deletion and a second allele in which a conditional allele of *Sox9* was deleted using *beta-actin*-Cre exhibited 23% residual expression of *Sox9* compared to wild-type XY gonads at E13.5, and presented with ovotestes. This provided evidence that about 25% of normal *Sox9* levels in the mouse gonad correspond to the 50% threshold level found in humans, at or below which XY-female sex reversal occurs (Gonen et al., 2017). Ovotestes can resolve by birth into either ovaries or testes, the latter tending to be smaller than normal, which can explain complete sex reversal seen in some cases of CD (Swain, Narvaez, Burgoyne, Camerino, & Lovell-Badge, 1998).

All of the above strongly indicated the importance of TES in regulating the gonadal expression of *Sox9*. However, the lack of sex reversal upon TES deletion in mice, combined with a failure to date to identify mutations affecting the equivalent human TES sequence, including a study examining 66 patients with idiopathic 46, XY gonadal dysgenesis (Georg et al., 2010), and data from human DSD patients identified with deletions and duplications far upstream of the *Sox9* gene (Symon & Harley, 2017), all indicated that there must be additional enhancers involved in the regulation of *Sox9* expression in the gonads.

3.2 The Enh13 enhancer and other novel enhancers of *Sox9*

To identify additional enhancers involved in the gonadal regulation of *Sox9* expression during embryonic stages, chromatin accessibility was examined in the 2MB gene desert upstream of *Sox9*. This initially made use of a DNaseI hypersensitivity dataset performed on sorted Sertoli cells at E13.5 and E15.5 as well as H3K27Ac ChIP-seq that mark active enhancers (Maatouk et al., 2017). ATAC-seq data, which similar to DNaseI-seq, marks chromatin accessible regions in the genome that can serve as regulatory elements, was also performed to compare E13.5, sorted Sertoli and granulosa cells. Thirty-three potential sites were identified and 16 of these were screened experimentally, *in vivo*, using transgenic mice. Four novel gonadal enhancers were identified apart from the previously known TES (Gonen et al., 2018) (Fig. 3). Enh13 and Enh14 showed very strong testicular expression while Enh32 showed a much weaker expression only on the ventral side of the testis. Surprisingly, Enh8 conferred strong β-Gal activity in the ovary, while it was barely present in the testis at E13.5.

Although *Sox9* is not expressed in the ovary at E13.5, it is expressed at low levels in both sexes from E10.5 until E11.5, before it is markedly upregulated in the male and all but lost in the female (Kent et al., 1996; Morais da Silva et al., 1996). However, *Sox9* can be expressed in the adult ovary upon either deleting *Foxl2* (Uhlenhaut et al., 2009) or overexpressing

Dmrt1 (Lindeman et al., 2015; Zhao et al., 2015), both of which encode transcription factors required to maintain ovary and testis fate, respectively. The onset of *Sox9* expression in granulosa cells leads to their transdifferentiation into Sertoli cells. Because repression of *Sox9* in granulosa cells is an active process throughout life, otherwise they transdifferentiate into Sertoli cells, the relevant enhancers also must be accessible (Jakob & Lovell-Badge, 2011; Uhlenhaut et al., 2009). Thus, it is perhaps not surprising to identify an enhancer that, when taken out of its original genomic context, can induce ovary expression, and we speculate that in the endogenous *Sox9* locus in females, Enh8 is usually sequestered away from the *Sox9* promoter by DNA looping. It will be interesting to explore whether the activity of this enhancer is involved in adult gonadal sex reversal as described above. Interestingly, and in support of this idea, ATAC-seq at E13.5 reveals a much stronger peak at Enh8 in granulosa cells compared to Sertoli cells. Additionally, this enhancer was found active and marked with H3K27ac in both male and female gonads at E13.5 (Gonen et al., 2018).

Further study of Enh13 revealed that it is critical for sex determination and upregulation of *Sox9* expression in the testis. Homozygous deletion of Enh13, either on a TES deleted or a C57BL/6J wild type genetic background, led to full XY female sex reversal (Gonen et al., 2018). This lowered *Sox9* mRNA levels in XY gonads at E11.5 to that of control XX gonads, explaining the complete and early sex reversal. SRY was found strongly bound to Enh13 *in vivo* using ChIP at E11.5, while by E13.5 SOX9 itself was bound to Enh13, demonstrating that this enhancer is involved both in the initiation as well as maintenance of *Sox9* expression in the testis (Gonen et al., 2018).

Enh13 would appear to be well conserved amongst many mammals, with consensus binding sites for SRY/SOX9, SF1 and other relevant transcription factors (Gonen et al., 2018). Interestingly, Enh13 is located toward the distal 5′end of a 25.7 kb region in mice showing conserved synteny with the 32.5 kb human XY SR locus (Kim et al., 2015) (Fig. 3). Ohnesorg et al. relying on published DNaseI hypersensitivity data in unsorted cells from human embryonic testis and ovaries, found 13 potential sites upstream of *SOX9* (Ohnesorg, Croft, Tan, & Sinclair, 2016). One of these, termed "SOX9 up 2", is 1773 bp-long, maps to the XY SR locus and would include the human equivalent of Enh13. This, as well as other sites, showed enhancer activity assessed by luciferase-assays, although these were carried out in heterologous cell lines (Ohnesorg et al., 2016). The data in Gonen et al. (2018) therefore suggest that Enh13 is the critical enhancer within the XY SR region, where it also plays a critical role in *Sox9* expression leading to Sertoli cell and testis differentiation in humans. In strong support of this, a recent study which utilized CRISPR genome-editing to delete different parts of the XY SR region found that removing the entire mouse XY SR region results in XY female sex reversal. By further refining the deleted fragment, they identified a 783 bp element (mm9 chr11: 112,078,499–112,079,281) that is able to induce XY female development when homozygously deleted. Enh13 (557 bp) is fully contained within this 783 bp element (Ogawa et al., 2018).

Remarkably, recent work by Croft and colleagues, which focused on identifying enhancers that regulate human *SOX9* expression in gonads, found a homologous enhancer to Enh13. They describe two new 46, XX testicular/ovo-testicular DSD patients carrying a micro-duplication with overlapping 5.2 kb. They were able to identify a 1514 bp enhancer, which they termed Sex Reversal Enhancer-A (eSR-A). Enh13 sequence is fully contained within the 1514 bp eSR-A enhancer. Interestingly, when they compare the human eSR-A and the mouse Enh13 using luciferase assays they show that whereas Enh13 is activated by both mSRY + hSF1 and mSOX9 + hSF1, the human eSR-A enhancer is only activated by hSOX9 + hSF1 but not hSRY + hSF1 (Croft, Ohnesorg, Hewitt, et al., 2018). This suggests that whereas in the mouse Enh13 is important for both activation and maintenance of *Sox9* expression, in the human context, eSR-A is more important for the maintenance of *SOX9* expression. However, there is a possible caveat in that, unlike mouse SRY that has a C-terminal activation domain, albeit one that appears weak, human SRY does not, and it is assumed to rely on an interacting co-activator (Zhao et al., 2014). It is possible that this is not present in the cell line used for the co-transfection experiments. If so, then it is possible that eSR-A and Enh13 have identical roles in activation as well as maintenance.

This work and the newly identified DSD patients further support the critical role of Enh13/eSR-A and show that both deletion of this enhancer in XY patients and duplication in XX patients, in the absence of SRY, are able to induce sex reversal (Croft, Ohnesorg, Hewitt, et al., 2018; Kim et al., 2015). This would also suggest that it is an early acting enhancer.

Croft et al. further identified two novel human enhancers. One of them is located within the XX SR region, termed Sex Reversal Enhancer-B (eSR-B) and the other is located slightly upstream to the human TES enhancer, termed enhancer Alternate Long-Distance Initiator (eALDI). The former seem to be activated by SF1 + SOX9, whereas the latter is

SRY-dependent. Unlike Enh13/eSR-A which seems to be highly conserved between human and mouse, the two other enhancers seem to be human-specific (Croft, Ohnesorg, Hewitt, et al., 2018).

It will be interesting to explore if some DSDs involve specific mutations, including Single Nucleotide Polymorphisms (SNPs), affecting Enh13/eSR-A or the other novel identified enhancers.

The notion of redundancy or "shadow enhancers" within a regulatory region is well established (Lagha, Bothma, & Levine, 2012; Visel, Rubin, & Pennacchio, 2009). Recently, Dickel et al. studied ultraconserved enhancers, showing perfect conservation across mammalian genomes, and found that deletion of single enhancers does not impact viability. They focused on eight ultraconserved enhancers of the *Arx* (*Aristaless related homeobox*) gene and showed that while single deletions do not affect *Arx* expression, deletion of pairs of enhancers can lead to a reduction in *Arx* expression levels. In accordance with this, phenotypes in the mouse brain are more severe upon deletion of pairs of enhancers compared to a deletion of single enhancer (Dickel et al., 2018). Similarly, another study created 23 mouse deletion lines including single and combinatorial deletions of 7 established enhancers required for limb development. Surprisingly, none of the individual deletions caused changes to limb morphology. It was only when a pair of limb enhancers near the same gene were deleted that a phenotype started to appear, strongly indicating that enhancers function redundantly in establishing normal cell fate and morphology (Osterwalder et al., 2018).

As discussed above, the expression of *Sox9* in cartilage also appears to rely on several redundantly acting enhancers (Mochizuki et al., 2018). Hence, knowing the major contribution of TES to maintaining *Sox9* expression levels, as well as the discovery of other novel gonadal enhancers, such as Enh14, it is remarkable to see that deleting a single enhancer phenocopies the loss of *Sox9* itself within the supporting cell lineage (Gonen et al., 2018; Lavery et al., 2011). Substantial evidence points to the time-dependent action of SRY on *Sox9* as the explanation. If *Sox9* fails to reach a critical threshold within a few hours then ovary-determining/anti-testis factors, such as Wnt signaling, accumulate to a sufficient level to repress *Sox9* and make it refractory to male promoting factors, including SRY, even though expression of the latter persists in XY gonads when Sertoli cells fail to differentiate (Hiramatsu et al., 2009). The most likely explanation for the critical role of Enh13, and one that is consistent with all current data, is that it is an early acting enhancer, such that without it *Sox9* transcription fails to increase to a level where the other enhancers can act before the gene is silenced.

Unlike the critical role of Enh13 during sex determination, deletion of Enh14 did not exhibit any phenotype and did not alter *Sox9* expression levels at the embryonic stages. It is likely that Enh14 functions in a redundant manner along with TES and Enh13 to ensure that high levels of *Sox9* are maintained (Gonen et al., 2018). Enh32 may also correspond to an enhancer used later. While the DNaseI-seq and ATAC-seq data showed it was open at E13.5 and E15.5 in Sertoli cells, transgene expression was low at these stages, but higher at postnatal stages (Gonen et al., 2018) and (unpublished data). Moreover, it should be stressed that because the ATAC-seq data was carried out in early gonads there could well be other enhancers involved at later stages and in the adult testis.

Interestingly, high-throughput methods, including chromosome capture confirmation assays (3C and derivatives 4C, 5C and HiC), indicate that a single gene promoter will interact on average with five different enhancers (Jin et al., 2013; Spitz, 2016). Our data agrees with this, with at least four different enhancers likely to act on *Sox9* expression during embryonic testis development (Gonen et al., 2018). Assuming it will be possible to perform 4C analysis on small number of cells isolated from the embryonic gonad, it will be interesting to verify these long-distance interactions between the remote enhancers and the *Sox9* promoter during sex determination, and to explore if there is a distal to proximal use of enhancers, showing either consecutive or additive roles in *Sox9* transcription in the testis. Finally, given the relevance for SOX9 in sex determination, cartilage development, *etc.*, throughout vertebrate evolution, it will be of interest to explore how well the enhancers and their arrangement are functionally conserved. Despite its obvious importance, sex determination mechanisms show rapid evolution. SRY itself is poorly conserved outside the HMG box, and it is the sex-determining gene only in eutherian and metatherian mammals, indeed it has even been lost in some species, such as mole voles (*Ellobius lutescens*) and spiny rat (*Tokudaia osimensis*) (Capel, 2017; Just et al., 2007; Kuroiwa, Ishiguchi, Yamada, Shintaro, & Matsuda, 2010; Soullier, Hanni, Catzeflis, Berta, & Laudet, 1998). While Enh13 shows good conservation across a range of mammals tested, it was not obvious in the chick (Gonen et al., 2018). Perhaps this reflects the very different mechanism of sex determination present in birds, including the absence of *Sry*, and the more critical involvement of DMRT1 and of estrogen signaling (Capel, 2017; Lambeth et al., 2014; Smith et al., 2009).

4. Higher order of chromatin organization at the *Sox9* locus

It is now well established that the genome is organized non-randomly into higher order chromatin domains (Andrey & Mundlos, 2017; de Laat & Duboule, 2013; Spitz, 2016). Chromosome conformation capture techniques (3C, 4C, 5C, HiC and derivatives) have revealed the presence of chromatin domains called topological association domains (TADs) that show high internal interactions while being separated by low interacting regions called boundary elements (Dixon et al., 2012; Nora et al., 2012). It is believed that TADs divide the genome into separate regulatory units and hence allow the binding of enhancers only to their target genes (Rao et al., 2014; Symmons et al., 2014). TADs are relatively conserved across different cell types and even between different species (Dixon et al., 2015, 2012).

Recently, Franke et al. (2016) studied the *Sox9* TAD in human and mice and explored how genomic duplications spanning this region affect the TAD formation and mediate disease. They performed 4C experiments on human fibroblasts carrying three different types of duplications. The first was a 0.5 Mb duplication covering the RevSex region but fully contained within the *SOX9* TAD (termed an intra-TAD duplication) in which the patient presented with XX male sex reversal. The second was a larger duplication that included the RevSex region, but extended further toward the neighboring TAD and contained the upstream *KCNJ2* and *KCNJ16* genes (termed an inter-TAD duplication) (Fig. 2). This duplication had no effect on sexual development, but resulted in Cooks syndrome, a congenital limb malformation (Kurth et al., 2009). The third duplication included the RevSex and the entire gene desert upstream of *SOX9* but not the *KCNJ2* and *KCNJ16* genes. These patients were phenotypically normal. Using 4C and capture HiC the authors demonstrate that the intra-TAD duplications within the *SOX9* TAD result in increased contacts of the duplicated regions within the TAD, without change in the overall TAD structure. In contrast, the inter-TAD duplications that did not include the *KCNJ2* and *KCNJ16* genes created a new TAD (neo-TAD) that was isolated from the rest of the genome and hence had no phenotypic effect. Conversely, the duplication that did contain the *KCNJ2* adjacent gene also showed the formation of a neo-TAD, but in this instance, the interactions within the neo-TAD resulted in the ectopic contacts of *KCNJ2* with the duplicated part of the

SOX9 regulatory region and led to the misexpression of *KCNJ2* and to a limb malformation phenotype (Andrey & Mundlos, 2017; Franke et al., 2016). In support of TADs being insulated chromatin environments, another study explored the *Shh* gene and its ZRS limb enhancer and demonstrated that as long as both the enhancer and the gene are located within the same TAD, the relative distance of the enhancer from the gene has no influence on its transcriptional activity (Symmons et al., 2016).

Interestingly, it is notable that in the mouse, we did not discover any putative enhancer in the RevSex region (Gonen et al., 2018). On the contrary, a study of gonadal *SOX9* enhancers in humans did find a putative 744 bp-long enhancer termed "SOX9 up 3" located within the RevSex region, however, luciferase assays did not show strong enhancer activity (Ohnesorg et al., 2016). It will be interesting to understand the underlying reason for the sex reversal observed in XX patients carrying duplications containing the RevSex region (Baetens et al., 2017). It is possible that the RevSex region contains a human-specific enhancer that is not conserved in mouse. This idea is strongly supported by the recent finding of a novel human enhancer located in that region termed eSR-B (Croft, Ohnesorg, Hewitt, et al., 2018). Alternatively, we hypothesize that the RevSex duplication within the *SOX9* TAD causes increased interactions and hence elevated expression of *SOX9,* leading to male development in XX individuals. This could involve an effect on Enh13, which is close by. Perhaps the strategy of looking in the shared minimal region that is missing in all patients is not the way forward, but instead the phenotype arises from changes in higher chromatin organization and amount of interactions occurring within a TAD as a result of the duplication.

5. Perspectives

Advances in recent years have strongly supported the presence of multiple enhancers regulating *Sox9* expression throughout gonadal differentiation. We showed that Enh13 is critical for sex determination at early stages (Gonen et al., 2018). However, apart from an important, albeit not critical, subsequent role for TES, we still do not know which enhancers are involved in controlling *Sox9* expression throughout gonadal development. For example, how is the initial, basal expression of *Sox9* in the genital ridge of both sexes controlled, and what are the enhancers that regulate the expression of *Sox9* in adulthood, which is needed to maintain testis identity

(Matson et al., 2011). Furthermore, it will be interesting to see whether *Sox9* expression at different stages is also controlled by a single, critical enhancer or mediated *via* the activity of several, redundant enhancers. In addition, much focus has been put on enhancers; however, from what we know based on genetic studies, sex determination and sex maintenance heavily rely on gene repression (Carre & Greenfield, 2016). Although more challenging to find, it will be important to understand if the pro-female transcription factors such as β-CATENIN and FOXL2 execute their repressive effect on *Sox9* expression *via* binding to the already known enhancers or to different, novel silencers.

The rapid development and improvements of techniques that allow the identification of potential regulatory elements such as DNaseI-Seq, ATAC-Seq, and ChIP-Seq for TFs and histone marks has allowed much progress and increased our understanding of the mechanism of activity of the critical transcription factors controlling sex determination (Garcia-Moreno, Futtner, et al., 2018; Gonen et al., 2018; Maatouk et al., 2017; Rahmoun et al., 2017). Efforts to map all the potential regulatory elements in a cell- and stage-specific manner in the developing gonads are proving to be very informative, although it is likely that the precise role of each will still need to be assessed in a focused way (Garcia-Moreno, Futtner, et al., 2018). However, it is still challenging to perform chromosome capture confirmation techniques on relatively small numbers of cells sorted from embryonic gonads. In addition, whereas ChIP-Seq is useful at detecting transcription factor binding to promoters at target genes, the ability to detect binding to remote regulatory elements can be very limited due to the small amount of starting material available, particularly in early gonads (Rahmoun et al., 2017). Therefore, further progress in downscaling those methods could allow the unbiased identification of novel regulatory elements that are bound by pro-male and pro-female critical transcription factors such as SRY, SOX9, DMRT1, SF1, FOXL2, β-CATENIN and others that mediate the activation as well as repression of downstream target genes. It will also allow identification, *in vivo*, of enhancer-promoter interactions throughout gonadal development.

Although much effort has been given to discover the genetic causes of Disorders of Sex Development (DSD), this is still not known for ~80% of the patients. Many cases have been analyzed by whole exome sequencing (Baetens et al., 2017; Croft, Ohnesorg, & Sinclair, 2018; Hyon et al., 2015). However, with the understanding that in some patients the mutation occurs in non-coding regions, and include copy number variations (CNV) of critical regulatory elements, whole genome sequencing is now the preferred method (Croft, Ohnesorg, & Sinclair, 2018). The discovery that Enh13 is responsible for XY female development in DSDs involving deletions of the 32.5 kb genomic region termed XY SR, strongly supports the above (Gonen et al., 2018) and advocates for the search for additional regulatory elements of other genes that may mediate sex reversal phenotypes. Adding Enh13, but also the other novel enhancers to genotyping screens may be informative for other cases of DSD.

Acknowledgments

We thank members of the Lovell-Badge lab for helpful discussion and critical reading of the manuscript. We are grateful to the editor, Blanche Capel for additional constructive comments. The Francis Crick Institute receives its core funding from Cancer Research UK (FC001107), the UK Medical Research Council (FC001107), the Wellcome (FC001107), and by the UK Medical Research Council (U117512772).

References

Andrey, G., & Mundlos, S. (2017). The three-dimensional genome: Regulating gene expression during pluripotency and development. *Development*, *144*, 3646–3658.

Arboleda, V. A., Fleming, A., Barseghyan, H., Delot, E., Sinsheimer, J. S., & Vilain, E. (2014). Regulation of sex determination in mice by a non-coding genomic region. *Genetics*, *197*, 885–897.

Baetens, D., Mendonca, B. B., Verdin, H., Cools, M., & De Baere, E. (2017). Non-coding variation in disorders of sex development. *Clinical Genetics*, *91*, 163–172.

Bagheri-Fam, S., Barrionuevo, F., Dohrmann, U., Gunther, T., Schule, R., Kemler, R., et al. (2006). Long-range upstream and downstream enhancers control distinct subsets of the complex spatiotemporal Sox9 expression pattern. *Developmental Biology*, *291*, 382–397.

Bagheri-Fam, S., Sim, H., Bernard, P., Jayakody, I., Taketo, M. M., Scherer, G., et al. (2008). Loss of Fgfr2 leads to partial XY sex reversal. *Developmental Biology*, *314*, 71–83.

Barrionuevo, F., Bagheri-Fam, S., Klattig, J., Kist, R., Taketo, M. M., Englert, C., et al. (2006). Homozygous inactivation of Sox9 causes complete XY sex reversal in mice. *Biology of Reproduction*, *74*, 195–201.

Barrionuevo, F., Georg, I., Scherthan, H., Lecureuil, C., Guillou, F., Wegner, M., et al. (2009). Testis cord differentiation after the sex determination stage is independent of Sox9 but fails in the combined absence of Sox9 and Sox8. *Developmental Biology*, *327*, 301–312.

Barrionuevo, F. J., Hurtado, A., Kim, G. J., Real, F. M., Bakkali, M., Kopp, J. L., et al. (2016). Sox9 and Sox8 protect the adult testis from male-to-female genetic reprogramming and complete degeneration. *eLife*, *5*, e15635. https://doi.org/10.7554/eLife.15635.

Barrionuevo, F., & Scherer, G. (2010). SOX E genes: SOX9 and SOX8 in mammalian testis development. *The International Journal of Biochemistry & Cell Biology*, *42*, 433–436.

Benko, S., Fantes, J. A., Amiel, J., Kleinjan, D. J., Thomas, S., Ramsay, J., et al. (2009). Highly conserved non-coding elements on either side of SOX9 associated with Pierre Robin sequence. *Nature Genetics*, *41*, 359–364.

Benko, S., Gordon, C. T., Mallet, D., Sreenivasan, R., Thauvin-Robinet, C., Brendehaug, A., et al. (2011). Disruption of a long distance regulatory region upstream

of SOX9 in isolated disorders of sex development. *Journal of Medical Genetics*, *48*, 825–830.
Bernard, P., Ryan, J., Sim, H., Czech, D. P., Sinclair, A. H., Koopman, P., et al. (2012). Wnt signaling in ovarian development inhibits Sf1 activation of Sox9 via the Tesco enhancer. *Endocrinology*, *153*, 901–912.
Bi, W., Huang, W., Whitworth, D. J., Deng, J. M., Zhang, Z., Behringer, R. R., et al. (2001). Haploinsufficiency of Sox9 results in defective cartilage primordia and premature skeletal mineralization. *Proceedings of the National Academy of Sciences of the United States of America*, *98*, 6698–6703.
Bishop, C. E., Whitworth, D. J., Qin, Y., Agoulnik, A. I., Agoulnik, I. U., Harrison, W. R., et al. (2000). A transgenic insertion upstream of sox9 is associated with dominant XX sex reversal in the mouse. *Nature Genetics*, *26*, 490–494.
Bullejos, M., & Koopman, P. (2005). Delayed Sry and Sox9 expression in developing mouse gonads underlies B6-Y(DOM) sex reversal. *Developmental Biology*, *278*, 473–481.
Capel, B. (2017). Vertebrate sex determination: Evolutionary plasticity of a fundamental switch. *Nature Reviews. Genetics*, *18*, 675–689.
Carre, G. A., & Greenfield, A. (2016). The gonadal supporting cell lineage and mammalian sex determination: The differentiation of Sertoli and granulosa cells. *Results and Problems in Cell Differentiation*, *58*, 47–66.
Chaboissier, M. C., Kobayashi, A., Vidal, V. I., Lutzkendorf, S., van de Kant, H. J., Wegner, M., et al. (2004). Functional analysis of Sox8 and Sox9 during sex determination in the mouse. *Development*, *131*, 1891–1901.
Chassot, A. A., Ranc, F., Gregoire, E. P., Roepers-Gajadien, H. L., Taketo, M. M., Camerino, G., et al. (2008). Activation of beta-catenin signaling by Rspo1 controls differentiation of the mammalian ovary. *Human Molecular Genetics*, *17*, 1264–1277.
Colvin, J. S., Green, R. P., Schmahl, J., Capel, B., & Ornitz, D. M. (2001). Male-to-female sex reversal in mice lacking fibroblast growth factor 9. *Cell*, *104*, 875–889.
Cox, J. J., Willatt, L., Homfray, T., & Woods, C. G. (2011). A SOX9 duplication and familial 46,XX developmental testicular disorder. *The New England Journal of Medicine*, *364*, 91–93.
Croft, B., Ohnesorg, T., Hewitt, J., Bowles, J., Quinn, A., Tan, J., et al. (2018). Human sex reversal is caused by duplication or deletion of core enhancers upstream of SOX9. *Nature Communications*, *9*, 5319.
Croft, B., Ohnesorg, T., & Sinclair, A. H. (2018). The role of copy number variants in disorders of sex development. *Sexual Development*, *12*, 19–29.
de Laat, W., & Duboule, D. (2013). Topology of mammalian developmental enhancers and their regulatory landscapes. *Nature*, *502*, 499–506.
Dickel, D. E., Ypsilanti, A. R., Pla, R., Zhu, Y., Barozzi, I., Mannion, B. J., et al. (2018). Ultraconserved enhancers are required for normal development. *Cell*, *172*, 491–499.e15.
Dixon, J. R., Jung, I., Selvaraj, S., Shen, Y., Antosiewicz-Bourget, J. E., Lee, A. Y., et al. (2015). Chromatin architecture reorganization during stem cell differentiation. *Nature*, *518*, 331–336.
Dixon, J. R., Selvaraj, S., Yue, F., Kim, A., Li, Y., Shen, Y., et al. (2012). Topological domains in mammalian genomes identified by analysis of chromatin interactions. *Nature*, *485*, 376–380.
Foster, J. W., Dominguez-Steglich, M. A., Guioli, S., Kwok, C., Weller, P. A., Stevanovic, M., et al. (1994). Campomelic dysplasia and autosomal sex reversal caused by mutations in an SRY-related gene. *Nature*, *372*, 525–530.
Franke, M., Ibrahim, D. M., Andrey, G., Schwarzer, W., Heinrich, V., Schopflin, R., et al. (2016). Formation of new chromatin domains determines pathogenicity of genomic duplications. *Nature*, *538*, 265–269.

Garcia-Moreno, S. A., Futtner, C. R., Salamone, I. M., Gonen, N., Lovell-Badge, R., & Maatouk, D. M. (2018). Gonadal supporting cells acquire sex-specific chromatin landscapes during mammalian sex determination. *Developmental Biology*. https://doi.org/10.1016/j.ydbio.2018.12.023 [Epub ahead of print].

Garcia-Moreno, S. A., Plebanek, M. P., & Capel, B. (2018). Epigenetic regulation of male fate commitment from an initially bipotential system. *Molecular and Cellular Endocrinology*, *468*, 19–30.

Georg, I., Bagheri-Fam, S., Knower, K. C., Wieacker, P., Scherer, G., & Harley, V. R. (2010). Mutations of the SRY-responsive enhancer of SOX9 are uncommon in XY gonadal dysgenesis. *Sexual Development*, *4*, 321–325.

Gonen, N., Futtner, C. R., Wood, S., Garcia-Moreno, S. A., Salamone, I. M., Samson, S. C., et al. (2018). Sex reversal following deletion of a single distal enhancer of Sox9. *Science*, *360*, 1469–1473.

Gonen, N., Quinn, A., O'Neill, H. C., Koopman, P., & Lovell-Badge, R. (2017). Normal levels of Sox9 expression in the developing mouse testis depend on the TES/TESCO enhancer, but this does not act alone. *PLoS Genetics*, *13*, e1006520.

Gordon, C. T., Tan, T. Y., Benko, S., Fitzpatrick, D., Lyonnet, S., & Farlie, P. G. (2009). Long-range regulation at the SOX9 locus in development and disease. *Journal of Medical Genetics*, *46*, 649–656.

Gubbay, J., Collignon, J., Koopman, P., Capel, B., Economou, A., Munsterberg, A., et al. (1990). A gene mapping to the sex-determining region of the mouse Y chromosome is a member of a novel family of embryonically expressed genes. *Nature*, *346*, 245–250.

Hediger, M., Burghardt, G., Siegenthaler, C., Buser, N., Hilfiker-Kleiner, D., Dubendorfer, A., et al. (2004). Sex determination in *Drosophila melanogaster* and *Musca domestica* converges at the level of the terminal regulator doublesex. *Development Genes and Evolution*, *214*, 29–42.

Hiramatsu, R., Matoba, S., Kanai-Azuma, M., Tsunekawa, N., Katoh-Fukui, Y., Kurohmaru, M., et al. (2009). A critical time window of Sry action in gonadal sex determination in mice. *Development*, *136*, 129–138.

Houston, C. S., Opitz, J. M., Spranger, J. W., Macpherson, R. I., Reed, M. H., Gilbert, E. F., et al. (1983). The campomelic syndrome: Review, report of 17 cases, and follow-up on the currently 17-year-old boy first reported by Maroteaux et al in 1971. *American Journal of Medical Genetics*, *15*, 3–28.

Huang, B., Wang, S., Ning, Y., Lamb, A. N., & Bartley, J. (1999). Autosomal XX sex reversal caused by duplication of SOX9. *American Journal of Medical Genetics*, *87*, 349–353.

Huang, S., Ye, L., & Chen, H. (2017). Sex determination and maintenance: The role of DMRT1 and FOXL2. *Asian Journal of Andrology*, *19*, 619–624.

Hyon, C., Chantot-Bastaraud, S., Harbuz, R., Bhouri, R., Perrot, N., Peycelon, M., et al. (2015). Refining the regulatory region upstream of SOX9 associated with 46,XX testicular disorders of sex development (DSD). *American Journal of Medical Genetics. Part A*, *167A*, 1851–1858.

Ikeda, T., Kamekura, S., Mabuchi, A., Kou, I., Seki, S., Takato, T., et al. (2004). The combination of SOX5, SOX6, and SOX9 (the SOX trio) provides signals sufficient for induction of permanent cartilage. *Arthritis and Rheumatism*, *50*, 3561–3573.

Jakob, S., & Lovell-Badge, R. (2011). Sex determination and the control of Sox9 expression in mammals. *The FEBS Journal*, *278*, 1002–1009.

Jakobsen, L. P., Ullmann, R., Christensen, S. B., Jensen, K. E., Molsted, K., Henriksen, K. F., et al. (2007). Pierre Robin sequence may be caused by dysregulation of SOX9 and KCNJ2. *Journal of Medical Genetics*, *44*, 381–386.

Jin, F., Li, Y., Dixon, J. R., Selvaraj, S., Ye, Z., Lee, A. Y., et al. (2013). A high-resolution map of the three-dimensional chromatin interactome in human cells. *Nature*, *503*, 290–294.

Jo, A., Denduluri, S., Zhang, B., Wang, Z., Yin, L., Yan, Z., et al. (2014). The versatile functions of Sox9 in development, stem cells, and human diseases. *Genes & Diseases*, *1*, 149–161.

Just, W., Baumstark, A., Suss, A., Graphodatsky, A., Rens, W., Schafer, N., et al. (2007). *Ellobius lutescens*: Sex determination and sex chromosome. *Sexual Development*, *1*, 211–221.

Kamachi, Y., & Kondoh, H. (2013). Sox proteins: Regulators of cell fate specification and differentiation. *Development*, *140*, 4129–4144.

Kamachi, Y., Uchikawa, M., & Kondoh, H. (2000). Pairing SOX off: With partners in the regulation of embryonic development. *Trends in Genetics*, *16*, 182–187.

Karl, J., & Capel, B. (1998). Sertoli cells of the mouse testis originate from the coelomic epithelium. *Developmental Biology*, *203*, 323–333.

Kent, J., Wheatley, S. C., Andrews, J. E., Sinclair, A. H., & Koopman, P. (1996). A male-specific role for SOX9 in vertebrate sex determination. *Development*, *122*, 2813–2822.

Kim, Y., Bingham, N., Sekido, R., Parker, K. L., Lovell-Badge, R., & Capel, B. (2007). Fibroblast growth factor receptor 2 regulates proliferation and Sertoli differentiation during male sex determination. *Proceedings of the National Academy of Sciences of the United States of America*, *104*, 16558–16563.

Kim, Y., Kobayashi, A., Sekido, R., DiNapoli, L., Brennan, J., Chaboissier, M. C., et al. (2006). Fgf9 and Wnt4 act as antagonistic signals to regulate mammalian sex determination. *PLoS Biology*, *4*, e187.

Kim, G. J., Sock, E., Buchberger, A., Just, W., Denzer, F., Hoepffner, W., et al. (2015). Copy number variation of two separate regulatory regions upstream of SOX9 causes isolated 46,XY or 46,XX disorder of sex development. *Journal of Medical Genetics*, *52*, 240–247.

Kohler, B., & Achermann, J. C. (2010). Update—Steroidogenic factor 1 (SF-1, NR5A1). *Minerva Endocrinologica*, *35*, 73–86.

Koopman, P., Sinclair, A., & Lovell-Badge, R. (2016). Of sex and determination: Marking 25 years of Randy, the sex-reversed mouse. *Development*, *143*, 1633–1637.

Kuroiwa, A., Ishiguchi, Y., Yamada, F., Shintaro, A., & Matsuda, Y. (2010). The process of a Y-loss event in an XO/XO mammal, the Ryukyu spiny rat. *Chromosoma*, *119*, 519–526.

Kurth, I., Klopocki, E., Stricker, S., van Oosterwijk, J., Vanek, S., Altmann, J., et al. (2009). Duplications of noncoding elements 5′ of SOX9 are associated with brachydactyly-anonychia. *Nature Genetics*, *41*, 862–863.

Lagha, M., Bothma, J. P., & Levine, M. (2012). Mechanisms of transcriptional precision in animal development. *Trends in Genetics*, *28*, 409–416.

Lambeth, L. S., Raymond, C. S., Roeszler, K. N., Kuroiwa, A., Nakata, T., Zarkower, D., et al. (2014). Over-expression of DMRT1 induces the male pathway in embryonic chicken gonads. *Developmental Biology*, *389*, 160–172.

Lavery, R., Lardenois, A., Ranc-Jianmotamedi, F., Pauper, E., Gregoire, E. P., Vigier, C., et al. (2011). XY Sox9 embryonic loss-of-function mouse mutants show complete sex reversal and produce partially fertile XY oocytes. *Developmental Biology*, *354*, 111–122.

Lee, C. H., & Taketo, T. (1994). Normal onset, but prolonged expression, of Sry gene in the B6.YDOM sex-reversed mouse gonad. *Developmental Biology*, *165*, 442–452.

Leung, V. Y., Gao, B., Leung, K. K., Melhado, I. G., Wynn, S. L., Au, T. Y., et al. (2011). SOX9 governs differentiation stage-specific gene expression in growth plate chondrocytes via direct concomitant transactivation and repression. *PLoS Genetics*, *7*, e1002356.

Lin, Y. T., & Capel, B. (2015). Cell fate commitment during mammalian sex determination. *Current Opinion in Genetics & Development, 32*, 144–152.

Lindeman, R. E., Gearhart, M. D., Minkina, A., Krentz, A. D., Bardwell, V. J., & Zarkower, D. (2015). Sexual cell-fate reprogramming in the ovary by DMRT1. *Current Biology, 25*, 764–771.

Lovell-Badge, R. (2010). The early history of the Sox genes. *The International Journal of Biochemistry & Cell Biology, 42*, 378–380.

Ludbrook, L. M., Bernard, P., Bagheri-Fam, S., Ryan, J., Sekido, R., Wilhelm, D., et al. (2012). Excess DAX1 leads to XY ovotesticular disorder of sex development (DSD) in mice by inhibiting steroidogenic factor-1 (SF1) activation of the testis enhancer of SRY-box-9 (Sox9). *Endocrinology, 153*, 1948–1958.

Maatouk, D. M., & Capel, B. (2008). Sexual development of the soma in the mouse. *Current Topics in Developmental Biology, 83*, 151–183.

Maatouk, D. M., DiNapoli, L., Alvers, A., Parker, K. L., Taketo, M. M., & Capel, B. (2008). Stabilization of beta-catenin in XY gonads causes male-to-female sex-reversal. *Human Molecular Genetics, 17*, 2949–2955.

Maatouk, D. M., Natarajan, A., Shibata, Y., Song, L., Crawford, G. E., Ohler, U., et al. (2017). Genome-wide identification of regulatory elements in Sertoli cells. *Development, 144*, 720–730.

Matson, C. K., Murphy, M. W., Sarver, A. L., Griswold, M. D., Bardwell, V. J., & Zarkower, D. (2011). DMRT1 prevents female reprogramming in the postnatal mammalian testis. *Nature, 476*, 101–104.

Mead, T. J., Wang, Q., Bhattaram, P., Dy, P., Afelik, S., Jensen, J., et al. (2013). A far-upstream (−70 kb) enhancer mediates Sox9 auto-regulation in somatic tissues during development and adult regeneration. *Nucleic Acids Research, 41*, 4459–4469.

Mochizuki, Y., Chiba, T., Kataoka, K., Yamashita, S., Sato, T., Kato, T., et al. (2018). Combinatorial CRISPR/Cas9 approach to elucidate a far-upstream enhancer complex for tissue-specific Sox9 expression. *Developmental Cell, 46*, 794–806.e6.

Moniot, B., Declosmenil, F., Barrionuevo, F., Scherer, G., Aritake, K., Malki, S., et al. (2009). The PGD2 pathway, independently of FGF9, amplifies SOX9 activity in Sertoli cells during male sexual differentiation. *Development, 136*, 1813–1821.

Morais da Silva, S., Hacker, A., Harley, V., Goodfellow, P., Swain, A., & Lovell-Badge, R. (1996). Sox9 expression during gonadal development implies a conserved role for the gene in testis differentiation in mammals and birds. *Nature Genetics, 14*, 62–68.

Nora, E. P., Lajoie, B. R., Schulz, E. G., Giorgetti, L., Okamoto, I., Servant, N., et al. (2012). Spatial partitioning of the regulatory landscape of the X-inactivation centre. *Nature, 485*, 381–385.

O'Bryan, M. K., Takada, S., Kennedy, C. L., Scott, G., Harada, S., Ray, M. K., et al. (2008). Sox8 is a critical regulator of adult Sertoli cell function and male fertility. *Developmental Biology, 316*, 359–370.

Ogawa, Y., Terao, M., Hara, S., Tamano, M., Okayasu, H., Kato, T., et al. (2018). Mapping of a responsible region for sex reversal upstream of Sox9 by production of mice with serial deletion in a genomic locus. *Scientific Reports, 8*, 17514.

Ohnesorg, T., Croft, B., Tan, J., & Sinclair, A. H. (2016). Using ROADMAP data to identify enhancers associated with disorders of sex development. *Sexual Development, 10*, 59–65.

Ohnesorg, T., van den Bergen, J. A., Belluoccio, D., Shankara-Narayana, N., Kean, A. M., Vasilaras, A., et al. (2017). A duplication in a patient with 46,XX ovo-testicular disorder of sex development refines the SOX9 testis-specific regulatory region to 24 kb. *Clinical Genetics, 92*, 347–349.

Osterwalder, M., Barozzi, I., Tissieres, V., Fukuda-Yuzawa, Y., Mannion, B. J., Afzal, S. Y., et al. (2018). Enhancer redundancy provides phenotypic robustness in mammalian development. *Nature*, *554*, 239–243.

Ottolenghi, C., Omari, S., Garcia-Ortiz, J. E., Uda, M., Crisponi, L., Forabosco, A., et al. (2005). Foxl2 is required for commitment to ovary differentiation. *Human Molecular Genetics*, *14*, 2053–2062.

Qin, Y., & Bishop, C. E. (2005). Sox9 is sufficient for functional testis development producing fertile male mice in the absence of Sry. *Human Molecular Genetics*, *14*, 1221–1229.

Qin, Y., Kong, L. K., Poirier, C., Truong, C., Overbeek, P. A., & Bishop, C. E. (2004). Long-range activation of Sox9 in Odd Sex (Ods) mice. *Human Molecular Genetics*, *13*, 1213–1218.

Racca, J. D., Chen, Y. S., Maloy, J. D., Wickramasinghe, N., Phillips, N. B., & Weiss, M. A. (2014). Structure–function relationships in human testis-determining factor SRY: An aromatic buttress underlies the specific DNA-bending surface of a high mobility group (HMG) box. *The Journal of Biological Chemistry*, *289*, 32410–32429.

Rahmoun, M., Lavery, R., Laurent-Chaballier, S., Bellora, N., Philip, G. K., Rossitto, M., et al. (2017). In mammalian foetal testes, SOX9 regulates expression of its target genes by binding to genomic regions with conserved signatures. *Nucleic Acids Research*, *45*, 7191–7211.

Rao, S. S., Huntley, M. H., Durand, N. C., Stamenova, E. K., Bochkov, I. D., Robinson, J. T., et al. (2014). A 3D map of the human genome at kilobase resolution reveals principles of chromatin looping. *Cell*, *159*, 1665–1680.

Schepers, G. E., Teasdale, R. D., & Koopman, P. (2002). Twenty pairs of sox: Extent, homology, and nomenclature of the mouse and human sox transcription factor gene families. *Developmental Cell*, *3*, 167–170.

Schepers, G., Wilson, M., Wilhelm, D., & Koopman, P. (2003). SOX8 is expressed during testis differentiation in mice and synergizes with SF1 to activate the Amh promoter in vitro. *The Journal of Biological Chemistry*, *278*, 28101–28108.

Schmahl, J., Eicher, E. M., Washburn, L. L., & Capel, B. (2000). Sry induces cell proliferation in the mouse gonad. *Development*, *127*, 65–73.

Schmidt, D., Ovitt, C. E., Anlag, K., Fehsenfeld, S., Gredsted, L., Treier, A. C., et al. (2004). The murine winged-helix transcription factor Foxl2 is required for granulosa cell differentiation and ovary maintenance. *Development*, *131*, 933–942.

Sekido, R. (2010). SRY: A transcriptional activator of mammalian testis determination. *The International Journal of Biochemistry & Cell Biology*, *42*, 417–420.

Sekido, R., Bar, I., Narvaez, V., Penny, G., & Lovell-Badge, R. (2004). SOX9 is up-regulated by the transient expression of SRY specifically in Sertoli cell precursors. *Developmental Biology*, *274*, 271–279.

Sekido, R., & Lovell-Badge, R. (2008). Sex determination involves synergistic action of SRY and SF1 on a specific Sox9 enhancer. *Nature*, *453*, 930–934.

She, Z. Y., & Yang, W. X. (2017). Sry and SoxE genes: How they participate in mammalian sex determination and gonadal development? *Seminars in Cell & Developmental Biology*, *63*, 13–22.

Sinclair, A. H., Berta, P., Palmer, M. S., Hawkins, J. R., Griffiths, B. L., Smith, M. J., et al. (1990). A gene from the human sex-determining region encodes a protein with homology to a conserved DNA-binding motif. *Nature*, *346*, 240–244.

Smith, C. A., Roeszler, K. N., Ohnesorg, T., Cummins, D. M., Farlie, P. G., Doran, T. J., et al. (2009). The avian Z-linked gene DMRT1 is required for male sex determination in the chicken. *Nature*, *461*, 267–271.

Sock, E., Schmidt, K., Hermanns-Borgmeyer, I., Bosl, M. R., & Wegner, M. (2001). Idiopathic weight reduction in mice deficient in the high-mobility-group transcription factor Sox8. *Molecular and Cellular Biology*, *21*, 6951–6959.

Soullier, S., Hanni, C., Catzeflis, F., Berta, P., & Laudet, V. (1998). Male sex determination in the spiny rat *Tokudaia osimensis* (Rodentia: Muridae) is not Sry dependent. *Mammalian Genome, 9*, 590–592.

Spitz, F. (2016). Gene regulation at a distance: From remote enhancers to 3D regulatory ensembles. *Seminars in Cell & Developmental Biology, 57*, 57–67.

Sreenivasan, R., Ludbrook, L., Fisher, B., Declosmenil, F., Knower, K. C., Croft, B., et al. (2018). Mutant NR5A1/SF-1 in patients with disorders of sex development shows defective activation of the SOX9 TESCO enhancer. *Human Mutation, 39*, 1861–1874.

Swain, A., Narvaez, V., Burgoyne, P., Camerino, G., & Lovell-Badge, R. (1998). Dax1 antagonizes Sry action in mammalian sex determination. *Nature, 391*, 761–767.

Symmons, O., Pan, L., Remeseiro, S., Aktas, T., Klein, F., Huber, W., et al. (2016). The Shh topological domain facilitates the action of remote enhancers by reducing the effects of genomic distances. *Developmental Cell, 39*, 529–543.

Symmons, O., Uslu, V. V., Tsujimura, T., Ruf, S., Nassari, S., Schwarzer, W., et al. (2014). Functional and topological characteristics of mammalian regulatory domains. *Genome Research, 24*, 390–400.

Symon, A., & Harley, V. (2017). SOX9: A genomic view of tissue specific expression and action. *The International Journal of Biochemistry & Cell Biology, 87*, 18–22.

Uda, M., Ottolenghi, C., Crisponi, L., Garcia, J. E., Deiana, M., Kimber, W., et al. (2004). Foxl2 disruption causes mouse ovarian failure by pervasive blockage of follicle development. *Human Molecular Genetics, 13*, 1171–1181.

Uhlenhaut, N. H., Jakob, S., Anlag, K., Eisenberger, T., Sekido, R., Kress, J., et al. (2009). Somatic sex reprogramming of adult ovaries to testes by FOXL2 ablation. *Cell, 139*, 1130–1142.

Vainio, S., Heikkila, M., Kispert, A., Chin, N., & McMahon, A. P. (1999). Female development in mammals is regulated by Wnt-4 signalling. *Nature, 397*, 405–409.

Velagaleti, G. V., Bien-Willner, G. A., Northup, J. K., Lockhart, L. H., Hawkins, J. C., Jalal, S. M., et al. (2005). Position effects due to chromosome breakpoints that map approximately 900 kb upstream and approximately 1.3 Mb downstream of SOX9 in two patients with campomelic dysplasia. *American Journal of Human Genetics, 76*, 652–662.

Vetro, A., Ciccone, R., Giorda, R., Patricelli, M. G., Della Mina, E., Forlino, A., et al. (2011). XX males SRY negative: A confirmed cause of infertility. *Journal of Medical Genetics, 48*, 710–712.

Vidal, V. P., Chaboissier, M. C., de Rooij, D. G., & Schedl, A. (2001). Sox9 induces testis development in XX transgenic mice. *Nature Genetics, 28*, 216–217.

Visel, A., Rubin, E. M., & Pennacchio, L. A. (2009). Genomic views of distant-acting enhancers. *Nature, 461*, 199–205.

Wagner, T., Wirth, J., Meyer, J., Zabel, B., Held, M., Zimmer, J., et al. (1994). Autosomal sex reversal and campomelic dysplasia are caused by mutations in and around the SRY-related gene SOX9. *Cell, 79*, 1111–1120.

Wegner, M. (2010). All purpose Sox: The many roles of Sox proteins in gene expression. *The International Journal of Biochemistry & Cell Biology, 42*, 381–390.

Wilhelm, D., Palmer, S., & Koopman, P. (2007). Sex determination and gonadal development in mammals. *Physiological Reviews, 87*, 1–28.

Wilkins, A. S. (1995). Moving up the hierarchy: A hypothesis on the evolution of a genetic sex determination pathway. *BioEssays, 17*, 71–77.

Wunderle, V. M., Critcher, R., Hastie, N., Goodfellow, P. N., & Schedl, A. (1998). Deletion of long-range regulatory elements upstream of SOX9 causes campomelic dysplasia. *Proceedings of the National Academy of Sciences of the United States of America, 95*, 10649–10654.

Xiao, B., Ji, X., Xing, Y., Chen, Y. W., & Tao, J. (2013). A rare case of 46, XX SRY-negative male with approximately 74-kb duplication in a region upstream of SOX9. *European Journal of Medical Genetics*, *56*, 695–698.

Yao, B., Wang, Q., Liu, C. F., Bhattaram, P., Li, W., Mead, T. J., et al. (2015). The SOX9 upstream region prone to chromosomal aberrations causing campomelic dysplasia contains multiple cartilage enhancers. *Nucleic Acids Research*, *43*, 5394–5408.

Zhao, L., Ng, E. T., Davidson, T. L., Longmuss, E., Urschitz, J., Elston, M., et al. (2014). Structure-function analysis of mouse Sry reveals dual essential roles of the C-terminal polyglutamine tract in sex determination. *Proceedings of the National Academy of Sciences of the United States of America*, *111*, 11768–11773.

Zhao, L., Svingen, T., Ng, E. T., & Koopman, P. (2015). Female-to-male sex reversal in mice caused by transgenic overexpression of Dmrt1. *Development*, *142*, 1083–1088.

Zhao, L., Wang, C., Lehman, M. L., He, M., An, J., Svingen, T., et al. (2018). Transcriptomic analysis of mRNA expression and alternative splicing during mouse sex determination. *Molecular and Cellular Endocrinology*, *478*, 84–96.

CHAPTER NINE

Sexually dimorphic germ cell identity in mammals

Cassy Spiller[a], Josephine Bowles[a,b,*]
[a]School of Biomedical Sciences, The University of Queensland, Brisbane, QLD, Australia
[b]Institute for Molecular Bioscience, The University of Queensland, Brisbane, QLD, Australia
*Corresponding author: e-mail address: jo.bowles@uq.edu.au

Contents

Abstract

Germ cells are the stem cells of the species. Thus, it is critical that we have a good understanding of how they are specified, how the somatic cells instruct and support them, how they commit to one or other sex, and how they ultimately develop into functional gametes. Here, we focus on specifics of how sexual fate is determined during fetal life. Because the majority of relevant experimental work has been done using the mouse model, we focus on that species. We review evidence regarding the identity of instructive signals from the somatic cells, and the molecular responses that occur in germ cells in response to those extrinsic signals. In this way we aim to clarify progress to date regarding the mechanisms underlying the mitotic to meiosis switch in germ cells of the fetal ovary, and those involved in adopting and securing male fate in germ cells of the fetal testis.

Current Topics in Developmental Biology, Volume 134
ISSN 0070-2153
https://doi.org/10.1016/bs.ctdb.2019.01.011

1. Introduction

Germ cells are the precursors of the gametes and are, therefore, critical for the survival of the species. Mature female (oocytes or "eggs") and male (spermatozoa or "sperm") germ cells are similar in that both are haploid cells derived from their diploid precursors by meiotic division, and in that both bear responsibility for transferring all genetic and epigenetic information from one generation to the next. Other than that, however, it is obvious that oocytes and spermatozoa are extremely different from each other, particularly in terms of their morphology and abundance—it is therefore perhaps surprising that oocytes and spermatozoa are both derived from the same progenitor cell type, the primordial germ cell (PGC). The PGCs are derived anew each generation during early embryonic development and, for the first few days after they arise, PGCs that are XX and XY are indistinguishable in terms of their morphology and behavior.

Sex-specific differentiation of the PGCs begins, as far as we know, only after they migrate into and colonize the embryonic gonads. The sexual fate of germline cells depends on the environment in which they find themselves rather than on their sex chromosome constitution (i.e., whether they are XX or XY) (Ford et al., 1975; Palmer & Burgoyne, 1991). For PGCs in the developing ovary, the challenge is timely initiation and progression through prophase of meiosis I, to the stage of dictyate arrest (Hilscher et al., 1974; McLaren, 1984). The requirements for PGC differentiation in the developing testis are perhaps more complicated, because it seems that controlling the totipotent potential of the germline is a major issue. In addition, meiotic initiation must be avoided, and the germ cells must ultimately cease proliferating and enter a state of quiescence prior to birth. The adoption of appropriate germ cell sexual fate during fetal life is the subject of this review.

2. Getting to the point of sexual divergence

2.1 PGC specification, survival and migration

In mammals, the founding germline population, the primordial germ cells (PGCs), are segregated anew from somatic cells during each generation, rather than being inherited as preformed "germ plasm" as is seen in some other species (Saitou et al., 2003). PGCs arise in the posterior proximal epiblast of post-implantation embryos at around 6.25 dpc (days post coitum), when a small group of multipotent cells located in that area is subject to inducing signals, including BMP and WNT3 (Lawson et al., 1999; Ohinata et al., 2009). These cells undergo what is essentially a de-differentiation event, downregulating characteristic somatic mesodermal genes such as those encoding homeobox proteins and re-expressing or maintaining expression of genes associated with pluripotency (such as *Oct4* (*Pou5f1*), *Sox2* and *Nanog*, please see Table 1 for gene details), and upregulating the expression of key germ cell-specific genes (Saitou, Barton, & Surani, 2002; Saitou et al., 2003; Saitou & Yamaji, 2010). *Prdm1* (also known as *Blimp1*) expression in the nascent PGCs is thought to be largely responsible for repressing the somatic mesodermal program and PRDM1-positive cells are considered the first definitive lineage-restricted PGCs (Kurimoto et al., 2008; Ohinata et al., 2005).

Initiation of expression of germline-specific genes begins shortly after PGC induction. NANOS3, an evolutionarily conserved RNA-binding protein, is expressed soon after germ cell specification at 7.5 dpc and appears to be required for germ cell survival during migration (Suzuki, Tsuda, Kiso, & Saga, 2008). Presumably onset of germ-cell-specific gene expression constricts developmental potency so that PGCs are unipotent, unlike the cells from which they arise. Once specified the nascent PGCs take on a polarized morphology with cytoplasmic extensions and, by about 8.0 dpc, they begin to migrate through the mouse embryo, proliferating as they go. The mechanisms involved in directing this migration are complex, as would be expected considering the distance to be traveled and the changing nature of the tissue terrain along their migratory route (reviewed by Cantu & Laird, 2017; Molyneaux & Wylie, 2004). Dynamic expression of the cell-cell adhesion protein E-cadherin (Bendel-Stenzel, Gomperts, Anderson, Heasman, & Wylie, 2000) and PGC-specific expression of the extracellular matrix molecule β1 integrin (Anderson et al., 1999) are involved. PGC survival, proliferation and directed migration seem to be largely dependent on two ligand/receptor pairs: c-Kit (a receptor tyrosine kinase) and Cxcr4 (a chemokine receptor) are expressed by migrating PGCs while their respective ligands, the membrane-bound form of Kit-ligand (also known as stem cell factor (Scf) or Steel factor (Sf)) and Sdf-1 (also called Cxcl12) are produced by somatic cells. Traveling *via* the hindgut endoderm and its mesentery, PGCs progress as a network of cells, eventually moving dorsally and laterally and separating into two populations to colonize the indifferent embryonic gonads (genital ridges) (Molyneaux, Stallock, Schaible, & Wylie, 2001). The first PGCs reach their target tissue at about 10.0 dpc

Table 1 Symbols, names and common aliases for genes involved in mouse germ cell pluripotency, sexual development and differentiation.

Gene symbol	Gene name	Common aliases	Protein name or symbol
Aldh1a1	*Aldehyde dehydrogenase family 1, subfamily A1*	*Raldh1*	ALDH1A1 RALDH1
Aldh1a2	*Aldehyde dehydrogenase family 1, subfamily A2*	*Raldh2*	ALDH1A2 RALDH2
Aldh1a3	*Aldehyde dehydrogenase family 1, subfamily A3*	*Raldh3*	ALDH1A3 RALDH3
Axin2	*Axin2*		AXIN2
Bax	*BCL2-associated X protein*		BAX
Bmp2	*Bone morphogenic protein 2*		BMP2
Bmp5	*Bone morphogenic protein 5*		BMP5
Ccne1	*Cyclin E1*		CCNE1
Ccne2	*Cyclin E2*		CCNE2
Cdh1	*Cadherin 1, E-cadherin*	*E-cad*	E-cadherin
Cdkn1a	*Cyclin-dependent kinase inhibitor 1A*	*p21, p21Cip1*	CDKN1A
Cdkn1b	*Cyclin-dependent kinase inhibitor 1B*	*p27, p27Kip1*	CDKN1B
Cdkn2a	*Cyclin-dependent kinase inhibitor 2A*	*p16, p16INK4a*	CDKN2A
Cdkn2b	*Cyclin-dependent kinase inhibitor 2B*	*p15, p15INK4b*	CDKN2B
Cdkn2c	*Cyclin-dependent kinase inhibitor 2C*	*p18, p18INK4c*	CDKN2C
Cdkn2d	*Cyclin-dependent kinase inhibitor 2B*	*p15, p19INK4d*	CDKN2D
Crebbp	*CREB-binding protein*		CBP
Cxcl12	*Chemokine (C-X-C motif) ligand 12, stromal cell-derived factor 1*	*Sdf1*	CXCL12, SDF1
Cxcr4	*Chemokine (C-X-C motif) receptor 4*	*Sdf1r*	SDF1R
Ctnnb1	*Cateinin (cadherin-associated protein), beta 1*	*β-catenin*	β-catenin
Cyp26b1	*Cytochrome P450, family 26, subfamily b, polypeptide 1*	*P450RAI-2*	CYP26B1
Dazl	*Deleted in azoospermia-like*	*Dazla*	DAZL
Ddx4	*DEAD (Asp-Glu-Ala-Asp) box polypeptide 4*	*Mvh*	DDX4 MVH

Table 1 Symbols, names and common aliases for genes involved in mouse germ cell pluripotency, sexual development and differentiation.—cont'd

Gene symbol	Gene name	Common aliases	Protein name or symbol
Dmc1	*DMC1 dosage suppressor of mck1 homolog, meiosis-specific homologous recombination (yeast)*	*Dmc1h*	DMC1
Dmrt1	*Doublesex and mab-3 related transcription factor 1*		DMRT1
Dnd1	*DND MicroRNA-Mediated Repression Inhibitor 1*		DND1
Dnmt3l	*DNA (cytosine-5-)-methyltransferase 3-like*		DNMT3L
Dnmt3c	*DNA (cytosine-5-)-methyltransferase 3-c*		
Fgf9	*Fibroblast growth factor 9*		FGF9
Fgfr2	*Fibroblast growth factor receptor 2*		FGFR2
Inhbb	*Inhibin beta-B*	*activin*	INHBB
Itgb1	*Beta 1 integrin*	*cd29*	ITGB1
Kit	*Kit oncogene*	*c-kit*	C-KIT
Kitl	*Kit ligand, steel factor, stem cell factor, mast cell growth factor*	*Scf*	KITL, SCF
Lefty1	*Left-right determination factor 1*	*Lefty B*	LEFTY1, LEFTYB
Lefty2	*Left-right determination factor 12*	*Lefty A*	LEFTY2, LEFTYA
Lgr4	*Leucine Rich Repeat Containing G Protein-Coupled Receptor 4*		LGR4
Lgr5	*Leucine Rich Repeat Containing G Protein-Coupled Receptor 5*		LGR5
Meioc	*Meiosis specific with coil-coil domain*		MEIOC
Msx1	*Msh homeobox 1*	*Hox7*	MSX1
Msx2	*Msh homeobox 2*	*Hox8*	MSX2
Nanog	*Nanog homeobox*		NANOG
Nanos2	*Nanos homolog 2 (Drosophila)*	*nos2*	NANOS2
Nanos3	*Nanos homolog 3 (Drosophila)*	*nos3*	NANOS3
Nodal	*Nodal growth differentiation factor*		NODAL

Continued

Table 1 Symbols, names and common aliases for genes involved in mouse germ cell pluripotency, sexual development and differentiation.—cont'd

Gene symbol	Gene name	Common aliases	Protein name or symbol
Notch1	*Notch 1*		NOTCH1
Piwil2	*Piwi-like RNA-mediated gene silencing 2*	*mili*	PIWIL2
Piwil4	*Piwi like RNA-mediated gene silencing 4*	*piwi*	PIWIL4
Pou5f1	*POU domain, class5, transcription factor 1*	*Oct4, Oct3*	POU5F1 OCT4
Prdm1	*PR domain containing 1*	*Blimp1*	PRDM1
Ptgds	*Prostaglandin D2 synthase*	*Pgds2*	PGD2
Rb1	*Retinoblastoma 1*	*Rb, pRb*	RB1, pRB
Rec8	*Meiotic recombination protein REC8 homolog*		REC8
Rnf2	*Ring finger protein 2*		RNF2
Rspo1	*R-spondin homolog (Xenopus laevis)*	*R-spondin*	R-spondin1
Sox2	*Sry-box containing gene 2*		SOX2
Sox9	*Sry-box containing gene 9*		SOX9
Spo11	*SPO11, Initiator Of Meiotic Double Stranded Breaks*		SPO11
Sry	*Sex determining region of Chromosome Y*		SRY
Stra8	*Stimulated by retinoic acid gene 8*		STRA8
Sycp3	*Synaptonemal complex protein 3*	*Scp3, Cor1*	SYCP3
Tdgf1	*Teratocarcinoma-derived growth factor 1*	*Cripto*	TDGF1
Wnt4	*Wingless-related MMTV integration site 4*		WNT4

and most have migrated into the gonad by 11.5 dpc (Gomperts, Garciacastro, Wylie, & Heasman, 1994). Gonadal germ cells continue to proliferate for several days after they colonize the genital ridges/gonads, their number reaching about 25,000 by 13.5 dpc (Tam & Snow, 1981). There is no evidence for sex-specific differences during specification or migration.

2.2 The new environment of the gonad

At the time of PGC colonization, the mouse genital ridges (the gonadal primordia) are in an "indifferent" or bipotential state, still possessing the capacity to differentiate into either ovarian or testicular tissue. This very quickly changes, however, because expression of the Y chromosomal testis-determining gene *Sry*, from about 10.5 dpc, triggers considerable transcriptional change by 11.5 dpc and testicular cords are clearly visible in XY gonads by 12.5 dpc (Jameson, Natarajan, et al., 2012; Munger, Natarajan, Looger, Ohler, & Capel, 2013; Nef et al., 2005). Therefore, PGCs enter what is essentially a "war zone" in which male and female programs of somatic development are being actively and rapidly determined, accompanied by production of extrinsic factors that instruct not only somatic, but also germ cell, development. A key direct downstream target of the transcription factor SRY is a related protein, SOX9, and it appears that essentially all effects of SRY are dependent on SOX9 function (Sekido, Bar, Narvaez, Penny, & Lovell-Badge, 2004; Vidal, Chaboissier, de Rooij, & Schedl, 2001). Critical functions of SOX9 include ensuring the upregulation of the growth factor FGF9, which antagonizes the ovarian pathway of development (Jameson, Lin, & Capel, 2012; Schmahl, Kim, Colvin, Ornitz, & Capel, 2004), and CYP26B1, which clears the gonadal tissue of the signaling molecule retinoic acid (RA) (Bowles et al., 2006; Koubova et al., 2006; MacLean, Li, Metzger, Chambon, & Petkovich, 2007). In addition to important roles in orchestrating gonadal somatic development (Bowles et al., 2018; Kim et al., 2006), both FGF9 and RA have key roles in determining germ cell sexual fate as will be discussed below. We would expect the environment of the indifferent gonad, at the time of germ cell colonization, to include substantial levels of the ovarian signaling molecules WNT4 and R-spondin 1 (RSPO1; Parma et al., 2006; Vainio, Heikkila, Kispert, Chin, & McMahon, 1999). By 11.5 dpc, FGF9 levels are rapidly increasing in the developing testis and falling in the developing ovary while RA is building in the ovary due to loss of CYP26B1 (Bowles et al., 2010, 2006; Kim et al., 2006; MacLean et al., 2007).

Both XX and XY germ cells acquire competence to undergo sex-specific differentiation once they reach the genital ridges. Shortly after their arrival, PGCs begin to express germ-cell-specific RNA-binding proteins DDX4 (also known as MVH), DAZL (Seligman & Page, 1998; Toyooka et al., 2000) and low levels of the meiosis-related protein SYCP3 (Chuma & Nakatsuji, 2001). Expression of these key germ-cell-specific genes is induced by the gonadal environment and, although the exact molecular mechanism involved is not understood, promoter demethylation seems to be involved (Hu et al., 2015; Maatouk et al., 2006). Expression of DAZL is necessary for sexual fate competence because *Dazl*-null XX germ cells fail to develop along either female or male pathways (although this effect is clear-cut only on the inbred C57BL/6 background) (Gill, Hu, Lin, & Page, 2011; Lin & Page, 2005). Hence, DAZL is considered a "licencing factor," allowing PGCs to enter a state whereby they are capable of responding to environmental signals and undergoing sexual differentiation (Gill et al., 2011; Lin, Gill, Koubova, & Page, 2008). DAZL promotes translation in germ cells, including for proteins DDX4 and SYCP3, a protein involved in synaptonemal complex formation in meiotic prophase 1 (Reynolds, Collier, Bingham, Gray, & Cooke, 2007; Reynolds et al., 2005).

2.3 Epigenetic reprogramming

The "chosen ones," the cells that de-differentiate to become the germline progenitors, undergo extensive epigenetic reprogramming; this is essential so that appropriate epigenetic information is transmitted to the next generation. In particular, it is critical that DNA methylation at control elements of imprinted loci is removed and appropriately reset so that correct sex-specific patterns are established in the progeny. Two phases of demethylation are recognized, a preliminary round during the highly proliferative migratory period, and a second phase of demethylation that occurs when PGCs colonize the gonad. The latter is particularly pertinent to sequences that carry long-term epigenetic memory and affects differentially-methylated regions of imprinted genes, and CpG islands in the promoters of germline-specific genes and on the X chromosome (Hajkova et al., 2002; Seisenberger et al., 2012). It is postulated that this near global demethylation does not result in widespread and promiscuous transcription because at all times either the core pluripotency or the germline-specific gene network is active (Seisenberger et al., 2012). Demethylation is accompanied by major changes in nuclear architecture, the result of extensive chromatin remodeling, erasure of histone modifications and changes in histone variant composition (Hajkova et al., 2008). The PGCs at this point are considered to be in an epigenetically naïve state, their genomes displaying the lowest amount of epigenetic information of any cell type at any time during the life cycle (Hajkova, 2011).

3. Adopting the female fate, in the ovary

Ovarian germ cells initiate the first round of meiosis before birth, while testicular germ cells do not embark on meiosis until postnatal life, commencing at puberty. It is believed that germ cell sexual fate is entirely dependent on influences of the somatic environment (Byskov & Saxen, 1976; Chiquoine, 1954; McLaren, 2003). Histologically, the first sign that germ cells have adopted a particular sexual fate is the appearance of characteristic condensed chromatin structures in ovarian germ cells, at about 13.5 dpc (McLaren, 1984; Zamboni, 1986). From this point the ovarian germ cells are referred to as "primary oocytes" and they progress through leptonema, zygonema, pachynema and diplonema until, at around birth, they enter a prolonged arrest stage known as dictyate (Speed, 1982). As the oocytes enter meiosis they stop proliferating and there is downregulation of *Oct4*, which presumably indicates of a loss of pluripotency (Bullejos & Koopman, 2004; Western, van den Bergen, Miles, & Sinclair, 2010). The first known genetic indication that germ cells have committed to meiosis is the expression of the pre-meiotic gene, *Stra8*, beginning at around 12.5 dpc (Baltus et al., 2006; Bowles et al., 2006; Koubova et al., 2006; Menke, Koubova, & Page, 2003).

3.1 Retinoic acid triggers the onset of meiosis

It is now well established that the main factor to orchestrate the switch from mitosis to meiosis in the fetal ovary is the pleiotropic signaling molecule, retinoic acid (RA) (Fig. 1). RA mediates its effects via binding to RA receptor heterodimers (RAR:RXR) that recognize RA response elements

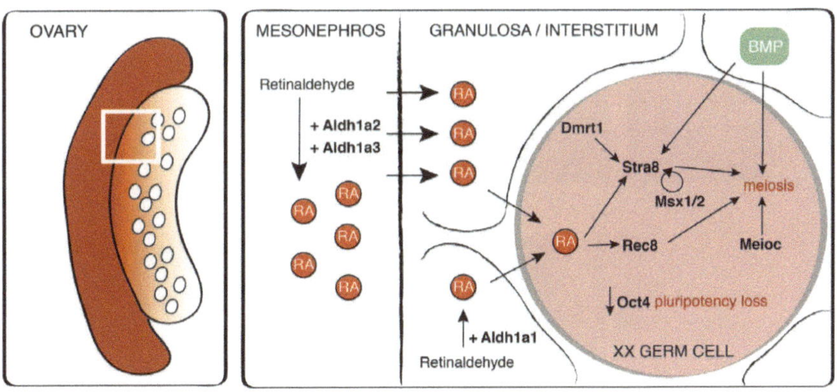

Fig. 1 Molecular regulators of female germ cell sex determination. A schematic of the developing ovary (pale), with attached mesonephros (dark) is shown on the left and the indicated area is expanded in the right panel. Retinoic acid (RA) is produced in the mesonephros and in the somatic tissue of the ovary. BMP is also produced by somatic cells. Both RA and BMP impact on female germ cell development, via key molecular regulators as indicated. For details see the text.

(RAREs) located upstream of target genes. RA has many widely-varying functions during development (Niederreither & Dolle, 2008; Rhinn & Dolle, 2012) and it seems that inducing meiotic onset is yet one more. When pregnant rats are made deficient in vitamin A, the precursor of RA, XX germ cells in the progeny do not upregulate *Stra8* and do not enter meiosis: this result supports the model that RA is required, and not merely sufficient, to trigger meiotic onset (Li & Clagett-Dame, 2009). Similarly, when RA receptors are prevented from functioning in the fetal gonad, ovarian germ cells do not express *Stra8* and do not enter meiosis (Bowles et al., 2016, 2006; Koubova et al., 2006). Recent studies have demonstrated that RA also acts to induce expression of another important meiotic marker, *Rec8*, independently of its role in triggering *Stra8* expression (Koubova et al., 2014). The presence of local RA is also key to the onset of meiosis in the pubertal male (Anderson et al., 2008; Koubova et al., 2006).

The onset of *Stra8* expression in germ cells of the developing ovary, and its avoidance in germ cells of the developing testis, is controlled by expression of the P450 enzyme Cyp26b1 (Bowles et al., 2016, 2006; MacLean et al., 2007) (Fig. 2). This enzyme, specifically expressed by somatic cells of the developing testis, efficiently degrades endogenous RA, effectively clearing the gonad of meiosis-inducing RA. In the *Cyp26b1*-null testis, where RA is not specifically degraded, testicular germ cells enter meiosis ectopically

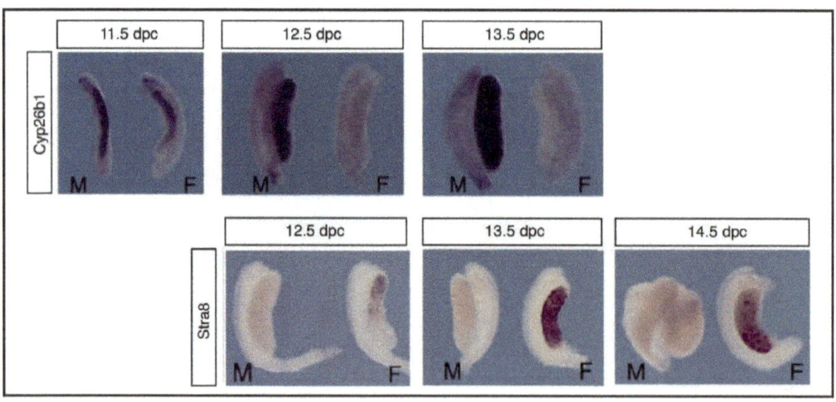

Fig. 2 Temporal expression of *Cyp26b1* and *Stra8* during mouse fetal gonad development. *Cyp26b1*, encoding a P450 enzyme that degrades RA, is expressed in both XY (M) and XX (F) gonads initially and is then upregulated in the testis and downregulated in the ovary. *Stra8* is the critical "gate-keeper" of meiosis and is expressed in ovarian germ cells in an anterior to posterior wave beginning at 12.5 dpc (in situ data first presented in Bowles et al., 2006). *dpc*, days post coitum.

and progress to the pachytene stage (Bowles et al., 2006; Koubova et al., 2014; MacLean et al., 2007; Saba, Wu, & Saga, 2014). Even prior to the discovery of the meiosis-inducing role of RA, it was appreciated that ovarian germ cells enter meiosis in an anterior-to-posterior (AP) wave (Bullejos & Koopman, 2004; Menke et al., 2003; Yao, DiNapoli, & Capel, 2003); this is now interpreted as reflective of the fact that a reservoir of *Stra8*-inducing RA exists at the anterior end of the adjacent mesonephric tissue (Bowles et al., 2016, 2006; Bowles & Koopman, 2007). It seems likely that the AP wave of entry into meiosis is merely an interesting phenomenon, reflecting RA availability, rather than being of any functional importance.

Evidence that ovarian germ cells are instructed to enter meiotic prophase, by a signaling molecule of the ovarian milieu, was somewhat of a surprise, because earlier observations of meiotic onset had led to the theory that germ cells autonomously enter meiosis unless specifically prevented from doing so (McLaren & Southee, 1997), although this view was not universally held (Byskov, 1974; Byskov & Saxen, 1976). The conclusion that germ cells did not require an extrinsic signal to instruct them to embark upon meiosis focused mainly on the observation that ectopic germ cells (lodged in the nearby adrenal or mesonephros rather than the gonad) enter meiosis with the same timing as those in the ovary (Zamboni & Upadhyay, 1983). The apparent discrepancies between that view and the current model of RA-induced meiosis are not difficult to understand if we take into consideration the fact that RA is widely present in the developing embryo and, hence, ectopic germ cells whether they are in the adrenal or the mesonephros, will quite likely be exposed to RA.

Nonetheless, there has been some ongoing controversy as to whether RA does actually induce meiosis in the fetal ovary: these concerns have focussed mainly on whether RA is actually present in the fetal ovarian tissue and whether RA is necessary for meiotic onset, given that at least some germ cells can initiate meiosis in the absence of certain RA-synthesizing enzymes, namely, *Aldh1a2* and *Aldh1a3* (Kumar et al., 2011; Kumar, Cunningham, & Duester, 2013). Although the levels of endogenous RA in the fetal ovary are evidently low, and hence its presence is difficult to detect, there is clearly more RA in the mouse fetal ovary than testis (Bowles et al., 2016). Moreover, the weak RA-synthesizing enzyme ALDH1A1 seems to play a role in providing RA in the mouse fetal testis (Bowles et al., 2016; Mu et al., 2013); this is in addition to the role of ALDH1A2, which is highly expressed in the mesonephric ducts and appears to produce the bulk of the RA detectable in the fetal ovary (Bowles et al., 2016) (Fig. 1). In the human fetus it appears that it is the ovarian somatic tissue that provides a source of meiosis-inducing RA (Childs, Cowan, Kinnell, Anderson, & Saunders, 2011; Jorgensen, Nielsen, Blomberg Jensen, Graem, & Rajpert-De Meyts, 2012; Le Bouffant et al., 2010).

Possibly because of their epigenetically-naïve state, fetal germ cells are highly sensitive to RA in the microenvironment. RA acts directly on germ cells to induce expression of *Stra8* (Bowles et al., 2010; Tedesco, Desimio, Klinger, De Felici, & Farini, 2013). Even very low levels of RA are sufficient to elicit this effect: when isolated germ cells, from either testis or ovary, are cultured in the presence of RA, at concentrations as low as 10nM, *Stra8* expression is rapidly induced (Bowles et al., 2010; Ohta, Lin, Hogg, Yamamoto, & Yamazaki, 2010; Tedesco et al., 2013). This exquisite sensitivity to RA is in accord with the low absolute levels of RA that can be detected in the intact fetal ovary (Bowles et al., 2016, 2006; Kumar et al., 2011). Even though RA availability in the fetal ovary is low, this contrasts with the apparent absence of RA in the fetal testis, due to the actions of CYP26B1 (Bowles et al., 2016, 2006; MacLean et al., 2007): it seems that this clear disparity orchestrates the sex-specific expression of *Stra8* and *Rec8* expression in fetal germ cells.

A different question relates to whether *Stra8* and *Rec8* are direct transcriptional targets of RA—that is, whether RA receptors (RARs) actually bind RA response elements (RAREs) in their promoters. Although it has been reported that RARs do not bind two RAREs identified upstream of the *Stra8* transcription start site when examined by chromatin immunoprecipitation (ChIP) (Kumar et al., 2011), a ChIP-seq study identified RAR-binding sites in both *Stra8* and *Rec8* promoter regions (Mahony et al., 2011). RA does seem to act directly to induce *Stra8* expression in ES (embryonic stem) cells (Giuili et al., 2002; Wang & Tilly, 2010) and direct transcriptional regulation by liganded RARs would be consistent with the rapidity of *Stra8* upregulation in ex vivo germ cell culture assays (Bowles et al., 2010; Gudas & Wagner, 2011; Tedesco et al., 2013).

Other non-RA factors are emerging as regulators of *Stra8* expression. There is some evidence that the transcription factor DMRT1 plays a role in ensuring the onset or augmentation of *Stra8* expression, though this effect appears strangely stochastic: some *Dmrt1*-null germ cells express *Stra8* and enter into meiosis apparently normally while other nearby germ cells fail to express *Stra8* at all (Krentz et al., 2011) (Fig. 1). It is difficult to imagine how DMRT1 could be instructive with respect to *Stra8* expression because *Dmrt1* is actually expressed more highly in testicular than ovarian germ cells

(Krentz et al., 2011; Lei et al., 2007). Intriguingly, DMRT1 only binds to the *Stra8* proximal promoter in ovarian germ cells (Krentz et al., 2011), suggesting that RA/RAR/RXR/RARE binding might ensure that DMRT1 is able to interact with the *Stra8* promoter region and, potentially, transactivate. There is also some evidence that related transcription factors MSX1 and MSX2 can directly bind to genomic regions upstream of *Stra8* (Le Bouffant et al., 2011). Although the absence of both of these factors is associated with lower than normal levels of *Stra8* transcription, the timing of the defect suggests that these may be involved in the maintenance, rather than the initiation, of *Stra8* expression. The meiosis-specific protein MEIOC is also involved in proper meiotic progression (Abby et al., 2016). When *Meioc* is deleted, *Stra8* is still upregulated but meiosis proceeds abnormally and germ cells undergo apoptosis. *Meioc* expression is not dependent on RA and its function appears to involve post-transcriptional stabilization of a number of meiosis-specific mRNAs.

3.2 Induction of Stra8 expression by RA is strictly time and cell type specific

RA is widely present during mammalian development, yet RA-induced initiation of *Stra8* only occurs in germ cells, and then only at specific times. Hence, one of the most puzzling outstanding questions relates to how *Stra8* expression is avoided in all somatic cell types and also during a large part of germ cell development. In the mouse, *Stra8* expression is only induced in fetal ovarian germ cells, and differentiated spermatogonia of the postnatal testis, just prior to meiotic entry (Bowles et al., 2006; Menke et al., 2003; Oulad-Abdelghani et al., 1996; Snyder, Small, & Griswold, 2010). Intriguingly, the in vitro response to RA is also highly specific: thus far only ES (embryonic stem) and EC (embryonal carcinoma) cells are known to be sensitive to RA in this respect (Bouillet et al., 1995; Oulad-Abdelghani et al., 1996). It seems likely that it is the characteristic epigenetic naivety of these cell types, including their flexible chromatin organization that makes them vulnerable to the *Stra8*-inducing effects of RA. Part of the answer to the time- and cell-specific expression of *Stra8* in response to RA may relate to the presence of Dazl: germ cells in *Dazl*-null gonads cultured ex vivo from 12.5 dpc are unable to respond to ectopic RA by upregulating *Stra8* (Lin et al., 2008). As noted above, the effect of Dazl is broader than simply bestowing RA-responsiveness, and may be essential to mature the germ cells, rendering them capable of responding to environmental influences more generally.

Chromatin dynamics, in addition to RA, also play a role in *Stra8* regulation. There is experimental evidence that regulatory regions of the *Stra8* promoter are maintained in a "closed," transcriptionally silent, status by histone deacetylation (Wang & Tilly, 2010) but this result is somewhat unsurprising because, in the absence of local RA, RARs are known to repress gene transcription by recruiting histone deacetylase enzymes (HDACs) as corepressors (Weston, Blumberg, & Underhill, 2003). Of relevance are results showing that RA induction of *Stra8* expression in ES cells is associated with an increase in histone acetylation at the *Stra8* promoter, mediated by the histone acetyltransferase CBP (Chen et al., 2013). Along similar lines, strong evidence supports a model whereby the transcriptional repressor PRC1 controls germ cell sensitivity to the *Stra8*-inducing effects of RA. It seems likely that PRC1 plays a long-term silencing role, in vivo, because when *RNF2* (encoding one of the central components of PRC1) is deleted, germ cells (in both ovarian and testicular environments) become sensitized to RA. Specifically, germ cells resident in the fetal ovary prematurely express *Stra8*, *Rec8* and other meiosis-related genes and precociously enter meiosis (Yokobayashi et al., 2013). These effects are observed about 1 day earlier than normal and are conditional on the presence of RA: they do not occur in the fetal ovary when RA signaling is prevented and they do not occur in the fetal testis unless ectopic RA is supplied. Chromatin immunoprecipitation (ChIP) analysis indicates that the RNF2 protein of the PRC1 multiprotein complex physically interacts with the *Stra8* promoter (Yokobayashi et al., 2013). Interestingly, ovarian germ cells in the *Cyp26b1*-null embryo also express *Stra8* about a day earlier than is normal (Bowles et al., 2006). Because we would expect endogenous RA levels to build up earlier in the *Cyp26b1*-null ovary than in the wildtype ovary, this result suggests that ovarian germ cells are capable of responding to RA for at least a short period prior to the normal *Stra8* expression onset time of 12.5 dpc. Hence it seems that the timing of *Stra8* onset relates to the interplay between openness or "looseness" of the chromatin conformation of the regulatory regions upstream of *Stra8*, regulated at least in part by PRC1 repression, and availability of sufficiently high RA levels.

3.3 The molecular function of STRA8 remains obscure

It is clear that STRA8 is critical for meiosis in both sexes because, in its absence, no functional gametes are produced in either sex (Anderson et al., 2008; Baltus et al., 2006). In *Stra8*-null ovarian germ cells the pre-meiotic DNA replication step does not occur, there is no meiotic chromosome condensation, cohesion, synapsis or recombination, and no meiotic progression (Baltus et al., 2006). Although the biological requirement for STRA8 seems quite clear, we know surprisingly little about its mechanism of action. *Stra8* was first identified in a screen looking for RA-responsive genes in P19 pluripotent embryonal carcinoma (EC) cells (Oulad-Abdelghani et al., 1996). The STRA8 protein was predicted to be 45-kD in size and to include one glutamic acid-rich domain, but to otherwise lack significant homology to protein databases. That initial report found that STRA8 expression was restricted to testis tissue in the adult mouse and in induced P19 EC cells, and differently phosphorylated forms were identified depending on whether all-*trans* or 9-*cis* RA was used for induction. It has been suggested that STRA8 shuttles between cytoplasm and nucleus (Tedesco, La Sala, Barbagallo, De Felici, & Farini, 2009) and that it binds to DNA (Choi et al., 2010).

3.4 Other extrinsic factors that may impact on meiotic progression

Recent studies suggest that BMP signaling may play an important role in instructing ovarian germ cell fate, acting in a complementary manner to RA signaling (Miyauchi et al., 2017). Although this study was done using induced PGC-like cells (PGCLCs), rather than bone fide fetal germ cells, the results are persuasive. In particular, it seems that RA can only induce PGCLCs to develop into primary oocytes if BMP is also provided: in vivo this would presumably come from the somatic cells because they are known to express high levels of both *Bmp2* and *Bmp5* (Jameson, Natarajan, et al., 2012; Yao et al., 2004). A requirement for some form of BMP/TGFβ signaling in oocyte development was suggested earlier, on the basis of the effects on germ cell development when the universal co-SMAD receptor (SMAD4) was inactivated, resulting in diminished expression of meiotic genes *Sycp3*, *Dmc1* and *Spo11* (Wu et al., 2016). In the more recent study (Miyauchi et al., 2017) it was found that BMP2 treatment of PGCLCs resulted in a loss of *Prdm1* expression and an upregulation of *Dazl* and *Mvh*, essentially mimicking the transcriptional changes observed in vivo when PGCs colonize the genital ridges. It was also shown that BMP signaling is necessary to achieve maximal *Stra8* expression, although BMP was not capable of inducing *Stra8* expression in the absence of RA (Fig. 1). It is tempting to speculate that the role of BMP signaling in his context is to mature the PGCLCs, as would normally occur in vivo upon entry into the gonad, and that it is not until this occurs that germ cells can respond normally to RA signaling. Counter to this idea is evidence that chemical inhibition of BMP signaling in vivo, at the appropriate timepoint, affects the sexual differentiation of ovarian, but not testicular, germ cells (Miyauchi et al., 2017). Interestingly, the combination of BMP and RA is not able to induce meiotic gene expression in PGCLCs until they have undergone a period of proliferative culture. Moreover, although RA/BMP cultured PGCLCs enter meiotic prophase, progression to the pachytene stage is inefficient, suggesting that additional extrinsic factors may be necessary in vivo.

There is also some evidence that additional non-retinoid non-BMP signaling molecules, produced by the ovarian soma, have a role in guiding ovarian germ cells through prophase of meiosis I. For example, the secreted factors WNT4 and its upstream regulator RSPO1 (R-spondin 1) both activate the canonical β-catenin signaling pathway during ovarian development (Chassot et al., 2008; Kim et al., 2006; Tomizuka et al., 2008; Vainio et al., 1999). In *Wnt4*- and *Rspo1*-null XX gonads, *Stra8* expression is delayed; it is unclear whether this reflects a direct effect on germ cells or a secondary (indirect) effect due to partial masculinization of the ovary (Chassot et al., 2011, 2008; Naillat et al., 2010; Tomizuka et al., 2008). The ectopic expression of *Cyp26b1* and reduced expression of *Aldh1a1* and *Aldh1a2* in *Wnt4*-null ovaries is consistent with an indirect effect (Naillat et al., 2010). An indirect effect is also supported by the fact that experimental overexpression of β-catenin in testicular germ cells does not induce them to enter meiosis (Chassot et al., 2011). Finally, neither LGR4 nor LGR5, receptors that enhance β-catenin signaling by virtue of their high affinity for RSPO1, are expressed by ovarian germ cells (Koizumi et al., 2015; Rastetter et al., 2014) suggesting that RSPO1, at least, cannot impact directly on germ cells. On the other hand, the WNT/β-catenin target gene *Axin2* is expressed in germ cells, and in XX more so than XY (Jameson, Natarajan, et al., 2012). Further work will be required to clarify the roles of RSPO1 and WNT4 in directing germ cell sexual fate and whether these are direct, indirect, or both.

It has also been reported that INHBA (activin-A) might co-operate with RA to promote meiosis in ovarian fetal germ cells (Liang et al., 2015). When activin A is added to cultured ovaries at 12.5 dpc, or delivered to 10.5 dpc litters by intraperitoneal injection of the dams, the expression of *Stra8* and meiotic genes was increased. The mechanism was suggested to be via activation of SMAD3 in pre-granulosa (somatic) cells, leading to a downregulation of *Cyp26b1* expression: an indirect effect on the germline.

It was also reported that inhibitors of Notch signaling, or siRNA-mediated gene knockdown of *Notch1*, diminish expression of *Stra8* and other meiosis-related genes in mouse fetal gonad cultures (Feng et al., 2014), however, it seems likely that these results are nonspecific. In particular, a direct role of Notch signaling in the instigation of meiotic entry is not consistent with evidence that Notch-responsive gene expression is not observed in pre-granulosa cells until 15.5 dpc (Vanorny, Prassasya, Chalpe, Kilen, & Mayo, 2014), well after the initiation of *Stra8* expression in ovarian germ cells at 12.5 dpc (Menke et al., 2003). A role for Notch signaling in the progression of pachytene oocytes through meiosis I does seem possible, though this is yet to be confirmed in vivo (Feng et al., 2014). There is clear evidence that Notch ligands are expressed by germ cells and that a Notch receptor, NOTCH2, is expressed by somatic pre-granulosa cells, and that these factors mediate cellular interactions during follicle assembly, which occurs around birth (Vanorny et al., 2014). The requirement for Notch and activin A signaling for normal meiotic onset and progression is yet to be studied in vivo.

4. Adopting the male fate in the testis

We recognize a number of distinct, but potentially interrelated, landmark features of fetal testicular germ cell development. Critically, XY germ cells in the fetal testis do not enter meiosis, unlike their contemporaries in the fetal ovary. Nonetheless, proliferation of testicular germ cells eventually slows and they enter a transient "mitotic arrest" in the G_0/G_1 phase, between 12.5 and 14.5 dpc. Simultaneously, expression of pluripotency-associated markers is diminished, characteristic male fate markers come to be expressed, and male-specific methylation patterns are re-established (Fig. 3). Presently it is unclear how these various processes are coordinated, nor how interdependent they are. We are gradually learning more about this interesting period of male germ cell development in the mammal; such knowledge is important because it is believed that human testicular germ cell tumors are "seeded" during fetal life when normal regulatory mechanisms go awry (Skakkebaek, Berthelsen, Giwercman, & Muller, 1987).

4.1 Avoidance of meiosis

Historically it was postulated that the female fate, or at least the meiotic initiation aspect of female fate, is the default state and that XY germ cells are

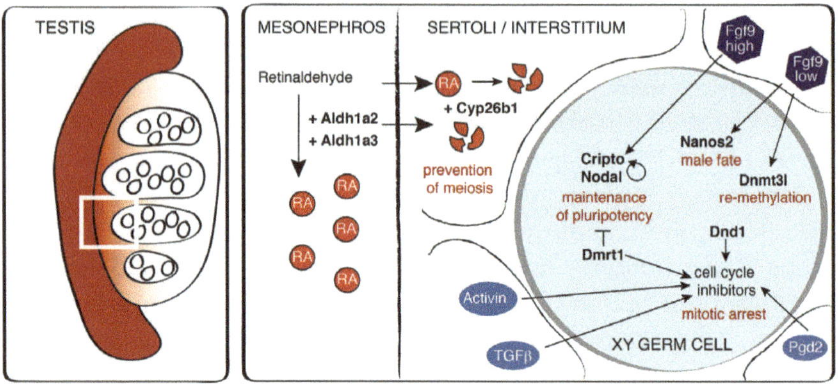

Fig. 3 Molecular regulators of male germ cell sex differentiation. A schematic of the developing testis (pale), with attached mesonephros (dark) is shown on the left and the indicated area is expanded in the right panel. Retinoic acid (RA) is produced in the mesonephros but this is degraded by CYP26B1, produced by somatic cells of the testis. FGF9 is produced by Sertoli cells and acts directly on germ cells to promote the maintenance of pluripotency as well as differentiation and re-methylation of the germline. Additional extrinsic factors including Activins, TGFβ and PGD2 also influence male germ cell development, particularly mitotic arrest. The key molecular regulators for each process are indicated. For details see the text.

only prevented from entering meiosis during fetal life by the actions of testicular somatic cells (Adams & McLaren, 2002; Chuma & Nakatsuji, 2001; Farini, Scaldaferri, Iona, La Sala, & De Felici, 2005; McLaren & Southee, 1997; Upadhyay & Zamboni, 1982). As discussed above, we now know that one action of those somatic cells is to produce the enzyme Cyp26b1 that efficiently degrades the secreted meiosis-promoting molecule RA, thereby preventing XY germ cells from expressing *Stra8* and *Rec8* and embarking on meiosis (Bowles et al., 2006; Koubova et al., 2014). When *Cyp26b1* is deleted, or inhibited, testicular germ cells strongly express *Stra8* and later meiotic markers, and progress through meiosis I. They reach at least the pachytene stage although they subsequently die, presumably because they receive conflicting molecular signals from the testicular soma (Bowles et al., 2010, 2006; MacLean et al., 2007).

Evidence that there needs to be more to the story—i.e., that a secreted testis-specific anti-meiotic factor must also be in play—has been provided by several studies. First, it was noted that some XY germ cells that lodge in the XY mesonephros are spared meiotic entry, even though XX germ cells that lodge in the XX mesonephros enter meiosis just as their ovarian counterparts do (McLaren, 1984); second, it was shown that if all cellular secretion is inhibited by use of the compound Brefeldin A some testicular germ cells do enter meiosis (Best, Sahlender, Walther, Peden, & Adams, 2008). Accordingly, it seems that the secreted growth factor FGF9, produced by early Sertoli cells shortly after the onset of *Sox9* expression, acts directly on germ cells to make them less responsive to RA, at least in terms of *Stra8* expression and meiotic entry (Barrios et al., 2010; Bowles et al., 2010; Ulu, Kim, Yokoyama, & Yamazaki, 2017). When isolated 11.5 or 12.5 dpc XX or XY germ cells are cultured in low doses of RA they rapidly upregulate *Stra8*, but if FGF9 is also present in the culture, this effect is attenuated (Bowles et al., 2010). It is difficult to design in vivo experiments that address this putative role for FGF9 because it is well established that FGF9 plays a key role in directing male fate in somatic cells of the testis: because *Fgf9*-null XY gonads develop as ovaries they are uninformative with respect to any direct effect of FGF9 on germ cell sexual fate (Colvin, Green, Schmahl, Capel, & Ornitz, 2001; Schmahl et al., 2004). Nonetheless, in vivo studies demonstrated that loss of even one copy of *Fgf9* was associated with higher propensity of germ cells in a *Cyp26b1*-null testis to respond to RA (Bowles et al., 2010). Importantly, treatment of XX or XY germ cells with an Fgfr antagonist alone does not augment *Stra8* expression suggesting that FGF9 does not affect the expression of *Stra8 per se* but, rather, that it makes germ cells less sensitive to RA, possibly by promoting the male fate including the upregulation of *Nanos2*, encoding an important specifier of male germ cell fate, as detailed below (Bowles et al., 2010). Therefore, it seems that RA is instructive and the absence of FGF9 is permissive in terms of meiotic onset.

It should be noted that a number of genetic deletion models are associated with a degree of ectopic meiosis in the fetal testis. These must be interpreted carefully because, as is well known, the soma instructs the sexual fate of germ cells and, hence, any perturbations of testicular soma development are likely to impact on germ cell behavior. For example, if testis development is delayed RA degradation would be expected to be delayed and if, for any reason, *Nanos2* expression is delayed, XY germ cells would be expected to remain sensitive to the meiosis-inducing effects of RA. In this category we would include the *Pgd2* deletion model, because overall testis development is delayed (Moniot et al., 2014), and the *Dnd1* mutant, because *Nanos2* expression is lost (Cook, Munger, Nadeau, & Capel, 2011). Similarly, ex vivo studies in which testis development is compromized, such as those in which all TGFβ signaling is antagonized (Wu et al., 2013), are also likely to result in ectopic meiosis that is not necessarily informative in terms of direct regulatory mechanisms. It is also important to recognize that ex vivo studies must be interpreted carefully because it is our experience that even control XY germ cell or gonadal cultures invariably exhibit "leaky" *Stra8* expression.

4.2 Controlling pluripotency regulators in testicular germ cells

Although PGCs and gonadal germ cells express factors associated with pluripotency (OCT4, SOX2, NANOG), they are not functionally pluripotent in that they do not self-renew and they do not contribute to germ line chimeras (Durcova-Hills, Adams, Barton, Surani, & McLaren, 2006). They can, however, be induced to de-differentiate and form embryonic germ (EG) cell colonies if cultured in the presence of LIF, FGF2 and SCF (Matsui, Zsebo, & Hogan, 1992; Resnick, Bixler, Cheng, & Donovan, 1992). EG cells are morphologically and functionally similar to ES cells and can be derived from germ cells of both male and female embryos, but only up until about 12.5 dpc (Durcova-Hills et al., 2006). This suggests that the underlying pluripotency of the germ cells is restricted at this time for both sexes. For the male mammal it seems that restraint of latent pluripotency during fetal life is particularly critical; it is postulated that human testicular germ cell tumors arise when germ cells do not differentiate normally during this time period (Skakkebaek et al., 1987; Sonne et al., 2009). Although genes such as *OCT4* and *SOX2* are not normally expressed in adult human germ cells, they are detected in testicular germ cell tumors (Oosterhuis & Looijenga, 2005).

Germ cells resident in the fetal testis maintain expression of pluripotency-associated genes for longer than do their counterparts in the fetal ovary (Bullejos & Koopman, 2004; Pesce, Wang, Wolgemuth, & Scholer, 1998; Western et al., 2005). OCT4, SOX2 and NANOG protein is lost in XY germ cells by 14.5 to 15.5 dpc (Western et al., 2010; Yamaguchi, Kimura, Tada, Nakatsuji, & Tada, 2005). Intriguingly, it seems that this is achieved by methylation of *Sox2* and *Nanog* promoters, which halts transcription, and by a post-transcriptional mechanism in the case of *Oct4*. Upregulation of *Dnmt3l*, encoding an enzyme that is critical for de novo methylation of the XY germ cell genome (Bourc'his & Bestor, 2004; La Salle et al., 2004), precedes the male-specific methylation of *Sox2* and *Nanog* promoter regions. Interestingly, loss of expression of pluripotency-associated genes as XX germ cells enter meiosis is not associated with promoter methylation, and loss of OCT4 is related to loss of the transcript rather than a post-translational mechanism (Western et al., 2010). The transcription factor DMRT1 appears to be an important germ cell intrinsic factor as regards the loss of pluripotency because in its absence, expression of *Oct4*, *Sox2* and *Nanog* is elevated; moreover, there is evidence that DMRT1 binds and directly represses the *Sox2* promoter (Krentz et al., 2009). In the teratoma-susceptible 129Sv genetic background depletion of DMRT1 specifically from germ cells greatly increases the likelihood of teratoma formation, indicating that functional pluripotency is retained in such germ cells (Krentz et al., 2009).

FGF9 seems to have a bearing on the ongoing expression of pluripotency-associated genes in testicular germ cells. In the *Cyp26b1*-null testis, XY germ cells embark on meiosis but this is not accompanied by loss of *Oct4* and *Sox2* expression, as is seen in the XX gonad, and in vivo studies suggest that this is because of the actions of FGF9 (Bowles et al., 2010). Accordingly, treatment of isolated 12.5 dpc germ cells with FGF9 upregulates *Oct4* and *Sox2* expression (Bowles et al., 2010). Another pluripotency-associated gene that is induced by FGF9 in XY germ cells is *Cripto* (also known as *Tdgf1*), encoding the obligate co-receptor for the TGFβ family signaling molecule NODAL (Spiller et al., 2012) (Fig. 2) and this is followed by an upregulation of *Nodal*, as well as genes encoding NODAL targets, such as *Lefty1* and *Lefty2* (Souquet et al., 2012; Spiller et al., 2012). It has been argued that the role of NODAL signaling is to allow testicular germ cells to avoid entering meiosis (Souquet et al., 2012; Wu et al., 2013) but this theory is not supported thus far in in vivo studies (Spiller et al., 2012). Some studies suggest that NODAL signaling ensures germ cells retain pluripotency and delay the onset of differentiation (Gustin, Stringer, Hogg, Sinclair, & Western, 2016; Spiller et al., 2012) although a role in inducing differentiation has also been proposed (Wu et al., 2013). CRIPTO/NODAL signaling is associated with an actual maintenance of pluripotency, not just maintenance of expression of pluripotency-related genes *Oct4*, *Sox2* and *Nanog*, because XY germ cells with diminished *Nodal* expression (approximately 30% of normal levels) are less capable of forming EG colonies than are wildtype XY germ cells (Spiller et al., 2012). It is interesting to note that teratoma-susceptible 129Sv fetal testes endogenously express higher levels of *Cripto*, *Nodal* and NODAL target genes than do fetal testes of the teratoma-resistant C57BL/6 strain (Krentz et al., 2009). The proposed role in maintaining functional pluripotency in testicular germ cells is in line with the well-defined role of NODAL signaling in the epiblast, in cancer stem cells and in human ES cells (reviewed in Brown et al., 2011; Vallier, Reynolds, & Pedersen, 2004). It is also in agreement with recent studies in human fetal testis xenografts in which it was demonstrated that stimulation of NODAL signaling increased the number of germ cells expressing pluripotency factors (Jorgensen et al., 2018). Definitive understanding of the role of CRIPTO/NODAL signaling will require analysis of a stringent germ cell-specific deletion of one or both genes.

4.3 Mitotic arrest

Although the mitotic cell cycle persists longer in germ cells of fetal testis than those of the fetal ovary, proliferative activity eventually ceases and the testicular germ cells become quiescent, remaining in G_0/G_1 until after birth (Hilscher et al., 1974; McLaren, 1984); once they mitotically arrest they are referred to as pro-spermatogonia (McCarrey, 2013). This phenomenon has been shown to occur gradually and in an unsynchronized manner between 12.5 and 14.5 dpc (Western, Miles, van den Bergen, Burton, & Sinclair, 2008). We have some knowledge of the germ cell intrinsic genes and proteins that are associated with the mitotic arrest in germ cells of the fetal testis. Onset of quiescence is marked by increased expression of genes encoding cell cycle inhibitors, such as $P27^{Kip1}$, $P21^{Cip1}$, $P15^{INK4B}$ and $P16^{INK4A}$, downregulation of genes encoding CYCLIN E1 and CYCLIN E2, and dephosphorylation (activation) of the G1/S checkpoint protein RB1 (Spiller, Wilhelm, & Koopman, 2009; Western et al., 2008). Deletion of RB1 in XY germ cells delays but does not completely derail mitotic arrest suggesting that considerable redundancy in the mechanisms of cell cycle is likely (Spiller, Wilhelm, & Koopman, 2010). Interestingly, mitotic arrest initiates before there is observable loss of pluripotency-associated proteins OCT4 and SOX2 (Western et al., 2010).

We know of two germ cell intrinsic factors that appear to have an impact on testicular germ cell arrest. When the gene encoding transcription factor DMRT1 is deleted, many XY germ cells were still cycling at 16.5 dpc and expression of $P18^{INK4C}$ and $P19^{INK4D}$ was decreased though, perhaps surprisingly, $P15^{INK4B}$ expression was increased. DMRT1 was found to bind the $p19^{INK4D}$ promoter, suggesting that it might directly regulate this key cell cycle inhibitor (Krentz et al., 2009). There is also strong evidence that the RNA-binding protein DND1 effects mitotic arrest by directly binding to, and protecting from degradation, a number of transcripts that encode negative regulators of the cell cycle, including $P27^{KIP1}$, $P21^{CIP1}$ and RB1 (Kedde et al., 2007). Further, there is some evidence that the RNA-binding male fate marker NANOS2 ensures mitotic arrest in a cell-autonomous manner (Saba, Kato, & Saga, 2014).

We are less clear regarding the somatic signals that instigate mitotic arrest. One report concluded that the presence of RA prevents XY germ cells from arresting, implying a second and independent role for the RA-degrading CYP26B1 enzyme (Trautmann et al., 2008); that RA can act in this manner is consistent with evidence that RA is a potent PGC mitogen in vitro (Koshimizu, Watanabe, & Nakatsuji, 1995). Accordingly, when ectopic RA is present in the developing testis but XY germ cells are unable to enter meiosis (*Cyp26b1*-null; *Stra8*-null model), mitotic arrest is moderately affected, suggesting that the loss of RA is important for cell cycle arrest, but that other factors must also be involved (Saba, Wu, et al., 2014). It seems likely that levels of the signaling molecule FGF9 need to fall before testicular germ cells stop proliferating. As mentioned above it is difficult to discern the role of FGF9 in vivo, but a high concentration of FGF9 can act *via* the ERK1/2 signal transduction pathway to induce proliferation in isolated germ cells (Ulu et al., 2017). There is some in vivo evidence that prostaglandin D_2 is involved in directing mitotic arrest in XY germ cells because when genes encoding both PDG2-producing enzymes are deleted, testicular germ cell mitotic arrest is inhibited (Moniot et al., 2014). However, because somatic development of the testis is also substantially delayed, it is not clear whether the effect is due to aberrant maturation of Sertoli cells (Moniot et al., 2009; Wilhelm et al., 2005) including, potentially, a secreted antagonist of germ cell proliferation. Nonetheless, it is possible that PGD2 directly mediates the cell cycle arrest because when it is added to isolated germ cell culture, expression of genes encoding $P21^{Cip1}$, $P15^{INK4b}$ and CYCLIN E2 is downregulated (Moniot et al., 2014). It has been proposed that a secreted product of P75-positive interstitial cells (which may include both peritubular myoid and fetal Leydig cells) triggers mitotic arrest (Guerquin et al., 2010). TGFβ and activin signaling have also been implicated in the instigation and maintenance of G_1/G_0 arrest in testicular germ cells (Godin & Wylie, 1991; Mendis, Meachem, Sarraj, & Loveland, 2011; Miles et al., 2013; Moreno et al., 2010). It is possible that CRIPTO expression on the surface of XY germ cells (see above) delays commitment to the male fate by making XY germ cells temporarily resistant to TGFβ and activin ligands (Spiller, Burnet, & Bowles, 2017; Spiller et al., 2012).

4.4 Active induction of the male fate

It has been argued, based on germ and somatic cell co-culture assays, that testicular germ cells commit to the male fate by 12.5 dpc (Adams &

McLaren, 2002), although some more recent studies have suggested that plasticity must remain for some additional days (Li, MacLean, Cameron, Clagett-Dame, & Petkovich, 2009; Ohta et al., 2010; Trautmann et al., 2008). Although classical studies strongly promoted the concept of a diffusible male-specific factor that is necessary to prevent germ cells from spontaneously entering meiosis, the notion that a male-specific factor(s) is required to direct germ cells to a spermatogenic fate was less strongly articulated (Adams & McLaren, 2002; Chuma & Nakatsuji, 2001; McLaren & Southee, 1997; Upadhyay & Zamboni, 1982). However, it is very clear that such a factor or factors must exist: perhaps this is most convincingly evident based on the observation that ovarian germ cells that cannot express *Stra8* and therefore do not enter meiotic prophase do not, instead, adopt a masculine fate (Baltus et al., 2006). Similarly, older studies showed that XX germ cells in a mouse ovary, if cultured without an attached mesonephros and therefore with limited exposure to RA, do not develop as pro-spermatogonia (Byskov, 1974). It has been reported that testicular germ cells that fail to become incorporated into testicular cords, possibly because they are "late arriving" to the genital ridges, develop along the male pathway, at least up until birth: this implies that any male fate-promoting signals in the testis are present not just within the cords but also in the interstitial tissue (Buehr, Gu, & Mclaren, 1993; Francavilla & Zamboni, 1985; McLaren, 1984). Hence, there is strong support for the theory that male fate in the testis must be either actively induced or, at least, that testis-specific factors must provide an environment that is permissive for progression to the male fate.

During the quiescent period prior to birth testicular germ cells begin to express certain genes that we now consider to signify an adoption of the masculine fate. These include *Nanos2* (Tsuda et al., 2003), *Dnmt3l* (La Salle et al., 2004), *Piwil4* (Aravin et al., 2008) and *Cdkn2b* (encoding *p15INK4b*; Western et al., 2008). Male imprints are also re-established at this time (Bourc'his & Bestor, 2004; Hata, Okano, Lei, & Li, 2002; La Salle et al., 2004). FGF9, mentioned above in the context of antagonizing meiosis in the testis, is thought to act directly on germ cells to push them into a male fate (Barrios et al., 2010; Bowles et al., 2010; Ulu et al., 2017). Isolated 12.5 dpc germ cells when treated with FGF9 upregulate *Nanos2*, *Dnmt3l* and *Cdkn2b* (Bowles et al., 2010; Ulu et al., 2017). There is evidence that it is only low levels of FGF9 that are associated with inducing the male fate markers and that FGF9 action employs the P38 MAPK signal transduction pathway in this context (Ulu et al., 2017). Intriguingly, an FGFR2 isoform switch appears to occur as germ cells colonize the gonad: migrating germ cells express *Fgfr2IIIb* transcripts but *Fgfr2IIIc* transcripts are detected in post-migration gonadal germ cells. As FGFR2-IIIc is a high affinity receptor for FGF9, and FGFR2-IIIb is not, this isoform switch would be expected to render germ cells highly sensitized to FGF9 upon entry into the gonad (Bowles et al., 2010).

The postulated requirement for FGF9 to directly induce male germ cell fate has been challenged by a study in which the effect of simultaneous deletion of both *Fgf9* and *Wnt4* was analysed (Jameson, Lin, et al., 2012). In terms of our understanding of somatic sex determination the results of that study were conclusive: XY gonads null for both *Fgf9* and *Wnt4* developed as normal testes thereby demonstrating that FGF9's critical function in fetal testis development is to antagonize the ovary-determining function of the signaling molecule WNT4. In terms of the hypothesis that FGF9 directly instructs the masculine germ cell fate, however, the results were unexpected. *Fgf9*-null; *Wnt4*-null XY testes were found to harbor XY germ cells that develop normally in the sense that they adopt the male fate, as assessed by expression DNMT3L. Despite these results it possible that another Fgf molecule acts redundantly to impose male fate, or that additional male-specific signaling molecules are involved and able to compensate for the loss of FGF9. A key experiment to address the issue of whether FGF signaling is required to instruct male germ cell fate would be to make germ cells insensitive to FGFs: such a germ cell-specific genetic deletion of FGF receptor(s) is yet to be reported.

The presence of ectopic RA impacts adversely on the adoption of the male fate as shown by study of *Cyp26b1*-null; *Stra8*-null testes (Saba, Wu, et al., 2014). In this experiment, XY germ cells were subjected to RA in the testis environment but were unable to enter meiosis, by virtue of their inability to produce STRA8. Although induction of *Nanos2* and was still observed, its expression was delayed and diminished in scale, suggesting that RA can partially inhibit male fate by an unknown mechanism and that this is independent of its role in inducing meiosis. FGF9 is produced by the Sertoli cells, but it has been reported that an as yet unidentified product of the interstitial cells can also act to induce expression of *Nanos2* (Guerquin et al., 2010). Prostaglandin D_2, produced by Sertoli cells can also induce *Nanos2* expression in isolated 12.5 dpc germ cells (Moniot et al., 2014).

Data thus far indicate that NANOS2, an RNA-binding protein, is responsible for "locking in" the masculine fate in fetal germ cells. *Nanos2* is first expressed in testicular germ cells at about 13.5 dpc (Tsuda et al., 2003).

When *Nanos2* is deleted, testicular germ cells die, but if rescued (by concomitant ablation of the pro-apoptotic protein BAX) they resume proliferation and start to express *Stra8*, albeit at levels substantially lower than is normal, and with a delay of about 2 days, compared with germ cells of the fetal ovary (Suzuki & Saga, 2008). XX germ cells that are forced to express ectopic *Nanos2* do not express *Stra8* or enter meiosis normally, suggesting that once NANOS2 protein is produced in a fetal germ cell, its window of sensitivity to the meiosis-inducing effects of RA closes. Presumably the delayed onset of *Stra8* expression in *Cyp26b1*-null testicular germ cells relates to the need for endogenous RA to build up, once *Cyp26b1* expression falls at later stages of fetal life (Suzuki & Saga, 2008).

Over and above an apparent role in making germ cells unresponsive to RA, NANOS2 is the key factor necessary to "lock in" male germ cell fate. This result was demonstrated by considering the effects of deleting both *Stra8* and *Nanos2*: even when testicular germ cells were prevented from ectopically upregulating *Stra8*, they still did not adopt the male fate normally (Saba, Kato, et al., 2014). NANOS2 interacts with the CCR4-NOT deadenylation complex and probably functions, in the context of XY germ cells, by degrading certain meiosis-related transcripts (Suzuki, Igarashi, Aisaki, Kanno, & Saga, 2010). This is interesting in light of evidence that *Sycp3* transcripts, which encode a key meiosis protein, are detectable in both XX and XY germ cells shortly after they colonize the fetal gonads (Chuma & Nakatsuji, 2001). More recent studies have shown that NANOS2 partners with DND1 in order to associate with its target RNAs (Suzuki et al., 2016) and that translation of *Dazl* mRNA must be suppressed by NANOS2/DND1 in order for normal male germ cell development to occur (Kato, Katsuki, Kokubo, Masuda, & Saga, 2016). It is possible that this effect on production of the RNA-binding protein DAZL is the mechanism by which NANOS2 renders germ cells unresponsive to RA, in terms of *Stra8* expression (Suzuki & Saga, 2008) and the means by which germ cells, resident in wildtype testes, become irreversibly committed to the male fate (McLaren & Southee, 1997).

The XY germ cell genome is almost completely demethylated shortly after gonadal colonization, but re-methylation begins at 14.5 dpc and by 16.5 dpc about 50% of the genome is methylated (Seisenberger et al., 2012). This de novo methylation plays an important role in establishing life-long epigenetic silencing of transposable elements (Aravin et al., 2008). Re-methylation requires the action of several DNA methyltransferases and, in the XY germ cells, the most dynamically-expressed is DNMT3L, a catalytically inactive protein that acts as a co-factor for related DNMT enzymes. *Dnmt3l* is robustly and male-specifically upregulated between 13.5 and 14.5 dpc (Western et al., 2010) while the newly-discovered and murid-specific *Dnmt3c* transcript is sharply upregulated at about 16.5 dpc (Barau et al., 2016). Deletion of either gene leads to impaired meiosis and male sterility (Barau et al., 2016; Bourc'his & Bestor, 2004).

5. Conclusions

Germ cells face a number of unique problems during their development. Uniquely, they must undergo meiosis to produce haploid gametes. Second, they need to retain latent totipotency, yet this requirement is sometimes problematic as evidenced by the propensity of male fetal germ cells to seed germ cell tumors. Finally, it is of course essential that germ cells, as the purveyor of all genetic and epigenetic information to the next generation, maintain a high level of genetic integrity.

It is well established that sex-specific development of the germline initiates during fetal life, and that it is the somatic tissue that instructs sexual fate. The first female- and male-specific transcripts reported thus far are *Stra8* and *Cripto*, respectively: these are crucial for launching meiosis in ovarian germ cells and, seemingly, for maintaining a state of transient pluripotency in testicular germ cells. STRA8 launches the meiotic program in ovarian germ cells, by an unknown mechanism, while NANOS2 is critical for "locking in" male fate and DNMT3L for progressively re-establishing paternal imprints in testicular germ cells.

In terms of signaling from the somatic environment, RA and FGF9 appear to be key, but additional molecules are also likely to have an impact on normal germ cell development. Contenders include WNT4 and RSPO1 in the ovary and PGD2 and TGFβ in the testis. Evidence that BMP signaling is likely necessary for robust meiotic entry in ovarian germ cells is an important recent development. Presumably the cumulative effects of at least some of these signaling molecules, and others yet to be discovered, converge to ensure that all aspects of fetal germ cell development occur in a coordinated fashion.

Our growing understanding of how murine fetal germ cells are instructed on the first steps of their transition to mature gametes has proven informative with respect to recent efforts to produce functional sperm and oocytes in vitro.

References

Abby, E., Tourpin, S., Ribeiro, J., Daniel, K., Messiaen, S., Moison, D., et al. (2016). Implementation of meiosis prophase I programme requires a conserved retinoid-independent stabilizer of meiotic transcripts. *Nature Communications*, 7, 10324.

Adams, I. R., & McLaren, A. (2002). Sexually dimorphic development of mouse primordial germ cells: Switching from oogenesis to spermatogenesis. *Development*, *129*, 1155–1164.

Anderson, E. L., Baltus, A. E., Roepers-Gajadien, H. L., Hassold, T. J., de Rooij, D. G., van Pelt, A. M. M., et al. (2008). Stra8 and its inducer, retinoic acid, regulate meiotic initiation in both spermatogenesis and oogenesis in mice. *Proceedings of the National Academy of Sciences of the United States of America*, *105*, 14976–14980.

Anderson, R., Fassler, R., Georges-Labouesse, E., Hynes, R. O., Bader, B. L., Kreidberg, J. A., et al. (1999). Mouse primordial germ cells lacking beta 1 integrins enter the germline but fail to migrate normally to the gonads. *Development*, *126*, 1655–1664.

Aravin, A. A., Sachidanandam, R., Bourc'his, D., Schaefer, C., Pezic, D., Toth, K. F., et al. (2008). A piRNA pathway primed by individual transposons is linked to de novo DNA methylation in mice. *Molecular Cell*, *31*, 785–799.

Baltus, A. E., Menke, D. B., Hu, Y. C., Goodheart, M. L., Carpenter, A. E., de Rooij, D. G., et al. (2006). In germ cells of mouse embryonic ovaries, the decision to enter meiosis precedes premeiotic DNA replication. *Nature Genetics*, *38*, 1430–1434.

Barau, J., Teissandier, A., Zamudio, N., Roy, S., Nalesso, V., Herault, Y., et al. (2016). The DNA methyltransferase DNMT3C protects male germ cells from transposon activity. *Science*, *354*, 909–912.

Barrios, F., Filipponi, D., Pellegrini, M., Paronetto, M. P., Di Siena, S., Geremia, R., et al. (2010). Opposing effects of retinoic acid and FGF9 on Nanos2 expression and meiotic entry of mouse germ cells. *Journal of Cell Science*, *123*, 871–880.

Bendel-Stenzel, M. R., Gomperts, M., Anderson, R., Heasman, J., & Wylie, C. (2000). The role of cadherins during primordial germ cell migration and early gonad formation in the mouse. *Mechanisms of Development*, *91*, 143–152.

Best, D., Sahlender, D. A., Walther, N., Peden, A. A., & Adams, I. R. (2008). Sdmg1 is a conserved transmembrane protein associated with germ cell sex determination and germline-soma interactions in mice. *Development*, *135*, 1415–1425.

Bouillet, P., Ouladabdelghani, M., Vicaire, S., Garnier, J. M., Schuhbaur, B., Dolle, P., et al. (1995). Efficient cloning of Cdnas of retinoic acid-responsive genes in P19 embryonal carcinoma-cells and characterization of a novel mouse gene, Stra1 (Mouse Lerk-2/Eplg2). *Developmental Biology*, *170*, 420–433.

Bourc'his, D., & Bestor, T. H. (2004). Meiotic catastrophe and retrotransposon reactivation in male germ cells lacking Dnmt3L. *Nature*, *431*, 96–99.

Bowles, J., Feng, C. W., Ineson, J., Miles, K., Spiller, C. M., Harley, V. R., et al. (2018). Retinoic acid antagonizes testis development in mice. *Cell Reports*, *24*, 1330–1341.

Bowles, J., Feng, C. W., Miles, K., Ineson, J., Spiller, C., & Koopman, P. (2016). ALDH1A1 provides a source of meiosis-inducing retinoic acid in mouse fetal ovaries. *Nature Communications*, 7, 10845.

Bowles, J., Feng, C. W., Spiller, C., Davidson, T. L., Jackson, A., & Koopman, P. (2010). FGF9 suppresses meiosis and promotes male germ cell fate in mice. *Developmental Cell*, *19*, 440–449.

Bowles, J., Knight, D., Smith, C., Wilhelm, D., Richman, J., Mamiya, S., et al. (2006). Retinoid signaling determines germ cell fate in mice. *Science*, *312*, 596–600.

Bowles, J., & Koopman, P. (2007). Retinoic acid, meiosis and germ cell fate in mammals. *Development*, *134*, 3401–3411.

Brown, S., Teo, A., Pauklin, S., Hannan, N., Cho, C. H., Lim, B., et al. (2011). Activin/nodal signaling controls divergent transcriptional networks in human embryonic stem cells and in endoderm progenitors. *Stem Cells*, *29*, 1176–1185.

Buehr, M., Gu, S., & Mclaren, A. (1993). Mesonephric contribution to testis differentiation in the fetal mouse. *Development*, *117*, 273–281.

Bullejos, M., & Koopman, P. (2004). Germ cells enter meiosis in a rostro-caudal wave during development of the mouse ovary. *Molecular Reproduction and Development*, *68*, 422–428.

Byskov, A. G. (1974). Does the rete ovarii act as a trigger for the onset of meiosis? *Nature*, *252*, 396–397.

Byskov, A. G., & Saxen, L. (1976). Induction of meiosis in fetal mouse testis invitro. *Developmental Biology*, *52*, 193–200.

Cantu, A. V., & Laird, D. J. (2017). A pilgrim's progress: Seeking meaning in primordial germ cell migration. *Stem Cell Research*, *24*, 181–187.

Chassot, A. A., Gregoire, E. P., Lavery, R., Taketo, M. M., de Rooij, D. G., Adams, I. R., et al. (2011). RSPO1/beta-catenin signaling pathway regulates oogonia differentiation and entry into meiosis in the mouse fetal ovary. *PLoS One*, *6*, e25641.

Chassot, A. A., Ranc, F., Gregoire, E. P., Roepers-Gajadien, H. L., Taketo, M. M., Camerino, G., et al. (2008). Activation of beta-catenin signaling by Rspo1 controls differentiation of the mammalian ovary. *Human Molecular Genetics*, *17*, 1264–1277.

Chen, W., Jia, W., Wang, K., Si, X., Zhu, S., Duan, T., et al. (2013). Distinct roles for CBP and p300 on the RA-mediated expression of the meiosis commitment gene Stra8 in mouse embryonic stem cells. *PLoS One*, *8*, e66076.

Childs, A. J., Cowan, G., Kinnell, H. L., Anderson, R. A., & Saunders, P. T. (2011). Retinoic acid signalling and the control of meiotic entry in the human fetal gonad. *PLoS One*, *6*, e20249.

Chiquoine, A. D. (1954). The identification, origin, and migration of the primordial germ cells in the mouse embryo. *The Anatomical Record*, *118*, 135–146.

Choi, Y. J., Yoon, J. W., Pyo, C. W., Kim, J. A., Bae, S. H., & Park, S. S. (2010). A possible role of STRA8 as a transcriptional factor. *Genes & Genomics*, *32*, 521–526.

Chuma, S., & Nakatsuji, N. (2001). Autonomous transition into meiosis of mouse fetal germ cells in vitro and its inhibition by gp130-mediated signaling. *Developmental Biology*, *229*, 468–479.

Colvin, J. S., Green, R. P., Schmahl, J., Capel, B., & Ornitz, D. M. (2001). Male-to-female sex reversal in mice lacking fibroblast growth factor 9. *Cell*, *104*, 875–889.

Cook, M. S., Munger, S. C., Nadeau, J. H., & Capel, B. (2011). Regulation of male germ cell cycle arrest and differentiation by DND1 is modulated by genetic background. *Development*, *138*, 23–32.

Durcova-Hills, G., Adams, I. R., Barton, S. C., Surani, M. A., & McLaren, A. (2006). The role of exogenous fibroblast growth factor-2 on the reprogramming of primordial germ cells into pluripotent stem cells. *Stem Cells*, *24*, 1441–1449.

Farini, D., Scaldaferri, M. L., Iona, S., La Sala, G., & De Felici, M. (2005). Growth factors sustain primordial germ cell survival, proliferation and entering into meiosis in the absence of somatic cells. *Developmental Biology*, *285*, 49–56.

Feng, Y. M., Liang, G. J., Pan, B., Qin, X. S., Zhang, X. F., Chen, C. L., et al. (2014). Notch pathway regulates female germ cell meiosis progression and early oogenesis events in fetal mouse. *Cell Cycle*, *13*, 782–791.

Ford, C. E., Evans, E. P., Burtenshaw, M. D., Clegg, H. M., Tuffrey, M., & Barnes, R. D. (1975). Functional sex-reversed oocyte in mouse. *Proceedings of the Royal Society of London - Series B: Biological Sciences*, *190*, 187–197.

Francavilla, S., & Zamboni, L. (1985). Differentiation of mouse ectopic germinal cells in intragonadal and perigonadal locations. *The Journal of Experimental Zoology*, *233*, 101–109.

Gill, M. E., Hu, Y. C., Lin, Y. F., & Page, D. C. (2011). Licensing of gametogenesis, dependent on RNA binding protein DAZL, as a gateway to sexual differentiation of fetal germ cells. *Proceedings of the National Academy of Sciences of the United States of America*, *108*, 7443–7448.

Giuili, G., Tomljenovic, A., Labrecque, N., Oulad-Abdelghani, M., Rassoulzadegan, M., & Cuzin, F. (2002). Murine spermatogonial stem cells: Targeted transgene expression and purification in an active state. *EMBO Reports*, *3*, 753–759.

Godin, I., & Wylie, C. C. (1991). TGF beta 1 inhibits proliferation and has a chemotropic effect on mouse primordial germ cells in culture. *Development*, *113*, 1451–1457.

Gomperts, M., Garciacastro, M., Wylie, C., & Heasman, J. (1994). Interactions between primordial germ-cells play a role in their migration in mouse embryos. *Development*, *120*, 135–141.

Gudas, L. J., & Wagner, J. A. (2011). Retinoids regulate stem cell differentiation. *Journal of Cellular Physiology*, *226*, 322–330.

Guerquin, M. J., Duquenne, C., Lahaye, J. B., Tourpin, S., Habert, R., & Livera, G. (2010). New testicular mechanisms involved in the prevention of fetal meiotic initiation in mice. *Developmental Biology*, *346*, 320–330.

Gustin, S. E., Stringer, J. M., Hogg, K., Sinclair, A. H., & Western, P. S. (2016). FGF9, activin and TGF beta promote testicular characteristics in an XX gonad organ culture model. *Reproduction*, *152*, 529–543.

Hajkova, P. (2011). Epigenetic reprogramming in the germline: Towards the ground state of the epigenome. *Philosophical Transactions of the Royal Society of London. Series B, Biological Sciences*, *366*, 2266–2273.

Hajkova, P., Ancelin, K., Waldmann, T., Lacoste, N., Lange, U. C., Cesari, F., et al. (2008). Chromatin dynamics during epigenetic reprogramming in the mouse germ line. *Nature*, *452*, 877–881.

Hajkova, P., Erhardt, S., Lane, N., Haaf, T., El-Maarri, O., Reik, W., et al. (2002). Epigenetic reprogramming in mouse primordial germ cells. *Mechanisms of Development*, *117*, 15–23.

Hata, K., Okano, M., Lei, H., & Li, E. (2002). Dnmt3L cooperates with the Dnmt3 family of de novo DNA methyltransferases to establish maternal imprints in mice. *Development*, *129*, 1983–1993.

Hilscher, B., Hilscher, W., Bulthoffohnolz, B., Kramer, U., Birke, A., Pelzer, H., et al. (1974). Kinetics of gametogenesis.1. Comparative histological and autoradiographic studies of oocytes and transitional prospermatogonia during oogenesis and prespermatogenesis. *Cell and Tissue Research*, *154*, 443–470.

Hu, Y. C., Nicholls, P. K., Soh, Y. Q., Daniele, J. R., Junker, J. P., van Oudenaarden, A., et al. (2015). Licensing of primordial germ cells for gametogenesis depends on genital ridge signaling. *PLoS Genetics*, *11*, e1005019.

Jameson, S. A., Lin, Y. T., & Capel, B. (2012). Testis development requires the repression of Wnt4 by Fgf signaling. *Developmental Biology*, *370*, 24–32.

Jameson, S. A., Natarajan, A., Cool, J., DeFalco, T., Maatouk, D. M., Mork, L., et al. (2012). Temporal transcriptional profiling of somatic and germ cells reveals biased lineage priming of sexual fate in the fetal mouse gonad. *PLoS Genetics*, *8*, e1002575.

Jorgensen, A., Macdonald, J., Nielsen, J. E., Kilcoyne, K. R., Perlman, S., Lundvall, L., et al. (2018). Nodal signaling regulates germ cell development and establishment of seminiferous cords in the human fetal testis. *Cell Reports*, *25*, 1924–1937.

Jorgensen, A., Nielsen, J. E., Blomberg Jensen, M., Graem, N., & Rajpert-De Meyts, E. (2012). Analysis of meiosis regulators in human gonads: A sexually dimorphic spatiotemporal expression pattern suggests involvement of DMRT1 in meiotic entry. *Molecular Human Reproduction*, *18*, 523–534.

Kato, Y., Katsuki, T., Kokubo, H., Masuda, A., & Saga, Y. (2016). Dazl is a target RNA suppressed by mammalian NANOS2 in sexually differentiating male germ cells. *Nature Communications*, *7*.

Kedde, M., Strasser, M. J., Boldajipour, B., Vrielink, J. A. F. O., Le Sage, C., Nagel, R., et al. (2007). RNA-binding protein Dnd1 inhibits microRNA access to target mRNA. *Cell*, *131*, 1273–1286.

Kim, Y., Kobayashi, A., Sekido, R., DiNapoli, L., Brennan, J., Chaboissier, M. C., et al. (2006). Fgf9 and Wnt4 act as antagonistic signals to regulate mammalian sex determination. *PLoS Biology*, *4*, e187.

Koizumi, M., Oyama, K., Yamakami, Y., Kida, T., Satoh, R., Kato, S., et al. (2015). Lgr4 controls specialization of female gonads in mice. *Biology of Reproduction*, *93*, 90.

Koshimizu, U., Watanabe, M., & Nakatsuji, N. (1995). Retinoic acid is a potent growth activator of mouse primordial germ-cells in-vitro. *Developmental Biology*, *168*, 683–685.

Koubova, J., Hu, Y. C., Bhattacharyya, T., Soh, Y. Q., Gill, M. E., Goodheart, M. L., et al. (2014). Retinoic acid activates two pathways required for meiosis in mice. *PLoS Genetics*, *10*, e1004541.

Koubova, J., Menke, D. B., Zhou, Q., Capel, B., Griswold, M. D., & Page, D. C. (2006). Retinoic acid regulates sex-specific timing of meiotic initiation in mice. *Proceedings of the National Academy of Sciences of the United States of America*, *103*, 2474–2479.

Krentz, A. D., Murphy, M. W., Kim, S., Cook, M. S., Capel, B., Zhu, R., et al. (2009). The DM domain protein DMRT1 is a dose-sensitive regulator of fetal germ cell proliferation and pluripotency. *Proceedings of the National Academy of Sciences of the United States of America*, *106*, 22323–22328.

Krentz, A. D., Murphy, M. W., Sarver, A. L., Griswold, M. D., Bardwell, V. J., & Zarkower, D. (2011). DMRT1 promotes oogenesis by transcriptional activation of Stra8 in the mammalian fetal ovary. *Developmental Biology*, *356*, 63–70.

Kumar, S., Chatzi, C., Brade, T., Cunningham, T. J., Zhao, X., & Duester, G. (2011). Sex-specific timing of meiotic initiation is regulated by Cyp26b1 independent of retinoic acid signalling. *Nature Communications*, *2*, 151.

Kumar, S., Cunningham, T. J., & Duester, G. (2013). Resolving molecular events in the regulation of meiosis in male and female germ cells. *Science Signaling*, *6*, pe25.

Kurimoto, K., Yabuta, Y., Ohinata, Y., Shigeta, M., Yamanaka, K., & Saitou, M. (2008). Complex genome-wide transcription dynamics orchestrated by Blimp1 for the specification of the germ cell lineage in mice. *Genes & Development*, *22*, 1617–1635.

La Salle, S., Mertineit, C., Taketo, T., Moens, P. B., Bestor, T. H., & Trasler, J. M. (2004). Windows for sex-specific methylation marked by DNA methyltransferase expression profiles in mouse germ cells. *Developmental Biology*, *268*, 403–415.

Lawson, K. A., Dunn, N. R., Roelen, B. A. J., Zeinstra, L. M., Davis, A. M., Wright, C. V. E., et al. (1999). Bmp4 is required for the generation of primordial germ cells in the mouse embryo. *Genes & Development*, *13*, 424–436.

Le Bouffant, R., Guerquin, M. J., Duquenne, C., Frydman, N., Coffigny, H., Rouiller-Fabre, V., et al. (2010). Meiosis initiation in the human ovary requires intrinsic retinoic acid synthesis. *Human Reproduction*, *25*, 2579–2590.

Le Bouffant, R., Souquet, B., Duval, N., Duquenne, C., Herve, R., Frydman, N., et al. (2011). Msx1 and Msx2 promote meiosis initiation. *Development*, *138*, 5393–5402.

Lei, N., Hornbaker, K. I., Rice, D. A., Karpova, T., Agbor, V. A., & Heckert, L. L. (2007). Sex-specific differences in mouse DMRT1 expression are both cell type- and stage-dependent during gonad development. *Biology of Reproduction*, *77*, 466–475.

Li, H., & Clagett-Dame, M. (2009). Vitamin A deficiency blocks the initiation of meiosis of germ cells in the developing rat ovary in vivo. *Biology of Reproduction*, *81*, 996–1001.

Li, H., MacLean, G., Cameron, D., Clagett-Dame, M., & Petkovich, M. (2009). Cyp26b1 expression in murine Sertoli cells is required to maintain male germ cells in an undifferentiated state during embryogenesis. *PLoS One*, *4*, e7501.

Liang, G. J., Zhang, X. F., Wang, J. J., Sun, Y. C., Sun, X. F., Cheng, S. F., et al. (2015). Activin A accelerates the progression of fetal oocytes throughout meiosis and early oogenesis in the mouse. *Stem Cells and Development*, *24*, 2455–2465.

Lin, Y. F., Gill, M. E., Koubova, J., & Page, D. C. (2008). Germ cell-intrinsic and -extrinsic factors govern meiotic initiation in mouse embryos. *Science*, *322*, 1685–1687.

Lin, Y., & Page, D. C. (2005). Dazl deficiency leads to embryonic arrest of germ cell development in XY C57BL/6 mice. *Developmental Biology*, *288*, 309–316.
Maatouk, D. M., Kellam, L. D., Mann, M. R. W., Lei, H., Li, E., Bartolomei, M. S., et al. (2006). DNA methylation is a primary mechanism for silencing postmigratory primordial germ cell genes in both germ cell and somatic cell lineages. *Development*, *133*, 3411–3418.
MacLean, G., Li, H., Metzger, D., Chambon, P., & Petkovich, M. (2007). Apoptotic extinction of germ cells in testes of Cyp26b1 knockout mice. *Endocrinology*, *148*, 4560–4567.
Mahony, S., Mazzoni, E. O., McCuine, S., Young, R. A., Wichterle, H., & Gifford, D. K. (2011). Ligand-dependent dynamics of retinoic acid receptor binding during early neurogenesis. *Genome Biology*, *12*, R2.
Matsui, Y., Zsebo, K., & Hogan, B. L. M. (1992). Derivation of Pluripotential embryonic stem-cells from murine primordial germ-cells in culture. *Cell*, *70*, 841–847.
McCarrey, J. R. (2013). Toward a more precise and informative nomenclature describing fetal and neonatal male germ cells in rodents. *Biology of Reproduction*, *89*, 1–9.
McLaren, A. (1984). Meiosis and differentiation of mouse germ cells. *Symposium of the Society of Experimental Biology*, *38*, 7–23.
McLaren, A. (2003). Primordial germ cells in the mouse. *Developmental Biology*, *262*, 1–15.
McLaren, A., & Southee, D. (1997). Entry of mouse embryonic germ cells into meiosis. *Developmental Biology*, *187*, 107–113.
Mendis, S. H., Meachem, S. J., Sarraj, M. A., & Loveland, K. L. (2011). Activin A balances sertoli and germ cell proliferation in the fetal mouse testis. *Biology of Reproduction*, *84*, 379–391.
Menke, D. B., Koubova, J., & Page, D. C. (2003). Sexual differentiation of germ cells in XX mouse gonads occurs in an anterior-to-posterior wave. *Developmental Biology*, *262*, 303–312.
Miles, D. C., Wakeling, S. I., Stringer, J. M., van den Bergen, J. A., Wilhelm, D., Sinclair, A. H., et al. (2013). Signaling through the TGF beta-activin receptors ALK4/5/7 regulates testis formation and male germ cell development. *PLoS One*, *8*, e54606.
Miyauchi, H., Ohta, H., Nagaoka, S., Nakaki, F., Sasaki, K., Hayashi, K., et al. (2017). Bone morphogenetic protein and retinoic acid synergistically specify female germ-cell fate in mice. *The EMBO Journal*, *36*, 3100–3119.
Molyneaux, K. A., Stallock, J., Schaible, K., & Wylie, C. (2001). Time-lapse analysis of living mouse germ cell migration. *Developmental Biology*, *240*, 488–498.
Molyneaux, K., & Wylie, C. (2004). Primordial germ cell migration. *International Journal of Developmental Biology*, *48*, 537–544.
Moniot, B., Declosmenil, F., Barrionuevo, F., Scherer, G., Aritake, K., Malki, S., et al. (2009). The PGD2 pathway, independently of FGF9, amplifies SOX9 activity in Sertoli cells during male sexual differentiation. *Development*, *136*, 1813–1821.
Moniot, B., Ujjan, S., Champagne, J., Hirai, H., Aritake, K., Nagata, K., et al. (2014). Prostaglandin D-2 acts through the Dp2 receptor to influence male germ cell differentiation in the foetal mouse testis. *Development*, *141*, 3561–3571.
Moreno, S. G., Attali, M., Allemand, I., Messiaen, S., Fouchet, P., Coffigny, H., et al. (2010). TGFbeta signaling in male germ cells regulates gonocyte quiescence and fertility in mice. *Developmental Biology*, *342*, 74–84.
Mu, X., Wen, J., Guo, M., Wang, J., Li, G., Wang, Z., et al. (2013). Retinoic acid derived from the fetal ovary initiates meiosis in mouse germ cells. *Journal of Cellular Physiology*, *228*, 627–639.
Munger, S. C., Natarajan, A., Looger, L. L., Ohler, U., & Capel, B. (2013). Fine time course expression analysis identifies cascades of activation and repression and maps a putative regulator of mammalian sex determination. *PLoS Genetics*, *9*, e1003630.

Naillat, F., Prunskaite-Hyyrylainen, R., Pietila, I., Sormunen, R., Jokela, T., Shan, J., et al. (2010). Wnt4/5a signalling coordinates cell adhesion and entry into meiosis during presumptive ovarian follicle development. *Human Molecular Genetics*, *19*, 1539–1550.

Nef, S., Schaad, O., Stallings, N. R., Cederroth, C. R., Pitetti, J. L., Schaer, G., et al. (2005). Gene expression during sex determination reveals a robust female genetic program at the onset of ovarian development. *Developmental Biology*, *287*, 361–377.

Niederreither, K., & Dolle, P. (2008). Retinoic acid in development: Towards an integrated view. *Nature Reviews. Genetics*, *9*, 541–553.

Ohinata, Y., Ohta, H., Shigeta, M., Yamanaka, K., Wakayama, T., & Saitou, M. (2009). A signaling principle for the specification of the germ cell lineage in mice. *Cell*, *137*, 571–584.

Ohinata, Y., Payer, B., O'Carroll, D., Ancelin, K., Ono, Y., Sano, M., et al. (2005). Blimp1 is a critical determinant of the germ cell lineage in mice. *Nature*, *436*, 207–213.

Ohta, K., Lin, Y., Hogg, N., Yamamoto, M., & Yamazaki, Y. (2010). Direct effects of retinoic acid on entry of fetal male germ cells into meiosis in mice. *Biology of Reproduction*, *83*, 1056–1063.

Oosterhuis, J. W., & Looijenga, L. H. J. (2005). Testicular germ-cell tumours in a broader perspective. *Nature Reviews. Cancer*, *5*, 210–222.

Oulad-Abdelghani, M., Bouillet, P., Decimo, D., Gansmuller, A., Heyberger, S., Dolle, P., et al. (1996). Characterization of a premeiotic germ cell-specific cytoplasmic protein encoded by Stra8, a novel retinoic acid-responsive gene. *The Journal of Cell Biology*, *135*, 469–477.

Palmer, S. J., & Burgoyne, P. S. (1991). Insitu analysis of fetal, prepuberal and adult Xx[—] Xy chimeric mouse testes—Sertoli cells are predominantly, but not exclusively, Xy. *Development*, *112*, 265–268.

Parma, P., Radi, O., Vidal, V., Chaboissier, M. C., Dellambra, E., Valentini, S., et al. (2006). R-spondin1 is essential in sex determination, skin differentiation and malignancy. *Nature Genetics*, *38*, 1304–1309.

Pesce, M., Wang, X. Y., Wolgemuth, D. J., & Scholer, H. (1998). Differential expression of the Oct-4 transcription factor during mouse germ cell differentiation. *Mechanisms of Development*, *71*, 89–98.

Rastetter, R. H., Bernard, P., Palmer, J. S., Chassot, A. A., Chen, H., Western, P. S., et al. (2014). Marker genes identify three somatic cell types in the fetal mouse ovary. *Developmental Biology*, *394*, 242–252.

Resnick, J. L., Bixler, L. S., Cheng, L. Z., & Donovan, P. J. (1992). Long-term proliferation of mouse primordial germ-cells in culture. *Nature*, *359*, 550–551.

Reynolds, N., Collier, B., Bingham, V., Gray, N. K., & Cooke, H. J. (2007). Translation of the synaptonemal complex component Sycp3 is enhanced in vivo by the germ cell specific regulator Dazl. *RNA*, *13*, 974–981.

Reynolds, N., Collier, B., Maratou, K., Bingham, V., Speed, R. M., Taggart, M., et al. (2005). Dazl binds in vivo to specific transcripts and can regulate the pre-meiotic translation of Mvh in germ cells. *Human Molecular Genetics*, *14*, 3899–3909.

Rhinn, M., & Dolle, P. (2012). Retinoic acid signalling during development. *Development*, *139*, 843–858.

Saba, R., Kato, Y., & Saga, Y. (2014). NANOS2 promotes male germ cell development independent of meiosis suppression. *Developmental Biology*, *385*, 32–40.

Saba, R., Wu, Q., & Saga, Y. (2014). CYP26B1 promotes male germ cell differentiation by suppressing STRA8-dependent meiotic and STRA8-independent mitotic pathways. *Developmental Biology*, *389*, 173–181.

Saitou, M., Barton, S. C., & Surani, M. A. (2002). A molecular programme for the specification of germ cell fate in mice. *Nature*, *418*, 293–300.

Saitou, M., Payer, B., Lange, U. C., Erhardt, S., Barton, S. C., & Surani, M. A. (2003). Specification of germ cell fate in mice. *Philosophical Transactions of the Royal Society of London. Series B, Biological Sciences*, *358*, 1363–1370.

Saitou, M., & Yamaji, M. (2010). Germ cell specification in mice: Signaling, transcription regulation, and epigenetic consequences. *Reproduction, 139*, 931–942.

Schmahl, J., Kim, Y., Colvin, J. S., Ornitz, D. M., & Capel, B. (2004). Fgf9 induces proliferation and nuclear localization of FGFR2 in Sertoli precursors during male sex determination. *Development, 131*, 3627–3636.

Seisenberger, S., Andrews, S., Krueger, F., Arand, J., Walter, J., Santos, F., et al. (2012). The dynamics of genome-wide DNA methylation reprogramming in mouse primordial germ cells. *Molecular Cell, 48*, 849–862.

Sekido, R., Bar, I., Narvaez, V., Penny, G., & Lovell-Badge, R. (2004). SOX9 is up-regulated by the transient expression of SRY specifically in sertoli cell precursors. *Developmental Biology, 274*, 271–279.

Seligman, J., & Page, D. C. (1998). The Dazh gene is expressed in male and female embryonic gonads before germ cell sex differentiation. *Biochemical and Biophysical Research Communications, 245*, 878–882.

Skakkebaek, N. E., Berthelsen, J. G., Giwercman, A., & Muller, J. (1987). Carcinoma-in-situ of the testis: Possible origin from gonocytes and precursor of all types of germ cell tumours except spermatocytoma. *International Journal of Andrology, 10*, 19–28.

Snyder, E. M., Small, C., & Griswold, M. D. (2010). Retinoic acid availability drives the asynchronous initiation of spermatogonial differentiation in the mouse. *Biology of Reproduction, 83*, 783–790.

Sonne, S. B., Almstrup, K., Dalgaard, M., Juncker, A. S., Edsgard, D., Ruban, L., et al. (2009). Analysis of gene expression profiles of microdissected cell populations indicates that testicular carcinoma in situ is an arrested gonocyte. *Cancer Research, 69*, 5241–5250.

Souquet, B., Tourpin, S., Messiaen, S., Moison, D., Habert, R., & Livera, G. (2012). Nodal signaling regulates the entry into meiosis in fetal germ cells. *Endocrinology, 153*, 2466–2473.

Speed, R. M. (1982). Meiosis in the fetal mouse ovary.1. An analysis at the light-microscope level using surface-spreading. *Chromosoma, 85*, 427–437.

Spiller, C., Burnet, G., & Bowles, J. (2017). Regulation of fetal male germ cell development by members of the TGF beta superfamily. *Stem Cell Research, 24*, 174–180.

Spiller, C. M., Feng, C. W., Jackson, A., Gillis, A. J., Rolland, A. D., Looijenga, L. H., et al. (2012). Endogenous nodal signaling regulates germ cell potency during mammalian testis development. *Development, 139*, 4123–4132.

Spiller, C., Wilhelm, D., & Koopman, P. (2009). Cell cycle analysis of fetal germ cells during sex differentiation in mice. *Biology of the Cell, 101*, 587–598.

Spiller, C. M., Wilhelm, D., & Koopman, P. (2010). Retinoblastoma 1 protein modulates XY germ cell entry into G(1)/G(0) arrest during fetal development in mice. *Biology of Reproduction, 82*, 433–443.

Suzuki, A., Igarashi, K., Aisaki, K., Kanno, J., & Saga, Y. (2010). NANOS2 interacts with the CCR4-NOT deadenylation complex and leads to suppression of specific RNAs. *Proceedings of the National Academy of Sciences of the United States of America, 107*, 3594–3599.

Suzuki, A., Niimi, Y., Shinmyozu, K., Zhou, Z., Kiso, M., & Saga, Y. (2016). Dead end1 is an essential partner of NANOS2 for selective binding of target RNAs in male germ cell development. *EMBO Reports, 17*, 37–46.

Suzuki, A., & Saga, Y. (2008). Nanos2 suppresses meiosis and promotes male germ cell differentiation. *Genes & Development, 22*, 430–435.

Suzuki, H., Tsuda, M., Kiso, M., & Saga, Y. (2008). Nanos3 maintains the germ cell lineage in the mouse by suppressing both Bax-dependent and -independent apoptotic pathways. *Developmental Biology, 318*, 133–142.

Tam, P. P. L., & Snow, M. H. L. (1981). Proliferation and migration of primordial germ-cells during compensatory growth in mouse embryos. *Journal of Embryology and Experimental Morphology, 64*, 133–147.

Tedesco, M., Desimio, M. G., Klinger, F. G., De Felici, M., & Farini, D. (2013). Minimal concentrations of retinoic acid induce stimulation by retinoic acid 8 and promote entry into meiosis in isolated pregonadal and gonadal mouse primordial germ cells. *Biology of Reproduction*, *88*, 145.

Tedesco, M., La Sala, G., Barbagallo, F., De Felici, M., & Farini, D. (2009). STRA8 shuttles between nucleus and cytoplasm and displays transcriptional activity. *The Journal of Biological Chemistry*, *284*, 35781–35793.

Tomizuka, K., Horikoshi, K., Kitada, R., Sugawara, Y., Iba, Y., Kojima, A., et al. (2008). R-spondin1 plays an essential role in ovarian development through positively regulating Wnt-4 signaling. *Human Molecular Genetics*, *17*, 1278–1291.

Toyooka, Y., Tsunekawa, N., Takahashi, Y., Matsui, Y., Satoh, M., & Noce, T. (2000). Expression and intracellular localization of mouse vasa-homologue protein during germ cell development. *Mechanisms of Development*, *93*, 139–149.

Trautmann, E., Guerquin, M. J., Duquenne, C., Lahaye, J. B., Habert, R., & Livera, G. (2008). Retinoic acid prevents germ cell mitotic arrest in mouse fetal testes. *Cell Cycle*, *7*, 656–664.

Tsuda, M., Sasaoka, Y., Kiso, M., Abe, K., Haraguchi, S., Kobayashi, S., et al. (2003). Conserved role of nanos proteins in germ cell development. *Science*, *301*, 1239–1241.

Ulu, F., Kim, S. M., Yokoyama, T., & Yamazaki, Y. (2017). Dose-dependent functions of fibroblast growth factor 9 regulate the fate of murine XY primordial germ cells. *Biology of Reproduction*, *96*, 122–133.

Upadhyay, S., & Zamboni, L. (1982). Ectopic germ-cells—Natural model for the study of germ-cell sexual-differentiation. *Proceedings of the National Academy of Sciences of the United States of America*, *79*, 6584–6588.

Vainio, S., Heikkila, M., Kispert, A., Chin, N., & McMahon, A. P. (1999). Female development in mammals is regulated by Wnt-4 signalling. *Nature*, *397*, 405–409.

Vallier, L., Reynolds, D., & Pedersen, R. A. (2004). Nodal inhibits differentiation of human embryonic stem cells along the neuroectodermal default pathway. *Developmental Biology*, *275*, 403–421.

Vanorny, D. A., Prassasya, R. D., Chalpe, A. J., Kilen, S. M., & Mayo, K. E. (2014). Notch signaling regulates ovarian follicle formation and coordinates follucular growth. *Molecular Endocrinology*, *28*, 499–511.

Vidal, V. P. I., Chaboissier, M. C., de Rooij, D. G., & Schedl, A. (2001). Sox9 induces testis development in XX transgenic mice. *Nature Genetics*, *28*, 216–217.

Wang, N., & Tilly, J. L. (2010). Epigenetic status determines germ cell meiotic commitment in embryonic and postnatal mammalian gonads. *Cell Cycle*, *9*, 339–349.

Western, P., Maldonado-Saldivia, J., Van den Bergen, J., Hajkova, P., Saitou, M., Barton, S., et al. (2005). Analysis of Esg1 expression in pluripotent cells and the germline reveals similarities with Oct4 and N2 and differences between human pluripotent cell lines. *Stem Cells*, *23*, 1436–1442.

Western, P. S., Miles, D. C., van den Bergen, J. A., Burton, M., & Sinclair, A. H. (2008). Dynamic regulation of mitotic arrest in fetal male germ cells. *Stem Cells*, *26*, 339–347.

Western, P. S., van den Bergen, J. A., Miles, D. C., & Sinclair, A. H. (2010). Male fetal germ cell differentiation involves complex repression of the regulatory network controlling pluripotency. *The FASEB Journal*, *24*, 3026–3035.

Weston, A. D., Blumberg, B., & Underhill, T. M. (2003). Active repression by unliganded retinoid receptors in development: Less is sometimes more. *The Journal of Cell Biology*, *161*, 223–228.

Wilhelm, D., Martinson, F., Bradford, S., Wilson, M. J., Combes, A. N., Beverdam, A., et al. (2005). Sertoli cell differentiation is induced both cell-autonomously and through prostaglandin signaling during mammalian sex determination. *Developmental Biology*, *287*, 111–124.

Wu, Q., Fukuda, K., Kato, Y., Zhou, Z., Deng, C. X., & Saga, Y. (2016). Sexual fate change of XX germ cells caused by the deletion of SMAD4 and STRA8 independent of somatic sex reprogramming. *PLoS Biology*, *14*, e1002553.

Wu, Q., Kanata, K., Saba, R., Deng, C. X., Hamada, H., & Saga, Y. (2013). Nodal/activin signaling promotes male germ cell fate and suppresses female programming in somatic cells. *Development*, *140*, 291–300.

Yamaguchi, S., Kimura, H., Tada, M., Nakatsuji, N., & Tada, T. (2005). Nanog expression in mouse germ cell development. *Gene Expression Patterns*, *5*, 639–646.

Yao, H. H., DiNapoli, L., & Capel, B. (2003). Meiotic germ cells antagonize mesonephric cell migration and testis cord formation in mouse gonads. *Development*, *130*, 5895–5902.

Yao, H. H. C., Matzuk, M. M., Jorgez, C. J., Menke, D. B., Page, D. C., Swain, A., et al. (2004). Follistatin operates downstream of Wnt4 in mammalian ovary organogenesis. *Developmental Dynamics*, *230*, 210–215.

Yokobayashi, S., Liang, C. Y., Kohler, H., Nestorov, P., Liu, Z., Vidal, M., et al. (2013). PRC1 coordinates timing of sexual differentiation of female primordial germ cells. *Nature*, *495*, 236–240.

Zamboni, L. (1986). Meiosis as a sexual dimorphic character of germinal cell differentiation. *The Tokai Journal of Experimental and Clinical Medicine*, *11*, 377–390.

Zamboni, L., & Upadhyay, S. (1983). Germ-cell differentiation in mouse adrenal-glands. *The Journal of Experimental Zoology*, *228*, 173–193.

CHAPTER TEN

Rethinking sex determination of non-gonadal tissues

Arthur P. Arnold*
Department of Integrative Biology & Physiology, Laboratory of Neuroendocrinology of the Brain Research Institute, University of California, Los Angeles, CA, United States
*Corresponding author: e-mail address: arnold@ucla.edu

Contents

Abstract

Evolution of genetic mechanisms of sex determination led to two processes causing sex differences in somatic phenotypes: gonadal differentiation and sex chromosome dosage inequality. In species with heteromorphic sex chromosomes, the sex of the individual is established at the time of formation of the zygote, leading to inherent sex differences in expression of sex chromosome genes beginning as soon as the embryonic transcriptome is activated. The inequality of sex chromosome gene expression causes sexual differentiation of the gonads and of non-gonadal tissues. The difference in gonad type in turn causes lifelong differences in gonadal hormones, which interact with unequal effects of X and Y genes acting within cells. Separating the effects of gonadal hormones and sex chromosomes has been possible using mouse models in which gonadal determination is separated from the sex chromosomes, allowing comparison of XX and XY mice with the same type of gonad. Sex differences caused by gonadal hormones and sex chromosomes affect basic physiology and disease mechanisms in most or all tissues.

Current Topics in Developmental Biology, Volume 134
ISSN 0070-2153
https://doi.org/10.1016/bs.ctdb.2019.01.003

1. Introduction

The evolution of sex chromosomes in mammals involved two major events affecting separate classes of molecular pathways causing sex differences in tissues. One event was the emergence of sex-chromosome-encoded genetic mechanisms determining the differentiation of a testis or ovary from the early gonad, leading to sex differences caused by different levels of gonadal hormones in the two sexes. The second event was the evolution of heteromorphic sex chromosomes resulting in an inherent XX *vs.* XY difference in expressed doses of X and Y genes, which are incompletely compensated by X inactivation and other balancing mechanisms. In particular, the evolution of sexual disparity of X chromosome number led to major sex-specific adjustments in gene expression and in cross talk between sex chromosomes and autosomes. Theories of sexual differentiation have long emphasized the first, gonad-specifying attribute of sex chromosomes, and underestimated the second mechanism based on difference in X and Y gene dosage. Moreover, although concepts of sexual differentiation focus on the inherent sexual inequality, the sex chromosomes also balance the effects of each other, making XX and XY cells more similar than they might otherwise be. The relative importance of the sex-differentiating roles of sex chromosomes, and of the sex-balancing roles, is not yet well understood, but bears significantly on the causes of sex differences in disease.

2. The classic *vs.* revised theories of sex determination and sexual differentiation

A widely-held tenet of sex determination is that "in most mammals, sex determination is initiated by transient expression of *Sry* ...," the Y gene initiating testis differentiation (Miura, Tomita, & Kanai, 2018; Zhao et al., 2010). Here, I challenge the thinking behind this narrative, because (1) ubiquitous sex differences in cell phenotypes emerge well before the expression of *Sry* and (2) after differentiation of gonads, significant sex differences in cells are not the result of the this "sex determination" process. The dominant narrative is flawed because "sex determination" excludes some components of sexual differentiation. The error can be dismissed as merely semantic, but if so, it is important to define terms to optimize communication among investigators. In my view, however, the issue is deeper, so that the classic narrative has limited the conceptualization of the evolution and processes of sexual differentiation, and therefore has undermined proper design and interpretation of studies of sexual differentiation.

The dominant classic tenet appears to derive from a definition of sex based exclusively on type of gonad: females have ovaries and make eggs, and males have testes and make sperm. This definition has the advantage of being simple, long-standing, and can be consistently applied across all sexually reproducing plants and animals. In animals with genetic sex determination (*e.g.*, mammals and birds), however, sex determination has been taken over by genetic mechanisms common to all cells, so that the onset of sexual differentiation of the gonads no longer defines sex because it does not encompass all sexual differentiation processes.

The central question of this chapter is: "What are the biological forces that make females and males different?" The focus is on sex as a set of diverse forces, not as an identity or simply defined condition. The discussion is about biological forces, not social forces, even though social forces have profound sex-biasing effects on phenotypes, which are nearly impossible to disentangle from the effects of biological factors with which they are strongly correlated, especially for human phenotypes that are difficult to model in animals. To answer the central question, a definition of sex based on gonadal type is not fundamental enough to encompass all forces that differentiate the two sexes. We explore the more expansive idea that all biological sex differences stem originally from the inherent imbalance of sex chromosome factors, present in the zygote, and acting in the embryo and throughout life.

In this sex-chromosome-centric view, the expression of *Sry* is not the sex determinant, but the testis-determining factor, one of several genes on the Y chromosome that make males different from females in diverse tissues (Fig. 1) (Arnold, 2012, 2017a). *Sry* is arguably the most important of sex-differentiating factors, because its expression leads to lifelong secretion of testicular hormones, rather than ovarian hormones. The different effects of gonadal hormones account for many sex differences in non-gonadal phenotypes, including the external and internal genitalia that are fundamental to classifying humans as male or female. XX and XY non-gonadal cells also experience the inherently sex-biased effects of X or Y genes, stemming from the inequality of representation of these genes within the male and female genome, which we call "sex chromosome effects" causing sex differences. Importantly, many sex-biased phenotypes are impacted by both gonadal hormones and sex chromosome effects, although the interaction of these different molecular mechanisms to date is poorly studied.

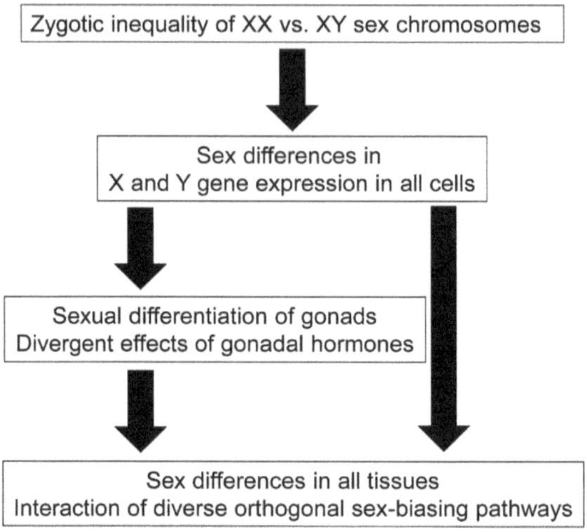

Fig. 1 The sex-chromosome-centric view of sexual differentiation. All phenotypic sex differences stem from the difference in number and type of sex chromosomes present in the zygote. In the undifferentiated gonadal ridge, *Sry* expression in XY cells induces differentiation of testes. Ovaries differentiate in XX animals in the absence of *Sry*, because of the expression of autosomal or X-linked genes. Differences in gonadal hormone levels contribute to sex differences in tissue function. In parallel, sex differences are also caused in all non-gonadal tissues by differences in effects of X and Y genes. Hormonal and sex chromosome factors interact to promote or reduce sex differences in phenotypes.

3. Evolution of sex chromosomes: Balance and inequality of X and Y

The inherent imbalance in XX *vs.* XY genomes can best be understood based on sex chromosome evolution. Early in the therian radiation, an autosomal mutation led to the emergence of the dominant testis-determining gene, *Sry*, which is expressed in the bipotential gonad to commit that tissue to a testicular fate (Sutton et al., 2011). The evolution of *Sry* led to several important consequences. (1) The genomic region near *Sry* was inherited only by gonadal males and it acquired genes that were good for males (Burgoyne & Mitchell, 2007; Rice, 1992). The near-*Sry* region lost ability to recombine with its partner proto-X chromosome (Bachtrog, 2013; Cortez et al., 2014; Hughes & Page, 2015). (2) Recombination of the X and Y chromosome was further reduced because of subsequent major inversions on the Y chromosome (Lahn & Page, 1999), leading to expansion of the male-only region to encompass nearly the entire Y chromosome, except for a small pseudoautosomal region (PAR) shared between the X and Y. The asexual inheritance (father to son, with no recombination) led to loss of most Y genes (Charlesworth, 1996; Graves, 2016; Rice, 1994; Vallender & Lahn, 2004). (3) The progressive loss of Y genes left X genes expressed at a lower dose in XY than XX cells, causing a sex difference in stoichiometric balance of X and autosomal genes that cooperate in gene networks (Veitia, Veyrunes, Bottani, & Birchler, 2015). For some X genes, the reduced balance with autosomal genes was critical, and favored evolution of dosage compensation mechanisms to adjust the X to autosomal expression ratios to within an acceptable range in both sexes. In XX cells, one X chromosome is transcriptionally silenced, so that both XX and XY cells express genes mostly from one X chromosome (Disteche, 2016). In addition, some evidence suggests that the expression of many X genes may be upregulated in both XX and XY cells to match the general level of expression of autosomal genes, with which they interact in gene networks (Disteche, 2016; Nguyen & Disteche, 2006; Sangrithi et al., 2017).

These dosage compensation mechanisms, X-inactivation and upregulation of X gene expression to better match autosomal expression, have the general consequence of making XX and XY cells phenotypically more similar to each other because they offset the effect of the difference in X chromosomal dose. The compensation is incomplete, however, so that XX and XY genomes have inherent functional differences caused by difference in X and Y gene dose. Some X genes escape inactivation and are expressed from all X chromosomes, and therefore at constitutively higher levels in XX cells than XY cells in many tissues and developmental stages (Carrel, Cottle, Goglin, & Willard, 1999; Tukiainen et al., 2017). In humans, about 23% of non-PAR X genes are expressed higher in XX than XY tissues (Fig. 2). Genes appear be left out of, or protected from, the inactivation mechanism either because they are recently added to the X chromosome (Graves, 2016) or because they occupy positions in gene networks that are insensitive to dose, where one copy is as effective as two (Naqvi, Bellott, Lin, & Page, 2018).

The dosage sensitivity of genes was apparently a major factor influencing evolution of sex chromosome gene content. Dosage sensitivity determined, in part, which Y genes were lost or retained during evolution, and affected which X genes were subject to inactivation (Bellott et al., 2014; Cortez et al., 2014; Naqvi et al., 2018; Veitia et al., 2015). One example is X-Y gene pairs, which have similar DNA sequence to each other, and evolved from a common autosomal ancestor gene. X-Y gene pairs are involved in basic cellular functions such as transcription factors and regulators of

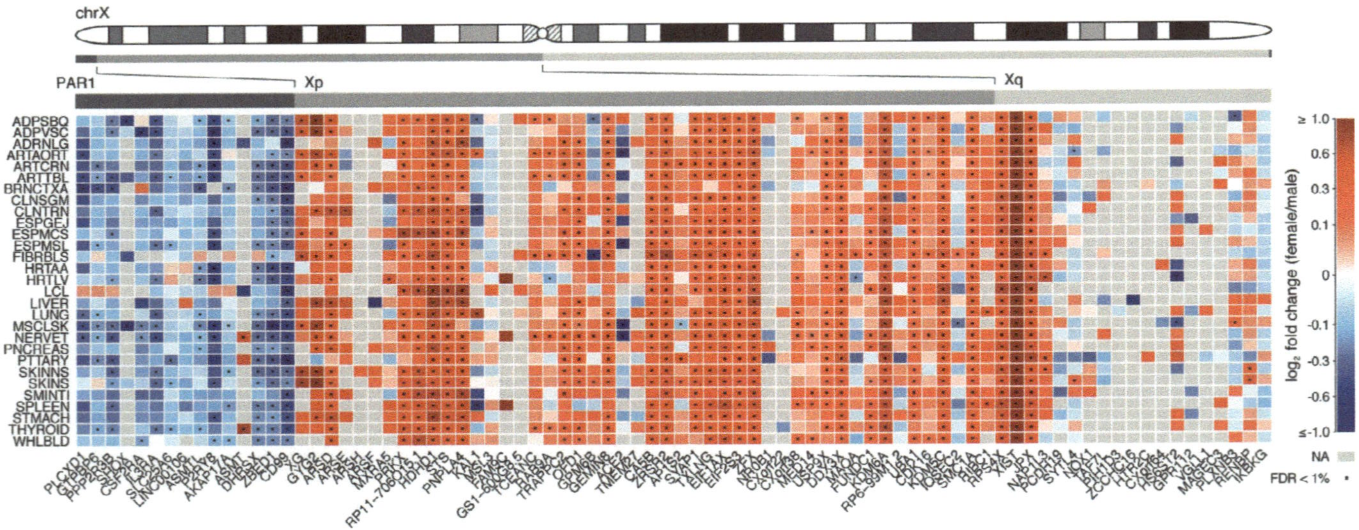

Fig. 2 Inherent bias in human X chromosome gene expression across tissues. Analysis of over 5500 transcriptomes from 29 human tissues from GTEx consortium data shows the level of sex bias in expression of X genes reported to escape inactivation. Rows indicate different tissues, and columns represent X genes reported to escape X inactivation, mapped by chromosomal position. Red indicates expression higher in females and blue indicates expression higher in males. Gray indicates expression too low to be analyzed. In the non-pseudoautosomal (non-PAR) region, about 23% of X genes overall showed higher expression in females, and in PAR1 region most genes were expressed higher in males. This general pattern can be attributed to escape from X inactivation in the non-PAR region in XX cells, and to some inactivation of PAR genes in XX cells but not XY cells. However, other sex-biasing factors (gonadal hormones, parental imprint) could contribute to sex differences as well, in either direction, and may account for colors that deviate from the pattern of adjacent genes (*e.g.*, red squares in the PAR, and blue squares in the non-PAR). *Reprinted from Tukiainen, T., Villani. A.C., Yen. A., Rivas. M.A., Marshall. J.L., Satija. R., et al. (2017) Landscape of X chromosome inactivation across human tissues.* Nature, 550, *244–248 under terms of Creative Commons Attribution 4.0 International license.*

RNA splicing and translation, and are thus expressed widely across tissues (Bellott et al., 2014). The Y partner genes have survived on the Y chromosome despite hundreds of millions of years of degrading influences that eliminated most other Y genes, probably because of the particularly high dosage sensitivity of the X-Y pair that made deletion of the Y gene lethal or highly disadvantageous. The X and Y genes often share high sequence similarity, attesting to their common origin and shared functions. The X partner gene typically escapes X inactivation and is expressed higher in XX cells than XY cells (Disteche, 2016). The Y partner gene likely compensates for the lower expression of the X partner gene in XY cells, if two X copies in XX cells have similar function compared to that of one X copy plus one Y copy in XY cells.

Paradoxically, although the evolutionary survival of the Y partner genes of X-Y gene pairs is explained by a balance of effects of the X and Y partners, X-Y gene pairs are considered a prime source of sexual inequality of XX and XY cells because their balance is variable across tissues, ages, or environmental conditions such as disease. For example, the Y genes are typically expressed at lower levels than the X partner in many tissues (Deng, Berletch, Nguyen, & Disteche, 2014). Moreover, the partner genes sometimes show different patterns of expression across tissues (Xu, Deng, Watkins, & Disteche, 2008). These inequalities can be explained if the balance of effects of the two genes might be critical in only some tissues or at specific developmental periods. Elsewhere, the Y gene was free to evolve male-specific functions, different from that of the X partner. An interesting case is the X-Y gene pair *Utx-Uty*, which is found in sex chromosomes of many mammalian species. *Utx* (a.k.a. *Kdm6a*) is a histone demethylase. Knock out of *Utx* is lethal in XX mouse embryos, but the presence of a functional copy of *Uty* prevents some of the mortality, illustrating the overlapping function of *Uty* with *Utx* (Shpargel, Sengoku, Yokoyama, & Magnuson, 2012). Nevertheless, the histone demethylase activity of *Utx* is not found in *Uty*, illustrating divergence of function of the two orthologous genes. In the brain, the pattern of expression of *Utx* and *Uty* is somewhat different (Xu et al., 2008). Thus, functional differences in XX and XY cells likely result from different levels of expression of both *Utx* and *Uty*.

Another illustrative example of X-Y gene pairs involves mouse *Sly* and its X partners, *Slx* and *Slxl1*. These genes participate in meiotic drive, an example of intergenomic conflict. Factors (drivers) arise on the Y chromosome that promote its inheritance in preference to the X chromosome. That process sets up counteracting pressures to select for suppressor sequences on the X chromosomes that favor its inheritance. Drivers and suppressors were duplicated on each chromosome, in successive waves that offset and overcame each other in evolutionary time. In mice, 95% of the Y chromosome represents massive duplications of only three gene families, and the X chromosome also contains extensive duplicated regions (Soh et al., 2014). The duplicated X-Y gene pair *Sly-Slx* participates in meiotic drive, because deletions of *Sly* undermine sperm viability, and deletions of *Slx* rescue these sperm differentiation defects. *Sly* deficiency also corrects the enhanced production of male offspring produced by *Slx* deficiency (Cocquet et al., 2012, 2009).

Further evidence supports the idea that the balance of X and Y gene effects is tenuous and varies across tissues, genotypes, age, and evolutionary time. XY male mice, of many strains, have greater body weight than XX female mice. In C57BL/6 mice, the number of X chromosomes causes sex differences in body weight and fat, but the Y chromosome has little effect (Chen et al., 2012) (Fig. 3). In MF1 strain mice, however, XO mice weigh less than XX or XY mice, so that the addition of a second sex chromosome of either type to the XO genotype has the same effect (Chen, McClusky, Itoh, Reue, & Arnold, 2013). In MF1 mice, the X and Y chromosomes have balanced effects on body weight and fat, but the two chromosomes have unequal effects in C57BL/6, illustrating that the balance of the two chromosomes can come and go depending on factors that interact with X and Y genes.

The tension between the gonad-centric and sex-chromosome-centric views of sexual determination uncovers a fundamental difference in thinking about evolutionary forces that lead to sex differences in phenotype. Within the gonad-centric view, sex differences are related predominantly to the evolution of sexual reproduction. Animals with testes or ovaries have evolved functional differences throughout the body that support the two reproductive roles. The evolution of sex differences in internal and external genitalia are obvious examples, because the different anatomical structures support the different roles. Sex differences in other tissues, such as the brain, have also long been rationalized as the result of separate selection pressures on males and females to make the brain support sex-biased reproductive functions such as ovulation, sex behavior, and aggression. In this view, sex differences are often seen as adaptive, rather than side-products of other evolutionary processes. The sex-chromosome-centric view emphasizes a different set of selection pressures. The inherent sexual inequality in genomic dose and effect of X and Y genes is the result of evolutionary factors

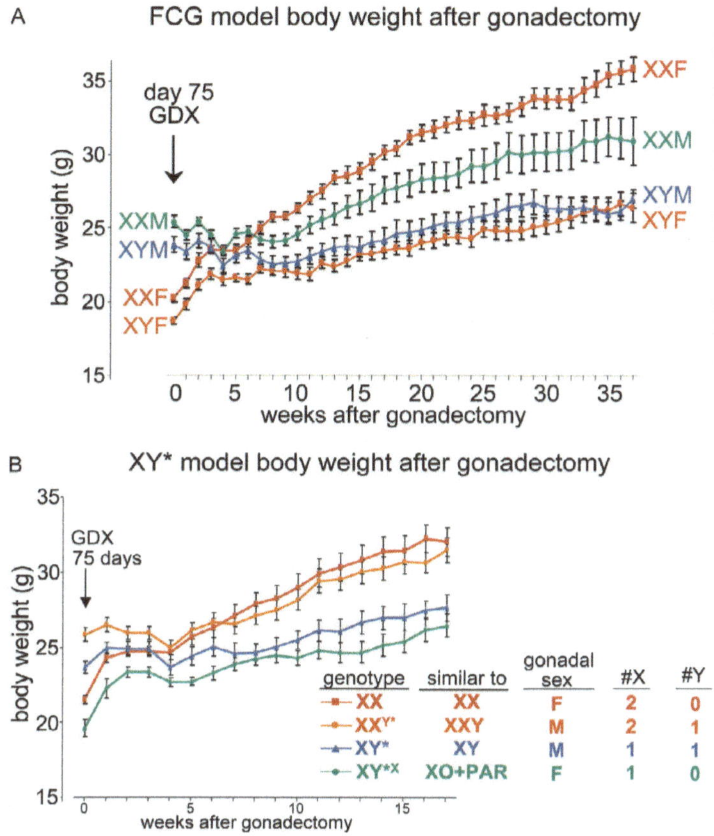

Fig. 3 The number of X chromosomes influences body weight in FCG and XY* mice. (A) In FCG mice, gonadal males at postnatal day 75 (week 0) weighed about 25% more than gonadal females. When gonads were removed on that day, the sex differences disappeared by 4 weeks after gonadectomy (GDX). After that, XX mice slowly gained much more body weight than XY mice, irrespective of previous gonadal type. (B) In the XY* model, the same experiment replicates the larger body size of two groups with testes, relative to two groups with ovaries. After gonadectomy, the groups with two X chromosomes gained more weight than those with one X chromosome, illustrating that the body weight is affected by the number of X chromosomes. M = gonadal male, F = gonadal female, PAR = pseudoautosomal region. *Reprinted from Chen, X., McClusky, R., Chen, J., Beaven, S.W., Tontonoz, P., Arnold, A.P., et al. (2012) The number of X chromosomes causes sex differences in adiposity in mice.* PLoS Genetics, 8, *e1002709 under terms of the Creative Commons Attribution License.*

generally unrelated to sexual function. As discussed above, sex differences caused by imbalance of X and Y genes occur because of loss of recombination of the X and Y chromosomes, loss of genomic regions on the Y chromosome, dosage compensation on the X and Y chromosomes, intergenomic conflict, *etc.* Thus, gonadal females get stuck with having two X chromosomes, and gonadal males with one X and one Y, for evolutionary reasons that do not primarily enhance reproduction. Moreover, the inequality of chromosome size affects many genes lacking a direct influence on reproduction. Of course, the different effects of X and Y genes in the two sexes must be compatible with their reproductive roles, so that both types of evolutionary pressure operate simultaneously within the constraints of tissue requirements related to sex roles. The evolution of different reproductive roles of cells and tissues, controlled dominantly by hormonal effects of gonadal secretions, occurs on a background of inherent sex differences in the genome.

4. Categories of X and Y genes causing phenotypic differences between XX and XY cells

It is useful to list all of the types of sexual inequality that are currently known or suspected in XX *vs.* XY cells because of the inherent imbalance of representation of sex chromosome genes (Arnold, 2017a). (1) The male-specific expression of Y genes causes XY cells to differ from XX cells (Arnold, 2017b; Case & Teuscher, 2015). (2) The constitutive difference in number of X chromosomes leads to four classes of X genes that underlie sex differences in phenotype. (2A) X genes escaping X inactivation are expressed higher in XX than XY cells (Carrel et al., 1999; Tukiainen et al., 2017). For example, sex differences in trophoblast expression of the putative X escapee gene *Ogt* are reported to cause sex differences in resilience to prenatal insults (Nugent, O'Donnell, Epperson, & Bale, 2018). (2B) Genes in the pseudoautosomal regions (PAR) of the sex chromosomes are often expressed at higher levels in XY cells, relative to XX cells, because some of them are subject to X inactivation in XX cells (Tukiainen et al., 2017) (Fig. 2). (2C) X genes that experience a parental imprint can be expressed at a higher or lower level in XX *vs.* XY cells because a paternal imprint affects only XX cells (Arnold, 2017a). (2D) The difference in epigenetic regulation of the X and Y chromosomes could have ramifications for epigenetic status of the autosomes. In *Drosophila*, for example, the large heterochromatic Y chromosome has indirect epigenetic effects on autosomal gene expression, which are not mediated by expression of Y genes (Silkaitis & Lemos, 2014). In mammals, it is conceivable that *Xist*, which initiates X inactivation *in cis* in XX cells, also has *trans* effects on autosomes. For example, silencing of *Xist* in adult XX mice leads to hematologic cancer

(Yildirim et al., 2013), although this could be the result of reactivation of the inactive X chromosome instead of direct effects of *Xist* on autosomes. Moreover, a large heterochromatic X chromosome, present only in XX cells, is postulated to sequester heterochromatizing factors and reduce their availability to regulate autosomal heterochromatin, affecting autosomal gene expression (Wijchers & Festenstein, 2011).

The molecular differences between XX and XY cells, just listed, operate during ontogeny of all individuals, and thus can often be measured in inbred laboratory strains of animals, especially mice. Other forces cause sex differences in populations of humans and other animals, because of natural genetic heterogeneity not modeled in inbred strains (reviewed by Arnold, 2017a). These factors cause average differences between groups of males and females. (1) XY males have hemizygous exposure of X alleles, so that they express X allelic variations more prominently than XX females (Midgeon, 2007). XY individuals with X-linked lethal alleles are removed from the population, and shift the mean phenotype of males *vs.* females. (2) XX individuals are mosaic for X alleles and X gene imprints, because the parental origin of the active X varies among XX cells but not among XY cells. The mosaicism itself can have a protective effect for XX but not XY tissues. For example, in XX mice with a null mutation in one allele of the X-linked *Hccs* gene, cells expressing the null allele are selectively removed from the heart during embryonic development (Drenckhahn et al., 2008). (3) Some autosomal alleles are likely to be better adapted to one sex than the other, and thus will be found in greater frequency in one sex, creating an average genetic difference in autosomes of males and females. (4) The different modes of inheritance of the mitochondrial and nuclear genomes lead to sex-biasing effects. Because the mitochondrial genome is inherited in the female lineage, mitochondrial alleles might arise that are beneficial to females but harmful to males. Such a mismatch of mitochondrial genomes of the mother and nuclear genomes of her sons can have differential harmful effects in males relative to females (Camus, Clancy, & Dowling, 2012; Frank, 2012; Innocenti, Morrow, & Dowling, 2011).

5. Mouse models and methods that separate effects of sex chromosomes *vs.* gonadal hormones

For well over a century, investigators have been able manipulate gonadal hormone levels effectively to test their roles in causing sex differences in physiological and behavioral traits. The classic method is to reduce hormone levels by gonadectomy, and increase them by treating with hormones. Hormone effects are also manipulated by pharmacologic or genetic manipulations of steroid receptors or synthetic enzymes. In contrast, manipulating the number of sex chromosomes has been much more difficult. Thus, the first reports of sex chromosome effects causing sex difference relied on naturalistic observations rather than experimental manipulation. In the marsupial tammar wallaby, sex differences in the differentiation of the pouch or scrotum precede the differentiation of gonads (Renfree & Short, 1988). In mice and other mammalian embryos, the size of XX and XY embryos differs long before differentiation of gonads (Burgoyne et al., 1995). The embryonic mouse transcriptome differs in XX and XY cells as soon as the embryonic transcriptome is activated at the eight-cell stage or earlier (Bramble et al., 2016; Lowe, Gemma, Rakyan, & Holland, 2015; Werner et al., 2017), and sex differences in the transcriptome are found at later embryonic stages prior to gonadal differentiation (Dewing, Shi, Horvath, & Vilain, 2003; Nef et al., 2005). In other cases, sex differences in phenotype are found in adult intersex animals such as avian gynandromorphs in which the sex difference is attributable to sex chromosome effects rather than hormonal effects (Agate et al., 2003; Witschi, 1939; Zhao et al., 2010).

The investigation of sex chromosome effects has been advanced by the production of mouse models in which whole mice with the same gonadal sex are engineered to have different complements of sex chromosomes (Arnold, 2014; Burgoyne & Arnold, 2016; Cox, Bonthuis, & Rissman, 2014). In these models, the hormones are held relatively "constant" across groups, while sex chromosome complement is varied. In the Four Core Genotypes (FCG) model, the *Sry* gene is deleted from the Y chromosome, producing the Y^- chromosome. XY^- gonadal females are compared to XX gonadal females. In addition, a *Sry* transgene is introduced onto chromosome 3, so that testis-determination is specified independently of the sex chromosome complement. In the FCG model, XY^-(*Sry*+) fathers mate with XX females to produce four types of offspring: XX and XY gonadal males, and XX and XY gonadal females (Fig. 3). The model separates sex-biasing factors into two groups: those that differ in XX *vs.* XY mice (comparing XX and XY mice with the same type of gonad), and those that differ in gonadal males and females (comparing gonadal males and females with the same sex chromosome complement) (De Vries et al., 2002). The model is widely applicable to assess the causes of sex differences in virtually any somatic trait. As expected, many phenotypes differ in gonadal males and females, confirming the long-standing concept that gonadal hormones cause many or most sex differences in somatic phenotypes throughout the body. However, an increasing number of XX *vs.* XY differences have also been discovered (Arnold, Cassis, Eghbali, Reue, & Sandberg, 2017; Arnold & Chen, 2009; Arnold, Chen, Link, Itoh, & Reue, 2013; Arnold et al., 2016; Cox et al., 2014).

The discovery of a phenotypic difference in XX *vs.* XY FCG mice is the first step in discovering the mechanisms of action of X or Y genes causing sex differences in phenotype. A second mouse model, the XY* model, is useful for figuring out whether the XX *vs.* XY phenotypic difference is attributable to X or Y genes. The XY* model produces genotypes very similar to XO, XX, XY, and XXY (Burgoyne & Arnold, 2016) (Fig. 3). The XY* model allows measurement of the effects of X chromosome number (XO *vs.* XX gonadal females, and XY *vs.* XXY gonadal males). It also allows measurement of effects of presence/absence of the Y chromosome: XO *vs.* XY, and XX *vs.* XXY. Because *Sry* is present on the Y chromosome in this model, the comparisons of mice with and without the Y chromosome are confounded by differences in the type of gonad and levels of gonadal hormones. However, if such gonadal effects are found not to cause the sex difference in phenotype in FCG mice, the effects of the Y chromosome in the XY* model can be attributed to non-hormonal (direct) effects of Y genes. Using the XY* model, sex differences have been attributed more often to the effects of X genes, but Y chromosome effects have been found as well (Arnold et al., 2016; Umar et al., 2018).

6. Sex differences caused by sex chromosome effects

The FCG and XY* mouse models have been used to discover sex differences in various mouse phenotypes that are caused by sex chromosome complement, as previously reviewed (Arnold, 2014, 2017ab; Cox et al., 2014). A few examples are selected for discussion here.

In the aggregate, the studies indicate that XX *vs.* XY differences participate in causing sex differences in a wide variety of tissues and traits, and affect multiple mouse models of disease. The XX *vs.* XY differences affect autoimmune disease and susceptibility to infection, brain and behavior, metabolic traits including adiposity and fat metabolism, and numerous cardiovascular traits. Although sex chromosome effects are now widely recognized, numerous questions remain. For most traits, it is not known which X or Y genes (or non-genic regions of the sex chromosomes) account for the sex chromosome effect. This question is particularly of translational importance. The identification of specific X or Y genes causing a sex chromosome effect will greatly facilitate research about the role of such genes in sex differences in human disease. For example, one can study effects of variation of that gene in human populations, and cellular pathways influenced by the gene in human cell systems. The FCG and XY* models can be used to discover effects of X or Y genes contributing to sex differences in mouse models, after which manipulation of specific candidate genes leads to identification of the genes responsible (reviewed by Burgoyne & Arnold, 2016).

A second major outstanding question is how the sex differences caused by sex chromosome complement interact with sex differences caused by gonadal hormonal effects. Typically, both of these factors contribute to sex differences of the same phenotypes in mice. In what molecular pathways do diverse sex-biasing factors interact to affect the same disease process?

6.1 Independent sex-biasing factors interact to cause sex differences in physiology and disease

6.1.1 Metabolism and body weight

The greater body weight of male mice, relative to females, is reduced by gonadectomy of adult mice, indicating that gonadal hormones largely are responsible for this difference. However, after gonadectomy of FCG mice, the body weight of XX mice increases slowly over weeks until it is much greater than that of XY mice, irrespective of their type of gonad (Fig. 3A) (Chen et al., 2012). The sex chromosome effect is eventually as large as the effect of gonadal hormones in young mice at the outset of the experiment. As discussed above, use of the XY* model allows us to test mice with different numbers of X or Y chromosomes. These studies show that the XX-XY difference is caused by X genes, not Y genes (Fig. 3B). The greater body weight is mostly the result of an X chromosome effect on the amount of body fat. In mice eating a high fat diet, sex chromosomes cause sex differences in the level of plasma cholesterol, and accumulation of liver triglycerides (Chen et al., 2012; Link et al., 2015). In the case of body weight, the gonadal hormonal effect counteracts the sex chromosome effect: male hormones are associated with greater body weight, but male sex chromosomes (XY) with lower body weight (Fig. 3). To date, little is known about the molecular site of interaction of the two effects, and whether they influence the same molecular pathways or distinct pathways that both impinge on the same emergent phenotype.

6.1.2 Sex differences in volumes of brain regions

Concepts of sexual differentiation in brain and behavior have stemmed disproportionately from study of a few phenotypes with large sex differences. In animals, a few brain regions are three to six times larger in one sex than the other (Arnold & Gorski, 1984; Forger, 2009), where it has been easiest to measure the effects of variables that cause the sex differences. These large sex differences are related to sexual function, such as sexual behavior, ovulation, aggression, and courtship. The principles learned for these brain regions have been thought to represent general principles by which sex variables (*e.g.*, gonadal hormones) influence brain development and function (Breedlove, Cooke, & Jordan, 1999). The advent of high-resolution brain-wide structural MRI has allowed investigators to ask if these principles also apply more generally across all brain regions. Importantly, study of FCG mice shows an unexpected incidence of sex chromosome effects not previously found in the most sexually dimorphic brain regions (Corre et al., 2016). Of 62 brain regions examined, 30 were significantly different among FCG groups, with 20 showing effects of gonadal hormones and 14 effects of sex chromosomes (Fig. 4). The brain regions affected by gonadal hormones or sex chromosomes tended to be distinct. When gonadal hormones were removed before puberty and brain anatomy was measured in adulthood, most differences between XX and XY brain regions persisted, although on average they were modestly attenuated (Vousden et al., 2018). In some brain regions, the effects of gonadal hormones and sex chromosomes were synergistic, such that male-female hormonal effects where in the same direction as XY-XX sex chromosome effects. In other regions, the two sets of variables were antagonistic or compensatory, such that removal of hormones caused an increase in effect of sex chromosome complement. These complex interactions of hormonal and sex chromosome factors illustrate the orthogonal actions of the two sets of factors in a brain-region dependent fashion (Vousden et al., 2018).

6.1.3 Sex chromosome regulation of brain sensitivity to gonadal hormones

The potential interaction of sex chromosome and sex hormone effects is also implied by the recent discovery in FCG mice that sex chromosomes influence the metabolism of gonadal hormones in the brain, and regulate the sensitivity of specific brain regions to sex steroid hormones. The amygdala has long been recognized as a site of action of gonadal steroids, to cause sex differences in function. In the anterior amygdala of E16 mouse embryos,

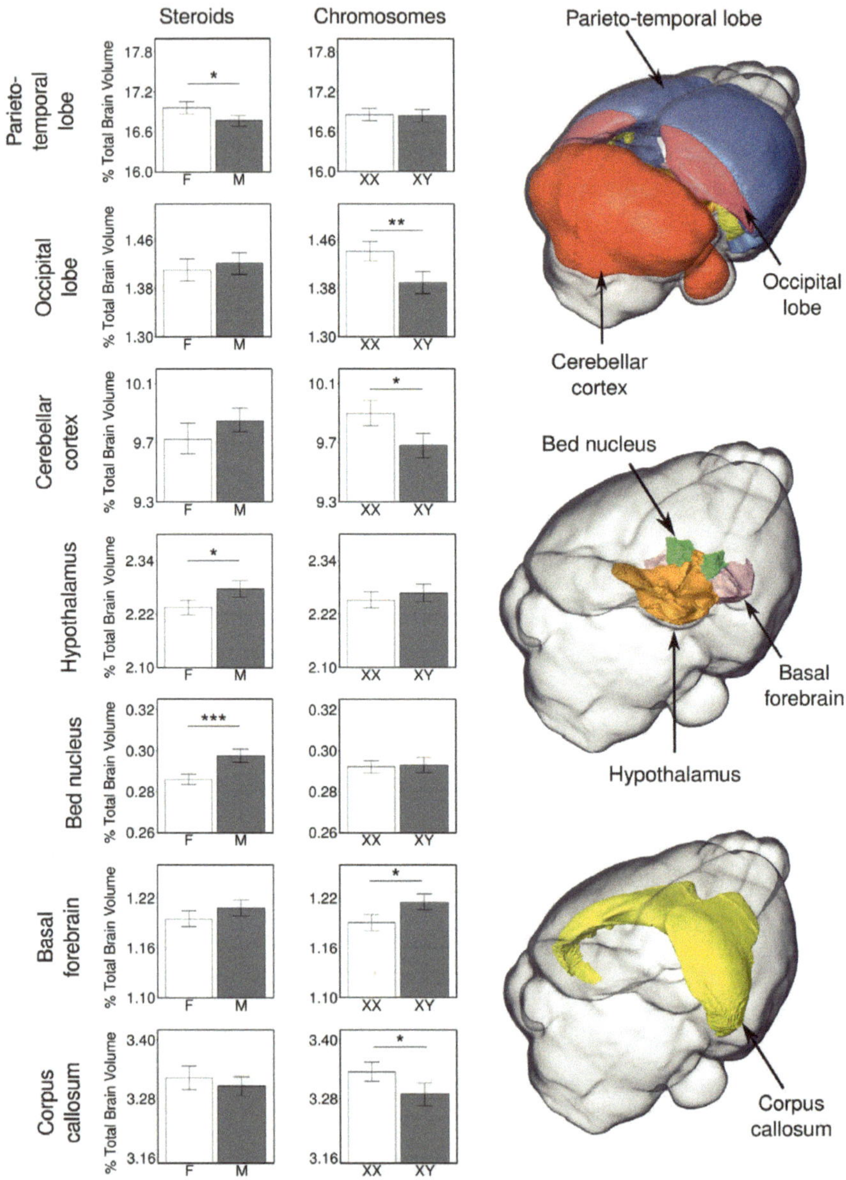

Fig. 4 Sex chromosome and hormonal effects on brain volumes from FCG mice. Whole-brain MRI was used to measure volumetric variation in gonad-intact FCG mice. Among seven brain regions illustrated, some showed significant effects of gonadal hormones (parieto-temporal lobe of the cerebral cortex, the hypothalamus, and the bed nucleus of the stria terminalis). Others showed effects of sex chromosomes (basal forebrain, occipital lobe of the cerebral cortex, cerebellar cortex, corpus callosum). Asterisks indicate changes were significant at false discovery rate of 5% or less. Error bars represent 95% confidence intervals. *Reprinted by permission of Springer Nature from Corre, C., Friedel, M., Vousden, D.A., Metcalf, A., Spring, S., Qiu, L.R., et al. (2016) Separate effects of sex hormones and sex chromosomes on brain structure and function revealed by high-resolution magnetic resonance imaging and spatial navigation assessment of the Four Core Genotype mouse model.* Brain Structure & Function, 221, *997–1016.*

XY mice, relative to XX, have greater expression of aromatase, the enzyme converting androgens to estrogens, irrespective of gonadal type (Cisternas et al., 2015). The same sex chromosome effect on aromatase is found in cultures of embryonic amygdala neurons (Cisternas, Cabrera Zapata, Arevalo, Garcia-Segura, & Cambiasso, 2017; Cisternas et al., 2015). *In vitro*, estrogen receptor beta (*Esr2*) is expressed higher in XY than XX cells (Cisternas et al., 2017). Treatment of neurons with estradiol or dihydrotestosterone increases expression of aromatase and *Esr2* in XX cells only, and abolishes the difference in *Esr2* expression caused by sex chromosomes. *Esr1* expression *in vitro* is also regulated by both sex chromosome complement and by gonadal hormones. This example is particularly interesting because it demonstrates a regulation by sex chromosomes both of local synthesis of estradiol, and of the sensitivity to estradiol, to regulate amygdalar phenotypes. The regulation of aromatase by sex chromosomes could shift the balance of action of estrogens and their androgenic precursors, contributing to hormonally driven sex differences in amygdala development. It will be exciting to discover the molecular pathways by which X or Y genes regulate steroid sensitivity of this and other brain regions (Cambiasso et al., 2017; Cisternas, Garcia-Segura, & Cambiasso, 2018).

6.1.4 Cardiovascular and pulmonary disease

Women and men differ in the susceptibility and progression of numerous cardiovascular diseases, including atherosclerosis, aneurysms, ischemia/reperfusion injury, and systemic and pulmonary hypertension (Arnold et al., 2017). In animal models, the sex differences appear to be caused by a combination of sex chromosome and hormonal effects. These diseases have quite diverse molecular mechanisms, indicating that a wide range of cellular processes are influenced by sex chromosome complement. In models of abdominal aortic aneurysms, which affect men much more than women, testosterone exacerbates disease, and XY sex chromosome complement is worse than XX (Alsiraj et al., 2016). Thus, both male hormones and male sex chromosomes appear to promote disease. In a model of ischemia/reperfusion injury, estrogens are generally protective, but XX sex chromosome complement is worse that XY, indicating compensatory or counteracting effects of estrogens and sex chromosomes (Li et al., 2014). Hypertension affects men more than women. Ovarian hormones appear to be protective, but XX mice are more affected in mouse models than XY mice, again suggesting counteracting hormonal/sex chromosomal effects (Ji et al., 2010). The sex differences in humans can often be complex, depending on changes in age, hormonal status (*e.g.*, pre- *vs.* postmenopausal), which might imply an age-related change in balance of hormonal and sex chromosome effects.

Pulmonary arterial hypertension affects women more than men. In a mouse model of experimental pulmonary hypertension using FCG mice, XY mice were less affected by hypoxia than XX mice. Further study of XY* mice indicated that the sex chromosome effects were caused by the presence of the Y chromosome, rather than the number of X chromosomes. Four Y genes, members X-Y gene pairs, are expressed in mouse lung, making them interesting candidates for the protective effect of the Y chromosome in this disease model (Umar et al., 2018).

6.1.5 Immune and autoimmune function

In Experimental Autoimmune Encephalomyelitis (EAE), a mouse model of multiple sclerosis (MS), different sex chromosome factors interact to produce sex differences in the disease. Mice are immunized with autoantigens that trigger an immune response resembling MS. Disease progression and neuropathology is worse in XX than XY mice (Palaszynski et al., 2005; Smith-Bouvier et al., 2008). However, when bone marrow chimeric mice are produced in which the sex chromosome complement (XX *vs.* XY) in the immune system is independent of that in the brain, XY brains are found to respond worse in EAE than XX brains (Du et al., 2014). The different direction of the sex chromosome effects may be related to the complex sex difference in humans, in which incidence of MS is greater in women than men, but disease progression is worse in men than women. Conceivably different sex chromosome genes influence the immune system and the response of brain tissue to the immunological attack. The sex chromosome effect is found also in models of other autoimmune diseases such as lupus erythematosus (Moore, Patel, Hannsun, Yang, & Tiwari-Woodruff, 2013; Smith-Bouvier et al., 2008). Sex chromosome effects also affect immunological response to some viruses (Robinson et al., 2011).

6.1.6 Interaction of sex chromosome effects and age

An experimental model of stroke involves temporary occlusion of the middle cerebral artery followed by reperfusion of the brain. Estrogens are known to protect from stroke damage in this model, causing a sex difference. In young FCG mice, no sex chromosome effect was discovered (Manwani et al., 2015). In aged mice, however, sex chromosome complement is the primary contributor to sex differences in this model. XX mice had greater lesions than XY mice, irrespective of gonadal sex, as well as greater activation of microglia and secretion of pro-inflammatory cytokines (McCullough et al., 2016). Among the aging processes that could contribute to an increase in sex chromosome effects is an age-related increase in escape from X inactivation (Disteche, 2016).

6.2 Sex differences in the germline

Germ cells have long been known to show sex differences in behavior, including onset of meiosis during embryonic life. Although some evidence pointed to the possibility that XX and XY cells are sexually differentiated before they migrate into the undifferentiated gonadal ridge (Chuva de Sousa Lopes et al., 2008), most evidence supported the idea that germline sex differences were imposed by the testicular *vs.* ovarian environment in which the cells differentiate (Spiller, Koopman, & Bowles, 2017). Thus, the sex differences were perceived to be downstream of the processes causing differentiation of gonadal somatic cells, induced by *Sry* in males *vs.* ovary-determining genes in females. Recently studies demonstrate, however, that the phenotypic differentiation of XX and XY germ cells depends also on sex chromosome complement. During the specification of germ cells in the epiblast and migration into the gonads, the active X chromosome of XX and XY germ cells is upregulated to be on par with autosomes. At early stages of differentiation of the gonads (E12.5–14.5), however, XX cells show X to autosome expression ratios above one, and XY cells have ratios below one [(Sangrithi et al., 2017); see also (Chuva de Sousa Lopes et al., 2008)]. This sex difference depends on the number of X chromosomes, not gonadal sex, and is another illustration that sex differences in development are not only dependent on gonadal type.

6.3 Y chromosome-mediated sex differences in physiology and disease

One method for testing the involvement of the Y chromosome in traits is the use of Y chromosome consomic mice, in which Y chromosomes from different mouse strains are crossed onto the same genetic background. Phenotypic differences among strains reveal the effects of variation of Y genes. The use of Y consomic strains has been expanded impressively recently in studies of EAE in which males differing only in their Y chromosomes show quite substantially different disease progression (Case & Teuscher, 2015; Case et al., 2013, 2015; Spach et al., 2009). Consomic Y chromosome strains also differ in their response to influenza A infection (Krementsov et al., 2017).The strain differences do not appear to be caused by different levels of testosterone. These studies allow the confident conclusion that genetic elements on the Y chromosome modulate disease, but may not necessarily imply that Y genes normally contribute to sex differences in disease mechanisms (Arnold, 2017b). On the one hand, Y genetic elements may have a "masculinizing" function in autoimmune disease in the sense that they make males different from females. But because the Y chromosome also balances the effects of the X chromosome, the variation in Y consomic strains may represent different mismatches of Y elements with X and autosomal genes with which they are artificially paired during breeding of consomic strains. The balancing role of the Y chromosome does, however, suggest a contribution of Y genes to sex differences more generally in human populations. In populations with different Y chromosome haplotypes, specific Y variants may be more or less compatible with certain X chromosomes or autosomes. In the same way that mismatches of the mitochondrial and nuclear genomes have a different effect on the two sexes, discussed above, the Y chromosome mismatches with the rest of the genome may induce male-specific variation in natural populations.

7. Comparison to birds

The study of sex differences in birds has contributed substantially to concepts of sex determination and sexual differentiation, and evolution of sex chromosomes. Males are homogametic (ZZ) and females are heterogametic (ZW). In the absence of easy transgenic methods in birds, two methods have historically been used to investigate a causal role for sex chromosomes in sex differences. One method is to transplant tissues from one sex to the other, to determine if the sexual phenotype of the donor is maintained or modified in the hormonal and tissue environment of the other sex. Using this approach, several studies found that some tissues retained their sex-specific character when transplanted to the other sex (Kozelka, 1932; Mueller, 1977). Although the suggestion was that genetic factors within the transplanted tissue might cause the sex differences, there remained the possibility that hormones might have caused permanent differences in the tissue before transplantation (Lillie, 1939). The second line of evidence came from study of lateral gynandromorphs, which are intersex birds with male plumage and a testis on one side of the body and female plumage and an ovary on the other. Early investigators realized that the presence of tissues typical of both sexes, within the same body, was unlikely to have an endocrine explanation, because hormones are expected to affect both sides (Macklin, 1923; Witschi, 1939).

Avian gynandromorphs have been analyzed since the advent of molecular tools that allow assessment of genotype and Z or W gene expression (Agate et al., 2003; Morris et al., 2018; Zhao et al., 2010). In a gynandromorphic zebra finch, brain regions controlling male song were larger on the genetically male side than on the genetically female side, implying sex chromosome effects on sexual development of the brain (Agate et al., 2003). In chickens, sexually dimorphic traits such as comb, wattles, leg spurs, and body size were male-like on the genetically male side, and more female-like on the genetically female side. Although the genotype of cells on the two sides is mixed ZZ-ZW (Zhao et al., 2010), the male side had more ZZ cells than the female side, and the female side had more ZW cells (Agate et al., 2003; Morris et al., 2018; Zhao et al., 2010). Interestingly, cells of one sex, introduced into chick embryos of the opposite sex, took up residence in the gonads and expressed markers typical of the donor sex, confirming that sex-specific development of gonadal somatic tissue is specified by sex chromosome complement (Zhao et al., 2010).

Zhao et al. (2010) suggested a distinction between birds and mammals, in which sexual phenotype is specified cell-autonomously in birds, in contrast to hormonal specification in mammals. This distinction is unwarranted, because of extensive evidence that gonadal hormones cause sex differences in avian tissues (Arnold & Itoh, 2011; Balthazart, Arnold, & Adkins-Regan, 2009; Witschi, 1961), and because of extensive evidence reviewed here that sex chromosomes have cell-autonomous sex-differentiating effects in mammals. However, the lack of a Z chromosome-wide mechanism of dosage compensation in birds (Itoh et al., 2007; Mank, 2009) means that most Z genes are expressed higher in every ZZ cell relative to ZW cells, making these genes available as drivers of sex differences. In mammals, the more effective X dosage compensation means that fewer X genes may be inherently sex-biased in their expression, so they may be less likely to acquire the ability to induce sex differences in cell function (Arnold, Itoh, & Melamed, 2008).

8. Conclusions

It is now widely recognized that the sex chromosomes of mammals are a major source of inherent sex bias in the genome, because of the unavoidable sex difference in copy number of the two chromosomes. Nevertheless, much work is needed to identify the specific genes and mechanisms by which the two chromosomes cause sex differences in tissue function. Moreover, it will be imperative to understand how diverse forces, from different sex chromosome genes and from gonadal hormones, interact with each other to cause sex differences. Discovering these sex-biasing pathways is important for better understanding of sex differences in a wide variety of diseases.

References

Agate, R. J., Grisham, W., Wade, J., Mann, S., Wingfield, J., Schanen, C., et al. (2003). Neural not gonadal origin of brain sex differences in a gynandromorphic finch. *Proceedings of the National Academy of Sciences of the United States of America, 100*, 4873–4878.

Alsiraj, Y., Thatcher, S. E., Charnigo, R., Kuey, C., Blalock, E., Daugherty, A., et al. (2016). Female mice with an XY sex chromosome complement develop severe angiotensin II-induced abdominal aortic aneurysms. *Circulation, 135*, 379–391.

Arnold, A. P. (2012). The end of gonad-centric sex determination in mammals. *Trends in Genetics, 28*, 55–61.

Arnold, A. P. (2014). Conceptual frameworks and mouse models for studying sex differences in physiology and disease: Why compensation changes the game. *Experimental Neurology, 259*, 2–9.

Arnold, A. P. (2017a). A general theory of sexual differentiation. *Journal of Neuroscience Research, 95*, 291–300.

Arnold, A. P. (2017b). Y chromosome's roles in sex differences in disease. *Proceedings of the National Academy of Sciences of the United States of America, 114*, 3787–3789.

Arnold, A. P., Cassis, L. A., Eghbali, M., Reue, K., & Sandberg, K. (2017). Sex hormones and sex chromosomes cause sex differences in the development of cardiovascular diseases. *Arteriosclerosis, Thrombosis, and Vascular Biology, 37*, 746–756.

Arnold, A. P., & Chen, X. (2009). What does the "four core genotypes" mouse model tell us about sex differences in the brain and other tissues? *Frontiers in Neuroendocrinology, 30*, 1–9.

Arnold, A. P., Chen, X., Link, J. C., Itoh, Y., & Reue, K. (2013). Cell-autonomous sex determination outside of the gonad. *Developmental Dynamics, 242*, 371–379.

Arnold, A. P., & Gorski, R. A. (1984). Gonadal steroid induction of structural sex differences in the CNS. *Annual Review of Neuroscience, 7*, 413–442.

Arnold, A. P., & Itoh, Y. (2011). Factors causing sex differences in birds. *Avian Biology Research, 4*, 44–51.

Arnold, A. P., Itoh, Y., & Melamed, E. (2008). A bird's-eye view of sex chromosome dosage compensation. *Annual Review of Genomics and Human Genetics, 9*, 109–127.

Arnold, A. P., Reue, K., Eghbali, M., Vilain, E., Chen, X., Ghahramani, N., et al. (2016). The importance of having two X chromosomes. *Philosophical Transactions of the Royal Society of London. Series B, Biological Sciences, 371*, 20150113.

Bachtrog, D. (2013). Y-chromosome evolution: Emerging insights into processes of Y-chromosome degeneration. *Nature Reviews. Genetics, 14*, 113–124.

Balthazart, J., Arnold, A. P., & Adkins-Regan, E. (2009). Sexual differentiation of brain and behavior in birds. In D. W. Pfaff, A. P. Arnold, A. M. Etgen, S. E. Fahrbach, & R. T. Rubin (Eds.), *Hormones, brain, and behavior* (pp. 1745–1788). San Diego: Elsevier.

Bellott, D. W., et al. (2014). Mammalian Y chromosomes retain widely expressed dosage-sensitive regulators. *Nature, 508*, 494–499.

Bramble, M. S., Roach, L., Lipson, A., Vashist, N., Eskin, A., Ngun, T., et al. (2016). Sex-specific effects of testosterone on the sexually dimorphic transcriptome and epigenome of embryonic neural stem/progenitor cells. *Scientific Reports*, *6*, 36916.

Breedlove, S. M., Cooke, B. M., & Jordan, C. L. (1999). The orthodox view of brain sexual differentiation. *Brain, Behavior and Evolution*, *54*, 8–14.

Burgoyne, P. S., & Arnold, A. P. (2016). A primer on the use of mouse models for identifying direct sex chromosome effects that cause sex differences in non-gonadal tissues. *Biology of Sex Differences*, 7, 68.

Burgoyne, P. S., & Mitchell, M. J. (2007). The role of mouse Y chromosome genes in spermatogenesis. In Y. F. C. Lau & W. Y. Chan (Eds.), *Y chromosome and male germ cell biology* (pp. 27–45). Hackensack NJ: World Scientific Publishers.

Burgoyne, P. S., Thornhill, A. R., Boudrean, S. K., Darling, S. M., Bishop, C. E., & Evans, E. P. (1995). The genetic basis of XX-XY differences present before gonadal sex differentiation in the mouse. *Philosophical Transactions of the Royal Society of London. Series B, Biological Sciences*, *350*, 253–260.

Cambiasso, M. J., Cisternas, C. D., Ruiz-Palmero, I., Scerbo, M. J., Arevalo, M. A., Azcoitia, I., et al. (2017). Interaction of sex chromosome complement, gonadal hormones and neuronal steroid synthesis on the sexual differentiation of mammalian neurons. *Journal of Neurogenetics*, *31*, 300–306.

Camus, M. F., Clancy, D. J., & Dowling, D. K. (2012). Mitochondria, maternal inheritance, and male aging. *Current Biology*, *22*, 1717–1721.

Carrel, L., Cottle, A. A., Goglin, K. C., & Willard, H. F. (1999). A first-generation X-inactivation profile of the human X chromosome. *Proceedings of the National Academy of Sciences of the United States of America*, *96*, 14440–14444.

Case, L. K., & Teuscher, C. (2015). Y genetic variation and phenotypic diversity in health and disease. *Biology of Sex Differences*, *6*, 6.

Case, L. K., Wall, E. H., Dragon, J. A., Saligrama, N., Krementsov, D. N., Moussawi, M., et al. (2013). The Y chromosome as a regulatory element shaping immune cell transcriptomes and susceptibility to autoimmune disease. *Genome Research*, *23*, 1474–1485.

Case, L. K., Wall, E. H., Osmanski, E. E., Dragon, J. A., Saligrama, N., Zachary, J. F., et al. (2015). Copy number variation in Y chromosome multicopy genes is linked to a paternal parent-of-origin effect on CNS autoimmune disease in female offspring. *Genome Biology*, *16*, 28.

Charlesworth, B. (1996). The evolution of chromosomal sex determination and dosage compensation. *Current Biology*, *6*, 149–162.

Chen, X., McClusky, R., Chen, J., Beaven, S. W., Tontonoz, P., Arnold, A. P., et al. (2012). The number of X chromosomes causes sex differences in adiposity in mice. *PLoS Genetics*, *8*, e1002709.

Chen, X., McClusky, R., Itoh, Y., Reue, K., & Arnold, A. P. (2013). X and Y chromosome complement influence adiposity and metabolism in mice. *Endocrinology*, *154*, 1092–1104.

Chuva de Sousa Lopes, S. M., Hayashi, K., Shovlin, T. C., Mifsud, W., Surani, M. A., & McLaren, A. (2008). X chromosome activity in mouse XX primordial germ cells. *PLoS Genetics*, *4*, e30.

Cisternas, C. D., Cabrera Zapata, L. E., Arevalo, M. A., Garcia-Segura, L. M., & Cambiasso, M. J. (2017). Regulation of aromatase expression in the anterior amygdala of the developing mouse brain depends on ERbeta and sex chromosome complement. *Scientific Reports*, 7, 5320.

Cisternas, C. D., Garcia-Segura, L. M., & Cambiasso, M. J. (2018). Hormonal and genetic factors interact to control aromatase expression in the developing brain. *Journal of Neuroendocrinology*, *30*, e12535.

Cisternas, C. D., Tome, K., Caeiro, X. E., Dadam, F. M., Garcia-Segura, L. M., & Cambiasso, M. J. (2015). Sex chromosome complement determines sex differences in aromatase expression and regulation in the stria terminalis and anterior amygdala of the developing mouse brain. *Molecular and Cellular Endocrinology*, *414*, 99–110.

Cocquet, J., Ellis, P. J., Mahadevaiah, S. K., Affara, N. A., Vaiman, D., & Burgoyne, P. S. (2012). A genetic basis for a postmeiotic X versus Y chromosome intragenomic conflict in the mouse. *PLoS Genetics*, *8*, e1002900.

Cocquet, J., Ellis, P. J., Yamauchi, Y., Mahadevaiah, S. K., Affara, N. A., Ward, M. A., et al. (2009). The multicopy gene Sly represses the sex chromosomes in the male mouse germline after meiosis. *PLoS Biology*, *7*, e1000244.

Corre, C., Friedel, M., Vousden, D. A., Metcalf, A., Spring, S., Qiu, L. R., et al. (2016). Separate effects of sex hormones and sex chromosomes on brain structure and function revealed by high-resolution magnetic resonance imaging and spatial navigation assessment of the Four Core Genotype mouse model. *Brain Structure & Function*, *221*, 997–1016.

Cortez, D., Marin, R., Toledo-Flores, D., Froidevaux, L., Liechti, A., Waters, P. D., et al. (2014). Origins and functional evolution of Y chromosomes across mammals. *Nature*, *508*, 488–493.

Cox, K. H., Bonthuis, P. J., & Rissman, E. F. (2014). Mouse model systems to study sex chromosome genes and behavior: Relevance to humans. *Frontiers in Neuroendocrinology*, *35*, 405–419.

De Vries, G. J., Rissman, E. F., Simerly, R. B., Yang, L. Y., Scordalakes, E. M., Auger, C. J., et al. (2002). A model system for study of sex chromosome effects on sexually dimorphic neural and behavioral traits. *The Journal of Neuroscience*, *22*, 9005–9014.

Deng, X., Berletch, J. B., Nguyen, D. K., & Disteche, C. M. (2014). X chromosome regulation: Diverse patterns in development, tissues and disease. *Nature Reviews. Genetics*, *15*, 367–378.

Dewing, P., Shi, T., Horvath, S., & Vilain, E. (2003). Sexually dimorphic gene expression in mouse brain precedes gonadal differentiation. *Brain Research. Molecular Brain Research*, *118*, 82–90.

Disteche, C. M. (2016). Dosage compensation of the sex chromosomes and autosomes. *Seminars in Cell & Developmental Biology*, *56*, 9–18.

Drenckhahn, J. D., Schwarz, Q. P., Gray, S., Laskowski, A., Kiriazis, H., Ming, Z., et al. (2008). Compensatory growth of healthy cardiac cells in the presence of diseased cells restores tissue homeostasis during heart development. *Developmental Cell*, *15*, 521–533.

Du, S., Itoh, N., Askarinam, S., Hill, H., Arnold, A. P., & Voskuhl, R. R. (2014). XY sex chromosome complement, compared with XX, in the CNS confers greater neurodegeneration during experimental autoimmune encephalomyelitis. *Proceedings of the National Academy of Sciences of the United States of America*, *111*, 2806–2811.

Forger, N. G. (2009). Control of cell number in the sexually dimorphic brain and spinal cord. *Journal of Neuroendocrinology*, *21*, 393–399.

Frank, S. A. (2012). Evolution: Mitochondrial burden on male health. *Current Biology*, *22*, R797–R799.

Graves, J. A. (2016). Evolution of vertebrate sex chromosomes and dosage compensation. *Nature Reviews. Genetics*, *17*, 33–46.

Hughes, J. F., & Page, D. C. (2015). The biology and evolution of mammalian Y chromosomes. *Annual Review of Genetics*, *49*, 507–527.

Innocenti, P., Morrow, E. H., & Dowling, D. K. (2011). Experimental evidence supports a sex-specific selective sieve in mitochondrial genome evolution. *Science*, *332*, 845–848.

Itoh, Y., Melamed, E., Yang, X., Kampf, K., Wang, S., Yehya, N., et al. (2007). Dosage compensation is less effective in birds than in mammals. *Journal of Biology*, *6*, 2.

Ji, H., Zheng, W., Wu, X., Liu, J., Ecelbarger, C. M., Watkins, R., et al. (2010). Sex chromosome effects unmasked in angiotensin II-induced hypertension. *Hypertension, 55*, 1275–1282.

Kozelka, A. W. (1932). Integumental grafting as a means of analyzing the factors determining the secondary sexual characteristics of leghorn fowl. *The Journal of Experimental Zoology, 61*, 431–495.

Krementsov, D. N., Case, L. K., Dienz, O., Raza, A., Fang, Q., Ather, J. L., et al. (2017). Genetic variation in chromosome Y regulates susceptibility to influenza A inflection. *Proceedings of the National Academy of Sciences of the United States of America, 114*, 3491–3496.

Lahn, B. T., & Page, D. C. (1999). Four evolutionary strata on the human X chromosome. *Science, 286*, 964–967.

Li, J., Chen, X., McClusky, R., Ruiz-Sundstrom, M., Itoh, Y., Umar, S., et al. (2014). The number of X chromosomes influences protection from cardiac ischaemia/reperfusion injury in mice: One X is better than two. *Cardiovascular Research, 102*, 375–394.

Lillie, F. R. (1939). General biological introduction. In E. Allen, C. H. Danforth, & E. A. Doisy (Eds.), *Sex and internal secretions* (pp. 3–14). Baltimore: Williams and Wilkins Co.

Link, J. C., Chen, X., Prien, C., Borja, M. S., Hammerson, B., Oda, M. N., et al. (2015). Increased high-density lipoprotein cholesterol levels in mice with XX versus XY sex chromosomes. *Arteriosclerosis, Thrombosis, and Vascular Biology, 35*, 1778–1786.

Lowe, R., Gemma, C., Rakyan, V. K., & Holland, M. L. (2015). Sexually dimorphic gene expression emerges with embryonic genome activation and is dynamic throughout development. *BMC Genomics, 16*, 295.

Macklin, M. T. (1923). A description of material from a gynandromorph fowl. *The Journal of Experimental Zoology, 38*, 355–375.

Mank, J. E. (2009). The W, X, Y and Z of sex-chromosome dosage compensation. *Trends in Genetics, 25*, 226–233.

Manwani, B., Bentivegna, K., Benashski, S. E., Venna, V. R., Xu, Y., Arnold, A. P., et al. (2015). Sex differences in ischemic stroke sensitivity are influenced by gonadal hormones, not by sex chromosome complement. *Journal of Cerebral Blood Flow and Metabolism, 35*, 221–229.

McCullough, L. D., Mirza, M. A., Xu, Y., Bentivegna, K., Steffens, E. B., Ritzel, R., et al. (2016). Stroke sensitivity in the aged: Sex chromosome complement *vs.* gonadal hormones. *Aging (Albany NY), 8*, 1432–1441.

Midgeon, B. R. (2007). *Females are mosaics*. Oxford: Oxford University Press.

Miura, K., Tomita, A., & Kanai, Y. (2018). Sex determination and differentiation in mammals. In K. Kobayashi, T. Kitano, Y. Iwao, & M. Kondo (Eds.), *Reproductive and developmental strategies: The continuity of life* (pp. 407–432). Japan: Springer.

Moore, S., Patel, R., Hannsun, G., Yang, J., & Tiwari-Woodruff, S. K. (2013). Sex chromosome complement influences functional callosal myelination. *Neurosciences, 245*, 166–178.

Morris, K. R., Hirst, C. E., Major, A. T., Ezaz, T., Ford, M., Bibby, S., et al. (2018). Gonadal and endocrine analysis of a gynandromorphic chicken. *Endocrinology, 159*, 3492–3502.

Mueller, N. (1977). Control of sex difference in the plumage of the house sparrow, *Passer domesticus. The Journal of Experimental Zoology, 202*, 45–48.

Naqvi, S., Bellott, D. W., Lin, K. S., & Page, D. C. (2018). Conserved microRNA targeting reveals preexisting gene dosage sensitivities that shaped amniote sex chromosome evolution. *Genome Research, 28*, 474–483.

Nef, S., Schaad, O., Stallings, N. R., Cederroth, C. R., Pitetti, J. L., Schaer, G., et al. (2005). Gene expression during sex determination reveals a robust female genetic program at the onset of ovarian development. *Developmental Biology, 287*, 361–377.

Nguyen, D. K., & Disteche, C. M. (2006). Dosage compensation of the active X chromosome in mammals. *Nature Genetics*, *38*, 47–53.

Nugent, B. M., O'Donnell, C. M., Epperson, C. N., & Bale, T. L. (2018). Placental H3K27me3 establishes female resilience to prenatal insults. *Nature Communications*, *9*, 2555.

Palaszynski, K. M., Smith, D. L., Kamrava, S., Burgoyne, P. S., Arnold, A. P., & Voskuhl, R. R. (2005). A Yin-Yang effect between sex chromosome complement and sex hormones on the immune response. *Endocrinology*, *146*, 3280–3285.

Renfree, M. B., & Short, R. V. (1988). Sex determination in marsupials: Evidence for a marsupial-eutherian dichotomy. *Philosophical Transactions of the Royal Society of London B: Biological Sciences*, *322*, 41–53.

Rice, W. R. (1992). Sexually antagonistic genes—Experimental evidence. *Science*, *256*, 1436–1439.

Rice, W. R. (1994). Degeneration of a nonrecombining chromosome. *Science*, *263*, 230–232.

Robinson, D. P., Huber, S. A., Moussawi, M., Roberts, B., Teuscher, C., Watkins, R., et al. (2011). Sex chromosome complement contributes to sex differences in coxsackievirus B3 but not influenza A virus pathogenesis. *Biology of Sex Differences*, *2*, 8.

Sangrithi, M. N., Royo, H., Mahadevaiah, S. K., Ojarikre, O., Bhaw, L., Sesay, A., et al. (2017). Non-canonical and sexually dimorphic X dosage compensation states in the mouse and human germline. *Developmental Cell*, *40*, 289–301.

Shpargel, K. B., Sengoku, T., Yokoyama, S., & Magnuson, T. (2012). UTX and UTY demonstrate histone demethylase-independent function in mouse embryonic development. *PLoS Genetics*, *8*, e1002964.

Silkaitis, K., & Lemos, B. (2014). Sex-biased chromatin and regulatory cross-talk between sex chromosomes, autosomes, and mitochondria. *Biology of Sex Differences*, *5*, 2.

Smith-Bouvier, D. L., Divekar, A. A., Sasidhar, M., Du, S., Tiwari-Woodruff, S. K., King, J. K., et al. (2008). A role for sex chromosome complement in the female bias in autoimmune disease. *The Journal of Experimental Medicine*, *205*, 1099–1108.

Soh, Y. Q., et al. (2014). Sequencing the mouse Y chromosome reveals convergent gene acquisition and amplification on both sex chromosomes. *Cell*, *159*, 800–813.

Spach, K. M., Blake, M., Bunn, J. Y., McElvany, B., Noubade, R., Blankenhorn, E. P., et al. (2009). Cutting edge: the Y chromosome controls the age-dependent experimental allergic encephalomyelitis sexual dimorphism in SJL/J mice. *Journal of Immunology*, *182*, 1789–1793.

Spiller, C., Koopman, P., & Bowles, J. (2017). Sex determination in the mammalian germline. *Annual Review of Genetics*, *51*, 265–285.

Sutton, E., et al. (2011). Identification of SOX3 as an XX male sex reversal gene in mice and humans. *The Journal of Clinical Investigation*, *121*, 328–341.

Tukiainen, T., Villani, A. C., Yen, A., Rivas, M. A., Marshall, J. L., Satija, R., et al. (2017). Landscape of X chromosome inactivation across human tissues. *Nature*, *550*, 244–248.

Umar, S., Cunningham, C. M., Itoh, Y., Moazeni, S., Vaillancourt, M., Sarji, S., et al. (2018). The Y chromosome plays a protective role in experimental hypoxic pulmonary hypertension. *American Journal of Respiratory and Critical Care Medicine*, *197*, 952–955.

Vallender, E. J., & Lahn, B. T. (2004). How mammalian sex chromosomes acquired their peculiar gene content. *BioEssays*, *26*, 159–169.

Veitia, R. A., Veyrunes, F., Bottani, S., & Birchler, J. A. (2015). X chromosome inactivation and active X upregulation in therian mammals: Facts, questions, and hypotheses. *Journal of Molecular Cell Biology*, *7*, 2–11.

Vousden, D. A., Corre, C., Spring, S., Qiu, L. R., Metcalf, A., Cox, E., et al. (2018). Impact of X/Y genes and sex hormones on mouse neuroanatomy. *NeuroImage*, *173*, 551–563.

Werner, R. J., Schultz, B. M., Huhn, J. M., Jelinek, J., Madzo, J., & Engel, N. (2017). Sex chromosomes drive gene expression and regulatory dimorphisms in mouse embryonic stem cells. *Biology of Sex Differences*, *8*, 28.

Wijchers, P. J., & Festenstein, R. J. (2011). Epigenetic regulation of autosomal gene expression by sex chromosomes. *Trends in Genetics*, *27*, 132–140.

Witschi, E. (1939). Modification of the development of sex in lower vertebrates and in mammals. In E. Allen, C. H. Danforth, & E. A. Doisy (Eds.), *Sex and internal secretions* (pp. 145–226). Baltimore: Williams & Wilkins.

Witschi, E. (1961). Sex and secondary characteristics. In A. J. Marshall (Ed.), *Biology and comparative physiology of birds* (pp. 115–168). New York: Academic Press.

Xu, J., Deng, X., Watkins, R., & Disteche, C. M. (2008). Sex-specific differences in expression of histone demethylases Utx and Uty in mouse brain and neurons. *The Journal of Neuroscience*, *28*, 4521–4527.

Yildirim, E., Kirby, J. E., Brown, D. E., Mercier, F. E., Sadreyev, R. I., Scadden, D. T., et al. (2013). Xist RNA is a potent suppressor of hematologic cancer in mice. *Cell*, *152*, 727–742.

Zhao, D., McBride, D., Nandi, S., McQueen, H. A., McGrew, M. J., Hocking, P. M., et al. (2010). Somatic sex identity is cell autonomous in the chicken. *Nature*, *464*, 237–242.

CHAPTER ELEVEN

Translating genomics to the clinical diagnosis of disorders/differences of sex development

Abhinav Parivesh[a], Hayk Barseghyan[a,b], Emmanuèle Délot[a,b,*], Eric Vilain[a,b,*]

[a]Center for Genetic Medicine Research, Children's National Medical Center, Washington, DC, United States
[b]Department of Genomics and Precision Medicine, The George Washington University, Washington, DC, United States
*Corresponding authors: e-mail address: edelot@gwu.edu; evilain@gwu.edu

Contents

Abstract

The medical and psychosocial challenges faced by patients living with Disorders/Differences of Sex Development (DSD) and their families can be alleviated by a rapid and accurate diagnostic process. Clinical diagnosis of DSD is limited by a lack of standardization of anatomical and endocrine phenotyping and genetic testing, as well as poor genotype/phenotype correlation. Historically, DSD genes have been identified through positional cloning of disease-associated variants segregating in families and validation of candidates in animal and *in vitro* modeling of variant pathogenicity. Owing to the complexity of conditions grouped under DSD, genome-wide scanning methods are better suited for identifying disease causing gene variant(s) and providing a clinical diagnosis. Here, we review a number of established genomic tools (karyotyping,

Current Topics in Developmental Biology, Volume 134
ISSN 0070-2153
https://doi.org/10.1016/bs.ctdb.2019.01.005

chromosomal microarrays and exome sequencing) used in clinic for DSD diagnosis, as well as emerging genomic technologies such as whole-genome (short-read) sequencing, long-read sequencing, and optical mapping used for novel DSD gene discovery. These, together with gene expression and epigenetic studies can potentiate the clinical diagnosis of DSD diagnostic rates and enhance the outcomes for patients and families.

1. Introduction

Sexual development is one of the most elegant and fascinating mechanisms in developmental biology where spatiotemporally orchestrated molecular signaling events determine the sexual fate of a developing embryo. Sexual development can be broadly divided into two sequential steps: sex determination and sex differentiation (Fig. 1). Sex determination in a developing embryo essentially begins with establishment of the sex chromosome complement, inherited from the parental gametes, which in turn triggers a sequence of mutually exclusive molecular signaling events in the presence or absence of the Y chromosome driving the bipotential

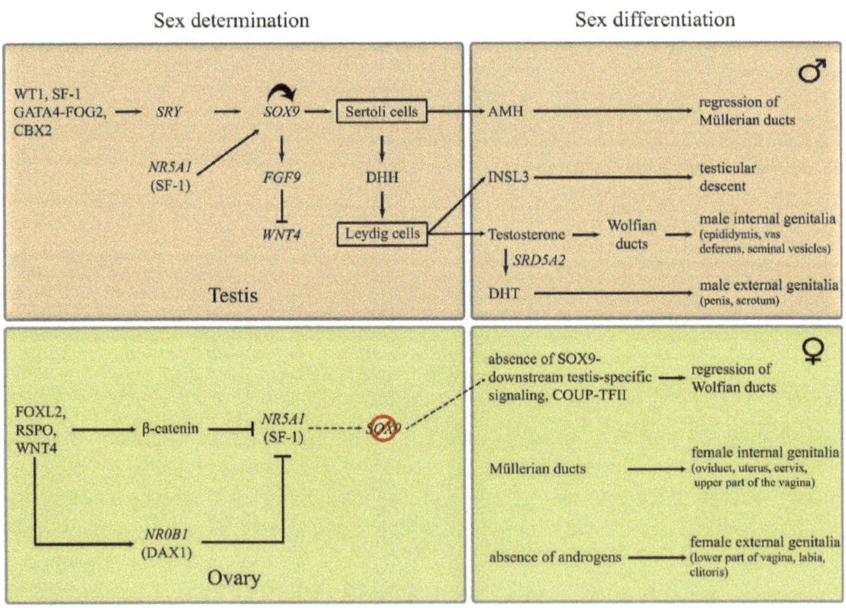

Fig. 1 Simplified overview of sex determination and sex differentiation in humans. The figure has been modified from Arbodela and Vilain (2009) using Barseghyan, Delot, and Vilain (2015), Biason-Lauber (2012), Croft, Ayers, Sinclair, and Ohnesorg (2016), Kyriakou, Lucas-Herald, McGowan, Tobias, and Ahmed (2015), Ohnesorg, Vilain, and Sinclair (2014), Ono and Harley (2013), Zhao et al. (2017); and references therein.

gonad in the developing embryo toward a male or female fate, respectively. Expression of Y-chromosome gene *SRY* in the somatic cells of the bipotential gonad of an XY embryo kickstarts a cascade of signaling events that lead to testicular organogenesis. In the absence of *SRY*, female-specific pathways are initiated, triggering ovarian development under the control of genes such as *RSPO1*, *WNT4* and DAX1/*NR0B1*.

The development of a testis or an ovary from a bipotential gonad is followed by secretion of the sex-specific milieu of hormones from those organs, which drives the second step of sexual development—sex differentiation, and results in sex-specific morphogenesis of the reproductive organs. Disruption in any of these developmental processes and the molecular networks driving them, as well as the enzymatic cascades involved in the synthesis of steroid hormones in the adrenal or central control of these processes, can lead to Disorders/Differences of Sex Development (DSD). The molecular mechanisms involved have been expertly reviewed elsewhere (Croft et al., 2016; Délot & Vilain, 2018; Kyriakou et al., 2015; Ohnesorg et al., 2014; Ono & Harley, 2013).

2. Disorders/differences of sex development, challenges in diagnosis and management

The term DSD was adopted following the 2005 consensus meeting of the Lawson Wilkins Pediatric Endocrine Society (LWPES) and the European Society for Paediatric Endocrinology (ESPE) to replace controversial and potentially confusing terms such as hermaphroditism, pseudohermaphroditism, and intersex. DSD was formally defined as "congenital conditions in which development of chromosomal, gonadal, or anatomical sex is atypical" (Hughes et al., 2006; Lee et al., 2006). As an umbrella term, it encompasses a range of rare and not so rare anatomical phenotypes (e.g., ambiguous genitalia, discordance of external genitalia appearance and chromosomal sex, isolated hypospadias, …); hormonal phenotypes (e.g., androgen insensitivity, over/under secretion of androgens); gonadal phenotypes (e.g., gonadal dysgenesis, ovotestis); and chromosomal complements (e.g., 46,XY, 46,XX, 47,XXY, 45,X). DSD can occur in isolated or syndromic forms (the latter reviewed in Délot et al., 2017; Hutson, Grover, O'Connell, & Pennell, 2014).

Management of DSD is complex as the conditions may be associated with increased infertility, cancer, and gender dysphoria risks, as well as pervasive challenges to psychosocial adaptation for patients and families

(Abacl, Çatll, & Berberoǀlu, 2015; Arboleda, Fleming, et al., 2014; Guercio & Rey, 2014; Sandberg, Callens, & Wisniewski, 2015; Sandberg & Mazur, 2014; Słowikowska-Hilczer et al., 2017). Correlation between treatment interventions (surgical, hormonal, psychosocial) and their clinical outcomes is poorly understood. This suggests further research in DSD pathophysiology is needed to provide optimal evidence-based clinical management. Major challenges in DSD management include lack of standardization in descriptions of phenotype and medical/surgical interventions, resulting in variations in diagnostic and treatment practices across medical, surgical and behavioral aspects of health (Sandberg et al., 2015). Substantial unexplained variance in outcomes is revealed by studies examining the relationship between molecular diagnosis, gender development and psychosocial trajectories (Rolston et al., 2017). These challenges complicate the principles of clinical management of DSDs with disagreements among and between healthcare providers, advocacy groups and patient communities regarding the definition of optimal care (Adam & Vilain, 2017). International efforts are underway to standardize practice and diagnosis (e.g., Audí et al., 2018; Délot et al., 2017).

From a diagnostic standpoint too, DSDs are challenging. For example, 46,XY disorders of androgen biosynthesis, which include *SRD5A2, HSD17B2, HSD3B2, StAR, CYP17A1,* and *POR* deficiencies, are often difficult to distinguish clinically. However, an early genetic diagnosis is critical as these conditions have drastically different natural histories requiring different management (recently reviewed in Délot & Vilain, 2018).

3. Genetics of DSD

The earliest characterized cases of DSD were cases of sex chromosome aneuploidies such as 45,X (Turner syndrome) and its mosaic variants (e.g., 45,X/46,XY), 47,XXY (Klinefelter syndrome), or 46,XX/46,XY chimerism. These conditions account for about 15% of DSDs (Cox et al., 2014; De Paula et al., 2016; Délot et al., 2017; Demirhan, Tanrıverdi, Süleymanova, Cetinel, & Yasar, 2015; Heeley et al., 2018). Many of the early identifications of individual DSD genes were achieved by positional cloning and linkage analysis and identifying microscopically discernable structural changes in the karyotype using cytogenetic techniques. A number of genes such as *SRY, SOX9, NR0B1*/DAX1, *DMRT1, WT1, MAP3K1* and *RSPO1* were discovered using these approaches.

However, family-based methods have reduced power for conditions where family size is limited by infertility.

Copy number variants (CNV) affecting coding sequences or regulatory elements of critical dosage-sensitive genes, as well as single nucleotide variants (SNVs) in at least 75 genes involved in gonadal development and/or sex hormone biosynthesis/action, are known causes of DSD (Ahmed & Hughes, 2002; Arboleda, Fleming, & Vilain, 2010; Auchus & Miller, 2012; Buonocore & Achermann, 2016; Croft, Ohnesorg, & Sinclair, 2018; Délot et al., 2017; Eggers & Sinclair, 2012; Ono & Harley, 2013). Genetic etiology has been well established for many sex differentiation conditions. For example, recessive inheritance of variants in *CYP21A2* is the most common cause of Congenital Adrenal Hyperplasia (CAH), which can lead to external genital masculinization in 46,XX patients (Nimkarn, Gangishetti, Yau, & New, 2016; Speiser et al., 2010). In 46,XY individuals, sex differentiation anomalies can result from complete or partial androgen insensitivity or defects in biosynthesis of steroid hormones. Androgen insensitivity in its complete form (CAIS) is caused, in >95% of patients, by loss-of-function mutations in *AR*, the androgen receptor. However, only a minority (~10%) of partial forms of androgen insensitivity (PAIS) are associated with *AR* variants. For the most difficult to diagnose DSD conditions, such as the non-syndromic forms of 46,XY gonadal dysgenesis, evaluation of historical data suggests that approximately ~15% are due to *SRY* defects, ~15% to mutations in SF1/*NR5A1*, and a few cases due to other rare genetic events (in *SOX9*, DAX1/*NR0B1*, ...) (Mohnach, Fechner, & Keegan, 2016). More recent data suggest that *MAP3K1* variants may explain an additional 10–18% (Baxter et al., 2015; Granados et al., 2017; Pearlman et al., 2010), still leaving nearly half of cases without a genetic diagnosis. In 46,XX individuals, the majority (80–90%) of non-syndromic Testicular DSD are explained by *SRY* translocations; however, only ~10% of Ovotesticular DSD have detectable Y material (Délot & Vilain, 2015) and etiology remains elusive. In XX individuals, abnormal ovarian development leads to primary ovarian insufficiency, an important medical challenge that affects over 1% of women over the age of 30 (Rossetti, Ferrari, Bonomi, & Persani, 2017). The majority of ovarian dysgenesis cases are linked to abnormal complements of the X chromosome (such as Turner Syndrome). In addition, many genes coding for oocyte-specific transcription factors (e.g., FOXL2), folliculogenesis growth factors (e.g., BMP15), proteins that control ovarian steroidogenesis (e.g., receptor for FSH) or proteins involved in DNA replication or integrity (e.g., STAG3) have now been shown to underlie ovarian dysgenesis or premature ovarian failure, though each explain only a minority of cases and many cases remain unexplained (Aittomäki et al., 1995; Arboleda, Sandberg, & Vilain, 2014; Caburet et al., 2014; Chapman, Cree, & Shelling, 2015).

The majority of DSD patients do not receive an accurate diagnosis, which may be due to lack of, or imperfect access to, state-of-the-art genetic testing. We recently showed that 97% of undiagnosed patients in the US-based DSD-TRN registry have not exhausted clinically available diagnostic testing (Délot et al., 2017). In this review, we focus on how the existing and upcoming high-throughput genomic technologies are used to help identify novel DSD-associated genes and how these can be incorporated into genetic diagnostic practice to improve the diagnosis of DSD.

4. Chromosomal microarrays (CMA)

Karyotyping can typically detect chromosomal rearrangement(s) with 5–10 Mb resolution. Hence, newer genome-wide scanning technologies were developed to achieve a better resolution in identifying gain or loss of genetic material. The two main types of chromosomal microarrays (CMA) used for genetic testing are Comparative Genomic Hybridization (array CGH) and Single-Nucleotide Polymorphism (SNP) arrays. Array CGH involves labeling the patient and control DNA samples with two different fluorophores and hybridizing them to an array containing 50–70 bp long oligonucleotides. The relative fluorescent signal shows regions of gain or loss of genetic material in the patient sample when directly compared with the control sample. It is useful for detecting CNVs, but it is not capable of detecting regions of homozygosity. Single Nucleotide Polymorphism (SNP) arrays, on the other hand, use a single fluorescently labeled (patient) DNA sample and hybridize it with an array containing oligonucleotides representing SNPs and the relative intensity is compared with that of a pool of controls available as online reference. In addition to detecting loss or gain of genetic material, new SNP arrays have incorporated aspects of the array CGH technology to be able to identify homozygosity, heterozygosity, regions of Loss of Homozygosity (LOH) as well as consanguinity (Sund & Rehder, 2014).

CMAs have been instrumental in genetic diagnosis in cases of DSD by identifying genetic changes in regions of the genome housing known DSD gene(s) as well as finding novel variants, potentiating discovery of novel

DSD genes (e.g., Aboura et al., 2009). For example, clinical CMA diagnosed CNV of the X-chromosome dosage-sensitive gene *NR0B1*/DAX1, whose deletions and duplications cause Adrenal Hypoplasia Congenita and XY gonadal dysgenesis, respectively (Bardoni et al., 1994; Ledig et al., 2010; Muscatelli et al., 1994; White et al., 2011). Array CGH analysis for a large cohort of 116 patients with idiopathic DSD, compared with 8951 controls without urogenital defects, was able to detect clinically relevant CNVs in 21.5% of the patients (Tannour-Louet et al., 2010). A number of chromosomal rearrangements were detected that not only encompassed known DSD genes such as *SRY* or *DMRT1* but also novel ones such as *FGFR2*, *KANK1*, *ADCY2*, or *ZEB2*. In another study of a cohort of 87 patients, array CGH identified CNVs comprising chromosomal deletions and duplications encompassing known DSD and novel genes for 5 out of 16 cases of syndromic DSD and 21 out of 71 cases of non-syndromic DSD (Ledig et al., 2010).

CMA has also been used to identify novel variants in the non-coding regulatory regions of DSD-associated genes. CMA identified a 250 kb deletion upstream of the *NR0B1* locus resulting in its increased expression and consequent sex reversal in an XY woman whose Y chromosome harbored an intact *SRY* (Smyk et al., 2007). Deletions and duplications upstream of *SOX9* have been associated with ovotesticular DSD and 46,XY gonadal dysgenesis (Benko et al., 2011; Cox, Willatt, Homfray, & Woods, 2011; Lecointre et al., 2009; Vetro et al., 2011; White et al., 2011). CMA was also used to identify the breakpoint of a 12:17 translocation replacing part of the cis-regulatory region upstream of *SOX9* with that of a chromosome 12 pseudogene, causing a gain of SOX9 function and a male phenotype in an XX child (Refai et al., 2010).

Correlation of CMA results from patients with animal model experiments can help validate novel SNVs. For example, a mouse knockout model of *Wwox* showed abnormal gonadal development (Aqeilan et al., 2009; Ludes-Meyers et al., 2007), but the gene was not associated with human DSD until CMA analysis on an XY DSD patient with abnormal gonadal development revealed an exonic deletion in the *WWOX* gene (White et al., 2012). SOX3 has a high amino acid sequence similarity (90%) and protein identity (67%) to SRY but is not normally expressed in developing gonads of humans or mice (Bowles, Schepers, & Koopman, 2000). While loss-of-function of *SOX3* doesn't seem to affect sex determination in mice or humans (Weiss et al., 2003), ectopic expression in XX mouse gonads results in a male phenotype (Sutton et al., 2011). Subsequent CMA analysis of human *SRY*-negative XX male individuals detected *SOX3* duplications in several patients (Moalem et al., 2012; Sutton et al., 2011; Vetro et al., 2015), supporting the hypothesis that increased dosage of *SOX3* is capable of inducing male sex determination in the absence of *SRY*.

DSD-relevant CNVs have been recently reviewed (Croft et al., 2018). Many candidates and chromosomal loci with potential roles in sex determination and DSD are awaiting further validation. These include deletions in the 9p23-24 chromosomal region encompassing *DMRT1* and *KANK1* in cases of XY gonadal dysgenesis (Igarashi et al., 2013; Ledig et al., 2010; Quinonez et al., 2013; Raymond et al., 1999), deletions at 12p13 and 16p11.2 in cases of hypospadias, deletions at 10p14 and Xq28 in cryptorchidism, deletions at 1p36.3, 9p24.3 and 19q12-13.1 in cases of ambiguous genitalia (Tannour-Louet et al., 2010), copy number gains at Xq28 encompassing the *VAMP7* gene in cases of newborns with congenital genitourinary tract masculinization disorders (Tannour-Louet et al., 2014).

In addition, clinical CMA typically only reports CNVs above a 25–50 kb cutoff, leaving a diagnostic gap for smaller CNVs. Previous studies in DSD cohorts have identified smaller CNVs, but interpretation remains a challenge (Kon & Fukami, 2015; Mizuno et al., 2014; Norling et al., 2013; Tannour-Louet et al., 2010; White et al., 2011). High-resolution CNV maps (with CNVs as small as 50 bp) are increasingly becoming available and are poised to bridge the gap with the smaller INDELs that are detectable using short-read sequencing. They have been recently used to investigate the potential consequences of small CNVs as well as genomic copy-neutral regions of homozygosity in DSD genes (Amarillo et al., 2016). This study identified novel potentially pathogenic mechanisms in well-established DSD genes such as *SOX9* or *PTPN11,* and strengthened the genetic evidence for pathogenicity of recently uncovered DSD genes such as *WWOX*. It also revealed potential novel recurrent mechanisms of pathogenicity involving imprinted genes, as well as regions that might modify the epigenetic landscape and RNA expression in the developing gonad.

5. Exome sequencing

Endocrine tests, anatomical studies, imaging studies, and the experience of the clinician have historically guided single candidate gene testing. However, in DSD, poor genotype-phenotype correlation and overlapping phenotypes pose a challenge to single-gene testing approaches.

In consequence, clinical testing has gradually evolved from single-gene testing (typically using Sanger sequencing) to multi-gene testing allowed by the development and cost decrease of massively parallel next-generation (short-read) sequencing. Exome sequencing (WES) captures and sequences only exons and the immediately adjacent intronic canonical splice sites, covering about 45 million base pairs (or ~1.5% of the genome). Current options include extended coverage to ~ 60 Mbp to include functional non-protein-coding elements, e.g., micro-RNA, long intergenic noncoding RNA. Alternatively a reduced coverage approach (~35 Mbp, limited to clinically relevant regions) has been used to decrease cost and increase read depth in the regions that are sequenced. WES was originally used to identify single nucleotide variants, but with the advancement of bioinformatic tools, it is now also capable of detecting some CNVs (D'Aurizio et al., 2016).

Exome sequencing involves capture of protein-coding regions, enrichment, sequencing and bioinformatic analysis. The major commercially available exome sequencing platforms differ in the capture and enrichment techniques. Exome capture can be solution-based or array-based. In the solution-based method, the DNA is fragmented; biotinylated oligonucleotide probes selectively hybridizing to a targeted genomic region are immobilized on magnetic streptavidin beads. Hybridized DNA targets are enriched using PCR, at a risk of introducing PCR-associated GC bias, and sequenced. Array-based enrichment is based on the same principles except that it utilizes a library of oligonucleotide probes immobilized on a high-density microarray (Warr et al., 2015). There are several commercial enrichment platforms available for exome sequencing, which differ in target region selection, bait length or density, molecule used for capture, and genomic fragmentation method (Shigemizu et al., 2015; Tobias & McElreavey, 2014).

WES allows for diagnosis of DSD not only by identifying known DSD pathogenic variants in patients but also the discovery of novel DSD genes. For patients without a diagnosis at a given time, reanalysis of the sequencing data using ever-evolving software and new publications about disease-gene associations can lead to almost a 10% increase in diagnostic yield after a 2–3 years time period (Wenger, Guturu, Bernstein, & Bejerano, 2017).

Clinical diagnosis by WES is challenged by factors such as incomplete gene coverage, platform variability, difficulties of variant interpretation, and the well-documented ethical issues linked to incidental findings (e.g., ACMG Board of Directors, 2014; Allyse & Michie, 2013; Green et al., 2013; Richards, Aziz, Bale, Bick, & Das, 2015; Tobias & McElreavey, 2014).

Since one of the important early steps in WES is exome enrichment by PCR technology, some parts of the exome can be difficult to access and enrich because of the regional nucleotide composition (such as GC content) leading to limitations in exome coverage. In fact, ~400 disease-causing exonic variants reported in the Human Genome Mutation Database (HGMD) are not covered by the current methods (Caspar et al., 2017). Commercially available enrichment kits using high concentration of capture probes that cover difficult-to-enrich exome regions can improve yields (Lelieveld, Spielmann, Mundlos, Veltman, & Gilissen, 2015; Meienberg et al., 2015). Evolution of capture techniques has impacted DSD testing. Before 2014, less than half of the genes in our typical DSD-specific list used for exome variants filtering had their entire sequence captured by exome (i.e., had 100% coverage). A newer capture method allowed to reach full coverage for 83% of genes on the list, with another seven genes having upward of 97% sequence coverage (Délot et al., 2017).

On average exome sequencing identifies 21,000 coding variants in each case when compared to the reference genome (Lee et al., 2014), out of which ~500 are rare variants present in <1% of the population (Kernohan et al., 2018). Increasing availability of large ethnically and geographically diverse public databases of healthy controls (such as ExAC (Lek et al., 2016) and now gnomAD) is critical to accurately assess allele frequency. Under-representation of certain populations in such reference databases can lead to variant calling errors for patients belonging to such populations (e.g., Petrovski & Goldstein, 2016). Manual curation of variants is highly labor intensive and automated in silico variant-prioritization platforms based on amino acid conservation and biochemical properties have been developed to interpret identified variants (Adzhubei et al., 2010; Kircher et al., 2014; Kumar, Henikoff, & Ng, 2009; Schwarz et al., 2014; Smedley et al., 2015).

Sequencing of the two parents along with the proband helps prioritize variants (Wenger et al., 2017): Trio exome sequencing has been reported to increase the diagnostic yields from 22% to 33% (Lee et al., 2014) and 23.6% to 31% (Retterer et al., 2016) in comparison to proband-only sequencing. In the case of DSD, though, this brings additional interpretation challenges. For example, *de novo* variants in *NR5A1* (the gene coding for SF1) are well known to cause developmental defects in gonads of both 46,XX and 46,XY individuals (reviewed in Ferraz-de-Souza, Lin, & Achermann, 2011). If the same variant was found in parents, automated algorithms would likely eliminate this variant as not pathogenic. However, many publications report as pathogenic novel heterozygous

NR5A1 variants inherited from the (fertile) parent of the same karyotype as the patient (e.g., Philibert et al., 2011; Swartz et al., 2017), including the well-established, recurrent p.Arg92Trp variant (Baetens et al., 2017; Bashamboo et al., 2016). It has been suggested that the parent may present a milder phenotype, such as early menopause in the mother or progressive gonadal dysgenesis in the father, allowing them to reproduce and pass on the variant to their more severely affected child. Whether modifier genes and/or environmental influences are responsible for the variable expressivity of the phenotype remains to be determined. A recent publication suggests genotype/phenotype variability may be explained by an oligogenic mode of inheritance, a mechanism clinical exome sequencing is poorly equipped to detect at this time (Camats, Fernández-Cancio, Audi, Schaller, & Flück, 2018).

Identified variants are classified for reporting using standardized nomenclature, following the American College of Medical Genetics and Genomics (ACMG) and the Association for Molecular Pathology recommendations (Richards et al., 2015). Variants are classified as pathogenic (previously reported in humans as the recognized cause of the specific DSD condition), likely pathogenic (e.g., previously unreported in a known DSD gene), benign (previously reported and recognized as a neutral variant), likely benign (e.g., previously unreported in DSD and is not predicted to affect protein functions). However, the vast majority remain classified as variants of uncertain clinical significance (VUS).

The early adoption of exome sequencing by academic institutions in the United States has helped to identify the genetic diagnosis in approximately 30% of cases for whom traditional molecular diagnostic methods were uninformative (Lee et al., 2014; Yang et al., 2014), a yield that has proven remarkably similar across countries, sequencing platforms, and conditions studied (Dong et al., 2016; Farwell et al., 2015; Sawyer et al., 2016; Soden et al., 2014; Yang et al., 2013). In many cases, the identified genetic diagnosis provided guidance for medical treatment and management. When performed early, exome sequencing is capable of not only increasing the diagnostic yields in DSD but also guiding clinical management of patients (e.g., Barseghyan et al., 2015; Baxter et al., 2015).

6. Gene-testing panels

Because interpretation of exome sequencing results beyond well-established variants in known DSD genes is challenging, clinical providers have developed gene panels. Gene panels are typically based on targeted next-generation sequencing (NGS) through either targeted capture or full exome capture followed by limited interpretation. They can provide high-throughput and accurate screening of multiple genes and different types of mutations in a highly efficient manner. Additional Sanger sequencing may be performed to fill in for genes with missing or insufficient read depth coverage. As gene panels test a limited set of genes, the utilization of NGS results in high read-depths enabling detection of non-reference alleles present at very low frequencies. This method facilitates detection of mosaics hypothesized to underlie variable phenotypic expression in DSD.

Comprehensive gene panels have been designed that report a high rate of diagnosis for 46,XY DSD: e.g., 43% using 64 genes (Eggers et al., 2016); 45% using 56 genes (Özen et al., 2017). Diagnosis rates are typically lower if undiagnosed 46,XX DSD cases are included (e.g., 29.5% of likely pathogenic or pathogenic variants found using a 67-gene panel (Kim et al., 2017)). A study using a panel of 80 DSD genes demonstrated the superiority of the approach over single-gene testing performed on the same patient samples (Fan et al., 2017). Panels are also time- and cost- effective, with one study quoting 3 days vs. 6 years to diagnosis on average and reduction of costs by 2/3rds (Özen et al., 2017).

Many commercial providers offer clinical genetic testing of DSD and many institutions have started developing their in-house tests. Prevention Genetics (www.preventiongenetics.com) provides a gene panel of 65 genes covering both syndromic and non-syndromic DSD accounting for ~35% of cases of 46,XY DSD. Using this panel they have reported deletions and duplications in known DSD genes such as *SOX3, LHCGR, SRY, NR0B1, DMRT1, NR5A1, GATA4, WT1, WNT4,* and *FGFR2*. Invitae (www.invitae.com) offers a panel of 8 genes (*AR, DHH, MAP3K1, NR0B1, NR5A1, SRD5A2, SRY* and *WT1*), expandable to 15, for genetic diagnosis of cases of 46,XY DSD, 46,XY complete gonadal dysgenesis and 46,XX testicular DSD. It combines full-gene sequencing and deletion/duplication analysis via NGS in clinically important genetic regions including exons and adjacent ±10 bp of intronic sequences. GeneDx (www.genedx.com) provides a 19-gene panel (*AR, ARX, ATRX, CHD7, CYP11A1, CYP17A1, DHCR7, DHH, DYNC2H1, HSD17B3, HSD3B2, NEK1, NR5A1, POR, SOX9, SRD5A2, SRY, STAR, WT1*) for diagnosis of prenatal and neonatal 46,XY DSD through NGS and deletion/duplication analysis. Blueprint genetics (www.blueprintgenetics.com) offers a gene panel covering 49 genes that assesses coding as well as non-coding variants for patients suspected to harbor DSD or congenital adrenal hyperplasia (CAH).

As is evident from the examples above, these panels are not standardized. With constantly evolving methods (capture kits, gene coverage percentage, sequencing platforms, and analysis pipelines) and variability of options, it is becoming increasingly difficult to ascertain what method was actually used to diagnose a patient and whether a different test might have found a different result.

7. Evaluation of genes on the DSD gene list

A challenge of both whole-exome and limited panels is deciding which genes make the gene list. A variety of criteria and resources are used, such as reported cases in PubMed (http://www.ncbi.nim.nih.gov/pubmed), the Online Mendelian Inheritance in Man (OMIM) database (http://www.ncbi.nlm.nih.gov/omim), GeneReviews, or clinical variant databases such as HGMD (http://www.hgmd.cf.ac.uk/ac/index.php) or ClinVar (https://www.ncbi.nlm.nih.gov/clinvar/). A typical clinical list includes 60–75 DSD genes. For research purposes, the list can be lengthened (to ~250) to include genes implicated in gonadal development, known molecular pathways (such as Hedgehog or WNT signaling, or AR-interacting proteins). However, the burden of proof available for assessing that a reported gene is indeed causative of DSD in humans varies greatly by gene. Some have had replication in multiple families and populations and are supported by extensive animal models and *in vitro* demonstration of the pathogenicity of specific variants, while others have been reported in single publications. As a proxy to assessing genetic evidence, we examined the variants reported in ClinVar, the widely used repository of clinical variants maintained and curated by the NIH. We extracted the variants classified as pathogenic and likely pathogenic in 69 genes, in cases reported with a DSD phenotype (Table 1). For each, we looked at the type of variant (missense, nonsense, frameshift, larger INDELs and rearrangements) and the publications quoted in support of the variant classification. We did not include the large CNVs (typically reported in batch from genome-wide scanning experiments) because whether deletion of the particular gene was indeed causative could not be assessed.

For several of the long-known DSD genes, the available proof of pathogenicity was obvious in the database, with the highest numbers of variants

Table 1 Pathogenic and likely pathogenic variants in 69 DSD genes listed in the ClinVar database.

Gene name	Number of variants	Type of variants	Example references
Sex determination			
BMP15 (POF4)	3	Missense, nonsense	Di Pasquale, Beck-Peccoz, and Persani (2004), Dixit et al. (2006), and Rossetti et al. (2009)
CBX2	2	Missense	Biason-Lauber, Konrad, Meyer, deBeaufort, and Schoenle (2009)
DHH	9	FS Del, nonsense, missense	Canto, Munguia, Soderlund, Castro, and Mendez (2008), Tajouri et al. (2018), Umehara et al. (2000), and Werner et al. (2015)
DMRT1	1	Missense	—
DMRT2	0	—	—
FSHR	17	Missense	Aittomäki et al. (1995), Beau et al. (1998), De Leener et al. (2008, 2006), Di Carlo et al. (1997), Doherty et al. (2002), Gromoll, Simoni, and Nieschlag (1996), Kuechler et al. (2010), Meduri et al. (2003), Montanelli et al. (2004), Smits et al. (2003), and Vasseur et al. (2003)
GATA4	1	Missense	Lourenco et al. (2011)
HHAT	0	—	—
MAP3K1	9	Missense, Splice site	Baxter et al. (2015), Granados et al. (2017), Loke and Ostrer (2012), and Pearlman et al. (2010)
NR0B1 (DAX1)	75	FS del/ins, rearrangement, gene deletion, gene duplication, missense, nonsense, splice site	Barbaro et al. (2007), Muscatelli et al. (1994), Nakae, Abe, et al. (1997), Schwartz, Blichfeldt, and Müller (1997), and Yanase et al. (1996)
NR5A1 (SF1)	26	Missense, nonsense, in-frame deletion, FS del	Achermann, Ito, Ito, Hindmarsh, and Jameson (1999), Baetens et al. (2017), Bashamboo et al. (2010), Colson et al. (2017), Correa et al. (2004), Igarashi et al. (2017), Lin et al. (2007), and Lourenço et al. (2009)

Table 1 Pathogenic and likely pathogenic variants in 69 DSD genes listed in the ClinVar database.—cont'd

Gene name	Number of variants	Type of variants	Example references
RSPO1	3	Exon deletion, FS ins, Splice site	Parma et al. (2006) and Tomaselli et al. (2008)
SOX3	3	Gene duplication, missense, in-frame duplication	Karaca et al. (2015) and Woods et al. (2005)
SOX9	14	Promoter Del, Promoter Dup/ Triplication, FS ins, FS del, missense, nonsense	Benko et al. (2011), Cox et al. (2011), Foster et al. (1994), Kim et al. (2015), Kwok et al. (1995), Pop, Zaragoza, Gaudette, Dohrmann, and Scherer (2005), and Vetro et al. (2011)
SRY	27	Missense, FS del, in-frame del, nonsense	Berta et al. (1990), Hawkins et al. (1992), Jäger, Anvret, Hall, and Scherer (1990), McElreavey et al. (1992), Nykamp et al. (2017), and Vilain et al. (1992)
STAG3 (POF8)	4	FS del, FS dup, Splice site, nonsense	Caburet et al. (2014), He et al. (2018), and Le Quesne Stabej et al. (2016)
WNT4	4	Missense	Biason-Lauber et al. (2007); Biason-Lauber, Konrad, Navratil, and Schoenle (2004), Mandel et al. (2008), and Philibert et al. (2008)
WT1	23	Splice site, missense, nonsense, FS del, FS dup	Barbaux et al. (1997), Barrera et al. (2016), Bruening et al. (1992), Little and Wells (1997), Royer-Pokora et al. (2004), and Wang, Zhang, et al. (2017)
WWOX	0	—	—
ZFPM2	3	Missense	Bashamboo et al. (2014) and Royer-Pokora et al. (2004)
Sex differentiation			
AKR1C2	4	Missense	Flück et al. (2011)
AKR1C4	0	—	—
AMH	4	FS Del, FS Ins, nonsense	Carré-Eusèbe et al. (1992), Knebelmann et al. (1991), and Lang-Muritano et al. (2001)

Continued

Table 1 Pathogenic and likely pathogenic variants in 69 DSD genes listed in the ClinVar database.—cont'd

Gene name	Number of variants	Type of variants	Example references
AMHR2	4	FS Del, in-frame del, Splice site, missense	Belville et al. (2009), Imbeaud et al. (1996, 1995), and Messika-Zeitoun et al. (2001)
AR	109	Splice site, missense, FS Ins/del, nonsense, multi exon deletion	Brown et al. (1988), Brüggenwirth et al. (1997), Choong, Quigley, French, and Wilson (1996), MacLean, Chu, Warne, and Zajac (1993), Sammarco et al. (2000), Vilchis et al. (2003), Zhu et al. (1999), and Zoppi et al. (1993)
ARX	2	Nonsense, FS Del	Kato et al. (2004) and Kitamura et al. (2002)
ATRX	24	Splice site, Nonsense, FS Del, missense	Gibbons, Picketts, Villard, and Higgs (1995), Mitson, Kelley, Sternberg, Higgs, & Gibbons, 2011, and Stevenson (1993)
CYP11A1	10	FS Del, FS Ins, in-frame insertion, missense	Hiort et al. (2005), Katsumata et al. (2002), Kim, Lin, et al. (2008), and Tajima, Fujieda, Kouda, Nakae, and Miller (2001)
CYP17A1	35	Splice site, in-frame deletion, FS del, FS Dup, rearrangement, missense, nonsense	Ahlgren, Yanase, Simpson, Winter, and Waterman (1992), Imai, Yanase, Waterman, Simpson, and Pratt (1992), Kagimoto, Winter, Kagimoto, Simpson, and Waterman (1988), Oshiro et al. (1995), Schwab, Moisan, Homoki, Peter, and Simard (2005), van den Akker et al. (2002), Yamaguchi, Nakazato, Miyazato, Kangawa, and Matsukura (1997), Yanase et al. (1990), and Yang et al. (2006)
CYP19A1	13	fusion, insertion, deletion, missense SNV, intron SNV, missense SNV, nonsense SNV	Deladoëy et al. (1999), Herrmann et al. (2002), Ito, Fisher, Conte, Grumbach, and Simpson (1993), Shozu et al. (2003), and Tiulpakov et al. (2005)

Table 1 Pathogenic and likely pathogenic variants in 69 DSD genes listed in the ClinVar database.—cont'd

Gene name	Number of variants	Type of variants	Example references
CYP21A2	49	Deletion, duplication, SNVs, insertion, intron SNV, missense SNV	Billerbeck et al. (2002), Hayashi et al. (2013), Kaupert et al. (2016), Kirac et al. (2014), Livadas et al. (2015), Wang, Pallan, et al. (2017), Wang et al. (2016), Welzel et al. (2010), and Yoo et al. (2013)
DHCR7	90	Missense, FS Ins, in-frame deletion, nonsense, Splice site	Lanthaler et al. (2015), Sparks et al. (2014), Wassif et al. (1998), Waterham and Hennekam (2012), Waye et al. (2002), and Yu et al. (2000)
FGFR2	3	Intragenic deletion, missense	Bagheri-Fam et al. (2015), Fonseca, Costa-Lima, Cosentino, and Orioli (2008), Przylepa et al. (1996), and Slavotinek et al. (2009)
FOXL2 (POF 3)	5	FS Del, FS Ins, in-frame deletion, missense	Harris et al. (2002) and Yang et al. (2018)
HSD17B3	23	Missense, Splice site, FS Del, SNV deleting a stop codon	Castro et al. (2012), Geissler et al. (1994), Lindqvist, Hughes, and Andersson (2001), and Phelan et al. (2015)
HSD3B2	11	FS ins, FS del, nonsense, missense	Rhéaume et al. (1992), Simard et al. (1994), and Welzel et al. (2008)
KDM5D	0	—	—
LHCGR	32	FS del, in-frame deletion, in-frame insertion, missense, nonsense, Splice site	Boot et al. (2011), Gromoll, Eiholzer, Nieschlag, and Simoni (2000), Laue et al. (1995), Misrahi et al. (1997), and Wu, Hallermeier, et al. (1998)
MAMLD1	4	Nonsense	Fukami et al. (2006)
POR	14	Splice site, missense, FS Ins, FS Dup, deletion	Arlt et al. (2004), Flück et al. (2004), Fukami et al. (2005), Huang et al. (2005), and Oh et al. (2017)

Continued

Table 1 Pathogenic and likely pathogenic variants in 69 DSD genes listed in the ClinVar database.—cont'd

Gene name	Number of variants	Type of variants	Example references
SRD5A2	41	FS Del, missense, nonsense, Splice site, exon loss deletion, in-frame deletion	Bertelloni et al. (2016), Cai et al. (1996), Can et al. (1998), Eggers et al. (2016), Ko et al. (2010), and Thigpen, Davis, Gautier, Imperato-McGinley, and Russell (1992)
STAR	33	FS Del, FS Ins, Splice site, missense, nonsense	Abdulhadi-Atwan et al. (2007), Bose et al. (1996), Flück et al. (2011), Huang et al. (2016), Nakae, Tajima, et al. (1997), Okuyama et al. (1997), and Tee et al. (1995)
VAMP7	0	—	—
Central causes of hypogonadism			
ARL6 (BBS3)	10	Missense, nonsense, gene deletion, FS Del, Splice site	Chiang et al. (2004), Fan et al. (2004), and Lindstrand et al. (2016)
CHD7	187	Multi-exon deletion, FS Del, FS Ins, nonsense, Splice site, rearrangement, a few missense	Jongmans et al. (2008), Kim, Kurth, et al. (2008), Moccia et al. (2018), and Udaka et al. (2007)
FGF8	7	FS Del, Missense, 5′UTR SNV	Costa-Barbosa et al. (2013), and Falardeau et al. (2008)
FGFR1	27	Missense, nonsense, FS Del, Splice site	Dodé et al. (2003), Pitteloud et al. (2006), Trarbach et al. (2006), and Xu et al. (2007)
FRAS1	17	FS Ins, FS Del, nonsense splice site, in-frame deletion	Cavalcanti, Matejas, Luquetti, Mello, and Zenker (2007), McGregor et al. (2003), and Slavotinek, Li, Sherr, and Chudley (2006)
FREM2	6	Missense, FS Del, FS Ins, nonsense, Splice site	Jadeja et al. (2005), Shafeghati, Kniepert, Vakili, and Zenker (2008), and van Haelst et al. (2008)
GNRH1	1	FS Ins	Bouligand et al. (2009)

Table 1 Pathogenic and likely pathogenic variants in 69 DSD genes listed in the ClinVar database.—cont'd

Gene name	Number of variants	Type of variants	Example references
GNRHR	20	Missense, nonsense, Splice site	Caron et al. (1999), Gianetti et al. (2012), Kottler et al. (2000), Layman et al. (1998), and Meysing et al. (2004)
GRIP1	2	Splice site, FS Del	Vogel et al. (2012)
HESX1	14	FS Del, FS Ins, missense, splice site, nonsense	Brickman et al. (2001), Cohen et al. (2003), Dattani et al. (1998), Sobrier et al. (2006), Tajima et al. (2003), and Thomas et al. (2001)
HFE	11	Missense, nonsense, FS Del	Beutler, Felitti, Koziol, Ho, and Gelbart (2002), de Villiers, Hillermann, Loubser, and Kotze (1999), Gochee et al. (2002), and Piperno et al. (2000)
KAL1/ ANOS1	25	Multi-exon deletion, FS Dup, FS Del, missense, nonsense, splice site	Albuisson et al. (2005), Canto et al. (2008), Massin et al. (2003), Söderlund, Canto, and Méndez (2002), and Trarbach et al. (2006)
KISS1R	7	Intragenic deletion, missense, nonsense, FS SNV	Brioude et al. (2013), Seminara et al. (2003), and Tenenbaum-Rakover et al. (2007)
LEP	4	FS Del, missense	Gibson et al. (2004), Strobel, Issad, Camoin, Ozata, and Strosberg (1998), and Wabitsch et al. (2015)
LEPR	2	Splice site, FS Ins	Clément et al. (1998)
LHX3	9	Gene deletion, intragenic deletion, FS del, nonsense, missense	Netchine et al. (2000), Pfaeffle et al. (2007), and Rajab et al. (2008)
PCSK1	5	Splice site, missense, nonsense, in-frame deletion	Farooqi et al. (2007) and Jackson et al. (2003, 1997)
PROK2	6	FS Ins, FS Del, missense	Cole et al. (2008), Dodé et al. (2006), Leroy et al. (2008), and Pitteloud et al. (2007)

Continued

Table 1 Pathogenic and likely pathogenic variants in 69 DSD genes listed in the ClinVar database.—cont'd

Gene name	Number of variants	Type of variants	Example references
PROKR2	12	Missense, FS Del	Cole et al. (2008), Dodé et al. (2006), Dodé and Rondard (2013), McCabe et al. (2013), Monnier et al. (2009), Reynaud et al. (2012), Ruiz-Ferrer et al. (2011), and Sarfati et al. (2013, 2010)
PROP1	31	Nonsense, missense, FS Ins, FS Del, Splice site	Cole et al. (2008), Deladoëy et al. (1999), Fofanova et al. (1998), Kelberman et al. (2009), Lemos et al. (2006), Reynaud et al. (2005), Sarfati et al. (2010), and Wu, Cogan, et al. (1998)
PTPN11	80	Missense, in-frame deletion	Sobreira et al. (2010), Tartaglia et al. (2006), Thiel et al. (2009), Yoshida et al. (2004), and Zenker et al. (2004)
SOS1	27	Missense	Lepri et al. (2011), Li, Ann Thomas, and Haber (2013), and Roberts et al. (2007)
TAC3	4	Missense, FS Del	Gianetti et al. (2010) and Topaloglu et al. (2009)
TACR3	7	Missense, nonsense	Francou et al. (2011), Gianetti et al. (2010), and Topaloglu et al. (2009)
TRIM32	6	Missense, FS Del	Chiang et al. (2006), Locke, Tinsley, Benson, and Blake (2009), and Van Goor, Yu, Burton, and Hoffman (2014)
TTC8	4	In-frame deletion, Splice site, 5′UTR deletion	Ansley et al. (2003) and Stoetzel et al. (2006)

The variants in 69 human DSD genes (column 1) deposited in the NIH-maintained ClinVar database and classified as pathogenic or likely pathogenic for a DSD condition were compiled. The total number (column 2) and type (column 3) of variants are shown (FS Del/Ins: deletion/insertion resulting in frameshift). Copy number variants that included multiple genes were not included, as pathogenicity could not be easily attributed to the single gene. If the genes were involved in several phenotypes, only the variants relevant to DSD were included. Variants for which phenotypic information was not available were not included. Selected references cited as evidence in the database are shown in column 4 (the complete list for each gene entry can be found at https://www.ncbi.nlm.nih.gov/clinvar/).

reported for *AR* (109), *NR0B1*/DAX1 (75), *CYP21A2* (49), *SRD5A2* (41), *SRY* (27), *NR5A1*/SF1 (26), *WT1* (23). For example, many types of severe variants (splice site, multi-exon deletion, nonsense, frameshift-causing insertions or deletions) support the pathogenicity of potentially milder (missense) variants in *AR* causing Androgen Insensitivity Syndrome. The extensive number of publications in support of these variants is complemented by data uploads from reliable clinical testing labs such as GeneDx, Johns Hopkins or Invitae. However, even these can be underreported. The *AR* mutation database created at McGill University (last updated in 2014) had >550 variants (Gottlieb & Trifiro, 2017), while only 109 are found in ClinVar. Similarly, the number of variants reported in ClinVar for *CYP21A2* is relatively low for such a frequent condition, and better resources exist elsewhere. A recent effort (Simonetti et al., 2018) curated existing variants from multiple sources (e.g., CYPAlleles and dbSNP databases) and compiled them in a new gene-specific database to support genetic counseling to families with Congenital Adrenal Hyperplasia.

Other "usual suspects" genes artificially show a low number of variants in ClinVar. In some cases, many more are actually in the database but, without a phenotype reported, they were excluded from the analysis, as "pathogenic" can only refer to a specific phenotype. In others cases, the explanation was less obvious. For example, *AKR1C2* seems to have a similar amount of evidence as *AMHR2*-4 variants each-with possible slant in favor of *AMHR2* due to more severe and diverse variants (e.g., frameshift, splice site, against only missense for *AKR1C2*). The *AKR1C2* variants, however, are all from the same publication, and the finding has not been replicated as widely. For *AMHR2*, one of the supporting publications quoted in the database (Belville et al., 2009) reports that 38 different mutations had been identified in patients with Persistent Müllerian Duct Syndrome by their team alone at that point, and describes the molecular effects of 10 of those variants. Those are yet to be included in ClinVar. Similarly, *ARX*, which causes an X-linked syndromic form of DSD in males, associating lissencephaly and abnormal genitalia, has only two ClinVar variants, one nonsense (Kato et al., 2004) and one deletion leading to a frameshift, but without a supporting reference. The quoted reference is actually the second paper from that team, which had reported at least another 15 variants in this gene. In these cases, the existing genetic proof is actually much stronger than reflected in the database.

This approach was also limited by the precision of phenotypic description in the database. For example, while pathogenicity of loss-of-function mutations in *ATRX* in the X-linked dominant alpha-thalassemia/MR syndrome is well documented with 24 variants in ClinVar, which of these mutations actually lead to DSD was more difficult to evaluate. The genital phenotype only affects males, while females have normal genitalia. However, variants are reported here without indication of whether they were found in XX or XY individuals. This is also an issue for syndromes that may include genital anomalies as a non-fully penetrant trait. Variants are reported in the databases under the umbrella of the syndrome, without specification of genital involvement. One striking example is *DHCR7*, the causative gene for CHARGE syndrome, which has one of the highest number of variants in the list (90). Clarifying which variants are associated with genital anomalies requires analyzing each original publication and is not possible in the case of variants reported by clinical testing labs, not linked to a publication. In these cases, genotype-phenotype correlations, if any exist, have to be manually curated.

Six genes in the table have no DSD-associated variants reported in ClinVar: *AKR1C4*, *DMRT2*, *HHAT*, *KDM5D*, *VAMP7* and *WWOX*. We examined the existing evidence for including those in a primary gene list for exome filtering in human DSD diagnosis. Three of these genes have in common to have been identified as causative on the basis of CNVs. As in the case for other dosage-sensitive genes involved in DSD (such as *SOX3*, *NR0B1*/DAX1 or *WNT4*), variants are found in very few patients each and are difficult to both identify and validate.

Duplication of the distal region of the X chromosome Xq28, including *VAMP7*, was found using array CGH in 2/116 patients with genitourinary tract masculinization disorder (Tannour-Louet et al., 2010). That paper identified 11 other cases in the literature, 3 more by quantitative PCR in a different cohort, and demonstrated that transgenic mice overexpressing human *VAMP7* mimicked the phenotype, possibly via potentiation of the estrogen signaling pathway through ESR1. The experimental evidence is solid, and awaits replication by others and identification of other types of variants in *VAMP7*.

A SNP array identified a 767-kb in-frame deletion of exons 6–8 of the *WWOX* gene in a patient with 46,XY DSD (White et al., 2012), and the corresponding truncated transcript was found in the patient's cells. The paper reported that two different knock-out mouse strains showed Leydig cell function defects and gonadal abnormalities. Validation of *WWOX* as a human DSD gene awaits experimental demonstration of protein dysfunction and identification of variants in other patients.

Multiple reports have associated 46,XY gonadal dysgenesis with deletion of chromosome 9p24.3, without the other features of terminal 9p deletion syndrome (cataloged as 46,XY Sex Reversal Type 4 in OMIM). This led to the hypothesis that haploinsufficiency for *DMRT1* and/or *DMRT2* is causative of the genital phenotype. Embryonic expression of *DMRT1* is gonad-specific and sexually dimorphic; *DMRT2* shows only weak expression in the developing gonad. Both genes share homology with sex-determining genes in Drosophila (*Doublesex*) and *Caenorhabditis elegans* (*Mab3*) but confirmatory evidence in humans is weak, with one missense (without supportive reference) in *DMRT1* in ClinVar and none in *DMRT2*. However, at least two other types of variants have now been identified in *DMRT1*: a missense mutation (Chauhan, Jyotsna, Jain, Khadgawat, & Dada, 2016) and a deletion of exons 3–4 (Ledig, Hiort, Wünsch, & Wieacker, 2012).

A short peptide derived from the KDM5D (lysine demethylase 5D) protein is the Y-chromosome-linked H-Y/HLA-B7 (aka HYA) minor histocompatibility antigen. GeneCards links *KDM5D* to "Spermatogenic Failure, Y-linked, Type 2." UniProt/SwissProt data suggests that it may regulate transcriptional activity of *AR*. However, OMIM does not report a link between this gene and the phenotype and no clear evidence was found in a literature search. Since the azoospermia AZF locus has been shown to not include HYA, this gene should probably be removed from primary DSD gene lists.

A homozygous missense variant in *HHAT*, the Hedgehog acetyl transferase, was identified by exome sequencing as the cause of a familial autosomal recessive syndromic form of 46,XY DSD associated with chondrodysplasia and severe brain abnormalities (Callier et al., 2014). An XX sibling with normal ovaries shared the non-genital phenotype and the variant. Evidence included prediction of protein disruption by multiple algorithms and confirmation by *in vitro* experiments, expression of the protein in the developing gonad, and demonstration that palmitoylation of Hedgehog ligands is required for testes organogenesis in mouse. The experimental evidence is elegant, but awaits identification of further cases.

AKR1C4 is part of a complex encoded on chromosome 10p15.1 that plays a role in metabolism of steroid hormones. Evidence for its involvement in human DSD rests on a single family with testicular and adrenal 17,20-desmolase deficiency resulting in a defect in androgen biosynthesis. In this family, a splice site variant in *AKR1C4* segregated together with a missense in related gene *AKR1C2* (Flück et al., 2011); another patient in the same report harbored a complex rearrangement of the region. Therefore *AKR1C4* is likely best considered a modifier (as OMIM classifies it for

46,XY Sex Reversal Type 8) or in a digenic model (e.g., with *AKR1C2*) and searching for such a mode of inheritance in exome results might help identify further patients.

8. Whole genome sequencing

While exome sequencing is a clear improvement over single-gene testing in providing clinical diagnosis for DSD, it leads to a definitive molecular diagnosis in only 30% of all cases (Baxter et al., 2015; Lee et al., 2014; Yang et al., 2014). About 85% of disease-associated variants are found in coding regions (Abecasis et al., 2010), but this is likely to be an ascertainment bias. Metaanalysis of hundreds of genome-wide association studies showed that >90% of genetic variants associated with complex multigenic disorders lie within non-coding regions of the genome instead of the regions captured by exome sequencing (Maurano et al., 2012). Genome-wide screens of DSD patients have discovered variants in the regulatory non-coding regions of DSD associated genes, in particular *SOX9* and *SOX3* (Croft et al., 2018; Harrison et al., 2013; Sutton et al., 2011; White et al., 2011). Therefore a technique sequencing the whole genome rather than only the protein-coding regions has the potential to provide a much higher diagnostic yield than WES.

Whole-Genome Sequencing (WGS), which is also based on short-read sequencing but is not constrained by the exome capture step, overcomes limitations of WES such as lack of coverage of some GC rich parts of the exome because of the PCR enrichment step, or the preferential capture of reference sequence allele leading to an allele distribution bias (Heinrich et al., 2012; Meynert, Ansari, FitzPatrick, & Taylor, 2014; Meynert, Bicknell, Hurles, Jackson, & Taylor, 2013; Quail et al., 2008). WGS is more powerful in detecting disease-causing SNVs than WES, even in exomic regions. For example, a study comparing WES and WGS results for six unrelated individuals showed that both techniques identified a similar number of the total as well as high quality SNVs and INDELs; however, the distributions of coverage depth, genotype quality and minor read ratio were more uniform for WGS than for WES (Belkadi et al., 2015). Confirmation using Sanger sequencing showed a similar percentage of false positives for INDELs (44% vs. 46%). However, SNV false positives were dramatically higher (78%) in WES than in WGS (17%).

Studies comparing the diagnostic yield of WGS with that of clinically conventional genetic diagnostic techniques typically report higher yield.

For example, a diagnostic yield of 34% in a cohort of children with various congenital anomalies, including developmental delay, represents a fourfold increase in diagnostic rate over CMA and twofold increase over targeted gene sequencing (Stavropoulos et al., 2016). NGS-based targeted sequencing is being increasingly used in conjunction with CMA and WES to improve clinical diagnostics of DSD cases. For example, a combination of targeted NGS, WES and CMA was able to detect genetic defects in 85% of the families in a cohort of 63 Japanese patients with biochemically uncharacterized pediatric onset primary adrenal insufficiency, with some of them exhibiting 46,XY DSD phenotypes (Amano et al., 2017). In another report, array CGH and WGS were utilized to determine a structural variant caused by a ~774 kb insertional translocation from chromosome 1 into a palindromic sequence 84 kb distal of the *SOX3* locus on chromosome X, leading to its upregulation in a 46,XX phenotypically male patient (Haines et al., 2015).

WGS is slowly being incorporated into the clinical diagnosis process throughout the world with the hope that providing physicians and patients with an interpreted genome sequence will reduce the uncertainty of clinical decision-making and improve clinical outcomes. However, the wide availability of WGS for clinical purposes is currently limited by several factors. With about 3 million variants compared to the reference genome in each individual, the complexity of variant filtering and prioritizing is exponentially amplified compared to exome sequencing. Diagnostic capacity and variant classification for non-coding variants is also limited by the availability of comprehensive, ethnically rich, reference databases of control and disease sequences. In addition, WGS has the potential to detect large SVs, CNVs, INDELs as well as SNPs across the genome, when *de novo* assembly becomes more routinely feasible. However, due to the utilization of short-reads, the discovery of SVs by WGS is compromised. Many bioinformatic tools have been developed for SV calling from NGS data using a combination of read-depth, read-pair, and split-read methods, but the number and type of SVs identified by each of the tools vary significantly with low sensitivity, low concordance rates, and large false discovery (English et al., 2015; Tattini, D'Aurizio, & Magi, 2015).

9. Long-read sequencing

The human genome is arguably the most complete mammalian reference genome assembly, but it still contained over 160 euchromatic gaps encompassing long runs of degenerate short tandem repeats, often several kilobases in length, embedded within GC-rich genomic regions

(Chaisson et al., 2015; Sedlazeck, Lee, Darby, & Schatz, 2018). Although NGS significantly reduced the cost of sequencing, the utilization of short-reads and amplification biases led to difficulties in identification of larger SVs, as well as problems in sequencing certain genomic regions (Chaisson et al., 2015). This limitation is being remedied with the advent of "third-generation," long-read sequencing (LRS) methods such as Single Molecule Real Time (SMRT) sequencing (Pacific Biosciences, PacBio) and nanopore sequencing (Oxford Nanopore Technologies, ONT). These also promise to revolutionize transcriptome analysis by allowing the sequencing of full-length RNA/cDNA molecules in which exon skipping and isoform expression can be determined.

PacBio's SMRT sequencing utilizes "sequencing by synthesis method" where incorporation of differentially fluorescently labeled nucleotides into immobilized individual DNA templates is recorded in real time, resulting in long reads that end only when the DNA polymerase dissociates from the template (Rank et al., 2009). The average read length from the current PacBio's Sequel System is approximately 10,000 bp, but new improved chemistry promises to increase the read lengths to an average of 85,000 bp (Kosicki, Tomberg, & Bradley, 2018; Sedlazeck, Rescheneder, et al., 2018).

Nanopore sequencing passes a single strand of DNA through a pore where the electric signal changes generated by the base-by-base movement of the molecule across the nanopore are converted into base-calls (Deamer, Akeson, & Branton, 2016; Jain et al., 2015; Jain, Olsen, Paten, & Akeson, 2016). Some users have observed read-lengths in excess of 2 Mb (www.nanoporetech.com).

LRS has been employed for a number of purposes to overcome challenges faced by NGS technologies. For example, it was successfully used to sequence regions of microsatellite repeats in cases of neurodegenerative diseases such as ALS and frontotemporal dementia (FTD) as well as polyglutamine expansion in Huntington disease, dentatorubropallidoluysian atrophy, spinal and bulbar muscular atrophy (Ebbert et al., 2018; Liu, Zhang, Wang, Gu, & Wang, 2017). PacBio long-range RNA sequencing was used to sequence the chicken transcriptome, which helped resolve regions encompassing multiple-mapping loci, repeat-regions and ambiguous splice junctions across the coding as well as non-coding transcriptome (Kuo et al., 2017). PacBio and Nanopore sequencing were also utilized in conjunction with Illumina SRS to create *de novo* genome assembly of *Saccharomyces cerevisiae* (Jenjaroenpun et al., 2018). Application of LRS in clinical diagnosis in a case of two patients with congenital abnormalities detected novel chromothripsis rearrangements with details of breakpoints, parental origin and structure, as well as novel variants originating from retrotransposon insertions at higher efficiencies in comparison to Illumina SRS (Cretu Stancu et al., 2017).

In 2018, both Nanopore (Jain et al., 2018) and PacBio (pacbio.com press release, Oct. 2018) have reported complete *de novo* assembly of a human genome. However both techniques have severe limitations that will need to be overcome before widespread use in clinical practice is possible. Both require important compute time and bioinformatic resources. Nanopore sequencing is fast but shows a lack of base-calling accuracy; Pac-Bio is very accurate, but slow. Methods are being developed to combine the strengths of each to create accurate, isoform-level transcriptome and genome annotation (Volden et al., 2018), promising to fully exploit the power of LRS technologies to discover large SV and sequence variation in a single molecule, which can be the next paradigm for DSDs diagnosis.

10. Optical genome mapping

Many SVs are caused by non-allelic homologous recombination events between duplicated DNA sequences as large as 300 kbp in size. The repetitive nature of these regions makes them all but invisible to short-read sequencing platforms (Conrad et al., 2010; Sharp et al., 2005) or traditional cytogenetics methods such as CMAs, which can elucidate gain or loss of genetic material, but fail to detect the order/orientation of the SV and are virtually blind for detection of balanced events such as inversions and translocations. Recently, a novel method—optical genome mapping has been proposed as the "go to" method for identification of large/complex structural variants due to its high specificity and sensitivity (Hastie et al., 2017). SVs account for the largest portion of human-to-human genome variation (Sudmant et al., 2015; Weischenfeldt, Symmons, Spitz, & Korbel, 2013) and have been implicated in a variety of disease etiologies including DSD (Barseghyan et al., 2015), neuropsychiatric disorders (Brand et al., 2014) and heart disease (Brand et al., 2014).

Optical genome mapping (supported by the Bionano Genomics platform) uses specialized nanochannel arrays capable of housing immobilized, labeled megabase-size DNA molecules for optical imaging (Hastie et al., 2017). Prior to imaging, the long DNA molecules are fluorescently labeled at specific sequence motifs throughout the genome, using restriction

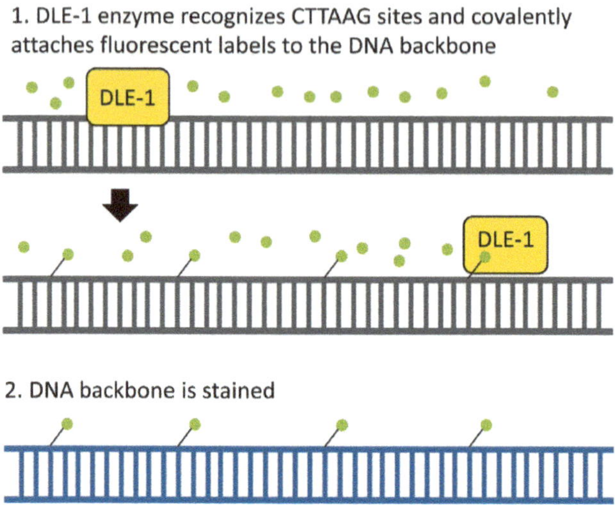

Fig. 2 Preparation of long DNA molecules for optical mapping: High molecular weight DNA is incubated with fluorescent labels and Direct Labeling Enzyme 1 (DLE1). The enzyme recognizes CTTAAG sequences throughout the genome and covalently attaches a fluorescent label (green). Subsequently, the DNA backbone is stained overnight and loaded onto the Bionano Genomics chips containing millions of nanochannels for imaging. The green fluorescent labels are used to map DLE1 enzyme recognition patterns, whereas the blue staining of the DNA backbone is used to identify molecule sizes. Both are used for *de novo* genome assembly.

enzymes (reviewed in Barseghyan, Délot, & Vilain, 2018) or other proprietary enzymes (Fig. 2). The resultant pattern of fluorescent tags within long DNA molecules is used for *de novo* assembly of each allele of the sample genome with scaffold length N50 values exceeding 60 Mbps (to compare with ~10.5 Mbps for Nanopore) (Jain et al., 2018). Comparison of the sample-specific maps with the in silico-digested maps of the human genome reference allows for identification of large deletions, insertions, inversions, translocations and complex rearrangements. Optical mapping has been used for identification of potentially pathogenic SVs in cancer genomes (Dixon et al., 2018; Du et al., 2018; Jaratlerdsiri et al., 2017), population-specific genome assemblies (McCaffrey et al., 2017; Seo et al., 2016; Shi et al., 2016), as well as muscle-wasting disorders such as Duchenne and facioscapulohumeral muscular dystrophies (Barseghyan et al., 2017; Dai et al., 2018). In Duchenne Muscular Dystrophy, we showed the ability of whole-genome optical mapping to identify large deletions, insertions, and inversions in hemizygous and heterozygous states with 100% clinical diagnosis concordance rates (Barseghyan et al., 2017).

Although optical mapping brings key advantages over the clinically available cytogenetic technologies such as karyotype and microarray, mainly increased resolution and detection of balanced events, its adoption in clinic is still pending. Our group is working on bringing optical mapping to DSD diagnosis (Barseghyan et al., 2018). Integration of WGS and whole-genome optical mapping allows researchers to survey all of the different types of genomic variations: SNVs, INDELs, CNVs and large SVs. Recent, unpublished work suggests that integration of the two technologies tremendously improves WGS SV identification rates, while also decreasing optical mapping's SV breakpoint uncertainty to <140 bp (originally ~3–5 kbp (Porat et al., 2018)). Combined integrated optical mapping and genome sequencing analysis is poised to provide higher diagnostic rates for DSD and other genetic disorders.

Identification of clinically relevant SVs will require integration of data sets obtained through all current technologies to ascertain both what is a true SV (variant calling) and which are pathogenic (variant classification). An early global map of 270 individuals using SNP arrays and array CGH identified ~1500 copy number variable regions over 1 kb, covering 12% of the genome (Redon et al. 2006). Integration of SRS, LRS and optical mapping on a genome identified 10,000 putative SVs compared to the hg19 reference assembly (English et al., 2015). The PacBio method identified >20,000 unique SVs (>50 bp) and ~400,000 1–49 bp INDELs, 80% of which could not be resolved with current SRS techniques (Ameur, Che, et al., 2018, Ameur, Kloosterman, & Hestand, 2018, Sedlazeck, Rescheneder, et al., 2018). A major difficulty remains assembling enough genomes to understand the complexity of pathogenic and non-pathogenic human genomic SV diversity.

11. Epigenetics

Finally, epigenetic variation may also represent a significant etiology in DSD. One of the most compelling pieces of evidence of the role of epigenetic changes governing sex determination is the promoter DNA methylation of the master regulator *Sry* (Nishino, Hattori, Tanaka, & Shiota, 2004). *Sry* expression in male mice happens in a subset of gonadal somatic cells under tight spatio-temporal regulation around embryonic days 10.5–12.5 (DiNapoli & Capel, 2008; Hiramatsu et al., 2009). The promoter region of *Sry* was observed to be hypermethylated (associated with transcriptional repression) in gonadal somatic cells at 8.5 dpc and hypomethylated (associated with transcriptional activation) at 11.5 dpc while remaining hypermethylated in tissue not expressing Sry at the same time-point (Nishino et al., 2004). There have also been reports of differences in methylation patterns of key sex determination genes in cases of animal DSD that have a high congruence with human DSD genes. Persistent abnormal hypermethylation of the *Sry* gene (resulting in down-regulated mRNA and protein expression) was established as the cause of ovotestis and female phenotype in cloned *Sry*-positive XY dogs generated through somatic cell nuclear transfer (Jeong et al., 2016). In another example involving XX DSD dogs with ovotestis or testis phenotype, candidate gene promoter methylation bisulfite sequencing revealed hypomethylation in *Sox9*, hypermethylation in *Wnt4* and hypermethylation of the *Sox3* promoter at levels similar to that of control XY male dogs (Salamon, Flisikowski, & Switonski, 2017). Both these examples show that epigenetic promoter methylation of important sex determination genes can be determinants of DSD in mammals.

Genomic DNA methylation at CpG sites can be detected by several technologies, including bisulfite conversion followed by SRS or SNP arrays (Kurdyukov & Bullock, 2016; Laird, 2010). Both PacBio and Oxford Nanopore LRS technologies potentially provide improved methylome sequencing capabilities (Fang et al., 2012; Murray et al., 2012) and Bionano is also developing a labeling method that will differentially detect unmethylated vs. methylated cytosines at the recognition site. All three have the advantage over current techniques to identify epigenetic marks and genomic sequence (PacBio, Nanopore) or mapping (Bionano) on the same, single molecule of DNA. These developing technologies bring the exciting prospect of enabling exploration of epigenetic modifications as a new and upcoming paradigm in diagnosis of DSDs.

References

Abacl, A., Çatll, G., & Berberołlu, M. (2015). Gonadal malignancy risk and prophylactic gonadectomy in disorders of sexual development. *Journal of Pediatric Endocrinology & Metabolism*, *28*, 1019–1027. https://doi.org/10.1515/jpem-2014-0522.

Abdulhadi-Atwan, M., Jean, A., Chung, W. K., Meir, K., Ben Neriah, Z., Stratigopoulos, G., et al. (2007). Role of a founder c.201_202delCT mutation and new phenotypic features of congenital lipoid adrenal hyperplasia in Palestinians. *The Journal of Clinical Endocrinology and Metabolism*, *92*, 4000–4008. https://doi.org/10.1210/jc.2007-1306.

Abecasis, G. R., Altshuler, D. L., Durbin, R. M., Bentley, D. R., Chakravarti, A., Clark, A. G., et al. (2010). A map of human genome variation from population-scale sequencing. *Nature*, *467*, 1061–1073. https://doi.org/10.1038/nature09534.

Aboura, A., Dupas, C., Tachdjian, G., Portnoï, M.-F., Bourcigaux, N., Dewailly, D., et al. (2009). Array comparative genomic hybridization profiling analysis reveals deoxyribonucleic acid copy number variations associated with premature ovarian failure. *The Journal of Clinical Endocrinology and Metabolism*, *94*, 4540–4546. https://doi.org/10.1210/jc.2009-0186.

Achermann, J. C., Ito, M., Ito, M., Hindmarsh, P. C., & Jameson, J. L. (1999). A mutation in the gene encoding steroidogenic factor-1 causes XY sex reversal and adrenal failure in humans. *Nature Genetics*, *22*, 125–126. https://doi.org/10.1038/9629.

ACMG Board of Directors. (2014). ACMG policy statement: Updated recommendations regarding analysis and reporting of secondary findings in clinical genome-scale sequencing. *Genetics in Medicine*, *17*, 68.

Adam, M. P., & Vilain, E. (2017). Emerging issues in disorders/differences of sex development (DSD). *American Journal of Medical Genetics. Part C, Seminars in Medical Genetics*, *175*, 249–252. https://doi.org/10.1002/ajmg.c.31564.

Adzhubei, I. A., Schmidt, S., Peshkin, L., Ramensky, V. E., Gerasimova, A., Bork, P., et al. (2010). A method and server for predicting damaging missense mutations. *Nature Methods*, *7*, 248–249. https://doi.org/10.1038/nmeth0410-248.

Ahlgren, R., Yanase, T., Simpson, E. R., Winter, J. S., & Waterman, M. R. (1992). Compound heterozygous mutations (Arg 239——stop, Pro 342——Thr) in the CYP17 (P45017 alpha) gene lead to ambiguous external genitalia in a male patient with partial combined 17 alpha-hydroxylase/17,20-lyase deficiency. *The Journal of Clinical Endocrinology and Metabolism*, *74*, 667–672. https://doi.org/10.1210/jcem.74.3.1740503.

Ahmed, S. F., & Hughes, I. A. (2002). The genetics of male undermasculinization. *Clinical Endocrinology*, *56*, 1–18.

Aittomäki, K., Lucena, J. L., Pakarinen, P., Sistonen, P., Tapanainen, J., Gromoll, J., et al. (1995). Mutation in the follicle-stimulating hormone receptor gene causes hereditary hypergonadotropic ovarian failure. *Cell*, *82*, 959–968.

Albuisson, J., Pêcheux, C., Carel, J.-C., Lacombe, D., Leheup, B., Lapuzina, P., et al. (2005). Kallmann syndrome: 14 Novel mutations in *KAL1* and *FGFR1* (*KAL2*). *Human Mutation*, *25*, 98–99. https://doi.org/10.1002/humu.9298.

Allyse, M., & Michie, M. (2013). Not-so-incidental findings: The ACMG recommendations on the reporting of incidental findings in clinical whole genome and whole exome sequencing. *Trends in Biotechnology*, *31*, 439–441. https://doi.org/10.1016/j.tibtech.2013.04.006.

Amano, N., Narumi, S., Hayashi, M., Takagi, M., Imai, K., Nakamura, T., et al. (2017). Genetic defects in pediatric-onset adrenal insufficiency in Japan. *European Journal of Endocrinology*, *177*, 187–194. https://doi.org/10.1530/EJE-17-0027.

Amarillo, I. E., Nievera, I., Hagan, A., Huchthagowder, V., Heeley, J., Hollander, A., et al. (2016). Integrated small copy number variations and epigenome maps of disorders of sex development. *Human Genome Variation*, *3*, 16012. https://doi.org/10.1038/hgv.2016.12.

Ameur, A., Che, H., Martin, M., Bunikis, I., Dahlberg, J., Höijer, I., et al. (2018). De novo assembly of two Swedish genomes reveals missing segments from the human GRCh38 reference and improves variant calling of population-scale sequencing data. *Genes (Basel)*, *9*, 1–16. https://doi.org/10.3390/genes9100486.

Ameur, A., Kloosterman, W. P., & Hestand, M. S. (2018). Single-molecule sequencing: Towards clinical applications. *Trends in Biotechnology*, *37*, 72–85. https://doi.org/10.1016/j.tibtech.2018.07.013.

Ansley, S. J., Badano, J. L., Blacque, O. E., Hill, J., Hoskins, B. E., Leitch, C. C., et al. (2003). Basal body dysfunction is a likely cause of pleiotropic Bardet–Biedl syndrome. *Nature*, *425*, 628–633. https://doi.org/10.1038/nature02030.

Aqeilan, R. I., Hagan, J. P., de Bruin, A., Rawahneh, M., Salah, Z., Gaudio, E., et al. (2009). Targeted ablation of the WW domain-containing oxidoreductase tumor suppressor leads to impaired steroidogenesis. *Endocrinology*, *150*, 1530–1535. https://doi.org/10.1210/en.2008-1087.

Arboleda, V. A., Fleming, A., Barseghyan, H., Delot, E., Sinsheimer, J. S., & Vilain, E. (2014). Regulation of sex determination in mice by a non-coding genomic region. *Genetics*, *197*, 885–897. https://doi.org/10.1534/genetics.113.160259.

Arboleda, V. A., Fleming, A. A., & Vilain, E. (2010). Disorders of sex development. In R. E. Weiss & S. Refetoff (Eds.), *Genetic diagnosis of endocrine disorders* (1st ed., pp. 227–243). Science Direct.

Arbodela, V. A., & Vilain, E. (2009). Disorders of sex development. In J. Strauss & R. Barbieri (Eds.), *Yen and Jaffe's reproductive endocrinology* (6th ed., pp. 367–393). Saunders Elsevier.

Arboleda, V. A., Sandberg, D. E., & Vilain, E. (2014). DSDs: Genetics, underlying pathologies and psychosexual differentiation. *Nature Reviews. Endocrinology*, *10*, 603–615. https://doi.org/10.1038/nrendo.2014.130.

Arlt, W., Walker, E. A., Draper, N., Ivison, H. E., Ride, J. P., Hammer, F., et al. (2004). Congenital adrenal hyperplasia caused by mutant P450 oxidoreductase and human androgen synthesis: Analytical study. *Lancet*, *363*, 2128–2135. https://doi.org/10.1016/S0140-6736(04)16503-3.

Auchus, R. J., & Miller, W. L. (2012). Defects in androgen biosynthesis causing 46,XY disorders of sexual development. *Seminars in Reproductive Medicine*, *30*, 417–426. https://doi.org/10.1055/s-0032-1324726.

Audí, L., Ahmed, S. F., Krone, N., Cools, M., McElreavey, K., Holterhus, P.-M., et al. (2018). GENETICS IN ENDOCRINOLOGY: Approaches to molecular genetic diagnosis in the management of differences/disorders of sex development (DSD): position paper of EU COST Action BM 1303 "DSDnet" *European Journal of Endocrinology*, *179*, R197–R206. https://doi.org/10.1530/EJE-18-0256.

Baetens, D., Stoop, H., Peelman, F., Todeschini, A.-L., Rosseel, T., Coppieters, F., et al. (2017). NR5A1 is a novel disease gene for 46,XX testicular and ovotesticular disorders of sex development. *Genetics in Medicine*, *19*, 367–376. https://doi.org/10.1038/gim.2016.118.

Bagheri-Fam, S., Ono, M., Li, L., Zhao, L., Ryan, J., Lai, R., et al. (2015). *FGFR2* mutation in 46,XY sex reversal with craniosynostosis. *Human Molecular Genetics*, *24*, 6699–6710. https://doi.org/10.1093/hmg/ddv374.

Barbaro, M., Oscarson, M., Schoumans, J., Staaf, J., Ivarsson, S. A., & Wedell, A. (2007). Isolated 46,XY gonadal dysgenesis in two sisters caused by a Xp21.2 interstitial duplication containing the DAX1 gene. *The Journal of Clinical Endocrinology and Metabolism*, *92*, 3305–3313. https://doi.org/10.1210/jc.2007-0505.

Barbaux, S., Niaudet, P., Gubler, M. C., Grünfeld, J. P., Jaubert, F., Kuttenn, F., et al. (1997). Donor splice-site mutations in WT1 are responsible for Frasier syndrome. *Nature Genetics*, *17*, 467–470. https://doi.org/10.1038/ng1297-467.

Bardoni, B., Zanaria, E., Guioli, S., Floridia, G., Worley, K. C., Tonini, G., et al. (1994). A dosage sensitive locus at chromosome Xp21 is involved in male to female sex reversal. *Nature Genetics*, *7*, 497–501. https://doi.org/10.1038/ng0894-497.

Barrera, L. A., Vedenko, A., Kurland, J. V., Rogers, J. M., Gisselbrecht, S. S., Rossin, E. J., et al. (2016). Survey of variation in human transcription factors reveals prevalent DNA binding changes. *Science*, *351*, 1450–1454. https://doi.org/10.1126/science.aad2257.

Barseghyan, H., Delot, E., & Vilain, E. (2015). New genomic technologies: An aid for diagnosis of disorders of sex development. *Hormone and Metabolic Research*, *47*, 312–320.

Barseghyan, H., Délot, E. C., & Vilain, E. (2018). New technologies to uncover the molecular basis of disorders of sex development. *Molecular and Cellular Endocrinology*, *468*, 60–69. https://doi.org/10.1016/j.mce.2018.04.003.

Barseghyan, H., Tang, W., Wang, R. T., Almalvez, M., Segura, E., Bramble, M. S., et al. (2017). Next-generation mapping: A novel approach for detection of pathogenic structural variants with a potential utility in clinical diagnosis. *Genome Medicine*, *9*, 90. https://doi.org/10.1186/s13073-017-0479-0.

Bashamboo, A., Brauner, R., Bignon-Topalovic, J., Lortat-Jacob, S., Karageorgou, V., Lourenco, D., et al. (2014). Mutations in the FOG2/ZFPM2 gene are associated with anomalies of human testis determination. *Human Molecular Genetics*, *23*, 3657–3665. https://doi.org/10.1093/hmg/ddu074.

Bashamboo, A., Ferraz-de-Souza, B., Lourenço, D., Lin, L., Sebire, N. J., Montjean, D., et al. (2010). Human male infertility associated with mutations in NR5A1 encoding steroidogenic factor 1. *American Journal of Human Genetics, 87*, 505–512. https://doi.org/10.1016/j.ajhg.2010.09.009.

Bashamboo, A., Donohoue, P. A., Vilain, E., Rojo, S., Calvel, P., Seneviratne, S. N., et al. (2016). A recurrent p.Arg92Trp variant in steroidogenic factor-1 (NR5A1) can act as a molecular switch in human sex development. *Human Molecular Genetics, 25*, 3446–3453. https://doi.org/10.1093/hmg/ddw186.

Baxter, R. M., Arboleda, V. A., Lee, H., Barseghyan, H., Adam, M. P., Fechner, P. Y., et al. (2015). Exome sequencing for the diagnosis of 46,XY disorders of sex development. *The Journal of Clinical Endocrinology and Metabolism, 100*, E333–E344. https://doi.org/10.1210/jc.2014-2605.

Beau, I., Touraine, P., Meduri, G., Gougeon, A., Desroches, A., Matuchansky, C., et al. (1998). A novel phenotype related to partial loss of function mutations of the follicle stimulating hormone receptor. *The Journal of Clinical Investigation, 102*, 1352–1359. https://doi.org/10.1172/JCI3795.

Belkadi, A., Bolze, A., Itan, Y., Cobat, A., Vincent, Q. B., Antipenko, A., et al. (2015). Whole-genome sequencing is more powerful than whole-exome sequencing for detecting exome variants. *Proceedings of the National Academy of Sciences of the United States of America, 112*, 5473–5478. https://doi.org/10.1073/pnas.1418631112.

Belville, C., Marechal, J.-D., Pennetier, S., Carmillo, P., Masgrau, L., Messika-Zeitoun, L., et al. (2009). Natural mutations of the anti-Mullerian hormone type II receptor found in persistent Mullerian duct syndrome affect ligand binding, signal transduction and cellular transport. *Human Molecular Genetics, 18*, 3002–3013. https://doi.org/10.1093/hmg/ddp238.

Benko, S., Gordon, C. T., Mallet, D., Sreenivasan, R., Thauvin-Robinet, C., Brendehaug, A., et al. (2011). Disruption of a long distance regulatory region upstream of SOX9 in isolated disorders of sex development. *Journal of Medical Genetics, 48*, 825–830. https://doi.org/10.1136/jmedgenet-2011-100255.

Berta, P., Hawkins, J. R., Sinclair, A. H., Taylor, A., Griffiths, B. L., Goodfellow, P. N., et al. (1990). Genetic evidence equating SRY and the testis-determining factor. *Nature, 348*, 448–450. https://doi.org/10.1038/348448A0.

Bertelloni, S., Baldinotti, F., Russo, G., Ghirri, P., Dati, E., Michelucci, A., et al. (2016). 5α-Reductase-2 deficiency: Clinical findings, endocrine pitfalls, and genetic features in a large Italian cohort. *Sexual Development, 10*, 28–36. https://doi.org/10.1159/000445090.

Beutler, E., Felitti, V. J., Koziol, J. A., Ho, N. J., & Gelbart, T. (2002). Penetrance of 845G→A (C282Y) HFE hereditary haemochromatosis mutation in the USA. *The Lancet, 359*, 211–218. https://doi.org/10.1016/S0140-6736(02)07447-0.

Biason-Lauber, A. (2012). WNT4, RSPO1, and FOXL2 in sex development. *Seminars in Reproductive Medicine, 30*, 387–395. https://doi.org/10.1055/s-0032-1324722.

Biason-Lauber, A., De Filippo, G., Konrad, D., Scarano, G., Nazzaro, A., & Schoenle, E. J. (2007). WNT4 deficiency—A clinical phenotype distinct from the classic Mayer–Rokitansky–Kuster–Hauser syndrome: A case report. *Human Reproduction, 22*, 224–229. https://doi.org/10.1093/humrep/del360.

Biason-Lauber, A., Konrad, D., Meyer, M., deBeaufort, C., & Schoenle, E. J. (2009). Ovaries and female phenotype in a girl with 46,XY karyotype and mutations in the CBX2 gene. *American Journal of Human Genetics, 84*, 658–663. https://doi.org/10.1016/j.ajhg.2009.03.016.

Biason-Lauber, A., Konrad, D., Navratil, F., & Schoenle, E. J. (2004). A *WNT4* mutation associated with Müllerian-duct regression and virilization in a 46,XX woman. *The New England Journal of Medicine, 351*, 792–798. https://doi.org/10.1056/NEJMoa040533.

Billerbeck, A. E. C., Mendonca, B. B., Pinto, E. M., Madureira, G., Arnhold, I. J. P., & Bachega, T. A. S. S. (2002). Three novel mutations in CYP21 gene in Brazilian patients with the classical form of 21-hydroxylase deficiency due to a founder effect. *The Journal of Clinical Endocrinology and Metabolism, 87*, 4314–4317. https://doi.org/10.1210/jc.2001-011939.

Boot, A. M., Lumbroso, S., Verhoef-Post, M., Richter-Unruh, A., Looijenga, L. H. J., Funaro, A., et al. (2011). Mutation analysis of the LH receptor gene in Leydig cell adenoma and hyperplasia and functional and biochemical studies of activating mutations of the LH receptor gene. *The Journal of Clinical Endocrinology and Metabolism*, *96*, E1197–E1205. https://doi.org/10.1210/jc.2010-3031.

Bose, H. S., Sugawara, T., Strauss, J. F., Miller, W. L., & International Congenital Lipoid Adrenal Hyperplasia Consortium. (1996). The pathophysiology and genetics of congenital lipoid adrenal hyperplasia. *The New England Journal of Medicine*, *335*, 1870–1879. https://doi.org/10.1056/NEJM199612193352503.

Bouligand, J., Ghervan, C., Tello, J. A., Brailly-Tabard, S., Salenave, S., Chanson, P., et al. (2009). Isolated familial hypogonadotropic hypogonadism and a *GNRH1* mutation. *The New England Journal of Medicine*, *360*, 2742–2748. https://doi.org/10.1056/NEJMoa0900136.

Bowles, J., Schepers, G., & Koopman, P. (2000). Phylogeny of the SOX family of developmental transcription factors based on sequence and structural indicators. *Developmental Biology*, *227*, 239–255. https://doi.org/10.1006/dbio.2000.9883.

Brand, H., Pillalamarri, V., Collins, R. L., Eggert, S., O'Dushlaine, C., Braaten, E. B., et al. (2014). Cryptic and complex chromosomal aberrations in early-onset neuropsychiatric disorders. *American Journal of Human Genetics*, *95*, 454–461. https://doi.org/10.1016/j.ajhg.2014.09.005.

Brickman, J. M., Clements, M., Tyrell, R., McNay, D., Woods, K., Warner, J., et al. (2001). Molecular effects of novel mutations in Hesx1/HESX1 associated with human pituitary disorders. *Development*, *128*, 5189–5199.

Brioude, F., Bouligand, J., Francou, B., Fagart, J., Roussel, R., Viengchareun, S., et al. (2013). Two families with normosmic congenital hypogonadotropic hypogonadism and biallelic mutations in KISS1R (KISS1 receptor): Clinical evaluation and molecular characterization of a novel mutation. *PLoS One*, *8* e53896. https://doi.org/10.1371/journal.pone.0053896.

Brown, T. R., Lubahn, D. B., Wilson, E. M., Joseph, D. R., French, F. S., & Migeon, C. J. (1988). Deletion of the steroid-binding domain of the human androgen receptor gene in one family with complete androgen insensitivity syndrome: Evidence for further genetic heterogeneity in this syndrome. *Proceedings of the National Academy of Sciences of the United States of America*, *85*, 8151–8155.

Bruening, W., Bardeesy, N., Silverman, B. L., Cohn, R. A., Machin, G. A., Aronson, A. J., et al. (1992). Germline intronic and exonic mutations in the Wilms' tumour gene (WT1) affecting urogenital development. *Nature Genetics*, *1*, 144–148. https://doi.org/10.1038/ng0592-144.

Brüggenwirth, H. T., Boehmer, A. L. M., Ramnarain, S., Verleun-Mooijman, M. C. T., Satijn, D. P. E., Trapman, J., et al. (1997). Molecular analysis of the androgen-receptor gene in a family with receptor-positive partial androgen insensitivity: An unusual type of intronic mutation. *American Journal of Human Genetics*, *61*, 1067–1077. https://doi.org/10.1086/301605.

Buonocore, F., & Achermann, J. C. (2016). Human sex development: Targeted technologies to improve diagnosis. *Genome Biology*, *17*, 257. https://doi.org/10.1186/s13059-016-1128-4.

Caburet, S., Arboleda, V. A., Llano, E., Overbeek, P. A., Barbero, J. L., Oka, K., et al. (2014). Mutant cohesin in premature ovarian failure. *The New England Journal of Medicine*, *370*, 943–949. https://doi.org/10.1056/NEJMoa1309635.

Cai, L. Q., Zhu, Y. S., Katz, M. D., Herrera, C., Baéz, J., DeFillo-Ricart, M., et al. (1996). 5 Alpha-reductase-2 gene mutations in the Dominican Republic. *The Journal of Clinical Endocrinology and Metabolism*, *81*, 1730–1735. https://doi.org/10.1210/jcem.81.5.8626825.

Callier, P., Calvel, P., Matevossian, A., Makrythanasis, P., Bernard, P., Kurosaka, H., et al. (2014). Loss of function mutation in the palmitoyl-transferase HHAT leads to syndromic 46, XY disorder of sex development by impeding hedgehog protein palmitoylation and signaling. *PLoS Genetics*, *10*, e1004340. https://doi.org/10.1371/journal.pgen.1004340.

Camats, N., Fernández-Cancio, M., Audi, L., Schaller, A., & Flück, C. E. (2018). Broad phenotypes in heterozygous NR5A1 46,XY patients with a disorder of sex development: An oligogenic origin? *European Journal of Human Genetics*, *26*, 1329–1338.

Can, S., Zhu, Y.-S., Cai, L.-Q., Ling, Q., Katz, M. D., Akgun, S., et al. (1998). The identification of 5α-reductase-2 and 17β-hydroxysteroid dehydrogenase-3 gene defects in male pseudohermaphrodites from a Turkish Kindred. *The Journal of Clinical Endocrinology and Metabolism*, *83*, 560–569. https://doi.org/10.1210/jcem.83.2.4535.

Canto, P., Munguia, P., Soderlund, D., Castro, J. J., & Mendez, J. P. (2008). Genetic analysis in patients with kallmann syndrome: Coexistence of mutations in prokineticin receptor 2 and KAL1. *Journal of Andrology*, *30*, 41–45. https://doi.org/10.2164/jandrol.108.005314.

Caron, P., Chauvin, S., Christin-Maitre, S., Bennet, A., Lahlou, N., Counis, R., et al. (1999). Resistance of hypogonadic patients with mutated GnRH receptor genes to pulsatile GnRH administration. *The Journal of Clinical Endocrinology and Metabolism*, *84*, 990–996. https://doi.org/10.1210/jcem.84.3.5518.

Carré-Eusèbe, D., Imbeaud, S., Harbison, M., New, M. I., Josso, N., & Picard, J. Y. (1992). Variants of the anti-Müllerian hormone gene in a compound heterozygote with the persistent Müllerian duct syndrome and his family. *Human Genetics*, *90*, 389–394.

Caspar, S. M., Dubacher, N., Kopps, A. M., Meienberg, J., Henggeler, C., & Matyas, G. (2017). Clinical sequencing: From raw data to diagnosis with lifetime value. *Clinical Genetics*, *93*, 508–519. https://doi.org/10.1111/cge.13190.

Castro, C. C. T., Guaragna-Filho, G., Calais, F. L., Coeli, F. B., Leal, I. R. L., Cavalcante-Junior, E. F., et al. (2012). Clinical and molecular spectrum of patients with 17β-hydroxysteroid dehydrogenase type 3 (17-β-HSD3) deficiency. *Arquivos Brasileiros de Endocrinologia e Metabologia*, *56*, 533–539.

Cavalcanti, D. P., Matejas, V., Luquetti, D., Mello, M. F., & Zenker, M. (2007). Fraser and Ablepharon macrostomia phenotypes: Concurrence in one family and association with mutated FRAS1. *American Journal of Medical Genetics. Part A*, *143A*, 241–247. https://doi.org/10.1002/ajmg.a.31426.

Chaisson, M. J. P., Huddleston, J., Dennis, M. Y., Sudmant, P. H., Malig, M., Hormozdiari, F., et al. (2015). Resolving the complexity of the human genome using single-molecule sequencing. *Nature*, *517*, 608–611. https://doi.org/10.1038/nature13907.

Chapman, C., Cree, L., & Shelling, A. N. (2015). The genetics of premature ovarian failure: Current perspectives. *International Journal of Women's Health*, 7, 799–810. https://doi.org/10.2147/IJWH.S64024.

Chauhan, V., Jyotsna, V., Jain, V., Khadgawat, R., & Dada, R. (2016). Novel heterozygous genetic variants in patients with 46,XY gonadal dysgenesis. *Hormone and Metabolic Research*, *49*, 36–42. https://doi.org/10.1055/s-0042-114778.

Chiang, A. P., Beck, J. S., Yen, H.-J., Tayeh, M. K., Scheetz, T. E., Swiderski, R. E., et al. (2006). Homozygosity mapping with SNP arrays identifies TRIM32, an E3 ubiquitin ligase, as a Bardet-Biedl syndrome gene (BBS11). *Proceedings of the National Academy of Sciences of the United States of America*, *103*, 6287–6292. https://doi.org/10.1073/pnas.0600158103.

Chiang, A. P., Nishimura, D., Searby, C., Elbedour, K., Carmi, R., Ferguson, A. L., et al. (2004). Comparative genomic analysis identifies an ADP-ribosylation factor–like gene as the cause of Bardet-Biedl syndrome (BBS3). *American Journal of Human Genetics*, *75*, 475–484. https://doi.org/10.1086/423903.

Choong, C. S., Quigley, C. A., French, F. S., & Wilson, E. M. (1996). A novel missense mutation in the amino-terminal domain of the human androgen receptor gene in a family with partial androgen insensitivity syndrome causes reduced efficiency of protein translation. *The Journal of Clinical Investigation*, *98*, 1423–1431. https://doi.org/10.1172/JCI118930.

Clément, K., Vaisse, C., Lahlou, N., Cabrol, S., Pelloux, V., Cassuto, D., et al. (1998). A mutation in the human leptin receptor gene causes obesity and pituitary dysfunction. *Nature*, *392*, 398–401. https://doi.org/10.1038/32911.

Cohen, R. N., Cohen, L. E., Botero, D., Yu, C., Sagar, A., Jurkiewicz, M., et al. (2003). Enhanced repression by HESX1 as a cause of hypopituitarism and septooptic dysplasia. *The Journal of Clinical Endocrinology and Metabolism*, *88*, 4832–4839. https://doi.org/10.1210/jc.2002-021868.

Cole, L. W., Sidis, Y., Zhang, C., Quinton, R., Plummer, L., Pignatelli, D., et al. (2008). Mutations in prokineticin 2 and prokineticin receptor 2 genes in human gonadotrophin-releasing hormone deficiency: Molecular genetics and clinical spectrum. *The Journal of Clinical Endocrinology and Metabolism*, *93*, 3551–3559. https://doi.org/10.1210/jc.2007-2654.

Colson, C., Aubry, E., Cartigny, M., Rémy, A.-A., Franquet, H., Leroy, X., et al. (2017). SF1 and spleen development: New heterozygous mutation, literature review and consequences for *NR5A1*-mutated patient's management. *Clinical Genetics*, *92*, 99–103. https://doi.org/10.1111/cge.12957.

Conrad, D. F., Pinto, D., Redon, R., Feuk, L., Gokcumen, O., Zhang, Y., et al. (2010). Origins and functional impact of copy number variation in the human genome. *Nature*, *464*, 704–712. https://doi.org/10.1038/nature08516.

Correa, R. V., Domenice, S., Bingham, N. C., Billerbeck, A. E. C., Rainey, W. E., Parker, K. L., et al. (2004). A microdeletion in the ligand binding domain of human steroidogenic factor 1 causes XY sex reversal without adrenal insufficiency. *The Journal of Clinical Endocrinology and Metabolism*, *89*, 1767–1772. https://doi.org/10.1210/jc.2003-031240.

Costa-Barbosa, F. A., Balasubramanian, R., Keefe, K. W., Shaw, N. D., Al-Tassan, N., Plummer, L., et al. (2013). Prioritizing genetic testing in patients with Kallmann syndrome using clinical phenotypes. *The Journal of Clinical Endocrinology and Metabolism*, *98*, E943–E953. https://doi.org/10.1210/jc.2012-4116.

Cox, K., Bryce, J., Jiang, J., Rodie, M., Sinnott, R., Alkhawari, M., et al. (2014). Novel associations in disorders of sex development: Findings from the I-DSD registry. *The Journal of Clinical Endocrinology and Metabolism*, *99*, E348–E355. https://doi.org/10.1210/jc.2013-2918.

Cox, J. J., Willatt, L., Homfray, T., & Woods, C. G. (2011). A SOX9 duplication and familial 46,XX developmental testicular disorder. *The New England Journal of Medicine*, *364*, 91–93. https://doi.org/10.1056/NEJMc1010311.

Cretu Stancu, M., Van Roosmalen, M. J., Renkens, I., Nieboer, M. M., Middelkamp, S., De Ligt, J., et al. (2017). Mapping and phasing of structural variation in patient genomes using nanopore sequencing. *Nature Communications*, *8*, 1–13. https://doi.org/10.1038/s41467-017-01343-4.

Croft, B., Ayers, K., Sinclair, A., & Ohnesorg, T. (2016). Review disorders of sex development: The evolving role of genomics in diagnosis and gene discovery. *Birth Defects Research. Part C, Embryo Today: Reviews*, *108*, 337–350. https://doi.org/10.1002/bdrc.21148.

Croft, B., Ohnesorg, T., & Sinclair, A. H. (2018). The role of copy number variants in disorders of sex development. *Sexual Development*, *12*, 19–29. https://doi.org/10.1159/000481896.

Dai, Y., Li, P., Wang, Z., Liang, F., Yang, F., Fang, L., et al. (2018). Single-molecule optical mapping enables accurate molecular diagnosis of facioscapulohumeral muscular dystrophy (FSHD). *bioRxiv*. *286104*, 1–24. https://doi.org/10.1101/286104.

Dattani, M. T., Martinez-Barbera, J.-P., Thomas, P. Q., Brickman, J. M., Gupta, R., Mårtensson, I.-L., et al. (1998). Mutations in the homeobox gene HESX1/Hesx1 associated with septo-optic dysplasia in human and mouse. *Nature Genetics*, *19*, 125–133. https://doi.org/10.1038/477.

D'Aurizio, R., Pippucci, T., Tattini, L., Giusti, B., Pellegrini, M., & Magi, A. (2016). Enhanced copy number variants detection from whole-exome sequencing data using EXCAVATOR2. *Nucleic Acids Research, 44*, e154. https://doi.org/10.1093/nar/gkw695.

De Leener, A., Caltabiano, G., Erkan, S., Idil, M., Vassart, G., Pardo, L., et al. (2008). Identification of the first germline mutation in the extracellular domain of the follitropin receptor responsible for spontaneous ovarian hyperstimulation syndrome. *Human Mutation, 29*, 91–98. https://doi.org/10.1002/humu.20604.

De Leener, A., Montanelli, L., Van Durme, J., Chae, H., Smits, G., Vassart, G., et al. (2006). Presence and absence of follicle-stimulating hormone receptor mutations provide some insights into spontaneous ovarian hyperstimulation syndrome physiopathology. *The Journal of Clinical Endocrinology and Metabolism, 91*, 555–562. https://doi.org/10.1210/jc.2005-1580.

De Paula, G. B., Barros, B. A., Carpini, S., Tincani, B. J., Mazzola, T. N., Sanches Guaragna, M., et al. (2016). 408 cases of genital ambiguity followed by single multidisciplinary team during 23 years: Etiologic diagnosis and sex of rearing. *International Journal of Endocrinology, 2016*, 1–9. https://doi.org/10.1155/2016/4963574.

de Villiers, J. N., Hillermann, R., Loubser, L., & Kotze, M. J. (1999). Spectrum of mutations in the HFE gene implicated in haemochromatosis and porphyria. *Human Molecular Genetics, 8*, 1517–1522.

Deamer, D., Akeson, M., & Branton, D. (2016). Three decades of nanopore sequencing. *Nature Biotechnology, 34*, 518–524. https://doi.org/10.1038/nbt.3423.

Deladoëy, J., Flück, C., Bex, M., Yoshimura, N., Harada, N., & Mullis, P. E. (1999). Aromatase deficiency caused by a novel P450 $_{arom}$ gene mutation: Impact of absent estrogen production on serum gonadotropin concentration in a boy. *The Journal of Clinical Endocrinology and Metabolism, 84*, 4050–4054. https://doi.org/10.1210/jcem.84.11.6135.

Demirhan, O., Tanrıverdi, N., Süleymanova, D., Cetinel, N., & Yasar, Y. (2015). The frequency and types of chromosomal aberrations in the patients with Hypogonadism. *Human Genetics & Embryology, 05*, 7–8. https://doi.org/10.4172/2161-0436.1000124.

Délot, E. C., Papp, J. C., Délot, E. C., Fox, M., Grody, W., Lee, H., et al. (2017). Genetics of disorders of sex development: The DSD-TRN experience. *Endocrinology and Metabolism Clinics of North America, 46*, 519–537. https://doi.org/10.1016/j.ecl.2017.01.015.

Délot, E. C., & Vilain, E. J. (2015). Nonsyndromic 46,XX testicular disorders of sex development. In M. P. Adam, H. H. Ardinger, R. A. Pagon, S. E. Wallace, L. J. H. Bean, K. Stephens, & A. Amemiya (Eds.), *GeneReviews*. Seattle (WA): University of Washington.

Délot, E., & Vilain, E. (2018). Disorders of sex development. In J. Strauss & R. B. Antonio Gargiulo (Eds.), *Yen & Jaffe's reproductive endocrinology*. Philadelphia, PA: Elsevier.

Di Carlo, C., Bruno, P., Cirillo, D., Morgera, R., Pellicano, M., & Nappi, C. (1997). Increased concentrations of renin, aldosterone and Ca125 in a case of spontaneous, recurrent, familial, severe ovarian hyperstimulation syndrome. *Human Reproduction, 12*, 2115–2117.

Di Pasquale, E., Beck-Peccoz, P., & Persani, L. (2004). Hypergonadotropic ovarian failure associated with an inherited mutation of human bone morphogenetic protein-15 (BMP15) gene. *American Journal of Human Genetics, 75*, 106–111. https://doi.org/10.1086/422103.

DiNapoli, L., & Capel, B. (2008). SRY and the standoff in sex determination. *Molecular Endocrinology, 22*, 1–9. https://doi.org/10.1210/me.2007-0250.

Dixit, H., Rao, L. K., Padmalatha, V. V., Kanakavalli, M., Deenadayal, M., Gupta, N., et al. (2006). Missense mutations in the BMP15 gene are associated with ovarian failure. *Human Genetics, 119*, 408–415. https://doi.org/10.1007/s00439-006-0150-0.

Dixon, J. R., Xu, J., Dileep, V., Zhan, Y., Song, F., Le, V. T., et al. (2018). Integrative detection and analysis of structural variation in cancer genomes. *Nature Genetics*, *50*, 1388–1398. https://doi.org/10.1038/s41588-018-0195-8.

Dodé, C., Levilliers, J., Dupont, J.-M., De Paepe, A., Le Dû, N., Soussi-Yanicostas, N., et al. (2003). Loss-of-function mutations in FGFR1 cause autosomal dominant Kallmann syndrome. *Nature Genetics*, *33*, 463–465. https://doi.org/10.1038/ng1122.

Dodé, C., & Rondard, P. (2013). PROK2/PROKR2 signaling and Kallmann syndrome. *Frontiers in Endocrinology*, *4*, 1–8. https://doi.org/10.3389/fendo.2013.00019.

Dodé, C., Teixeira, L., Levilliers, J., Fouveaut, C., Bouchard, P., Kottler, M.-L., et al. (2006). Kallmann syndrome: Mutations in the genes encoding prokineticin-2 and prokineticin receptor-2. *PLoS Genetics*, *2*, 1648–1652. https://doi.org/10.1371/journal.pgen.0020175.

Doherty, E., Pakarinen, P., Tiitinen, A., Kiilavuori, A., Huhtaniemi, I., Forrest, S., et al. (2002). A novel mutation in the FSH receptor inhibiting signal transduction and causing primary ovarian failure. *The Journal of Clinical Endocrinology and Metabolism*, *87*, 1151–1155. https://doi.org/10.1210/jcem.87.3.8319.

Dong, Y., Yi, Y., Yao, H., Yang, Z., Hu, H., Liu, J., et al. (2016). Targeted next-generation sequencing identification of mutations in patients with disorders of sex development. *BMC Medical Genetics*, *17*, 23. https://doi.org/10.1186/s12881-016-0286-2.

Du, C., Mark, D., Wappenschmidt, B., Böckmann, B., Pabst, B., Chan, S., et al. (2018). A tandem duplication of BRCA1 exons 1-19 through DHX8 exon 2 in four families with hereditary breast and ovarian cancer syndrome. *Breast Cancer Research and Treatment*, *172*, 561–569. https://doi.org/10.1007/s10549-018-4957-x.

Ebbert, M. T. W., Farrugia, S. L., Sens, J. P., Jansen-West, K., Gendron, T. F., Prudencio, M., et al. (2018). Long-read sequencing across the C9orf72 "GGGGCC" repeat expansion: Implications for clinical use and genetic discovery efforts in human disease. *Molecular Neurodegeneration*, *13*, 1–17. https://doi.org/10.1186/s13024-018-0274-4.

Eggers, S., Sadedin, S., van den Bergen, J. A., Robevska, G., Ohnesorg, T., Hewitt, J., et al. (2016). Disorders of sex development: Insights from targeted gene sequencing of a large international patient cohort. *Genome Biology*, *17*, 243. https://doi.org/10.1186/s13059-016-1105-y.

Eggers, S., & Sinclair, A. (2012). Mammalian sex determination-insights from humans and mice. *Chromosome Research*, *20*, 215–238. https://doi.org/10.1007/s10577-012-9274-3.

English, A. C., Salerno, W. J., Hampton, O. A., Gonzaga-Jauregui, C., Ambreth, S., Ritter, D. I., et al. (2015). Assessing structural variation in a personal genome-towards a human reference diploid genome. *BMC Genomics*, *16*, 286. https://doi.org/10.1186/s12864-015-1479-3.

Falardeau, J., Chung, W. C. J., Beenken, A., Raivio, T., Plummer, L., Sidis, Y., et al. (2008). Decreased FGF8 signaling causes deficiency of gonadotropin-releasing hormone in humans and mice. *The Journal of Clinical Investigation*, *118*, 2822–2831. https://doi.org/10.1172/JCI34538.

Fan, Y., Esmail, M. A., Ansley, S. J., Blacque, O. E., Boroevich, K., Ross, A. J., et al. (2004). Mutations in a member of the Ras superfamily of small GTP-binding proteins causes Bardet-Biedl syndrome. *Nature Genetics*, *36*, 989–993. https://doi.org/10.1038/ng1414.

Fan, Y., Zhang, X., Wang, L., Wang, R., Huang, Z., Sun, Y., et al. (2017). Diagnostic application of targeted next-generation sequencing of 80 genes associated with disorders of sexual development. *Scientific Reports*, 7, 44536. https://doi.org/10.1038/srep44536.

Fang, G., Munera, D., Friedman, D. I., Mandlik, A., Chao, M. C., Banerjee, O., et al. (2012). Genome-wide mapping of methylated adenine residues in pathogenic Escherichia coli using single-molecule real-time sequencing. *Nature Biotechnology*, *30*, 1232–1239. https://doi.org/10.1038/nbt.2432.

Farooqi, I. S., Volders, K., Stanhope, R., Heuschkel, R., White, A., Lank, E., et al. (2007). Hyperphagia and early-onset obesity due to a novel homozygous missense mutation in prohormone convertase 1/3. *The Journal of Clinical Endocrinology and Metabolism*, *92*, 3369–3373. https://doi.org/10.1210/jc.2007-0687.

Farwell, K. D., Shahmirzadi, L., El-Khechen, D., Powis, Z., Chao, E. C., Tippin Davis, B., et al. (2015). Enhanced utility of family-centered diagnostic exome sequencing with inheritance model-based analysis: Results from 500 unselected families with undiagnosed genetic conditions. *Genetics in Medicine*, *17*, 578–586. https://doi.org/10.1038/gim.2014.154.

Ferraz-de-Souza, B., Lin, L., & Achermann, J. C. (2011). Steroidogenic factor-1 (SF-1, NR5A1) and human disease. *Molecular and Cellular Endocrinology*, *336*, 198–205. https://doi.org/10.1016/j.mce.2010.11.006.

Flück, C. E., Pandey, A. V., Dick, B., Camats, N., Fernández-Cancio, M., Clemente, M., et al. (2011). Characterization of novel StAR (steroidogenic acute regulatory protein) mutations causing non-classic lipoid adrenal hyperplasia. *PLoS One*, *6*, e20178. https://doi.org/10.1371/journal.pone.0020178.

Flück, C. E., Tajima, T., Pandey, A. V., Arlt, W., Okuhara, K., Verge, C. F., et al. (2004). Mutant P450 oxidoreductase causes disordered steroidogenesis with and without Antley-Bixler syndrome. *Nature Genetics*, *36*, 228–230. https://doi.org/10.1038/ng1300.

Flück, C. E., Meyer-Böni, M., Pandey, A. V., Kempná, P., Miller, W. L., Schoenle, E. J., et al. (2011). Why boys will be boys: Two pathways of fetal testicular androgen biosynthesis are needed for male sexual differentiation. *American Journal of Human Genetics*, *89*, 201–218. https://doi.org/10.1016/j.ajhg.2011.06.009.

Fofanova, Q., Takamura, N., Kinoshita, E., Parks, J. S., Brown, M. R., Peterkova, V. A., et al. (1998). Compound heterozygous deletion of the *PROP*-1 gene in children with combined pituitary hormone deficiency. *The Journal of Clinical Endocrinology and Metabolism*, *83*, 2601–2604. https://doi.org/10.1210/jcem.83.7.5094.

Fonseca, R., Costa-Lima, M. A., Cosentino, V., & Orioli, I. M. (2008). Second case of Beare-Stevenson syndrome with an FGFR2 Ser372Cys mutation. *American Journal of Medical Genetics. Part A*, *146A*, 658–660. https://doi.org/10.1002/ajmg.a.32176.

Foster, J. W., Dominguez-Steglich, M. A., Guioli, S., Kwok, C., Weller, P. A., Stevanović, M., et al. (1994). Campomelic dysplasia and autosomal sex reversal caused by mutations in an SRY-related gene. *Nature*, *372*, 525–530. https://doi.org/10.1038/372525a0.

Francou, B., Bouligand, J., Voican, A., Amazit, L., Trabado, S., Fagart, J., et al. (2011). Normosmic congenital hypogonadotropic hypogonadism due to TAC3/TACR3 mutations: Characterization of neuroendocrine phenotypes and novel mutations. *PLoS One*, *6*, e25614. https://doi.org/10.1371/journal.pone.0025614.

Fukami, M., Horikawa, R., Nagai, T., Tanaka, T., Naiki, Y., Sato, N., et al. (2005). Cytochrome P450 oxidoreductase gene mutations and Antley-Bixler syndrome with abnormal genitalia and/or impaired steroidogenesis: Molecular and clinical studies in 10 patients. *The Journal of Clinical Endocrinology and Metabolism*, *90*, 414–426. https://doi.org/10.1210/jc.2004-0810.

Fukami, M., Wada, Y., Miyabayashi, K., Nishino, I., Hasegawa, T., Nordenskjöld, A., et al. (2006). CXorf6 is a causative gene for hypospadias. *Nature Genetics*, *38*, 1369–1371. https://doi.org/10.1038/ng1900.

Geissler, W. M., Davis, D. L., Wu, L., Bradshaw, K. D., Patel, S., Mendonca, B. B., et al. (1994). Male pseudohermaphroditism caused by mutations of testicular 17β-hydroxysteroid dehydrogenase 3. *Nature Genetics*, *7*, 34–39. https://doi.org/10.1038/ng0594-34.

Gianetti, E., Hall, J. E., Au, M. G., Kaiser, U. B., Quinton, R., Stewart, J. A., et al. (2012). When genetic load does not correlate with phenotypic spectrum: Lessons from the GnRH receptor (*GNRHR*). *The Journal of Clinical Endocrinology and Metabolism*, *97*, E1798–E1807. https://doi.org/10.1210/jc.2012-1264.

Gianetti, E., Tusset, C., Noel, S. D., Au, M. G., Dwyer, A. A., Hughes, V. A., et al. (2010). *TAC3/TACR3* mutations reveal preferential activation of gonadotropin-releasing hormone release by Neurokinin B in neonatal life followed by reversal in adulthood. *The Journal of Clinical Endocrinology and Metabolism, 95*, 2857–2867. https://doi.org/10.1210/jc.2009-2320.

Gibbons, R. J., Picketts, D. J., Villard, L., & Higgs, D. R. (1995). Mutations in a putative global transcriptional regulator cause X-linked mental retardation with alpha-thalassemia (ATR-X syndrome). *Cell, 80*, 837–845.

Gibson, W. T., Farooqi, I. S., Moreau, M., DePaoli, A. M., Lawrence, E., O'Rahilly, S., et al. (2004). Congenital leptin deficiency due to homozygosity for the Δ133G mutation: Report of another case and evaluation of response to four years of leptin therapy. *The Journal of Clinical Endocrinology and Metabolism, 89*, 4821–4826. https://doi.org/10.1210/jc.2004-0376.

Gochee, P. A., Powell, L. W., Cullen, D. J., Du Sart, D., Rossi, E., & Olynyk, J. K. (2002). A population-based study of the biochemical and clinical expression of the H63D hemochromatosis mutation. *Gastroenterology, 122*, 646–651.

Gottlieb, B., & Trifiro, M. A. (2017). *Androgen insensitivity syndrome, GeneReviews*. Seattle: University of Washington.

Granados, A., Alaniz, V. I., Mohnach, L., Barseghyan, H., Vilain, E., Ostrer, H., et al. (2017). MAP3K1-related gonadal dysgenesis: Six new cases and review of the literature. *American Journal of Medical Genetics. Part C, Seminars in Medical Genetics, 175*, 253–259. https://doi.org/10.1002/ajmg.c.31559.

Green, R. C., Berg, J. S., Grody, W. W., Kalia, S. S., Korf, B. R., Martin, C. L., et al. (2013). ACMG recommendations for reporting of incidental findings in clinical exome and genome sequencing. *Genetics in Medicine, 15*, 565–574. https://doi.org/10.1038/gim.2013.73.

Gromoll, J., Eiholzer, U., Nieschlag, E., & Simoni, M. (2000). Male hypogonadism caused by homozygous deletion of exon 10 of the luteinizing hormone (lh) receptor: Differential action of human chorionic gonadotropin and LH. *The Journal of Clinical Endocrinology and Metabolism, 85*, 2281–2286. https://doi.org/10.1210/jcem.85.6.6636.

Gromoll, J., Simoni, M., & Nieschlag, E. (1996). An activating mutation of the follicle-stimulating hormone receptor autonomously sustains spermatogenesis in a hypophysectomized man. *The Journal of Clinical Endocrinology and Metabolism, 81*, 1367–1370. https://doi.org/10.1210/jcem.81.4.8636335.

Guercio, G., & Rey, R. A. (2014). Fertility issues in the management of patients with disorders of sex development. *Endocrine Development, 27*, 87–98. https://doi.org/10.1159/000363633.

Haines, B., Hughes, J., Corbett, M., Shaw, M., Innes, J., Patel, L., et al. (2015). Interchromosomal insertional translocation at Xq26.3 alters SOX3 expression in an individual with XX male sex reversal. *The Journal of Clinical Endocrinology and Metabolism, 100*, E815–E820. https://doi.org/10.1210/jc.2014-4383.

Harris, S. E., Chand, A. L., Winship, I. M., Gersak, K., Aittomäki, K., & Shelling, A. N. (2002). Identification of novel mutations in FOXL2 associated with premature ovarian failure. *Molecular Human Reproduction, 8*, 729–733.

Harrison, S. M., Campbell, I. M., Keays, M., Granberg, C. F., Villanueva, C., Tannin, G., et al. (2013). Screening and familial characterization of copy-number variations in *NR5A1* in 46,XY disorders of sex development and premature ovarian failure. *American Journal of Medical Genetics. Part A, 161*, 2487–2494. https://doi.org/10.1002/ajmg.a.36084.

Hastie, A. R., Lam, E. T., Pang, A. W. C., Zhang, X., Andrews, W., Lee, J., et al. (2017). Rapid automated large structural variation detection in a diploid genome by nanoChannel based next-generation mapping. *bioRxiv*, 102764. https://doi.org/10.1101/102764.

Hawkins, J. R., Taylor, A., Goodfellow, P. N., Migeon, C. J., Smith, K. D., & Berkovitz, G. D. (1992). Evidence for increased prevalence of SRY mutations in XY females with complete rather than partial gonadal dysgenesis. *American Journal of Human Genetics*, *51*, 979–984.

Hayashi, M., Kataoka, Y., Sugimura, Y., Kato, F., Fukami, M., Ogata, T., et al. (2013). A 68-year-old phenotypically male patient with 21-hydroxylase deficiency and concomitant adrenocortical neoplasm producing testosterone and cortisol. *The Tohoku Journal of Experimental Medicine*, *231*, 75–84.

He, W.-B., Banerjee, S., Meng, L.-L., Du, J., Gong, F., Huang, H., et al. (2018). Whole-exome sequencing identifies a homozygous donor splice-site mutation in *STAG3* that causes primary ovarian insufficiency. *Clinical Genetics*, *93*, 340–344. https://doi.org/10.1111/cge.13034.

Heeley, J. M., Hollander, A. S., Austin, P. F., Merritt, D. F., Wesevich, V. G., Amarillo, I. E., et al. (2018). Risk association of congenital anomalies in patients with ambiguous genitalia: A 22-year single-center experience. *Journal of Pediatric Urology*, *14*, 153.e1–153.e7. https://doi.org/10.1016/j.jpurol.2017.09.027.

Heinrich, V., Stange, J., Dickhaus, T., Imkeller, P., Krüger, U., Bauer, S., et al. (2012). The allele distribution in next-generation sequencing data sets is accurately described as the result of a stochastic branching process. *Nucleic Acids Research*, *40*, 2426–2431. https://doi.org/10.1093/nar/gkr1073.

Herrmann, B. L., Saller, B., Janssen, O. E., Gocke, P., Bockisch, A., Sperling, H., et al. (2002). Impact of estrogen replacement therapy in a male with congenital aromatase deficiency caused by a novel mutation in the CYP19 gene. *The Journal of Clinical Endocrinology and Metabolism*, *87*, 5476–5484. https://doi.org/10.1210/jc.2002-020498.

Hiort, O., Holterhus, P.-M., Werner, R., Marschke, C., Hoppe, U., Partsch, C.-J., et al. (2005). Homozygous disruption of P450 side-chain cleavage (*CYP11A1*) is associated with prematurity, complete 46,XY sex reversal, and severe adrenal failure. *The Journal of Clinical Endocrinology and Metabolism*, *90*, 538–541. https://doi.org/10.1210/jc.2004-1059.

Hiramatsu, R., Matoba, S., Kanai-Azuma, M., Tsunekawa, N., Katoh-Fukui, Y., Kurohmaru, M., et al. (2009). A critical time window of Sry action in gonadal sex determination in mice. *Development*, *136*, 129–138. https://doi.org/10.1242/dev.029587.

Huang, N., Pandey, A. V., Agrawal, V., Reardon, W., Lapunzina, P. D., Mowat, D., et al. (2005). Diversity and function of mutations in P450 oxidoreductase in patients with Antley-Bixler syndrome and disordered steroidogenesis. *American Journal of Human Genetics*, *76*, 729–749. https://doi.org/10.1086/429417.

Huang, Z., Ye, J., Han, L., Qiu, W., Zhang, H., Yu, Y., et al. (2016). Identification of five novel STAR variants in ten Chinese patients with congenital lipoid adrenal hyperplasia. *Steroids*, *108*, 85–91. https://doi.org/10.1016/j.steroids.2016.01.016.

Hughes, I. A., Houk, C., Ahmed, S. F., Lee, P. A., Group, L.C, & Group, E.C. (2006). Consensus statement on management of intersex disorders. *Archives of Disease in Childhood*, *91*, 554–563. https://doi.org/10.1136/adc.2006.098319.

Hutson, J. M., Grover, S. R., O'Connell, M., & Pennell, S. D. (2014). Malformation syndromes associated with disorders of sex development. *Nature Reviews. Endocrinology*, *10*, 476–487. https://doi.org/10.1038/nrendo.2014.83.

Igarashi, M., Dung, V. C., Suzuki, E., Ida, S., Nakacho, M., Nakabayashi, K., et al. (2013). Cryptic genomic rearrangements in three patients with 46,XY disorders of sex development. *PLoS One*, *8*, e68194. https://doi.org/10.1371/journal.pone.0068194.

Igarashi, M., Takasawa, K., Hakoda, A., Kanno, J., Takada, S., Miyado, M., et al. (2017). Identical *NR5A1* missense mutations in two unrelated 46,XX individuals with testicular tissues. *Human Mutation*, *38*, 39–42. https://doi.org/10.1002/humu.23116.

Imai, T., Yanase, T., Waterman, M. R., Simpson, E. R., & Pratt, J. J. (1992). Canadian mennonites and individuals residing in the Friesland region of The Netherlands share the same molecular basis of 17 alpha-hydroxylase deficiency. *Human Genetics*, *89*, 95–96.

Imbeaud, S., Belville, C., Messika-Zeitoun, L., Rey, R., di Clemente, N., Josso, N., et al. (1996). A 27 base-pair deletion of the anti-müllerian type II receptor gene is the most common cause of the persistent müllerian duct syndrome. *Human Molecular Genetics*, *5*, 1269–1277.

Imbeaud, S., Faure, E., Lamarre, I., Mattéi, M. G., di Clemente, N., Tizard, R., et al. (1995). Insensitivity to anti-müllerian hormone due to a mutation in the human anti-müllerian hormone receptor. *Nature Genetics*, *11*, 382–388. https://doi.org/10.1038/ng1295-382.

Ito, Y., Fisher, C. R., Conte, F. A., Grumbach, M. M., & Simpson, E. R. (1993). Molecular basis of aromatase deficiency in an adult female with sexual infantilism and polycystic ovaries. *Proceedings of the National Academy of Sciences of the United States of America*, *90*, 11673–11677.

Jackson, R. S., Creemers, J. W. M., Farooqi, I. S., Raffin-Sanson, M.-L., Varro, A., Dockray, G. J., et al. (2003). Small-intestinal dysfunction accompanies the complex endocrinopathy of human proprotein convertase 1 deficiency. *The Journal of Clinical Investigation*, *112*, 1550–1560. https://doi.org/10.1172/JCI18784.

Jackson, R. S., Creemers, J. W. M., Ohagi, S., Raffin-Sanson, M.-L., Sanders, L., Montague, C. T., et al. (1997). Obesity and impaired prohormone processing associated with mutations in the human prohormone convertase 1 gene. *Nature Genetics*, *16*, 303–306. https://doi.org/10.1038/ng0797-303.

Jadeja, S., Smyth, I., Pitera, J. E., Taylor, M. S., van Haelst, M., Bentley, E., et al. (2005). Identification of a new gene mutated in Fraser syndrome and mouse myelencephalic blebs. *Nature Genetics*, *37*, 520–525. https://doi.org/10.1038/ng1549.

Jäger, R. J., Anvret, M., Hall, K., & Scherer, G. (1990). A human XY female with a frame shift mutation in the candidate testis-determining gene SRY. *Nature*, *348*, 452–454. https://doi.org/10.1038/348452a0.

Jain, M., Fiddes, I. T., Miga, K. H., Olsen, H. E., Paten, B., & Akeson, M. (2015). Improved data analysis for the MinION nanopore sequencer. *Nature Methods*, *12*, 351–356. https://doi.org/10.1038/nmeth.3290.

Jain, M., Koren, S., Miga, K. H., Quick, J., Rand, A. C., Sasani, T. A., et al. (2018). Nanopore sequencing and assembly of a human genome with ultra-long reads. *Nature Biotechnology*, *36*, 338–345. https://doi.org/10.1038/nbt.4060.

Jain, M., Olsen, H. E., Paten, B., & Akeson, M. (2016). Erratum to: The Oxford Nanopore MinION: Delivery of nanopore sequencing to the genomics community. [Genome Biol. (2016), 17, 239]. https://doi.org/10.1186/s13059-016-1103-0. *Genome Biology*, *17*, 1–11. https://doi.org/10.1186/s13059-016-1122-x.

Jaratlerdsiri, W., Chan, E. K. F., Petersen, D. C., Yang, C., Croucher, P. I., Bornman, M. S. R., et al. (2017). Next generation mapping reveals novel large genomic rearrangements in prostate cancer. *Oncotarget*, *8*, 23588–23602. https://doi.org/10.18632/oncotarget.15802.

Jenjaroenpun, P., Wongsurawat, T., Pereira, R., Patumcharoenpol, P., Ussery, D. W., Nielsen, J., et al. (2018). Complete genomic and transcriptional landscape analysis using third-generation sequencing: A case study of Saccharomyces cerevisiae CEN.PK113-7D. *Nucleic Acids Research*, *46*, 1–15. https://doi.org/10.1093/nar/gky014.

Jeong, Y. H., Lu, H., Park, C. H., Li, M., Luo, H., Kim, J. J., et al. (2016). Stochastic anomaly of methylome but persistent SRY hypermethylation in disorder of sex development in canine somatic cell nuclear transfer. *Scientific Reports*, *6*, 1–11. https://doi.org/10.1038/srep31088.

Jongmans, M. C. J., Hoefsloot, L. H., van der Donk, K. P., Admiraal, R. J., Magee, A., van de Laar, I., et al. (2008). Familial CHARGE syndrome and theCHD7 gene: A recurrent missense mutation, intrafamilial recurrence and variability. *American Journal of Medical Genetics. Part A*, *146A*, 43–50. https://doi.org/10.1002/ajmg.a.31921.

Kagimoto, M., Winter, J. S. D., Kagimoto, K., Simpson, E. R., & Waterman, M. R. (1988). Structural characterization of normal and mutant human steroid 17α-hydroxylase genes: Molecular basis of one example of combined 17α-hydroxylase/17,20 lyase deficiency. *Molecular Endocrinology*, *2*, 564–570. https://doi.org/10.1210/mend-2-6-564.

Karaca, E., Harel, T., Pehlivan, D., Jhangiani, S. N., Gambin, T., Coban Akdemir, Z., et al. (2015). Genes that affect brain structure and function identified by rare variant analyses of mendelian neurologic disease. *Neuron*, *88*, 499–513. https://doi.org/10.1016/j.neuron.2015.09.048.

Kato, M., Das, S., Petras, K., Kitamura, K., Morohashi, K., Abuelo, D. N., et al. (2004). Mutations of ARX are associated with striking pleiotropy and consistent genotype-phenotype correlation. *Human Mutation*, *23*, 147–159. https://doi.org/10.1002/humu.10310.

Katsumata, N., Ohtake, M., Hojo, T., Ogawa, E., Hara, T., Sato, N., et al. (2002). Compound heterozygous mutations in the cholesterol side-chain cleavage enzyme gene (*CYP11A*) cause congenital adrenal insufficiency in humans. *The Journal of Clinical Endocrinology and Metabolism*, *87*, 3808–3813. https://doi.org/10.1210/jcem.87.8.8763.

Kaupert, L. C., Gomes, L. G., Brito, V. N., Lemos-Marini, S. H. V., de Mello, M. P., Longui, C. A., et al. (2016). A single nucleotide variant in the promoter region of 17β-HSD type 5 gene influences external genitalia virilization in females with 21-hydroxylase deficiency. *Hormone Research in Pædiatrics*, *85*, 333–338. https://doi.org/10.1159/000445684.

Kelberman, D., Turton, J. P. G., Woods, K. S., Mehta, A., Al-Khawari, M., Greening, J., et al. (2009). Molecular analysis of novel *PROP1* mutations associated with combined pituitary hormone deficiency (CPHD). *Clinical Endocrinology*, *70*, 96–103. https://doi.org/10.1111/j.1365-2265.2008.03326.x.

Kernohan, K. D., Hartley, T., Alirezaie, N., Robinson, P. N., Dyment, D. A., & Boycott, K. M. (2018). Evaluation of exome filtering techniques for the analysis of clinically relevant genes. *Human Mutation*, *39*, 197–201. https://doi.org/10.1002/humu.23374.

Kim, J. H., Kang, E., Heo, S. H., Kim, G. H., Jang, J. H., Cho, E. H., et al. (2017). Diagnostic yield of targeted gene panel sequencing to identify the genetic etiology of disorders of sex development. *Molecular and Cellular Endocrinology*, *444*, 19–25. https://doi.org/10.1016/j.mce.2017.01.037.

Kim, H.-G., Kurth, I., Lan, F., Meliciani, I., Wenzel, W., Eom, S. H., et al. (2008). Mutations in CHD7, encoding a chromatin-remodeling protein, cause idiopathic hypogonadotropic hypogonadism and Kallmann syndrome. *American Journal of Human Genetics*, *83*, 511–519. https://doi.org/10.1016/j.ajhg.2008.09.005.

Kim, C. J., Lin, L., Huang, N., Quigley, C. A., AvRuskin, T. W., Achermann, J. C., et al. (2008). Severe combined adrenal and gonadal deficiency caused by novel mutations in the cholesterol side chain cleavage enzyme, P450scc. *The Journal of Clinical Endocrinology and Metabolism*, *93*, 696–702. https://doi.org/10.1210/jc.2007-2330.

Kim, G.-J., Sock, E., Buchberger, A., Just, W., Denzer, F., Hoepffner, W., et al. (2015). Copy number variation of two separate regulatory regions upstream of *SOX9* causes isolated 46,XY or 46,XX disorder of sex development. *Journal of Medical Genetics*, *52*, 240–247. https://doi.org/10.1136/jmedgenet-2014-102864.

Kirac, D., Guney, A. I., Akcay, T., Guran, T., Ulucan, K., Turan, S., et al. (2014). The frequency and the effects of 21-hydroxylase gene defects in congenital adrenal hyperplasia patients. *Annals of Human Genetics*, *78*, 399–409. https://doi.org/10.1111/ahg.12083.

Kircher, M., Witten, D. M., Jain, P., O'Roak, B. J., Cooper, G. M., & Shendure, J. (2014). A general framework for estimating the relative pathogenicity of human genetic variants. *Nature Genetics*, *46*, 310–315. https://doi.org/10.1038/ng.2892.

Kitamura, K., Yanazawa, M., Sugiyama, N., Miura, H., Iizuka-Kogo, A., Kusaka, M., et al. (2002). Mutation of ARX causes abnormal development of forebrain and testes in mice and X-linked lissencephaly with abnormal genitalia in humans. *Nature Genetics*, *32*, 359–369. https://doi.org/10.1038/ng1009.

Knebelmann, B., Boussin, L., Guerrier, D., Legeai, L., Kahn, A., Josso, N., et al. (1991). Anti-Müllerian hormone Bruxelles: A nonsense mutation associated with the persistent Müllerian duct syndrome. *Proceedings of the National Academy of Sciences of the United States of America*, *88*, 3767–3771.

Ko, J. M., Cheon, C.-K., Kim, G.-H., Kim, S. H., Kim, K. S., & Yoo, H.-W. (2010). Clinical characterization and analysis of the SRD5A2 gene in six Korean patients with 5α-reductase type 2 deficiency. *Hormone Research in Pædiatrics*, *73*, 41–48. https://doi.org/10.1159/000271915.

Kon, M., & Fukami, M. (2015). Submicroscopic copy-number variations associated with 46, XY disorders of sex development. *Molecular and Cellular Pediatrics*, *2*, 7. https://doi.org/10.1186/s40348-015-0018-2.

Kosicki, M., Tomberg, K., & Bradley, A. (2018). Repair of double-strand breaks induced by CRISPR–Cas9 leads to large deletions and complex rearrangements. *Nature Biotechnology, 36*. https://doi.org/10.1038/nbt.4192.

Kottler, M.-L., Chauvin, S., Lahlou, N., Harris, C. E., Johnston, C. J., Lagarde, J.-P., et al. (2000). A new compound heterozygous mutation of the gonadotropin-releasing hormone receptor (L314X, Q106R) in a woman with complete hypogonadotropic hypogonadism: Chronic estrogen administration amplifies the gonadotropin defect. *The Journal of Clinical Endocrinology and Metabolism*, *85*, 3002–3008. https://doi.org/10.1210/jcem.85.9.6783.

Kuechler, A., Hauffa, B. P., Köninger, A., Kleinau, G., Albrecht, B., Horsthemke, B., et al. (2010). An unbalanced translocation unmasks a recessive mutation in the follicle-stimulating hormone receptor (FSHR) gene and causes FSH resistance. *European Journal of Human Genetics*, *18*, 656–661. https://doi.org/10.1038/ejhg.2009.244.

Kumar, P., Henikoff, S., & Ng, P. C. (2009). Predicting the effects of coding non-synonymous variants on protein function using the SIFT algorithm. *Nature Protocols*, *4*, 1073–1081. https://doi.org/10.1038/nprot.2009.86.

Kuo, R. I., Tseng, E., Eory, L., Paton, I. R., Archibald, A. L., & Burt, D. W. (2017). Normalized long read RNA sequencing in chicken reveals transcriptome complexity similar to human. *BMC Genomics*, *18*, 1–19. https://doi.org/10.1186/s12864-017-3691-9.

Kurdyukov, S., & Bullock, M. (2016). DNA methylation analysis: Choosing the right method. *Biology (Basel)*, *5*, 3. https://doi.org/10.3390/biology5010003.

Kwok, C., Weller, P. A., Guioli, S., Foster, J. W., Mansour, S., Zuffardi, O., et al. (1995). Mutations in SOX9, the gene responsible for Campomelic dysplasia and autosomal sex reversal. *American Journal of Human Genetics*, *57*, 1028–1036.

Kyriakou, A., Lucas-Herald, A. L., McGowan, R., Tobias, A. S., & Ahmed, S. F. (2015). Disorders of sex development: Advances in genetic diagnosis and challenges in management. *Advances in Genomis & Genetics*, *2015*, 165–177. https://doi.org/10.2147/AGG.S53226.

Laird, P. W. (2010). Principles and challenges of genome-wide DNA methylation analysis. *Nature Reviews. Genetics*, *11*, 191–203. https://doi.org/10.1038/nrg2732.

Lang-Muritano, M., Biason-Lauber, A., Gitzelmann, C., Belville, C., Picard, Y., & Schoenle, E. J. (2001). A novel mutation in the anti-müllerian hormone gene as cause of persistent müllerian duct syndrome. *European Journal of Pediatrics*, *160*, 652–654.

Lanthaler, B., Hinderhofer, K., Maas, B., Haas, D., Sawyer, H., Burton-Jones, S., et al. (2015). Characterization of large deletions in the *DHCR7* gene. *Clinical Genetics*, *88*, 149–154. https://doi.org/10.1111/cge.12454.

Laue, L., Wu, S. M., Kudo, M., Hsueh, A. J., Cutler, G. B., Griffin, J. E., et al. (1995). A nonsense mutation of the human luteinizing hormone receptor gene in Leydig cell hypoplasia. *Human Molecular Genetics*, *4*, 1429–1433.

Layman, L. C., Cohen, D. P., Jin, M., Xie, J., Li, Z., Reindollar, R. H., et al. (1998). Mutations in gonadotropin-releasing hormone receptor gene cause hypogonadotropic hypogonadism. *Nature Genetics*, *18*, 14–15. https://doi.org/10.1038/ng0198-14.

Le Quesne Stabej, P., Williams, H. J., James, C., Tekman, M., Stanescu, H. C., Kleta, R., et al. (2016). STAG3 truncating variant as the cause of primary ovarian insufficiency. *European Journal of Human Genetics*, *24*, 135–138. https://doi.org/10.1038/ejhg.2015.107.

Lecointre, C., Pichon, O., Hamel, A., Heloury, Y., Michel-Calemard, L., Morel, Y., et al. (2009). Familial acampomelic form of campomelic dysplasia caused by a 960 kb deletion upstream of SOX9. *American Journal of Medical Genetics. Part A*, *149A*, 1183–1189. https://doi.org/10.1002/ajmg.a.32830.

Ledig, S., Hiort, O., Scherer, G., Hoffmann, M., Wolff, G., Morlot, S., et al. (2010). Array-CGH analysis in patients with syndromic and non-syndromic XY gonadal dysgenesis: Evaluation of array CGH as diagnostic tool and search for new candidate loci. *Human Reproduction*, *25*, 2637–2646. https://doi.org/10.1093/humrep/deq167.

Ledig, S., Hiort, O., Wünsch, L., & Wieacker, P. (2012). Partial deletion of DMRT1 causes 46,XY ovotesticular disorder of sexual development. *European Journal of Endocrinology*, *167*, 119–124. https://doi.org/10.1530/EJE-12-0136.

Lee, H., Deignan, J. L., Dorrani, N., Strom, S. P., Kantarci, S., Quintero-Rivera, F., et al. (2014). Clinical exome sequencing for genetic identification of rare Mendelian disorders. *Journal of the American Medical Association*, *312*, 1880–1887. https://doi.org/10.1001/jama.2014.14604.

Lee, P. A., Houk, C. P., Ahmed, S. F., Hughes, I. A., & International Consensus Conference on Intersex organized by the Lawson Wilkins Pediatric Endocrine Society and the European Society for Paediatric Endocrinology (2006). Consensus statement on management of intersex disorders. International consensus conference on intersex. *Pediatrics*, *118*, e488–e500. https://doi.org/10.1542/peds.2006-0738.

Lek, M., Karczewski, K. J., Minikel, E. V., Samocha, K. E., Banks, E., Fennell, T., et al. (2016). Analysis of protein-coding genetic variation in 60, 706 humans. *Nature*, *536*, 285–291. https://doi.org/10.1038/nature19057.

Lelieveld, S. H., Spielmann, M., Mundlos, S., Veltman, J. A., & Gilissen, C. (2015). Comparison of exome and genome sequencing Technologies for the Complete Capture of protein-coding regions. *Human Mutation*, *36*, 815–822. https://doi.org/10.1002/humu.22813.

Lemos, M. C., Gomes, L., Bastos, M., Leite, V., Limbert, E., Carvalho, D., et al. (2006). PROP1 gene analysis in Portuguese patients with combined pituitary hormone deficiency. *Clinical Endocrinology*, *65*, 479–485. https://doi.org/10.1111/j.1365-2265.2006.02617.x.

Lepri, F., De Luca, A., Stella, L., Rossi, C., Baldassarre, G., Pantaleoni, F., et al. (2011). SOS1 mutations in Noonan syndrome: Molecular spectrum, structural insights on pathogenic effects, and genotype-phenotype correlations. *Human Mutation*, *32*, 760–772. https://doi.org/10.1002/humu.21492.

Leroy, C., Fouveaut, C., Leclercq, S., Jacquemont, S., Du Boullay, H., Lespinasse, J., et al. (2008). Biallelic mutations in the prokineticin-2 gene in two sporadic cases of Kallmann syndrome. *European Journal of Human Genetics*, *16*, 865–868. https://doi.org/10.1038/ejhg.2008.15.

Li, K., Ann Thomas, M., Haber, R.M., (2013). Ulerythema ophryogenes, a rarely reported cutaneous manifestation of noonan syndrome: Case report and review of the literature. Journal of Cutaneous Medicine and Surgery 17, 212–8. https://doi.org/10.2310/7750.2011.11017

Lin, L., Philibert, P., Ferraz-de-Souza, B., Kelberman, D., Homfray, T., Albanese, A., et al. (2007). Heterozygous missense mutations in steroidogenic factor 1 (SF1/Ad4BP, *NR5A1*) are associated with 46,XY disorders of sex development with normal adrenal function. *The Journal of Clinical Endocrinology and Metabolism, 92*, 991–999. https://doi.org/10.1210/jc.2006-1672.

Lindqvist, A., Hughes, I. A., & Andersson, S. (2001). Substitution mutation C268Y causes 17β-hydroxysteroid dehydrogenase 3 deficiency. *The Journal of Clinical Endocrinology and Metabolism, 86*, 921–923. https://doi.org/10.1210/jcem.86.2.7172.

Lindstrand, A., Frangakis, S., Carvalho, C. M. B., Richardson, E. B., McFadden, K. A., Willer, J. R., et al. (2016). Copy-number variation contributes to the mutational load of Bardet-Biedl syndrome. *American Journal of Human Genetics, 99*, 318–336. https://doi.org/10.1016/j.ajhg.2015.04.023.

Little, M., & Wells, C. (1997). A clinical overview of WT1 gene mutations. *Human Mutation, 9*, 209–225. https://doi.org/10.1002/(SICI)1098-1004(1997)9:3<209::AID-HUMU2>3.0.CO;2-2.

Liu, Q., Zhang, P., Wang, D., Gu, W., & Wang, K. (2017). Interrogating the "unsequenceable" genomic trinucleotide repeat disorders by long-read sequencing. *Genome Medicine, 9*, 1–16. https://doi.org/10.1186/s13073-017-0456-7.

Livadas, S., Dracopoulou, M., Dastamani, A., Sertedaki, A., Maniati-Christidi, M., Magiakou, A.-M., et al. (2015). The spectrum of clinical, hormonal and molecular findings in 280 individuals with nonclassical congenital adrenal hyperplasia caused by mutations of the *CYP21A2* gene. *Clinical Endocrinology, 82*, 543–549. https://doi.org/10.1111/cen.12543.

Locke, M., Tinsley, C. L., Benson, M. A., & Blake, D. J. (2009). TRIM32 is an E3 ubiquitin ligase for dysbindin. *Human Molecular Genetics, 18*, 2344–2358. https://doi.org/10.1093/hmg/ddp167.

Loke, J., & Ostrer, H. (2012). Rapidly screening variants of uncertain significance in the MAP3K1 gene for phenotypic effects. *Clinical Genetics, 81*, 272–277. https://doi.org/10.1111/j.1399-0004.2011.01834.x.

Lourenço, D., Brauner, R., Lin, L., De Perdigo, A., Weryha, G., Muresan, M., et al. (2009). Mutations in *NR5A1* associated with ovarian insufficiency. *The New England Journal of Medicine, 360*, 1200–1210. https://doi.org/10.1056/NEJMoa0806228.

Lourenco, D., Brauner, R., Rybczynska, M., Nihoul-Fekete, C., McElreavey, K., & Bashamboo, A. (2011). Loss-of-function mutation in GATA4 causes anomalies of human testicular development. *Proceedings of the National Academy of Sciences of the United States of America, 108*, 1597–1602. https://doi.org/10.1073/pnas.1010257108.

Ludes-Meyers, J. H., Kil, H., Nunez, M. I., Conti, C. J., Parker-Thornburg, J., Bedford, M. T., et al. (2007). WWOX hypomorphic mice display a higher incidence of B-cell lymphomas and develop testicular atrophy. *Genes, Chromosomes and Cancer, 46*, 1129–1136. https://doi.org/10.1002/gcc.20497.

MacLean, H. E., Chu, S., Warne, G. L., & Zajac, J. D. (1993). Related individuals with different androgen receptor gene deletions. *The Journal of Clinical Investigation, 91*, 1123–1128. https://doi.org/10.1172/JCI116271.

Mandel, H., Shemer, R., Borochowitz, Z. U., Okopnik, M., Knopf, C., Indelman, M., et al. (2008). SERKAL syndrome: An autosomal-recessive disorder caused by a loss-of-function mutation in WNT4. *American Journal of Human Genetics, 82*, 39–47. https://doi.org/10.1016/j.ajhg.2007.08.005.

Massin, N., Pêcheux, C., Eloit, C., Bensimon, J.-L., Galey, J., Kuttenn, F., et al. (2003). X chromosome-linked Kallmann syndrome: Clinical heterogeneity in three siblings carrying an intragenic deletion of the *KAL-1* gene. *The Journal of Clinical Endocrinology and Metabolism, 88*, 2003–2008. https://doi.org/10.1210/jc.2002-021981.

Maurano, M. T., Humbert, R., Rynes, E., Thurman, R. E., Haugen, E., Wang, H., et al. (2012). Systematic localization of common disease-associated variation in regulatory DNA. Science. *337*, 1190–1195. https://doi.org/10.1126/science.1222794.

McCabe, M. J., Gaston-Massuet, C., Gregory, L. C., Alatzoglou, K. S., Tziaferi, V., Sbai, O., et al. (2013). Variations in *PROKR2*, but Not *PROK2*, are associated with hypopituitarism and septo-optic dysplasia. *The Journal of Clinical Endocrinology and Metabolism, 98*, E547–E557. https://doi.org/10.1210/jc.2012-3067.

McCaffrey, J., Young, E., Lassahn, K., Sibert, J., Pastor, S., Riethman, H., et al. (2017). High-throughput single-molecule telomere characterization. *Genome Research, 27*, 1904–1915. https://doi.org/10.1101/gr.222422.117.

McElreavey, K. D., Vilain, E., Boucekkine, C., Vidaud, M., Jaubert, F., Richaud, F., et al. (1992). XY sex reversal associated with a nonsense mutation in SRY. *Genomics, 13*, 838–840.

McGregor, L., Makela, V., Darling, S. M., Vrontou, S., Chalepakis, G., Roberts, C., et al. (2003). Fraser syndrome and mouse blebbed phenotype caused by mutations in FRAS1/Fras1 encoding a putative extracellular matrix protein. *Nature Genetics, 34*, 203–208. https://doi.org/10.1038/ng1142.

Meduri, G., Touraine, P., Beau, I., Lahuna, O., Desroches, A., Vacher-Lavenu, M. C., et al. (2003). Delayed puberty and primary amenorrhea associated with a novel mutation of the human follicle-stimulating hormone receptor: Clinical, histological, and molecular studies. *The Journal of Clinical Endocrinology and Metabolism, 88*, 3491–3498. https://doi.org/10.1210/jc.2003-030217.

Meienberg, J., Zerjavic, K., Keller, I., Okoniewski, M., Patrignani, A., Ludin, K., et al. (2015). New insights into the performance of human whole-exome capture platforms. *Nucleic Acids Research, 43*, 1–14. https://doi.org/10.1093/nar/gkv216.

Messika-Zeitoun, L., Gouédard, L., Belville, C., Dutertre, M., Lins, L., Imbeaud, S., et al. (2001). Autosomal recessive segregation of a truncating mutation of anti-müllerian type II receptor in a family affected by the persistent müllerian duct syndrome contrasts with its dominant negative activity *in vitro*. *The Journal of Clinical Endocrinology and Metabolism, 86*, 4390–4397. https://doi.org/10.1210/jcem.86.9.7839.

Meynert, A. M., Ansari, M., FitzPatrick, D. R., & Taylor, M. S. (2014). Variant detection sensitivity and biases in whole genome and exome sequencing. *BMC Bioinformatics, 15*, 247. https://doi.org/10.1186/1471-2105-15-247.

Meynert, A. M., Bicknell, L. S., Hurles, M. E., Jackson, A. P., & Taylor, M. S. (2013). Quantifying single nucleotide variant detection sensitivity in exome sequencing. *BMC Bioinformatics, 14*, 195. https://doi.org/10.1186/1471-2105-14-195.

Meysing, A. U., Kanasaki, H., Bedecarrats, G. Y., Acierno, J. S., Conn, P. M., Martin, K. A., et al. (2004). *GNRHR* mutations in a woman with idiopathic hypogonadotropic hypogonadism highlight the differential sensitivity of luteinizing hormone and follicle-stimulating hormone to gonadotropin-releasing hormone. *The Journal of Clinical Endocrinology and Metabolism, 89*, 3189–3198. https://doi.org/10.1210/jc.2003-031808.

Misrahi, M., Meduri, G., Pissard, S., Bouvattier, C., Beau, I., Loosfelt, H., et al. (1997). Comparison of immunocytochemical and molecular features with the phenotype in a case of incomplete male pseudohermaphroditism associated with a mutation of the luteinizing hormone receptor. *The Journal of Clinical Endocrinology and Metabolism, 82*, 2159–2165. https://doi.org/10.1210/jcem.82.7.4039.

Mitson, M., Kelley, L. A., Sternberg, M. J. E., Higgs, D. R., & Gibbons, R. J. (2011). Functional significance of mutations in the Snf2 domain of ATRX. *Human Molecular Genetics, 20*, 2603–2610. https://doi.org/10.1093/hmg/ddr163.

Mizuno, K., Kojima, Y., Kamisawa, H., Moritoki, Y., Nishio, H., Nakane, A., et al. (2014). Elucidation of distinctive genomic DNA structures in patients with 46,XX testicular disorders of sex development using genome wide analyses. *The Journal of Urology, 192*, 535–541. https://doi.org/10.1016/j.juro.2014.02.044.

Moalem, S., Babul-Hirji, R., Stavropolous, D. J., Wherrett, D., Bagli, D. J., Thomas, P., et al. (2012). XX male sex reversal with genital abnormalities associated with a *de novo* SOX3 gene duplication. *American Journal of Medical Genetics. Part A, 158A*, 1759–1764. https://doi.org/10.1002/ajmg.a.35390.

Moccia, A., Srivastava, A., Skidmore, J. M., Bernat, J. A., Wheeler, M., Chong, J. X., et al. (2018). Genetic analysis of CHARGE syndrome identifies overlapping molecular biology. *Genetics in Medicine, 20*, 1022–1029. https://doi.org/10.1038/gim.2017.233.
Mohnach, L., Fechner, P. Y., & Keegan, C. E. (2016). *Nonsyndromic disorders of testicular development, GeneReviews®*. Seattle: University of Washington.
Monnier, C., Dodé, C., Fabre, L., Teixeira, L., Labesse, G., Pin, J.-P., et al. (2009). PROKR2 missense mutations associated with Kallmann syndrome impair receptor signalling activity. *Human Molecular Genetics, 18*, 75–81. https://doi.org/10.1093/hmg/ddn318.
Montanelli, L., Delbaere, A., Di Carlo, C., Nappi, C., Smits, G., Vassart, G., et al. (2004). A mutation in the follicle-stimulating hormone receptor as a cause of familial spontaneous ovarian hyperstimulation syndrome. *The Journal of Clinical Endocrinology and Metabolism, 89*, 1255–1258.
Murray, I. A., Clark, T. A., Morgan, R. D., Boitano, M., Anton, B. P., Luong, K., et al. (2012). The methylomes of six bacteria. *Nucleic Acids Research, 40*, 11450–11462. https://doi.org/10.1093/nar/gks891.
Muscatelli, F., Strom, T. M., Walker, A. P., Zanaria, E., Recan, D., Meindl, A., et al. (1994). Mutations in the DAX-1 gene give rise to both X-linked adrenal hypoplasia congenita and hypogonadotropic hypogonadism. *Nature, 372*, 672–676. https://doi.org/10.1038/372672a0.
Nakae, J., Abe, S., Tajima, T., Shinohara, N., Murashita, M., Igarashi, Y., et al. (1997). Three novel mutations and a *de novo* deletion mutation of the DAX-1 gene in patients with X-linked adrenal hypoplasia congenita. *The Journal of Clinical Endocrinology and Metabolism, 82*, 3835–3841. https://doi.org/10.1210/jcem.82.11.4342.
Nakae, J., Tajima, T., Sugawara, T., Arakane, F., Hanaki, K., Hotsubo, T., et al. (1997). Analysis of the steroidogenic acute regulatory protein (StAR) gene in Japanese patients with congenital lipoid adrenal hyperplasia. *Human Molecular Genetics, 6*, 571–576.
Netchine, I., Sobrier, M.-L., Krude, H., Schnabel, D., Maghnie, M., Marcos, E., et al. (2000). Mutations in LHX3 result in a new syndrome revealed by combined pituitary hormone deficiency. *Nature Genetics, 25*, 182–186. https://doi.org/10.1038/76041.
Nimkarn, S., Gangishetti, P. K., Yau, M., & New, M. I. (2016). *21-hydroxylase-deficient congenital adrenal hyperplasia, GeneReviews®*. Seattle: University of Washington.
Nishino, K., Hattori, N., Tanaka, S., & Shiota, K. (2004). DNA methylation-mediated Control of *Sry* gene expression in mouse gonadal development. *The Journal of Biological Chemistry, 279*, 22306–22313. https://doi.org/10.1074/jbc.M309513200.
Norling, A., Lindén Hirschberg, A., Iwarsson, E., Persson, B., Wedell, A., & Barbaro, M. (2013). Novel candidate genes for 46,XY gonadal dysgenesis identified by a customized 1 M array-CGH platform. *European Journal of Medical Genetics, 56*, 661–668. https://doi.org/10.1016/j.ejmg.2013.09.003.
Nykamp, K., Anderson, M., Powers, M., Garcia, J., Herrera, B., Ho, Y.-Y., et al. (2017). Sherloc: A comprehensive refinement of the ACMG–AMP variant classification criteria. *Genetics in Medicine, 19*, 1105–1117. https://doi.org/10.1038/gim.2017.37.
Oh, J., Song, J.-S., Park, J. E., Jang, S. Y., Ki, C.-S., & Kim, D.-K. (2017). A case of Antley-Bixler syndrome with a novel likely pathogenic variant (c.529G > C) in the *POR* gene. *Annals of Laboratory Medicine, 37*, 559. https://doi.org/10.3343/alm.2017.37.6.559.
Ohnesorg, T., Vilain, E., & Sinclair, A. H. (2014). The genetics of disorders of sex development in humans. *Sexual Development*. https://doi.org/10.1159/000357956.
Okuyama, E., Nishi, N., Onishi, S., Itoh, S., Ishii, Y., Miyanaka, H., et al. (1997). A novel splicing junction mutation in the gene for the steroidogenic acute regulatory protein causes congenital lipoid adrenal hyperplasia. *The Journal of Clinical Endocrinology and Metabolism, 82*, 2337–2342. https://doi.org/10.1210/jcem.82.7.4045.
Ono, M., & Harley, V. R. (2013). Disorders of sex development: New genes, new concepts. *Nature Reviews. Endocrinology, 9*, 79–91. https://doi.org/10.1038/nrendo.2012.235.

Oshiro, C., Takasu, N., Wakugami, T., Komiya, I., Yamada, T., Eguchi, Y., et al. (1995). Seventeen alpha-hydroxylase deficiency with one base pair deletion of the cytochrome P450c17 (CYP17) gene. *The Journal of Clinical Endocrinology and Metabolism*, *80*, 2526–2529. https://doi.org/10.1210/jcem.80.8.7629254.

Özen, S., Onay, H., Atik, T., Solmaz, A. E., Özklnay, F., Gökşen, D., et al. (2017). Rapid molecular genetic diagnosis with next-generation sequencing in 46,XY disorders of sex development cases: Efficiency and cost assessment. *Hormone Research in Pædiatrics*, *87*, 81–87. https://doi.org/10.1159/000452995.

Parma, P., Radi, O., Vidal, V., Chaboissier, M. C., Dellambra, E., Valentini, S., et al. (2006). R-spondin1 is essential in sex determination, skin differentiation and malignancy. *Nature Genetics*, *38*, 1304–1309. https://doi.org/10.1038/ng1907.

Pearlman, A., Loke, J., Le Caignec, C., White, S., Chin, L., Friedman, A., et al. (2010). Mutations in MAP3K1 cause 46,XY disorders of sex development and implicate a common signal transduction pathway in human testis determination. *American Journal of Human Genetics*, *87*, 898–904. https://doi.org/10.1016/j.ajhg.2010.11.003.

Petrovski, S., & Goldstein, D. B. (2016). Unequal representation of genetic variation across ancestry groups creates healthcare inequality in the application of precision medicine. *Genome Biology*, *17*, 157. https://doi.org/10.1186/s13059-016-1016-y.

Pfaeffle, R. W., Savage, J. J., Hunter, C. S., Palme, C., Ahlmann, M., Kumar, P., et al. (2007). Four novel mutations of the *LHX3* gene cause combined pituitary hormone deficiencies with or without limited neck rotation. *The Journal of Clinical Endocrinology and Metabolism*, *92*, 1909–1919. https://doi.org/10.1210/jc.2006-2177.

Phelan, N., Williams, E. L., Cardamone, S., Lee, M., Creighton, S. M., Rumsby, G., et al. (2015). Screening for mutations in 17β-hydroxysteroid dehydrogenase and androgen receptor in women presenting with partially virilised 46,XY disorders of sex development. *European Journal of Endocrinology*, *172*, 745–751. https://doi.org/10.1530/EJE-14-0994.

Philibert, P., Biason-Lauber, A., Rouzier, R., Pienkowski, C., Paris, F., Konrad, D., et al. (2008). Identification and functional analysis of a new WNT4 gene mutation among 28 adolescent girls with primary amenorrhea and müllerian duct abnormalities: A french collaborative study. *The Journal of Clinical Endocrinology and Metabolism*, *93*, 895–900. https://doi.org/10.1210/jc.2007-2023.

Philibert, P., Polak, M., Colmenares, A., Lortat-Jacob, S., Audran, F., Poulat, F., et al. (2011). Predominant Sertoli cell deficiency in a 46,XY disorders of sex development patient with a new NR5A1/SF-1 mutation transmitted by his unaffected father. *Fertility and Sterility*, *95*, 1788.e5–1788.e9. https://doi.org/10.1016/j.fertnstert.2010.11.035.

Piperno, A., Arosio, C., Fossati, L., Viganò, M., Trombini, P., Vergani, A., et al. (2000). Two novel nonsense mutations of HFE gene in five unrelated italian patients with hemochromatosis. *Gastroenterology*, *119*, 441–445.

Pitteloud, N., Acierno, J. S., Meysing, A., Eliseenkova, A. V., Ma, J., Ibrahimi, O. A., et al. (2006). Mutations in fibroblast growth factor receptor 1 cause both Kallmann syndrome and normosmic idiopathic hypogonadotropic hypogonadism. *Proceedings of the National Academy of Sciences of the United States of America*, *103*, 6281–6286. https://doi.org/10.1073/pnas.0600962103.

Pitteloud, N., Zhang, C., Pignatelli, D., Li, J.-D., Raivio, T., Cole, L. W., et al. (2007). Loss-of-function mutation in the prokineticin 2 gene causes Kallmann syndrome and normosmic idiopathic hypogonadotropic hypogonadism. *Proceedings of the National Academy of Sciences of the United States of America*, *104*, 17447–17452. https://doi.org/10.1073/pnas.0707173104.

Pop, R., Zaragoza, M. V., Gaudette, M., Dohrmann, U., & Scherer, G. (2005). A homozygous nonsense mutation in SOX9 in the dominant disorder campomelic dysplasia: A case of mitotic gene conversion. *Human Genetics*, *117*, 43–53. https://doi.org/10.1007/s00439-005-1295-y.

Porat, Y., Lev, O., Einhorn, M., Shani, O., Vilain, E., Barseghyan, H., et al. (2018). A novel approach for structural variant calling: Combining data from whole genome next-generation sequencing and optical mapping. In *68th Annual Meeting of the American Society of Human Genetics, (October 16-20, 2018, San Diego, California). San Diego, California.* Poster abstract/1643F.

Przylepa, K. A., Paznekas, W., Zhang, M., Golabi, M., Bias, W., Bamshad, M. J., et al. (1996). Fibroblast growth factor receptor 2 mutations in Beare-Stevenson cutis gyrata syndrome. *Nature Genetics*, *13*, 492–494. https://doi.org/10.1038/ng0896-492.

Quail, M. A., Kozarewa, I., Smith, F., Scally, A., Stephens, P. J., Durbin, R., et al. (2008). A large genome center's improvements to the Illumina sequencing system. *Nature Methods*, *5*, 1005–1010. https://doi.org/10.1038/nmeth.1270.

Quinonez, S. C., Park, J. M., Rabah, R., Owens, K. M., Yashar, B. M., Glover, T. W., et al. (2013). 9p partial monosomy and disorders of sex development: Review and postulation of a pathogenetic mechanism. *American Journal of Medical Genetics. Part A*, *161A*, 1882–1896. https://doi.org/10.1002/ajmg.a.36018.

Rajab, A., Kelberman, D., de Castro, S. C. P., Biebermann, H., Shaikh, H., Pearce, K., et al. (2008). Novel mutations in LHX3 are associated with hypopituitarism and sensorineural hearing loss. *Human Molecular Genetics*, *17*, 2150–2159. https://doi.org/10.1093/hmg/ddn114.

Rank, D., Baybayan, P., Bettman, B., Bibillo, A., Bjornson, K., Chaudhuri, B., et al. (2009). Single polymerase molecules. *Science*, *323*, 133–138. https://doi.org/10.1126/science.1162986.

Raymond, C. S., Parker, E. D., Kettlewell, J. R., Brown, L. G., Page, D. C., Kusz, K., et al. (1999). A region of human chromosome 9p required for testis development contains two genes related to known sexual regulators. *Human Molecular Genetics*, *8*, 989–996.

Refai, O., Friedman, A., Terry, L., Jewett, T., Pearlman, A., Perle, M. A., et al. (2010). De novo 12;17 translocation upstream of *SOX9* resulting in 46,XX testicular disorder of sex development. *American Journal of Medical Genetics. Part A*, *152A*, 422–426. https://doi.org/10.1002/ajmg.a.33201.

Retterer, K., Juusola, J., Cho, M. T., Vitazka, P., Millan, F., Gibellini, F., et al. (2016). Clinical application of whole-exome sequencing across clinical indications. *Genetics in Medicine*, *18*, 696–704. https://doi.org/10.1038/gim.2015.148.

Reynaud, R., Barlier, A., Vallette-Kasic, S., Saveanu, A., Guillet, M.-P., Simonin, G., et al. (2005). An uncommon phenotype with familial central hypogonadism caused by a Novel *PROP1* gene mutant truncated in the transactivation domain. *The Journal of Clinical Endocrinology and Metabolism*, *90*, 4880–4887. https://doi.org/10.1210/jc.2005-0119.

Reynaud, R., Jayakody, S. A., Monnier, C., Saveanu, A., Bouligand, J., Guedj, A.-M., et al. (2012). PROKR2 variants in multiple hypopituitarism with pituitary stalk interruption. *The Journal of Clinical Endocrinology and Metabolism*, *97*, E1068–E1073. https://doi.org/10.1210/jc.2011-3056.

Rhéaume, E., Simard, J., Morel, Y., Mebarki, F., Zachmann, M., Forest, M. G., et al. (1992). Congenital adrenal hyperplasia due to point mutations in the type II 3β-hydroxysteroid dehydrogenase gene. *Nature Genetics*, *1*, 239–245. https://doi.org/10.1038/ng0792-239.

Richards, S., Aziz, N., Bale, S., Bick, D., & Das, S. (2015). Standards and guidelines for the interpretation of sequence variants: A joint consensus recommendation of the American College of Medical Genetics and Genomics and the Association for Molecular Pathology. *Genetics in Medicine*, *17*, 1–20. https://doi.org/10.1038/gim.2015.30.

Roberts, A. E., Araki, T., Swanson, K. D., Montgomery, K. T., Schiripo, T. A., Joshi, V. A., et al. (2007). Germline gain-of-function mutations in SOS1 cause Noonan syndrome. *Nature Genetics*, *39*, 70–74. https://doi.org/10.1038/ng1926.

Rolston, A. M., Gardner, M., van Leeuwen, K., Mohnach, L., Keegan, C., Delot, E., et al. (2017). Disorders of sex development (DSD): Clinical service delivery in the United States. *American Journal of Medical Genetics. Part C, Seminars in Medical Genetics, 175*, 268–278. https://doi.org/10.1002/ajmg.c.31558.

Rossetti, R., Di Pasquale, E., Marozzi, A., Bione, S., Toniolo, D., Grammatico, P., et al. (2009). BMP15 mutations associated with primary ovarian insufficiency cause a defective production of bioactive protein. *Human Mutation, 30*, 804–810. https://doi.org/10.1002/humu.20961.

Rossetti, R., Ferrari, I., Bonomi, M., & Persani, L. (2017). Genetics of primary ovarian insufficiency. *Clinical Genetics, 91*, 183–198. https://doi.org/10.1111/cge.12921.

Royer-Pokora, B., Beier, M., Henzler, M., Alam, R., Schumacher, V., Weirich, A., et al. (2004). Twenty-four new cases ofWT1 germline mutations and review of the literature: Genotype/phenotype correlations for Wilms tumor development. *American Journal of Medical Genetics, 127A*, 249–257. https://doi.org/10.1002/ajmg.a.30015.

Ruiz-Ferrer, M., Torroglosa, A., Núñez-Torres, R., de Agustín, J. C., Antiñolo, G., & Borrego, S. (2011). Expression of PROKR1 and PROKR2 in human enteric neural precursor cells and identification of sequence variants suggest a role in HSCR. *PLoS One, 6*, e23475. https://doi.org/10.1371/journal.pone.0023475.

Salamon, S., Flisikowski, K., & Switonski, M. (2017). Methylation patterns of SOX3, SOX9, and WNT4 genes in gonads of dogs with XX (SRY -negative) disorder of sexual development. *Sexual Development, 11*, 86–93. https://doi.org/10.1159/000466712.

Sammarco, I., Grimaldi, P., Rossi, P., Cappa, M., Moretti, C., Frajese, G., et al. (2000). novel point mutation in the splice donor site of exon-intron junction 6 of the androgen receptor gene in a patient with partial androgen insensitivity syndrome. *The Journal of Clinical Endocrinology and Metabolism, 85*, 3256–3261. https://doi.org/10.1210/jcem.85.9.6815.

Sandberg, D. E., Callens, N., & Wisniewski, A. B. (2015). Disorders of sex development (DSD): Networking and standardization considerations. *Hormone and Metabolic Research, 47*, 387–393. https://doi.org/10.1055/s-0035-1548936.

Sandberg, D. E., & Mazur, T. (2014). A noncategorical approach to the psychosocial care of persons with DSD and their families. In B. P. C. Kreukels, T. D. Steensma, & A. L. C. de Vries (Eds.), *Gender dysphoria and disorders of sex development: Progress in Care and Knowledge* (pp. 93–114). Boston, MA: Springer US. https://doi.org/10.1007/978-1-4614-7441-8_5.

Sarfati, J., Fouveaut, C., Leroy, C., Jeanpierre, M., Hardelin, J.-P., & Dodé, C. (2013). Greater prevalence of PROKR2 mutations in Kallmann syndrome patients from the Maghreb than in European patients. *European Journal of Endocrinology, 169*, 805–809. https://doi.org/10.1530/EJE-13-0419.

Sarfati, J., Guiochon-Mantel, A., Rondard, P., Arnulf, I., Garcia-Piñero, A., Wolczynski, S., et al. (2010). A comparative phenotypic study of Kallmann syndrome patients carrying monoallelic and biallelic mutations in the prokineticin 2 or prokineticin receptor 2 genes. *The Journal of Clinical Endocrinology and Metabolism, 95*, 659–669. https://doi.org/10.1210/jc.2009-0843.

Sawyer, S. L., Hartley, T., Dyment, D. A., Beaulieu, C. L., Schwartzentruber, J., Smith, A., et al. (2016). Utility of whole-exome sequencing for those near the end of the diagnostic odyssey: Time to address gaps in care. *Clinical Genetics, 89*, 275–284. https://doi.org/10.1111/cge.12654.

Schwab, K. O., Moisan, A.-M., Homoki, J., Peter, M., & Simard, J. (2005). 17alpha-hydroxylase/17,20-Lyase deficiency due to novel compound heterozygote mutations: Treatment for tall stature in a female with male pseudohermaphroditism and spontaneous puberty in her affected sister. *Journal of Pediatric Endocrinology & Metabolism, 18*, 403–411.

Schwartz, M., Blichfeldt, S., & Müller, J. (1997). X-linked adrenal hypoplasia in a large Greenlandic family. Detection of a missense mutation (N4401) in the DAX-1 gene; implication for genetic counselling and carrier diagnosis. *Human Genetics*, *99*, 83–87.
Schwarz, R. F., Trinh, A., Sipos, B., Brenton, J. D., Goldman, N., & Markowetz, F. (2014). Phylogenetic quantification of intra-tumour heterogeneity. *PLoS Computational Biology*, *10*, e1003535. https://doi.org/10.1371/journal.pcbi.1003535.
Sedlazeck, F. J., Lee, H., Darby, C. A., & Schatz, M. C. (2018). Piercing the dark matter: Bioinformatics of long-range sequencing and mapping. *Nature Reviews. Genetics*, *19*, 329–346. https://doi.org/10.1038/s41576-018-0003-4.
Sedlazeck, F. J., Rescheneder, P., Smolka, M., Fang, H., Nattestad, M., Von Haeseler, A., et al. (2018). Accurate detection of complex structural variations using single-molecule sequencing. *Nature Methods*, *15*, 461–468. https://doi.org/10.1038/s41592-018-0001-7.
Seminara, S. B., Messager, S., Chatzidaki, E. E., Thresher, R. R., Acierno, J. S., Shagoury, J. K., et al. (2003). The *GPR54* gene as a regulator of puberty. *The New England Journal of Medicine*, *349*, 1614–1627. https://doi.org/10.1056/NEJMoa035322.
Seo, J.-S., Rhie, A., Kim, J., Lee, S., Sohn, M.-H., Kim, C.-U., et al. (2016). De novo assembly and phasing of a Korean human genome. *Nature*, *538*, 243–247. https://doi.org/10.1038/nature20098.
Shafeghati, Y., Kniepert, A., Vakili, G., & Zenker, M. (2008). Fraser syndrome due to homozygosity for a splice site mutation of FREM2. *American Journal of Medical Genetics. Part A*, *146A*, 529–531. https://doi.org/10.1002/ajmg.a.32091.
Sharp, A. J., Locke, D. P., McGrath, S. D., Cheng, Z., Bailey, J. A., Vallente, R. U., et al. (2005). Segmental duplications and copy-number variation in the human genome. *American Journal of Human Genetics*, 77, 78–88. https://doi.org/10.1086/431652.
Shi, L., Guo, Y., Dong, C., Huddleston, J., Yang, H., Han, X., et al. (2016). Long-read sequencing and *de novo* assembly of a Chinese genome. *Nature Communications*, 7, 12065. https://doi.org/10.1038/ncomms12065.
Shigemizu, D., Momozawa, Y., Abe, T., Morizono, T., Boroevich, K. A., Takata, S., et al. (2015). Performance comparison of four commercial human whole-exome capture platforms. *Scientific Reports*, *5*, 1–8. https://doi.org/10.1038/srep12742.
Shozu, M., Sebastian, S., Takayama, K., Hsu, W.-T., Schultz, R. A., Neely, K., et al. (2003). Estrogen excess associated with novel gain-of-function mutations affecting the aromatase gene. *The New England Journal of Medicine*, *348*, 1855–1865. https://doi.org/10.1056/NEJMoa021559.
Simard, J., Rhéaume, E., Leblanc, J. F., Wallis, S. C., Joplin, G. F., Gilbey, S., et al. (1994). Congenital adrenal hyperplasia caused by a novel homozygous frameshift mutation 273 delta AA in type II 3 beta-hydroxysteroid dehydrogenase gene (HSD3B2) in three male patients of Afghan/Pakistani origin. *Human Molecular Genetics*, *3*, 327–330.
Simonetti, L., Bruque, C. D., Fernández, C. S., Benavides-Mori, B., Delea, M., Kolomenski, J. E., et al. (2018). *CYP21A2* mutation update: Comprehensive analysis of databases and published genetic variants. *Human Mutation*, *39*, 5–22. https://doi.org/10.1002/humu.23351.
Slavotinek, A., Crawford, H., Golabi, M., Tao, C., Perry, H., Oberoi, S., et al. (2009). Novel *FGFR2* deletion in a patient with Beare-Stevenson-like syndrome. *American Journal of Medical Genetics. Part A*, *149A*, 1814–1817. https://doi.org/10.1002/ajmg.a.32947.
Slavotinek, A., Li, C., Sherr, E. H., & Chudley, A. E. (2006). Mutation analysis of theFRAS1 gene demonstrates new mutations in a propositus with Fraser syndrome. *American Journal of Medical Genetics. Part A*, *140A*, 1909–1914. https://doi.org/10.1002/ajmg.a.31399.
Słowikowska-Hilczer, J., Hirschberg, A. L., Claahsen-van der Grinten, H., Reisch, N., Bouvattier, C., Thyen, U., et al. (2017). Fertility outcome and information on fertility issues in individuals with different forms of disorders of sex development: Findings from the dsd-LIFE study. *Fertility and Sterility*, *108*, 822–831. https://doi.org/10.1016/j.fertnstert.2017.08.013.

Smedley, D., Jacobsen, J. O. B., Jäger, M., Köhler, S., Holtgrewe, M., Schubach, M., et al. (2015). Next-generation diagnostics and disease-gene discovery with the exomiser. *Nature Protocols, 10,* 2004–2015. https://doi.org/10.1038/nprot.2015.124.

Smits, G., Olatunbosun, O., Delbaere, A., Pierson, R., Vassart, G., & Costagliola, S. (2003). Ovarian hyperstimulation syndrome due to a mutation in the follicle-stimulating hormone receptor. *The New England Journal of Medicine, 349,* 760–766. https://doi.org/10.1056/NEJMoa030064.

Smyk, M., Berg, J. S., Pursley, A., Curtis, F. K., Fernandez, B. A., Bien-Willner, G. A., et al. (2007). Male-to-female sex reversal associated with an ~250 kb deletion upstream of NR0B1 (DAX1). *Human Genetics, 122,* 63–70. https://doi.org/10.1007/s00439-007-0373-8.

Sobreira, N. L. M., Cirulli, E. T., Avramopoulos, D., Wohler, E., Oswald, G. L., Stevens, E. L., et al. (2010). Whole-genome sequencing of a single proband together with linkage analysis identifies a mendelian disease gene. *PLoS Genetics, 6* e1000991 https://doi.org/10.1371/journal.pgen.1000991.

Sobrier, M.-L., Maghnie, M., Vié-Luton, M.-P., Secco, A., di Iorgi, N., Lorini, R., et al. (2006). Novel *HESX1* mutations associated with a life-threatening neonatal phenotype, pituitary aplasia, but normally located posterior pituitary and no optic nerve abnormalities. *The Journal of Clinical Endocrinology and Metabolism, 91,* 4528–4536. https://doi.org/10.1210/jc.2006-0426.

Soden, S. E., Saunders, C. J., Willig, L. K., Farrow, E. G., Smith, L. D., Petrikin, J. E., et al. (2014). Effectiveness of exome and genome sequencing guided by acuity of illness for diagnosis of neurodevelopmental disorders. *Science Translational Medicine, 6,* 265ra168. https://doi.org/10.1126/scitranslmed.3010076.

Söderlund, D., Canto, P., & Méndez, J. P. (2002). Identification of three novel mutations in the KAL1 gene in patients with Kallmann syndrome. *The Journal of Clinical Endocrinology and Metabolism, 87,* 2589–2592. https://doi.org/10.1210/jcem.87.6.8611.

Sparks, S. E., Wassif, C. A., Goodwin, H., Conley, S. K., Lanham, D. C., Kratz, L. E., et al. (2014). Decreased cerebral spinal fluid neurotransmitter levels in Smith-Lemli-Opitz syndrome. *Journal of Inherited Metabolic Disease, 37,* 415–420. https://doi.org/10.1007/s10545-013-9672-5.

Speiser, P. W., Azziz, R., Baskin, L. S., Ghizzoni, L., Hensle, T. W., Merke, D. P., et al. (2010). Congenital adrenal hyperplasia due to steroid 21-hydroxylase deficiency: An Endocrine Society clinical practice guideline. *The Journal of Clinical Endocrinology and Metabolism, 95,* 4133–4160. https://doi.org/10.1210/jc.2009-2631.

Stavropoulos, D. J., Merico, D., Jobling, R., Bowdin, S., Monfared, N., Thiruvahindrapuram, B., et al. (2016). Whole-genome sequencing expands diagnostic utility and improves clinical management in paediatric medicine. *NPJ Genomic Medicine, 1,* 15012. https://doi.org/10.1038/npjgenmed.2015.12.

Stoetzel, C., Laurier, V., Faivre, L., Mégarbané, A., Perrin-Schmitt, F., Verloes, A., et al. (2006). BBS8 is rarely mutated in a cohort of 128 Bardet–Biedl syndrome families. *Journal of Human Genetics, 51,* 81–84. https://doi.org/10.1007/s10038-005-0320-2.

Strobel, A., Issad, T., Camoin, L., Ozata, M., & Strosberg, A. D. (1998). A leptin missense mutation associated with hypogonadism and morbid obesity. *Nature Genetics, 18,* 213–215. https://doi.org/10.1038/ng0398-213.

Sudmant, P. H., Rausch, T., Gardner, E. J., Handsaker, R. E., Abyzov, A., Huddleston, J., et al. (2015). An integrated map of structural variation in 2,504 human genomes. *Nature, 526,* 75–81. https://doi.org/10.1038/nature15394.

Sund, K. L., & Rehder, C. W. (2014). Detection and reporting of homozygosity associated with consanguinity in the clinical laboratory. *Human Heredity,* 77, 217–224. https://doi.org/10.1159/000362448.

Sutton, E., Hughes, J., White, S., Sekido, R., Tan, J., Arboleda, V., et al. (2011). Identification of SOX3 as an XX male sex reversal gene in mice and humans. *The Journal of Clinical Investigation, 121,* 328–341. https://doi.org/10.1172/JCI42580.

Swartz, J. M., Ciarlo, R., Guo, M. H., Abrha, A., Diamond, D. A., Chan, Y.-M., et al. (2017). Two unrelated undervirilized 46,XY males with inherited NR5A1 variants identified by whole-exome sequencing. *Hormone Research in Pædiatrics*, *87*, 264–270. https://doi.org/10.1159/000448754.

Tajima, T., Fujieda, K., Kouda, N., Nakae, J., & Miller, W. L. (2001). Heterozygous mutation in the cholesterol side chain cleavage enzyme (P450scc) gene in a patient with 46, XY sex reversal and adrenal insufficiency. *The Journal of Clinical Endocrinology and Metabolism*, *86*, 3820–3825. https://doi.org/10.1210/jcem.86.8.7748.

Tajima, T., Hattorri, T., Nakajima, T., Okuhara, K., Sato, K., Abe, S., et al. (2003). Sporadic heterozygous frameshift mutation of *HESX1* causing pituitary and optic nerve hypoplasia and combined pituitary hormone deficiency in a Japanese patient. *The Journal of Clinical Endocrinology and Metabolism*, *88*, 45–50. https://doi.org/10.1210/jc.2002-020818.

Tajouri, A., Kharrat, M., Hizem, S., Zaghdoudi, H., M'rad, R., Simic-Schleicher, G., et al. (2018). *In vitro* functional characterization of the novel *DHH* mutations p.(Asn337Lysfs*24) and p.(Glu212Lys) associated with gonadal dysgenesis. *Human Mutation*, *39*, 2097–2109. https://doi.org/10.1002/humu.23664.

Tannour-Louet, M., Han, S., Corbett, S. T., Louet, J. F., Yatsenko, S., Meyers, L., et al. (2010). Identification of *de novo* copy number variants associated with human disorders of sexual development. *PLoS One*, *5*, e15392. https://doi.org/10.1371/journal.pone.0015392.

Tannour-Louet, M., Han, S., Louet, J. F., Zhang, B., Romero, K., Addai, J., et al. (2014). Increased gene copy number of VAMP7 disrupts human male urogenital development through altered estrogen action. *Nature Medicine*, *20*, 715–724. https://doi.org/10.1038/nm.3580.

Tartaglia, M., Martinelli, S., Stella, L., Bocchinfuso, G., Flex, E., Cordeddu, V., et al. (2006). Diversity and functional consequences of germline and somatic PTPN11 mutations in human disease. *American Journal of Human Genetics*, *78*, 279–290. https://doi.org/10.1086/499925.

Tattini, L., D'Aurizio, R., & Magi, A. (2015). Detection of genomic structural variants from next-generation sequencing data. *Frontiers in Bioengineering and Biotechnology*, *3*, 1–8. https://doi.org/10.3389/fbioe.2015.00092.

Tee, M. K., Lin, D., Sugawara, T., Holt, J. A., Guiguen, Y., Buckingham, B., et al. (1995). T→A transversion 11 bp from a splice acceptor site in the human gene for steroidogenic acute regulatory protein causes congenital lipoid adrenal hyperplasia. *Human Molecular Genetics*, *4*, 2299–2305.

Tenenbaum-Rakover, Y., Commenges-Ducos, M., Iovane, A., Aumas, C., Admoni, O., & de Roux, N. (2007). Neuroendocrine phenotype analysis in five patients with isolated hypogonadotropic hypogonadism due to a L102P inactivating mutation of GPR54. *The Journal of Clinical Endocrinology and Metabolism*, *92*, 1137–1144. https://doi.org/10.1210/jc.2006-2147.

Thiel, C., Wilken, M., Zenker, M., Sticht, H., Fahsold, R., Gusek-Schneider, G.-C., et al. (2009). Independent *NF1* and *PTPN11* mutations in a family with neurofibromatosis-Noonan syndrome. *American Journal of Medical Genetics. Part A*, *149A*, 1263–1267. https://doi.org/10.1002/ajmg.a.32837.

Thigpen, A. E., Davis, D. L., Gautier, T., Imperato-McGinley, J., & Russell, D. W. (1992). The molecular basis of steroid 5α-reductase deficiency in a large Dominican kindred. *The New England Journal of Medicine*, *327*, 1216–1219. https://doi.org/10.1056/NEJM199210223271706.

Thomas, P. Q., Dattani, M. T., Brickman, J. M., McNay, D., Warne, G., Zacharin, M., et al. (2001). Heterozygous HESX1 mutations associated with isolated congenital pituitary hypoplasia and septo-optic dysplasia. *Human Molecular Genetics*, *10*, 39–45.

Tiulpakov, A., Kalintchenko, N., Semitcheva, T., Polyakov, A., Dedov, I., Sverdlova, P., et al. (2005). A potential rearrangement between CYP19 and TRPM7 genes on chromosome 15q21.2 as a cause of aromatase excess syndrome. *The Journal of Clinical Endocrinology and Metabolism*, *90*, 4184–4190. https://doi.org/10.1210/jc.2004-2176.

Tobias, E. S., & McElreavey, K. (2014). Next generation sequencing for disorders of sex development. *Endocrine Development*, *27*, 53–62. https://doi.org/10.1159/000363615.

Tomaselli, S., Megiorni, F., De Bernardo, C., Felici, A., Marrocco, G., Maggiulli, G., et al. (2008). Syndromic true hermaphroditism due to an R-spondin1 (*RSPO1*) homozygous mutation. *Human Mutation*, *29*, 220–226. https://doi.org/10.1002/humu.20665.

Topaloglu, A. K., Reimann, F., Guclu, M., Yalin, A. S., Kotan, L. D., Porter, K. M., et al. (2009). TAC3 and TACR3 mutations in familial hypogonadotropic hypogonadism reveal a key role for Neurokinin B in the central control of reproduction. *Nature Genetics*, *41*, 354–358. https://doi.org/10.1038/ng.306.

Trarbach, E. B., Costa, E. M. F., Versiani, B., de Castro, M., Baptista, M. T. M., Garmes, H. M., et al. (2006). Novel fibroblast growth factor receptor 1 mutations in patients with congenital hypogonadotropic hypogonadism with and without anosmia. *The Journal of Clinical Endocrinology and Metabolism*, *91*, 4006–4012. https://doi.org/10.1210/jc.2005-2793.

Udaka, T., Okamoto, N., Aramaki, M., Torii, C., Kosaki, R., Hosokai, N., et al. (2007). An *Alu* retrotransposition-mediated deletion of *CHD7* in a patient with CHARGE syndrome. *American Journal of Medical Genetics. Part A*, *143A*, 721–726. https://doi.org/10.1002/ajmg.a.31441.

Umehara, F., Tate, G., Itoh, K., Yamaguchi, N., Douchi, T., Mitsuya, T., et al. (2000). A novel mutation of desert hedgehog in a patient with 46,XY partial gonadal dysgenesis accompanied by minifascicular neuropathy. *American Journal of Human Genetics*, *67*, 1302–1305. https://doi.org/10.1016/S0002-9297(07)62958-9.

van den Akker, E. L. T., Koper, J. W., Boehmer, A. L. M., Themmen, A. P. N., Verhoef-Post, M., Timmerman, M. A., et al. (2002). Differential inhibition of 17α-hydroxylase and 17,20-lyase activities by three novel missense cyp17 mutations identified in patients with P450c17 deficiency. *The Journal of Clinical Endocrinology and Metabolism*, *87*, 5714–5721. https://doi.org/10.1210/jc.2001-011880.

Van Goor, F., Yu, H., Burton, B., & Hoffman, B. J. (2014). Effect of ivacaftor on CFTR forms with missense mutations associated with defects in protein processing or function. *Journal of Cystic Fibrosis*, *13*, 29–36. https://doi.org/10.1016/j.jcf.2013.06.008.

van Haelst, M. M., Maiburg, M., Baujat, G., Jadeja, S., Monti, E., Bland, E., et al. (2008). Molecular study of 33 families with Fraser syndrome new data and mutation review. *American Journal of Medical Genetics. Part A*, *146A*, 2252–2257. https://doi.org/10.1002/ajmg.a.32440.

Vasseur, C., Rodien, P., Beau, I., Desroches, A., Gérard, C., de Poncheville, L., et al. (2003). A chorionic gonadotropin-sensitive mutation in the follicle-stimulating hormone receptor as a cause of familial gestational spontaneous ovarian hyperstimulation syndrome. *The New England Journal of Medicine*, *349*, 753–759. https://doi.org/10.1056/NEJMoa030065.

Vetro, A., Ciccone, R., Giorda, R., Patricelli, M. G., Della Mina, E., Forlino, A., et al. (2011). XX males SRY negative: A confirmed cause of infertility. *Journal of Medical Genetics*, *48*, 710–712. https://doi.org/10.1136/jmedgenet-2011-100036.

Vetro, A., Dehghani, M. R., Kraoua, L., Giorda, R., Beri, S., Cardarelli, L., et al. (2015). Testis development in the absence of SRY: Chromosomal rearrangements at SOX9 and SOX3. *European Journal of Human Genetics*, *23*, 1025–1032. https://doi.org/10.1038/ejhg.2014.237.

Vilain, E., McElreavey, K., Jaubert, F., Raymond, J. P., Richaud, F., & Fellous, M. (1992). Familial case with sequence variant in the testis-determining region associated with two sex phenotypes. *American Journal of Human Genetics*, *50*, 1008–1011.

Vilchis, F., Ramos, L., Kofman-Alfaro, S., Zenteno, J. C., Méndez, J. P., & Chávez, B. (2003). Extreme androgen resistance in a kindred with a novel insertion/deletion mutation in exon 5 of the androgen receptor gene. *Journal of Human Genetics*, *48*, 346–351. https://doi.org/10.1007/s10038-003-0036-0.

Vogel, M. J., van Zon, P., Brueton, L., Gijzen, M., van Tuil, M. C., Cox, P., et al. (2012). Mutations in *GRIP1* cause Fraser syndrome. *Journal of Medical Genetics*, *49*, 303–306. https://doi.org/10.1136/jmedgenet-2011-100590.

Volden, R., Palmer, T., Byrne, A., Cole, C., Schmitz, R. J., Green, R. E., et al. (2018). Improving nanopore read accuracy with the R2C2 method enables the sequencing of highly multiplexed full-length single-cell cDNA. *Proceedings of the National Academy of Sciences of the United States of America*, *115*, 9726–9731. https://doi.org/10.1073/pnas.1806447115.

Wabitsch, M., Funcke, J.-B., Lennerz, B., Kuhnle-Krahl, U., Lahr, G., Debatin, K.-M., et al. (2015). Biologically inactive leptin and early-onset extreme obesity. *The New England Journal of Medicine*, *372*, 48–54. https://doi.org/10.1056/NEJMoa1406653.

Wang, C., Pallan, P. S., Zhang, W., Lei, L., Yoshimoto, F. K., Waterman, M. R., et al. (2017). Functional analysis of human cytochrome P450 21A2 variants involved in congenital adrenal hyperplasia. *The Journal of Biological Chemistry*, *292*, 10767–10778. https://doi.org/10.1074/jbc.M117.792465.

Wang, R., Yu, Y., Ye, J., Han, L., Qiu, W., Zhang, H., et al. (2016). 21-Hydroxylase deficiency-induced congenital adrenal hyperplasia in 230 Chinese patients: Genotype–phenotype correlation and identification of nine novel mutations. *Steroids*, *108*, 47–55. https://doi.org/10.1016/j.steroids.2016.01.007.

Wang, F., Zhang, Y., Mao, J., Yu, Z., Yi, Z., Yu, L., et al. (2017). Spectrum of mutations in Chinese children with steroid-resistant nephrotic syndrome. *Pediatric Nephrology*, *32*, 1181–1192. https://doi.org/10.1007/s00467-017-3590-y.

Warr, A., Robert, C., Hume, D., Archibald, A., Deeb, N., & Watson, M. (2015). Exome sequencing: Current and future perspectives. *G3: Genes|Genomes|Genetics*, *5*, 1543–1550. https://doi.org/10.1534/g3.115.018564.

Wassif, C. A., Maslen, C., Kachilele-Linjewile, S., Lin, D., Linck, L. M., Connor, W. E., et al. (1998). Mutations in the human sterol Δ7-reductase gene at 11q12-13 cause Smith-Lemli-Opitz syndrome. *American Journal of Human Genetics*, *63*, 55–62. https://doi.org/10.1086/301936.

Waterham, H. R., & Hennekam, R. C. M. (2012). Mutational spectrum of Smith-Lemli-Opitz syndrome. *American Journal of Medical Genetics. Part C, Seminars in Medical Genetics*, *160C*, 263–284. https://doi.org/10.1002/ajmg.c.31346.

Waye, J. S., Nakamura, L. M., Eng, B., Hunnisett, L., Chitayat, D., Costa, T., et al. (2002). Smith-Lemli-Opitz syndrome: Carrier frequency and spectrum of DHCR7 mutations in Canada. *Journal of Medical Genetics*, *39*, E31.

Weischenfeldt, J., Symmons, O., Spitz, F., & Korbel, J. O. (2013). Phenotypic impact of genomic structural variation: Insights from and for human disease. *Nature Reviews. Genetics*, *14*, 125–138. https://doi.org/10.1038/nrg3373.

Weiss, J., Meeks, J. J., Hurley, L., Raverot, G., Frassetto, A., & Jameson, J. L. (2003). Sox3 is required for gonadal function, but not sex determination, in males and females. *Molecular and Cellular Biology*, *23*, 8084–8091.

Welzel, M., Schwarz, H.-P., Hedderich, J., Dörr, H. G., Binder, G., Brämswig, J. H., et al. (2010). No correlation between androgen receptor CAG and GGN repeat length and the degree of genital virilization in females with 21-hydroxylase deficiency. *The Journal of Clinical Endocrinology and Metabolism*, *95*, 2443–2450. https://doi.org/10.1210/jc.2009-1338.

Welzel, M., Wüstemann, N., Ŝimić-Schleicher, G., Dörr, H. G., Schulze, E., Shaikh, G., et al. (2008). Carboxyl-terminal mutations in 3β-hydroxysteroid dehydrogenase type II cause severe salt-wasting congenital adrenal hyperplasia. *The Journal of Clinical Endocrinology and Metabolism, 93*, 1418–1425. https://doi.org/10.1210/jc.2007-1874.

Wenger, A. M., Guturu, H., Bernstein, J. A., & Bejerano, G. (2017). Systematic reanalysis of clinical exome data yields additional diagnoses: Implications for providers. *Genetics in Medicine, 19*, 209–214. https://doi.org/10.1038/gim.2016.88.

Werner, R., Merz, H., Birnbaum, W., Marshall, L., Schröder, T., Reiz, B., et al. (2015). 46, XY gonadal dysgenesis due to a homozygous mutation in desert hedgehog (*DHH*) identified by exome sequencing. *The Journal of Clinical Endocrinology and Metabolism, 100*, E1022–E1029. https://doi.org/10.1210/jc.2015-1314.

White, S., Hewitt, J., Turbitt, E., Van Der Zwan, Y., Hersmus, R., Drop, S., et al. (2012). A multi-exon deletion within WWOX is associated with a 46,XY disorder of sex development. *European Journal of Human Genetics, 20*, 348–351. https://doi.org/10.1038/ejhg.2011.204.

White, S., Ohnesorg, T., Notini, A., Roeszler, K., Hewitt, J., Daggag, H., et al. (2011). Copy number variation in patients with disorders of sex development due to 46, XY gonadal dysgenesis. *PLoS One, 6*, e17793. https://doi.org/10.1371/journal.pone.0017793.

Woods, K. S., Cundall, M., Turton, J., Rizotti, K., Mehta, A., Palmer, R., et al. (2005). Over- and underdosage of SOX3 is associated with infundibular hypoplasia and hypopituitarism. *American Journal of Human Genetics, 76*, 833–849. https://doi.org/10.1086/430134.

Wu, W., Cogan, J. D., Pfäffle, R. W., Dasen, J. S., Frisch, H., O'Connell, S. M., et al. (1998). Mutations in PROP1 cause familial combined pituitary hormone deficiency. *Nature Genetics, 18*, 147–149. https://doi.org/10.1038/ng0298-147.

Wu, S. M., Hallermeier, K. M., Laue, L., Brain, C., Berry, A. C., Grant, D. B., et al. (1998). Inactivation of the luteinizing hormone/chorionic gonadotropin receptor by an insertional mutation in Leydig cell hypoplasia. *Molecular Endocrinology, 12*, 1651–1660. https://doi.org/10.1210/mend.12.11.0189.

Xu, N., Qin, Y., Reindollar, R. H., Tho, S. P. T., McDonough, P. G., & Layman, L. C. (2007). A mutation in the fibroblast growth factor receptor 1 gene causes fully penetrant normosmic isolated hypogonadotropic hypogonadism. *The Journal of Clinical Endocrinology and Metabolism, 92*, 1155–1158. https://doi.org/10.1210/jc.2006-1183.

Yamaguchi, H., Nakazato, M., Miyazato, M., Kangawa, K., & Matsukura, S. (1997). A 5′-splice site mutation in the cytochrome p450 steroid 17α-hydroxylase gene in 17α-hydroxylase deficiency. *The Journal of Clinical Endocrinology and Metabolism, 82*, 1934–1938. https://doi.org/10.1210/jcem.82.6.4027.

Yanase, T., Sanders, D., Shibata, A., Matsui, N., Simpson, E. R., & Waterman, M. R. (1990). Combined 17 α-hydroxylase/17,20-lyase deficiency due to a 7-basepair duplication in the N-terminal region of the cytochrome P450 17 α (CYP17) Gene*. *The Journal of Clinical Endocrinology and Metabolism, 70*, 1325–1329. https://doi.org/10.1210/jcem-70-5-1325.

Yanase, T., Takayanagi, R., Oba, K., Nishi, Y., Ohe, K., & Nawata, H. (1996). New mutations of DAX-1 genes in two Japanese patients with X-linked congenital adrenal hypoplasia and hypogonadotropic hypogonadism. *The Journal of Clinical Endocrinology and Metabolism, 81*, 530–535. https://doi.org/10.1210/jcem.81.2.8636263.

Yang, J., Cui, B., Sun, S., Shi, T., Zheng, S., Bi, Y., et al. (2006). Phenotype-genotype correlation in eight chinese 17α-hydroxylase/17,20 lyase-deficiency patients with five novel mutations of CYP17A1 gene. *The Journal of Clinical Endocrinology and Metabolism, 91*, 3619–3625. https://doi.org/10.1210/jc.2005-2283.

Yang, X.-W., He, W.-B., Gong, F., Li, W., Li, X.-R., Zhong, C.-G., et al. (2018). Novel *FOXL2* mutations cause blepharophimosis-ptosis-epicanthus inversus syndrome with premature ovarian insufficiency. *Molecular Genetics & Genomic Medicine, 6*, 261–267. https://doi.org/10.1002/mgg3.366.

Yang, Y., Muzny, D. M., Reid, J. G., Bainbridge, M. N., Willis, A., Ward, P. A., et al. (2013). Clinical whole-exome sequencing for the diagnosis of Mendelian disorders. *The New England Journal of Medicine, 369*, 1502–1511. https://doi.org/10.1056/NEJMoa1306555.

Yang, Y., Muzny, D. M., Xia, F., Niu, Z., Person, R., Ding, Y., et al. (2014). Molecular findings among patients referred for clinical whole-exome sequencing. *JAMA, 312*, 1870–1879. https://doi.org/10.1001/jama.2014.14601.

Yoo, Y., Chang, M. S., Lee, J., Cho, S. Y., Park, S. W., Jin, D.-K., et al. (2013). Genotype-phenotype correlation in 27 pediatric patients in congenital adrenal hyperplasia due to 21-hydroxylase deficiency in a single center. *Annals of Pediatric Endocrinology & Metabolism, 18*, 128. https://doi.org/10.6065/apem.2013.18.3.128.

Yoshida, R., Hasegawa, T., Hasegawa, Y., Nagai, T., Kinoshita, E., Tanaka, Y., et al. (2004). Protein-tyrosine phosphatase, nonreceptor type 11 mutation analysis and clinical assessment in 45 patients with Noonan syndrome. *The Journal of Clinical Endocrinology and Metabolism, 89*, 3359–3364. https://doi.org/10.1210/jc.2003-032091.

Yu, H., Lee, M. H., Starck, L., Elias, E. R., Irons, M., Salen, G., et al. (2000). Spectrum of delta(7)-dehydrocholesterol reductase mutations in patients with the Smith-Lemli-Opitz (RSH) syndrome. *Human Molecular Genetics, 9*, 1385–1391.

Zenker, M., Buheitel, G., Rauch, R., Koenig, R., Bosse, K., Kress, W., et al. (2004). Genotype-phenotype correlations in Noonan syndrome. *The Journal of Pediatrics, 144*, 368–374. https://doi.org/10.1016/j.jpeds.2003.11.032.

Zhao, F., Franco, H. L., Rodriguez, K. F., Brown, P. R., Tsai, M. J., Tsai, S. Y., et al. (2017). Elimination of the male reproductive tract in the female embryo is promoted by COUP-TFII in mice. *Science, 357*, 717–720. https://doi.org/10.1126/science.aai9136.

Zhu, Y.-S., Cai, L.-Q., Cordero, J. J., Canovatchel, W. J., Katz, M. D., & Imperato-McGinley, J. (1999). A novel mutation in the CAG triplet region of exon 1 of androgen receptor gene causes complete androgen insensitivity syndrome in a large kindred. *The Journal of Clinical Endocrinology and Metabolism, 84*, 1590–1594. https://doi.org/10.1210/jcem.84.5.5695.

Zoppi, S., Wilson, C. M., Harbison, M. D., Griffin, J. E., Wilson, J. D., McPhaul, M. J., et al. (1993). Complete testicular feminization caused by an amino-terminal truncation of the androgen receptor with downstream initiation. *The Journal of Clinical Investigation, 91*, 1105–1112. https://doi.org/10.1172/JCI116269.

Further reading

Chérot, E., Keren, B., Dubourg, C., Carré, W., Fradin, M., Lavillaureix, A., et al. (2018). Using medical exome sequencing to identify the causes of neurodevelopmental disorders: Experience of 2 clinical units and 216 patients. *Clinical Genetics, 93*, 567–576. https://doi.org/10.1111/cge.13102.

Espiner, E. A., Veale, A. M. O., Sands, V. E., & Fitzgerald, P. H. (1970). Familial syndrome of streak gonads and normal male karyotype in five phenotypic females. *The New England Journal of Medicine, 283*, 6–11. https://doi.org/10.1056/NEJM197007022830102.

Guerrier, D., Tran, D., Vanderwinden, J. M., Hideux, S., Van Outryve, L., Legeai, L., et al. (1989). The persistent mullerian duct syndrome: A molecular approach. *The Journal of Clinical Endocrinology and Metabolism, 68*, 46–52. https://doi.org/10.1210/jcem-68-1-46.

Harbison, M. D., Magid, M. L., Josso, N., Mininberg, D. T., & New, M. I. (1991). Anti-Müllerian hormone in three intersex conditions. *Annales de Génétique, 34*, 226–232.

Jawaheer, D., Juo, S.-H. H., Le Caignec, C., David, A., Petit, C., Gregersen, P., et al. (2003). Mapping a gene for 46,XY gonadal dysgenesis by linkage analysis. *Clinical Genetics, 63*, 530–535.

Kwitek-Black, A. E., Carmi, R., Duyk, G. M., Buetow, K. H., Elbedour, K., Parvari, R., et al. (1993). Linkage of Bardet–Biedl syndrome to chromosome 16q and evidence for non-allelic genetic heterogeneity. *Nature Genetics*, *5*, 392–396. https://doi.org/10.1038/ng1293-392.

Le Caignec, C., Baron, S., McElreavey, K., Joubert, M., Rival, J.-M., Mechinaud, F., et al. (2003). 46,XY gonadal dysgenesis: Evidence for autosomal dominant transmission in a large kindred. *American Journal of Medical Genetics*, *116A*, 37–43. https://doi.org/10.1002/ajmg.a.10820.

Loke, J., Pearlman, A., Radi, O., Zuffardi, O., Giussani, U., Pallotta, R., et al. (2014). Mutations in MAP3K1 tilt the balance from SOX9/FGF9 to WNT/β-catenin signaling. *Human Molecular Genetics*, *23*, 1073–1083. https://doi.org/10.1093/hmg/ddt502.

Marohnic, C. C., Huber, W. J., III, Patrick Connick, J., Reed, J. R., McCammon, K., Panda, S. P., et al. (2011). Mutations of human cytochrome P450 reductase differentially modulate heme oxygenase-1 activity and oligomerization. *Archives of Biochemistry and Biophysics*, *513*, 42–50. https://doi.org/10.1016/j.abb.2011.06.008.

Nordfors, K., Haapasalo, J., Afyounian, E., Tuominen, J., Annala, M., Häyrynen, S., et al. (2018). Whole-exome sequencing identifies germline mutation in *TP53* and *ATRX* in a child with genomically aberrant AT/RT and her mother with anaplastic astrocytoma. *Cold Spring Harbor Molecular Case Studies*, *4*, a002246. https://doi.org/10.1101/mcs.a002246.

Olatunbosun, O. A., Gilliland, B., Brydon, L. A., Chizen, D. R., & Pierson, R. A. (1996). Spontaneous ovarian hyperstimulation syndrome in four consecutive pregnancies. *Clinical and Experimental Obstetrics & Gynecology*, *23*, 127–132.

Stevenson, R. E. (2014). *Alpha-thalassemia X-linked intellectual disability syndrome, GeneReviews®*.

FSC
www.fsc.org
MIX
Papier | Fördert gute Waldnutzung
FSC® C083411

Kolibri 360 GmbH
Ferdinand-Jühlke-Straße 7
99095 Erfurt, Deutschland
produktsicherheit@kolibri360.de